STUDENT'S SOLUTIONS MANUAL

JUDITH A. PENNA

COLLEGE ALGEBRA
FIFTH EDITION

D0074445

Judith A. Beecher

Judith A. Penna

Marvin L. Bittinger
Indiana University Purdue University Indianapolis

PEARSON

Boston Columbus Hoboken Indianapolis New York San Francisco
Amsterdam Cape Town Dubai London Madrid Milan Munich Paris Montreal Toronto
Delhi Mexico City São Paulo Sydney Hong Kong Seoul Singapore Taipei Tokyo

The author and publisher of this book have used their best efforts in preparing this book. These efforts include the development, research, and testing of the theories and programs to determine their effectiveness. The author and publisher make no warranty of any kind, expressed or implied, with regard to these programs or the documentation contained in this book. The author and publisher shall not be liable in any event for incidental or consequential damages in connection with, or arising out of, the furnishing, performance, or use of these programs.

Reproduced by Pearson from electronic files supplied by the author.

ISBN-13: 978-0-321-96995-8
ISBN-10: 0-321-96995-2

1 2 3 4 5 6 OPM 18 17 16 15 14

www.pearsonhighered.com

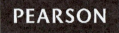

Contents

Chapter 1

Graphs, Functions, and Models

Exercise Set 1.1

1. Point A is located 5 units to the left of the y-axis and 4 units up from the x-axis, so its coordinates are $(-5, 4)$.

   Point B is located 2 units to the right of the y-axis and 2 units down from the x-axis, so its coordinates are $(2, -2)$.

   Point C is located 0 units to the right or left of the y-axis and 5 units down from the x-axis, so its coordinates are $(0, -5)$.

   Point D is located 3 units to the right of the y-axis and 5 units up from the x-axis, so its coordinates are $(3, 5)$.

   Point E is located 5 units to the left of the y-axis and 4 units down from the x-axis, so its coordinates are $(-5, -4)$.

   Point F is located 3 units to the right of the y-axis and 0 units up or down from the x-axis, so its coordinates are $(3, 0)$.

3. To graph $(4, 0)$ we move from the origin 4 units to the right of the y-axis. Since the second coordinate is 0, we do not move up or down from the x-axis.

   To graph $(-3, -5)$ we move from the origin 3 units to the left of the y-axis. Then we move 5 units down from the x-axis.

   To graph $(-1, 4)$ we move from the origin 1 unit to the left of the y-axis. Then we move 4 units up from the x-axis.

   To graph $(0, 2)$ we do not move to the right or the left of the y-axis since the first coordinate is 0. From the origin we move 2 units up.

   To graph $(2, -2)$ we move from the origin 2 units to the right of the y-axis. Then we move 2 units down from the x-axis.

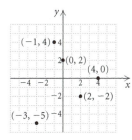

5. To graph $(-5, 1)$ we move from the origin 5 units to the left of the y-axis. Then we move 1 unit up from the x-axis.

   To graph $(5, 1)$ we move from the origin 5 units to the right of the y-axis. Then we move 1 unit up from the x-axis.

   To graph $(2, 3)$ we move from the origin 2 units to the right of the y-axis. Then we move 3 units up from the x-axis.

To graph $(2, -1)$ we move from the origin 2 units to the right of the y-axis. Then we move 1 unit down from the x-axis.

To graph $(0, 1)$ we do not move to the right or the left of the y-axis since the first coordinate is 0. From the origin we move 1 unit up.

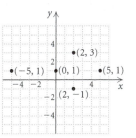

7. The first coordinate represents the year and the second coordinate represents the number of Sprint Cup Series races in which Tony Stewart finished in the top five. The ordered pairs are $(2008, 10)$, $(2009, 15)$, $(2010, 9)$, $(2011, 9)$, $(2012, 12)$, and $(2013, 5)$.

9. To determine whether $(-1, -9)$ is a solution, substitute -1 for x and -9 for y.

$$\begin{array}{c|c} \multicolumn{2}{c}{y = 7x - 2} \\ \hline -9 \ ? \ 7(-1) - 2 & \\ & -7 - 2 \\ -9 & -9 \qquad \text{TRUE} \end{array}$$

The equation $-9 = -9$ is true, so $(-1, -9)$ is a solution.

To determine whether $(0, 2)$ is a solution, substitute 0 for x and 2 for y.

$$\begin{array}{c|c} \multicolumn{2}{c}{y = 7x - 2} \\ \hline 2 \ ? \ 7 \cdot 0 - 2 & \\ & 0 - 2 \\ 2 & -2 \qquad \text{FALSE} \end{array}$$

The equation $2 = -2$ is false, so $(0, 2)$ is not a solution.

11. To determine whether $\left(\dfrac{2}{3}, \dfrac{3}{4}\right)$ is a solution, substitute $\dfrac{2}{3}$ for x and $\dfrac{3}{4}$ for y.

$$\begin{array}{c|c} \multicolumn{2}{c}{6x - 4y = 1} \\ \hline 6 \cdot \dfrac{2}{3} - 4 \cdot \dfrac{3}{4} \ ? \ 1 & \\ 4 - 3 & \\ 1 & 1 \quad \text{TRUE} \end{array}$$

The equation $1 = 1$ is true, so $\left(\dfrac{2}{3}, \dfrac{3}{4}\right)$ is a solution.

To determine whether $\left(1, \dfrac{3}{2}\right)$ is a solution, substitute 1 for x and $\dfrac{3}{2}$ for y.

$$6x - 4y = 1$$

$$
\begin{array}{c|c}
6 \cdot 1 - 4 \cdot \dfrac{3}{2} \;?\; 1 & \\
6 - 6 & \\
0 & 1 \quad \text{FALSE}
\end{array}
$$

The equation $0 = 1$ is false, so $\left(1, \dfrac{3}{2}\right)$ is not a solution.

13. To determine whether $\left(-\dfrac{1}{2}, -\dfrac{4}{5}\right)$ is a solution, substitute $-\dfrac{1}{2}$ for a and $-\dfrac{4}{5}$ for b.

$$2a + 5b = 3$$

$$
\begin{array}{c|c}
2\left(-\dfrac{1}{2}\right) + 5\left(-\dfrac{4}{5}\right) \;?\; 3 & \\
-1 - 4 & \\
-5 & 3 \quad \text{FALSE}
\end{array}
$$

The equation $-5 = 3$ is false, so $\left(-\dfrac{1}{2}, -\dfrac{4}{5}\right)$ is not a solution.

To determine whether $\left(0, \dfrac{3}{5}\right)$ is a solution, substitute 0 for a and $\dfrac{3}{5}$ for b.

$$2a + 5b = 3$$

$$
\begin{array}{c|c}
2 \cdot 0 + 5 \cdot \dfrac{3}{5} \;?\; 3 & \\
0 + 3 & \\
3 & 3 \quad \text{TRUE}
\end{array}
$$

The equation $3 = 3$ is true, so $\left(0, \dfrac{3}{5}\right)$ is a solution.

15. To determine whether $(-0.75, 2.75)$ is a solution, substitute -0.75 for x and 2.75 for y.

$$x^2 - y^2 = 3$$

$$
\begin{array}{c|c}
(-0.75)^2 - (2.75)^2 \;?\; 3 & \\
0.5625 - 7.5625 & \\
-7 & 3 \quad \text{FALSE}
\end{array}
$$

The equation $-7 = 3$ is false, so $(-0.75, 2.75)$ is not a solution.

To determine whether $(2, -1)$ is a solution, substitute 2 for x and -1 for y.

$$x^2 - y^2 = 3$$

$$
\begin{array}{c|c}
2^2 - (-1)^2 \;?\; 3 & \\
4 - 1 & \\
3 & 3 \quad \text{TRUE}
\end{array}
$$

The equation $3 = 3$ is true, so $(2, -1)$ is a solution.

17. Graph $5x - 3y = -15$.

To find the x-intercept we replace y with 0 and solve for x.

$$5x - 3 \cdot 0 = -15$$
$$5x = -15$$
$$x = -3$$

The x-intercept is $(-3, 0)$.

To find the y-intercept we replace x with 0 and solve for y.

$$5 \cdot 0 - 3y = -15$$
$$-3y = -15$$
$$y = 5$$

The y-intercept is $(0, 5)$.

We plot the intercepts and draw the line that contains them. We could find a third point as a check that the intercepts were found correctly.

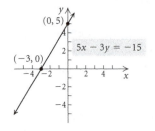

19. Graph $2x + y = 4$.

To find the x-intercept we replace y with 0 and solve for x.

$$2x + 0 = 4$$
$$2x = 4$$
$$x = 2$$

The x-intercept is $(2, 0)$.

To find the y-intercept we replace x with 0 and solve for y.

$$2 \cdot 0 + y = 4$$
$$y = 4$$

The y-intercept is $(0, 4)$.

We plot the intercepts and draw the line that contains them. We could find a third point as a check that the intercepts were found correctly.

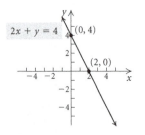

21. Graph $4y - 3x = 12$.

To find the x-intercept we replace y with 0 and solve for x.

$$4 \cdot 0 - 3x = 12$$
$$-3x = 12$$
$$x = -4$$

The x-intercept is $(-4, 0)$.

To find the y-intercept we replace x with 0 and solve for y.

$$4y - 3 \cdot 0 = 12$$
$$4y = 12$$
$$y = 3$$

The y-intercept is $(0, 3)$.

We plot the intercepts and draw the line that contains them. We could find a third point as a check that the intercepts were found correctly.

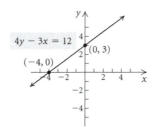

23. Graph $y = 3x + 5$.

We choose some values for x and find the corresponding y-values.

When $x = -3$, $y = 3x + 5 = 3(-3) + 5 = -9 + 5 = -4$.

When $x = -1$, $y = 3x + 5 = 3(-1) + 5 = -3 + 5 = 2$.

When $x = 0$, $y = 3x + 5 = 3 \cdot 0 + 5 = 0 + 5 = 5$

We list these points in a table, plot them, and draw the graph.

x	y	(x, y)
-3	-4	$(-3, -4)$
-1	2	$(-1, 2)$
0	5	$(0, 5)$

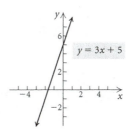

25. Graph $x - y = 3$.

Make a table of values, plot the points in the table, and draw the graph.

x	y	(x, y)
-2	-5	$(-2, -5)$
0	-3	$(0, -3)$
3	0	$(3, 0)$

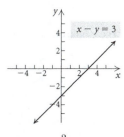

27. Graph $y = -\dfrac{3}{4}x + 3$.

By choosing multiples of 4 for x, we can avoid fraction values for y. Make a table of values, plot the points in the table, and draw the graph.

x	y	(x, y)
-4	6	$(-4, 6)$
0	3	$(0, 3)$
4	0	$(4, 0)$

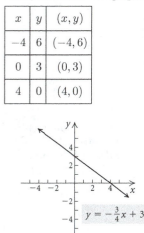

29. Graph $5x - 2y = 8$.

We could solve for y first.

$$5x - 2y = 8$$
$$-2y = -5x + 8 \quad \text{Subtracting } 5x \text{ on both sides}$$
$$y = \frac{5}{2}x - 4 \quad \text{Multiplying by } -\frac{1}{2} \text{ on both sides}$$

By choosing multiples of 2 for x we can avoid fraction values for y. Make a table of values, plot the points in the table, and draw the graph.

x	y	(x, y)
0	-4	$(0, -4)$
2	1	$(2, 1)$
4	6	$(4, 6)$

31. Graph $x - 4y = 5$.

Make a table of values, plot the points in the table, and draw the graph.

x	y	(x, y)
-3	-2	$(-3, -2)$
1	-1	$(1, -1)$
5	0	$(5, 0)$

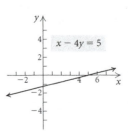

33. Graph $2x + 5y = -10$.

In this case, it is convenient to find the intercepts along with a third point on the graph. Make a table of values, plot the points in the table, and draw the graph.

x	y	(x, y)
-5	0	$(-5, 0)$
0	-2	$(0, -2)$
5	-4	$(5, -4)$

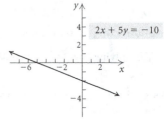

35. Graph $y = -x^2$.

Make a table of values, plot the points in the table, and draw the graph.

x	y	(x, y)
-2	-4	$(-2, -4)$
-1	-1	$(-1, -1)$
0	0	$(0, 0)$
1	-1	$(1, -1)$
2	-4	$(2, -4)$

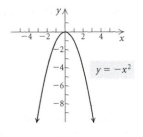

37. Graph $y = x^2 - 3$.

Make a table of values, plot the points in the table, and draw the graph.

x	y	(x, y)
-3	6	$(-3, 6)$
-1	-2	$(-1, -2)$
0	-3	$(0, -3)$
1	-2	$(1, -2)$
3	6	$(3, 6)$

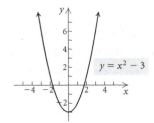

39. Graph $y = -x^2 + 2x + 3$.

Make a table of values, plot the points in the table, and draw the graph.

x	y	(x, y)
-2	-5	$(-2, -5)$
-1	0	$(-1, 0)$
0	3	$(0, 3)$
1	4	$(1, 4)$
2	3	$(2, 3)$
3	0	$(3, 0)$
4	-5	$(4, -5)$

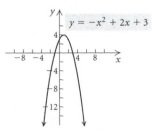

41. Either point can be considered as (x_1, y_1).
$$d = \sqrt{(4-5)^2 + (6-9)^2}$$
$$= \sqrt{(-1)^2 + (-3)^2} = \sqrt{10} \approx 3.162$$

43. Either point can be considered as (x_1, y_1).
$$d = \sqrt{(-13-(-8))^2 + (1-(-11))^2}$$
$$= \sqrt{(-5)^2 + 12^2} = \sqrt{169} = 13$$

45. Either point can be considered as (x_1, y_1).
$$d = \sqrt{(6-9)^2 + (-1-5)^2}$$
$$= \sqrt{(-3)^2 + (-6)^2} = \sqrt{45} \approx 6.708$$

47. Either point can be considered as (x_1, y_1).
$$d = \sqrt{(-8-8)^2 + \left(\frac{7}{11} - \frac{7}{11}\right)^2}$$
$$= \sqrt{(-16)^2 + 0^2} = 16$$

49. $d = \sqrt{\left[-\frac{3}{5} - \left(-\frac{3}{5}\right)\right]^2 + \left(-4 - \frac{2}{3}\right)^2}$
$$= \sqrt{0^2 + \left(-\frac{14}{3}\right)^2} = \frac{14}{3}$$

51. Either point can be considered as (x_1, y_1).
$$d = \sqrt{(-4.2-2.1)^2 + [3-(-6.4)]^2}$$
$$= \sqrt{(-6.3)^2 + (9.4)^2} = \sqrt{128.05} \approx 11.316$$

53. Either point can be considered as (x_1, y_1).
$$d = \sqrt{(0-a)^2 + (0-b)^2} = \sqrt{a^2 + b^2}$$

55. First we find the length of the diameter:
$$d = \sqrt{(-3-9)^2 + (-1-4)^2}$$
$$= \sqrt{(-12)^2 + (-5)^2} = \sqrt{169} = 13$$
The length of the radius is one-half the length of the diameter, or $\frac{1}{2}(13)$, or 6.5.

57. First we find the distance between each pair of points.
For $(-4, 5)$ and $(6, 1)$:
$$d = \sqrt{(-4-6)^2 + (5-1)^2}$$
$$= \sqrt{(-10)^2 + 4^2} = \sqrt{116}$$
For $(-4, 5)$ and $(-8, -5)$:
$$d = \sqrt{(-4-(-8))^2 + (5-(-5))^2}$$
$$= \sqrt{4^2 + 10^2} = \sqrt{116}$$
For $(6, 1)$ and $(-8, -5)$:
$$d = \sqrt{(6-(-8))^2 + (1-(-5))^2}$$
$$= \sqrt{14^2 + 6^2} = \sqrt{232}$$
Since $(\sqrt{116})^2 + (\sqrt{116})^2 = (\sqrt{232})^2$, the points could be the vertices of a right triangle.

59. First we find the distance between each pair of points.
For $(-4, 3)$ and $(0, 5)$:
$$d = \sqrt{(-4-0)^2 + (3-5)^2}$$
$$= \sqrt{(-4)^2 + (-2)^2} = \sqrt{20}$$
For $(-4, 3)$ and $(3, -4)$:
$$d = \sqrt{(-4-3)^2 + [3-(-4)]^2}$$
$$= \sqrt{(-7)^2 + 7^2} = \sqrt{98}$$
For $(0, 5)$ and $(3, -4)$:
$$d = \sqrt{(0-3)^2 + [5-(-4)]^2}$$
$$= \sqrt{(-3)^2 + 9^2} = \sqrt{90}$$
The greatest distance is $\sqrt{98}$, so if the points are the vertices of a right triangle, then it is the hypotenuse. But $(\sqrt{20})^2 + (\sqrt{90})^2 \neq (\sqrt{98})^2$, so the points are not the vertices of a right triangle.

61. We use the midpoint formula.
$$\left(\frac{4 + (-12)}{2}, \frac{-9 + (-3)}{2}\right) = \left(-\frac{8}{2}, -\frac{12}{2}\right) = (-4, -6)$$

63. We use the midpoint formula.
$$\left(\frac{0 + \left(-\frac{2}{5}\right)}{2}, \frac{\frac{1}{2} - 0}{2}\right) = \left(-\frac{\frac{2}{5}}{2}, \frac{\frac{1}{2}}{2}\right) = \left(-\frac{1}{5}, \frac{1}{4}\right)$$

65. We use the midpoint formula.
$$\left(\frac{6.1 + 3.8}{2}, \frac{-3.8 + (-6.1)}{2}\right) = \left(\frac{9.9}{2}, -\frac{9.9}{2}\right) =$$
$$(4.95, -4.95)$$

67. We use the midpoint formula.
$$\left(\frac{-6 + (-6)}{2}, \frac{5 + 8}{2}\right) = \left(-\frac{12}{2}, \frac{13}{2}\right) = \left(-6, \frac{13}{2}\right)$$

69. We use the midpoint formula.
$$\left(\frac{-\frac{1}{6} + \left(-\frac{2}{3}\right)}{2}, \frac{-\frac{3}{5} + \frac{5}{4}}{2}\right) = \left(-\frac{\frac{5}{6}}{2}, \frac{\frac{13}{20}}{2}\right) =$$
$$\left(-\frac{5}{12}, \frac{13}{40}\right)$$

71.

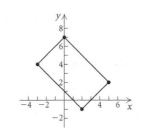

For the side with vertices $(-3, 4)$ and $(2, -1)$:
$$\left(\frac{-3 + 2}{2}, \frac{4 + (-1)}{2}\right) = \left(-\frac{1}{2}, \frac{3}{2}\right)$$
For the side with vertices $(2, -1)$ and $(5, 2)$:
$$\left(\frac{2 + 5}{2}, \frac{-1 + 2}{2}\right) = \left(\frac{7}{2}, \frac{1}{2}\right)$$

For the side with vertices $(5, 2)$ and $(0, 7)$:

$$\left(\frac{5+0}{2}, \frac{2+7}{2}\right) = \left(\frac{5}{2}, \frac{9}{2}\right)$$

For the side with vertices $(0, 7)$ and $(-3, 4)$:

$$\left(\frac{0+(-3)}{2}, \frac{7+4}{2}\right) = \left(-\frac{3}{2}, \frac{11}{2}\right)$$

For the quadrilateral whose vertices are the points found above, the diagonals have endpoints

$$\left(-\frac{1}{2}, \frac{3}{2}\right), \left(\frac{5}{2}, \frac{9}{2}\right) \text{ and } \left(\frac{7}{2}, \frac{1}{2}\right), \left(-\frac{3}{2}, \frac{11}{2}\right).$$

We find the length of each of these diagonals.

For $\left(-\frac{1}{2}, \frac{3}{2}\right), \left(\frac{5}{2}, \frac{9}{2}\right)$:

$$d = \sqrt{\left(-\frac{1}{2} - \frac{5}{2}\right)^2 + \left(\frac{3}{2} - \frac{9}{2}\right)^2}$$

$$= \sqrt{(-3)^2 + (-3)^2} = \sqrt{18}$$

For $\left(\frac{7}{2}, \frac{1}{2}\right), \left(-\frac{3}{2}, \frac{11}{2}\right)$:

$$d = \sqrt{\left(\frac{7}{2} - \left(-\frac{3}{2}\right)\right)^2 + \left(\frac{1}{2} - \frac{11}{2}\right)^2}$$

$$= \sqrt{5^2 + (-5)^2} = \sqrt{50}$$

Since the diagonals do not have the same lengths, the midpoints are not vertices of a rectangle.

73. We use the midpoint formula.

$$\left(\frac{\sqrt{7} + \sqrt{2}}{2}, \frac{-4 + 3}{2}\right) = \left(\frac{\sqrt{7} + \sqrt{2}}{2}, -\frac{1}{2}\right)$$

75. $\quad (x - h)^2 + (y - k)^2 = r^2$

$$(x - 2)^2 + (y - 3)^2 = \left(\frac{5}{3}\right)^2 \quad \text{Substituting}$$

$$(x - 2)^2 + (y - 3)^2 = \frac{25}{9}$$

77. The length of a radius is the distance between $(-1, 4)$ and $(3, 7)$:

$$r = \sqrt{(-1 - 3)^2 + (4 - 7)^2}$$

$$= \sqrt{(-4)^2 + (-3)^2} = \sqrt{25} = 5$$

$$(x - h)^2 + (y - k)^2 = r^2$$

$$[x - (-1)]^2 + (y - 4)^2 = 5^2$$

$$(x + 1)^2 + (y - 4)^2 = 25$$

79. The center is the midpoint of the diameter:

$$\left(\frac{7 + (-3)}{2}, \frac{13 + (-11)}{2}\right) = (2, 1)$$

Use the center and either endpoint of the diameter to find the length of a radius. We use the point $(7, 13)$:

$$r = \sqrt{(7 - 2)^2 + (13 - 1)^2}$$

$$= \sqrt{5^2 + 12^2} = \sqrt{169} = 13$$

$$(x - h)^2 + (y - k)^2 = r^2$$

$$(x - 2)^2 + (y - 1)^2 = 13^2$$

$$(x - 2)^2 + (y - 1)^2 = 169$$

81. Since the center is 2 units to the left of the y-axis and the circle is tangent to the y-axis, the length of a radius is 2.

$$(x - h)^2 + (y - k)^2 = r^2$$

$$[x - (-2)]^2 + (y - 3)^2 = 2^2$$

$$(x + 2)^2 + (y - 3)^2 = 4$$

83. $\quad\quad\quad x^2 + y^2 = 4$

$$(x - 0)^2 + (y - 0)^2 = 2^2$$

Center: $(0, 0)$; radius: 2

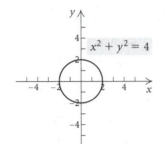

85. $\quad\quad x^2 + (y - 3)^2 = 16$

$$(x - 0)^2 + (y - 3)^2 = 4^2$$

Center: $(0, 3)$; radius: 4

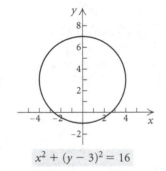

87. $\quad (x - 1)^2 + (y - 5)^2 = 36$

$$(x - 1)^2 + (y - 5)^2 = 6^2$$

Center: $(1, 5)$; radius: 6

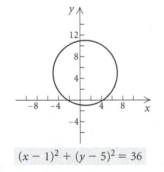

89. $\quad\quad (x + 4)^2 + (y + 5)^2 = 9$

$$[x - (-4)]^2 + [y - (-5)]^2 = 3^2$$

Center: $(-4, -5)$; radius: 3

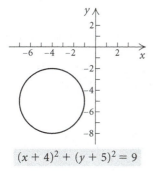

$$(x+4)^2 + (y+5)^2 = 9$$

91. From the graph we see that the center of the circle is $(-2, 1)$ and the radius is 3. The equation of the circle is $[x - (-2)]^2 + (y - 1)^2 = 3^2$, or $(x + 2)^2 + (y - 1)^2 = 3^2$.

93. From the graph we see that the center of the circle is $(5, -5)$ and the radius is 15. The equation of the circle is $(x - 5)^2 + [y - (-5)]^2 = 15^2$, or $(x - 5)^2 + (y + 5)^2 = 15^2$.

95. If the point (p, q) is in the fourth quadrant, then $p > 0$ and $q < 0$. If $p > 0$, then $-p < 0$ so both coordinates of the point $(q, -p)$ are negative and $(q, -p)$ is in the third quadrant.

97. Use the distance formula. Either point can be considered as (x_1, y_1).

$$d = \sqrt{(a + h - a)^2 + (\sqrt{a + h} - \sqrt{a})^2}$$
$$= \sqrt{h^2 + a + h - 2\sqrt{a^2 + ah} + a}$$
$$= \sqrt{h^2 + 2a + h - 2\sqrt{a^2 + ah}}$$

Next we use the midpoint formula.

$$\left(\frac{a + a + h}{2}, \frac{\sqrt{a} + \sqrt{a + h}}{2} \right) = \left(\frac{2a + h}{2}, \frac{\sqrt{a} + \sqrt{a + h}}{2} \right)$$

99. First use the formula for the area of a circle to find r^2:

$$A = \pi r^2$$
$$36\pi = \pi r^2$$
$$36 = r^2$$

Then we have:

$$(x - h)^2 + (y - k)^2 = r^2$$
$$(x - 2)^2 + [y - (-7)]^2 = 36$$
$$(x - 2)^2 + (y + 7)^2 = 36$$

101. Let $(0, y)$ be the required point. We set the distance from $(-2, 0)$ to $(0, y)$ equal to the distance from $(4, 6)$ to $(0, y)$ and solve for y.

$$\sqrt{[0 - (-2)]^2 + (y - 0)^2} = \sqrt{(0 - 4)^2 + (y - 6)^2}$$
$$\sqrt{4 + y^2} = \sqrt{16 + y^2 - 12y + 36}$$
$$4 + y^2 = 16 + y^2 - 12y + 36$$
$$\text{Squaring both sides}$$
$$-48 = -12y$$
$$4 = y$$

The point is $(0, 4)$.

103. a) When the circle is positioned on a coordinate system as shown in the text, the center lies on the y-axis and is equidistant from $(-4, 0)$ and $(0, 2)$.

Let $(0, y)$ be the coordinates of the center.

$$\sqrt{(-4-0)^2 + (0-y)^2} = \sqrt{(0-0)^2 + (2-y)^2}$$
$$4^2 + y^2 = (2 - y)^2$$
$$16 + y^2 = 4 - 4y + y^2$$
$$12 = -4y$$
$$-3 = y$$

The center of the circle is $(0, -3)$.

b) Use the point $(-4, 0)$ and the center $(0, -3)$ to find the radius.

$$(-4 - 0)^2 + [0 - (-3)]^2 = r^2$$
$$25 = r^2$$
$$5 = r$$

The radius is 5 ft.

105.

$$\frac{x^2 + y^2 = 1}{\left(\frac{\sqrt{3}}{2}\right)^2 + \left(-\frac{1}{2}\right)^2 \ ?\ 1}$$
$$\frac{3}{4} + \frac{1}{4} \Big|$$
$$1 \ \Big| \ 1 \quad \text{TRUE}$$

$\left(\dfrac{\sqrt{3}}{2}, -\dfrac{1}{2} \right)$ lies on the unit circle.

107.

$$\frac{x^2 + y^2 = 1}{\left(-\frac{\sqrt{2}}{2}\right)^2 + \left(\frac{\sqrt{2}}{2}\right)^2 \ ?\ 1}$$
$$\frac{2}{4} + \frac{2}{4} \Big|$$
$$1 \ \Big| \ 1 \quad \text{TRUE}$$

$\left(-\dfrac{\sqrt{2}}{2}, \dfrac{\sqrt{2}}{2} \right)$ lies on the unit circle.

109. a), b) See the answer section in the text.

Exercise Set 1.2

1. This correspondence is a function, because each member of the domain corresponds to exactly one member of the range.

3. This correspondence is a function, because each member of the domain corresponds to exactly one member of the range.

5. This correspondence is not a function, because there is a member of the domain (m) that corresponds to more than one member of the range (A and B).

7. This correspondence is a function, because each member of the domain corresponds to exactly one member of the range.

9. This correspondence is a function, because each car has exactly one license number.

11. This correspondence is a function, because each integer less than 9 corresponds to exactly one multiple of 5.

13. This correspondence is not a function, because at least one student will have more than one neighboring seat occupied by another student.

15. The relation is a function, because no two ordered pairs have the same first coordinate and different second coordinates.

The domain is the set of all first coordinates: $\{2, 3, 4\}$.

The range is the set of all second coordinates: $\{10, 15, 20\}$.

17. The relation is not a function, because the ordered pairs $(-2, 1)$ and $(-2, 4)$ have the same first coordinate and different second coordinates.

The domain is the set of all first coordinates: $\{-7, -2, 0\}$.

The range is the set of all second coordinates: $\{3, 1, 4, 7\}$.

19. The relation is a function, because no two ordered pairs have the same first coordinate and different second coordinates.

The domain is the set of all first coordinates: $\{-2, 0, 2, 4, -3\}$.

The range is the set of all second coordinates: $\{1\}$.

21. $g(x) = 3x^2 - 2x + 1$

a) $g(0) = 3 \cdot 0^2 - 2 \cdot 0 + 1 = 1$

b) $g(-1) = 3(-1)^2 - 2(-1) + 1 = 6$

c) $g(3) = 3 \cdot 3^2 - 2 \cdot 3 + 1 = 22$

d) $g(-x) = 3(-x)^2 - 2(-x) + 1 = 3x^2 + 2x + 1$

e) $g(1 - t) = 3(1 - t)^2 - 2(1 - t) + 1 =$
$3(1 - 2t + t^2) - 2(1 - t) + 1 = 3 - 6t + 3t^2 - 2 + 2t + 1 =$
$3t^2 - 4t + 2$

23. $g(x) = x^3$

a) $g(2) = 2^3 = 8$

b) $g(-2) = (-2)^3 = -8$

c) $g(-x) = (-x)^3 = -x^3$

d) $g(3y) = (3y)^3 = 27y^3$

e) $g(2 + h) = (2 + h)^3 = 8 + 12h + 6h^2 + h^3$

25. $g(x) = \dfrac{x - 4}{x + 3}$

a) $g(5) = \dfrac{5 - 4}{5 + 3} = \dfrac{1}{8}$

b) $g(4) = \dfrac{4 - 4}{4 + 7} = 0$

c) $g(-3) = \dfrac{-3 - 4}{-3 + 3} = \dfrac{-7}{0}$

Since division by 0 is not defined, $g(-3)$ does not exist.

d) $g(-16.25) = \dfrac{-16.25 - 4}{-16.25 + 3} = \dfrac{-20.25}{-13.25} = \dfrac{81}{53} \approx 1.5283$

e) $g(x + h) = \dfrac{x + h - 4}{x + h + 3}$

27. $g(x) = \dfrac{x}{\sqrt{1 - x^2}}$

$g(0) = \dfrac{0}{\sqrt{1 - 0^2}} = \dfrac{0}{\sqrt{1}} = \dfrac{0}{1} = 0$

$g(-1) = \dfrac{-1}{\sqrt{1 - (-1)^2}} = \dfrac{-1}{\sqrt{1 - 1}} = \dfrac{-1}{\sqrt{0}} = \dfrac{-1}{0}$

Since division by 0 is not defined, $g(-1)$ does not exist.

$g(5) = \dfrac{5}{\sqrt{1 - 5^2}} = \dfrac{5}{\sqrt{1 - 25}} = \dfrac{5}{\sqrt{-24}}$

Since $\sqrt{-24}$ is not defined as a real number, $g(5)$ does not exist as a real number.

$g\left(\dfrac{1}{2}\right) = \dfrac{\dfrac{1}{2}}{\sqrt{1 - \left(\dfrac{1}{2}\right)^2}} = \dfrac{\dfrac{1}{2}}{\sqrt{1 - \dfrac{1}{4}}} = \dfrac{\dfrac{1}{2}}{\sqrt{\dfrac{3}{4}}} =$

$\dfrac{\dfrac{1}{2}}{\dfrac{\sqrt{3}}{2}} = \dfrac{1}{2} \cdot \dfrac{2}{\sqrt{3}} = \dfrac{1 \cdot 2}{2\sqrt{3}} = \dfrac{1}{\sqrt{3}}, \text{ or } \dfrac{\sqrt{3}}{3}$

29. Graph $f(x) = \dfrac{1}{2}x + 3$.

We select values for x and find the corresponding values of $f(x)$. Then we plot the points and connect them with a smooth curve.

x	$f(x)$	$(x, f(x))$
-4	1	$(-4, 1)$
0	3	$(0, 3)$
2	4	$(2, 4)$

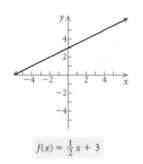

$f(x) = \frac{1}{2}x + 3$

31. Graph $f(x) = -x^2 + 4$.

We select values for x and find the corresponding values of $f(x)$. Then we plot the points and connect them with a smooth curve.

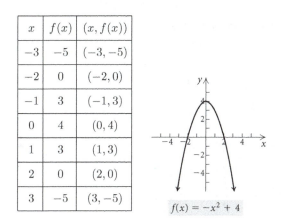

x	$f(x)$	$(x, f(x))$
-3	-5	$(-3, -5)$
-2	0	$(-2, 0)$
-1	3	$(-1, 3)$
0	4	$(0, 4)$
1	3	$(1, 3)$
2	0	$(2, 0)$
3	-5	$(3, -5)$

$$f(x) = -x^2 + 4$$

33. Graph $f(x) = \sqrt{x-1}$.

We select values for x and find the corresponding values of $f(x)$. Then we plot the points and connect them with a smooth curve.

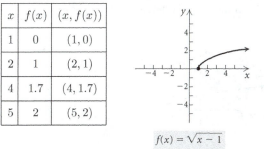

x	$f(x)$	$(x, f(x))$
1	0	$(1, 0)$
2	1	$(2, 1)$
4	1.7	$(4, 1.7)$
5	2	$(5, 2)$

$$f(x) = \sqrt{x-1}$$

35. From the graph we see that, when the input is 1, the output is -2, so $h(1) = -2$. When the input is 3, the output is 2, so $h(3) = 2$. When the input is 4, the output is 1, so $h(4) = 1$.

37. From the graph we see that, when the input is -4, the output is 3, so $s(-4) = 3$. When the input is -2, the output is 0, so $s(-2) = 0$. When the input is 0, the output is -3, so $s(0) = -3$.

39. From the graph we see that, when the input is -1, the output is 2, so $f(-1) = 2$. When the input is 0, the output is 0, so $f(0) = 0$. When the input is 1, the output is -2, so $f(1) = -2$.

41. This is not the graph of a function, because we can find a vertical line that crosses the graph more than once.

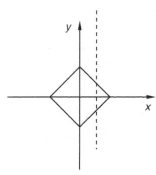

43. This is the graph of a function, because there is no vertical line that crosses the graph more than once.

45. This is the graph of a function, because there is no vertical line that crosses the graph more than once.

47. This is not the graph of a function, because we can find a vertical line that crosses the graph more than once.

49. We can substitute any real number for x. Thus, the domain is the set of all real numbers, or $(-\infty, \infty)$.

51. We can substitute any real number for x. Thus, the domain is the set of all real numbers, or $(-\infty, \infty)$.

53. The input 0 results in a denominator of 0. Thus, the domain is $\{x | x \neq 0\}$, or $(-\infty, 0) \cup (0, \infty)$.

55. We can substitute any real number in the numerator, but we must avoid inputs that make the denominator 0. We find these inputs.

$$2 - x = 0$$
$$2 = x$$

The domain is $\{x | x \neq 2\}$, or $(-\infty, 2) \cup (2, \infty)$.

57. We find the inputs that make the denominator 0:

$$x^2 - 4x - 5 = 0$$
$$(x - 5)(x + 1) = 0$$
$$x - 5 = 0 \ \ or \ \ x + 1 = 0$$
$$x = 5 \ \ or \ \ \ \ \ \ x = -1$$

The domain is $\{x | x \neq 5 \ and \ x \neq -1\}$, or $(-\infty, -1) \cup (-1, 5) \cup (5, \infty)$.

59. We can substitute any real number for x. Thus, the domain is the set of all real numbers, or $(-\infty, \infty)$.

61. We can substitute any real number in the numerator, but we must avoid inputs that make the denominator 0. We find these inputs.

$$x^2 - 7x = 0$$
$$x(x - 7) = 0$$
$$x = 0 \ or \ x - 7 = 0$$
$$x = 0 \ or \ \ \ \ \ \ x = 7$$

The domain is $\{x | x \neq 0 \ and \ x \neq 7\}$, or $(-\infty, 0) \cup (0, 7) \cup (7, \infty)$.

63. We can substitute any real number for x. Thus, the domain is the set of all real numbers, or $(-\infty, \infty)$.

65. The inputs on the x-axis that correspond to points on the graph extend from 0 to 5, inclusive. Thus, the domain is $\{x | 0 \leq x \leq 5\}$, or $[0, 5]$.

The outputs on the y-axis extend from 0 to 3, inclusive. Thus, the range is $\{y | 0 \leq y \leq 3\}$, or $[0, 3]$.

67. The inputs on the x-axis that correspond to points on the graph extend from -2π to 2π inclusive. Thus, the domain is $\{x | -2\pi \leq x \leq 2\pi\}$, or $[-2\pi, 2\pi]$.

The outputs on the y-axis extend from -1 to 1, inclusive. Thus, the range is $\{y | -1 \leq y \leq 1\}$, or $[-1, 1]$.

69. The graph extends to the left and to the right without bound. Thus, the domain is the set of all real numbers, or $(-\infty, \infty)$.

The only output is -3, so the range is $\{-3\}$.

71. The inputs on the x-axis extend from -5 to 3, inclusive. Thus, the domain is $[-5, 3]$.

The outputs on the y-axis extend from -2 to 2, inclusive. Thus, the range is $[-2, 2]$.

73.

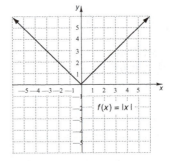

To find the domain we look for the inputs on the x-axis that correspond to a point on the graph. We see that each point on the x-axis corresponds to a point on the graph so the domain is the set of all real numbers, or $(-\infty, \infty)$.

To find the range we look for outputs on the y-axis. The number 0 is the smallest output, and every number greater than 0 is also an output. Thus, the range is $[0, \infty)$.

75.

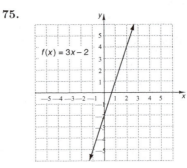

We see that each point on the x-axis corresponds to a point on the graph so the domain is the set of all real numbers, or $(-\infty, \infty)$. We also see that each point on the y-axis corresponds to an output so the range is the set of all real numbers, or $(-\infty, \infty)$.

77.

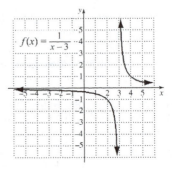

Since the graph does not touch or cross either the vertical line $x = 3$ or the x-axis, $y = 0$, 3 is excluded from the domain and 0 is excluded from the range.

Domain: $(-\infty, 3) \cup (3, \infty)$

Range: $(-\infty, 0) \cup (0, \infty)$

79.

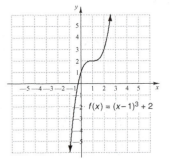

Each point on the x-axis corresponds to a point on the graph, so the domain is the set of all real numbers, or $(-\infty, \infty)$.

Each point on the y-axis also corresponds to a point on the graph, so the range is the set of all real numbers, $(-\infty, \infty)$.

81.

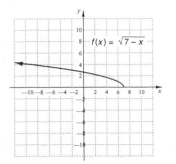

The largest input on the x-axis is 7 and every number less than 7 is also an input. Thus, the domain is $(-\infty, 7]$.

The number 0 is the smallest output, and every number greater than 0 is also an output. Thus, the range is $[0, \infty)$.

83.

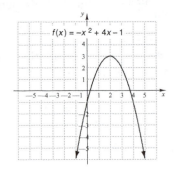

Each point on the x-axis corresponds to a point on the graph, so the domain is the set of all real numbers, or $(-\infty, \infty)$.

The largest output is 3 and every number less than 3 is also an output. Thus, the range is $(-\infty, 3]$.

85. a) $V(33) = 0.4306(33) + 11.0043 \approx \$25.21$
$V(40) = 0.4306(40) + 11.0043 \approx \$28.23$

b) Substitute 32 for $V(x)$ and solve for x.

$$32 = 0.4306x + 11.0043$$
$$20.9957 = 0.4306x$$
$$49 \approx x$$

It will take approximately \$32 to equal the value of \$1 in 1913 about 49 years after 1985, or in 2034.

87. $E(t) = 1000(100 - t) + 580(100 - t)^2$

a) $E(99.5) = 1000(100 - 99.5) + 580(100 - 99.5)^2$
$$= 1000(0.5) + 580(0.5)^2$$
$$= 500 + 580(0.25) = 500 + 145$$
$$= 645 \text{ m above sea level}$$

b) $E(100) = 1000(100 - 100) + 580(100 - 100)^2$
$$= 1000 \cdot 0 + 580(0)^2 = 0 + 0$$
$$= 0 \text{ m above sea level, or at sea level}$$

89. For $(-3, -2)$:
$$\underline{\quad y^2 - x^2 = -5 \quad}$$
$$(-2)^2 - (-3)^2 \ ? \ -5$$
$$4 - 9 \ \Big|$$
$$-5 \ \Big| \ -5 \ \text{ TRUE}$$

The equation $-5 = -5$ is true, so $(-3, -2)$ is a solution.

For $(2, -3)$:
$$\underline{\quad y^2 - x^2 = -5 \quad}$$
$$(-3)^2 - 2^2 \ ? \ -5$$
$$9 - 4 \ \Big|$$
$$5 \ \Big| \ -5 \ \text{ FALSE}$$

The equation $5 = -5$ is false, so $(2, -3)$ is not a solution.

91. For $\left(\dfrac{4}{5}, -2\right)$:
$$\underline{\quad 15x - 10y = 32 \quad}$$
$$15 \cdot \frac{4}{5} - 10(-2) \ ? \ 32$$
$$12 + 20 \ \Big|$$
$$32 \ \Big| \ 32 \quad \text{TRUE}$$

The equation $32 = 32$ is true, so $\left(\dfrac{4}{5}, -2\right)$ is a solution.

For $\left(\dfrac{11}{5}, \dfrac{1}{10}\right)$:
$$\underline{\quad 15x - 10y = 32 \quad}$$
$$15 \cdot \frac{11}{5} - 10 \cdot \frac{1}{10} \ ? \ 32$$
$$33 - 1 \ \Big|$$
$$32 \ \Big| \ 32 \quad \text{TRUE}$$

The equation $32 = 32$ is true, so $\left(\dfrac{11}{5}, \dfrac{1}{10}\right)$ is a solution.

93. Graph $y = \dfrac{1}{3}x - 6$.

Make a table of values, plot the points in the table, and draw the graph. If we choose values of x that are multiples of 3, we can avoid adding or subtracting fractions.

x	y	(x, y)
-3	-7	$(-3, -7)$
0	-6	$(0, -6)$
3	-5	$(0, -5)$

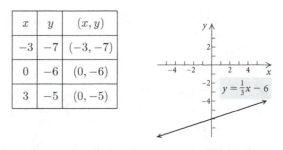

95. Graph $(x - 3)^2 + y^2 = 4$.

This is the equation of a circle. Writing it in standard form, we have

$$(x - 3)^2 + (y - 0)^2 = 2^2.$$

The circle has center $(3, 0)$ and radius 2.

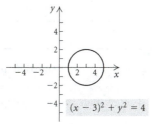

97. In the numerator we can substitute any real number for which the radicand is nonnegative. We see that $x + 1 \geq 0$ for $x \geq -1$. The denominator is 0 when $x = 0$, so 0 cannot be an input. Thus the domain is $\{x | x \geq -1 \ and \ x \neq 0\}$, or $[-1, 0) \cup (0, \infty)$.

99. $\sqrt{x}$ is defined for $x \geq 0$.

We find the inputs for which $4 - x$ is nonnegative.

$$4 - x \geq 0$$
$$4 \geq x, \text{ or } x \leq 4$$

The domain is $\{x | 0 \leq x \leq 4\}$, or $[0, 4]$.

101.

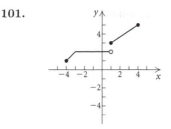

103. $f(x) = |x + 3| - |x - 4|$

a) If x is in the interval $(-\infty, -3)$, then $x + 3 < 0$ and $x - 4 < 0$. We have:

$$f(x) = |x + 3| - |x - 4|$$
$$= -(x + 3) - [-(x - 4)]$$
$$= -(x + 3) - (-x + 4)$$
$$= -x - 3 + x - 4$$
$$= -7$$

b) If x is in the interval $[-3, 4)$, then $x + 3 \geq 0$ and $x - 4 < 0$. We have:

$$
\begin{aligned}
f(x) &= |x + 3| - |x - 4| \\
&= x + 3 - [-(x - 4)] \\
&= x + 3 - (-x + 4) \\
&= x + 3 + x - 4 \\
&= 2x - 1
\end{aligned}
$$

c) If x is in the interval $[4, \infty)$, then $x + 3 > 0$ and $x - 4 \geq 0$. We have:

$$
\begin{aligned}
f(x) &= |x + 3| - |x - 4| \\
&= x + 3 - (x - 4) \\
&= x + 3 - x + 4 \\
&= 7
\end{aligned}
$$

Exercise Set 1.3

1. a) Yes. Each input is 1 more than the one that precedes it.

   b) Yes. Each output is 3 more than the one that precedes it.

   c) Yes. Constant changes in inputs result in constant changes in outputs.

3. a) Yes. Each input is 15 more than the one that precedes it.

   b) No. The change in the outputs varies.

   c) No. Constant changes in inputs do not result in constant changes in outputs.

5. Two points on the line are $(-4, -2)$ and $(1, 4)$.

$$
m = \frac{y_2 - y_1}{x_2 - x_1} = \frac{4 - (-2)}{1 - (-4)} = \frac{6}{5}
$$

7. Two points on the line are $(0, 3)$ and $(5, 0)$.

$$
m = \frac{y_2 - y_1}{x_2 - x_1} = \frac{0 - 3}{5 - 0} = \frac{-3}{5}, \text{ or } -\frac{3}{5}
$$

9. $m = \dfrac{y_2 - y_1}{x_2 - x_1} = \dfrac{3 - 3}{3 - 0} = \dfrac{0}{3} = 0$

11. $m = \dfrac{y_2 - y_1}{x_2 - x_1} = \dfrac{2 - 4}{-1 - 9} = \dfrac{-2}{-10} = \dfrac{1}{5}$

13. $m = \dfrac{y_2 - y_1}{x_2 - x_1} = \dfrac{6 - (-9)}{4 - 4} = \dfrac{15}{0}$

Since division by 0 is not defined, the slope is not defined.

15. $m = \dfrac{y_2 - y_1}{x_2 - x_1} = \dfrac{-0.4 - (-0.1)}{-0.3 - 0.7} = \dfrac{-0.3}{-1} = 0.3$

17. $m = \dfrac{y_2 - y_1}{x_2 - x_1} = \dfrac{-2 - (-2)}{4 - 2} = \dfrac{0}{2} = 0$

19. $m = \dfrac{y_2 - y_1}{x_2 - x_1} = \dfrac{\frac{3}{5} - \left(-\frac{3}{5}\right)}{-\frac{1}{2} - \frac{1}{2}} = \dfrac{\frac{6}{5}}{-1} = -\dfrac{6}{5}$

21. $m = \dfrac{y_2 - y_1}{x_2 - x_1} = \dfrac{-5 - (-13)}{-8 - 16} = \dfrac{8}{-24} = -\dfrac{1}{3}$

23. $m = \dfrac{7 - (-7)}{-10 - (-10)} = \dfrac{14}{0}$

Since division by 0 is not defined, the slope is not defined.

25. We have the points $(4, 3)$ and $(-2, 15)$.

$$
m = \frac{y_2 - y_1}{x_2 - x_1} = \frac{15 - 3}{-2 - 4} = \frac{12}{-6} = -2
$$

27. We have the points $\left(\dfrac{1}{5}, \dfrac{1}{2}\right)$ and $\left(-1, -\dfrac{11}{2}\right)$.

$$
m = \frac{y_2 - y_1}{x_2 - x_1} = \frac{-\frac{11}{2} - \frac{1}{2}}{-1 - \frac{1}{5}} = \frac{-6}{-\frac{6}{5}} = -6 \cdot \left(-\frac{5}{6}\right) = 5
$$

29. We have the points $\left(-6, \dfrac{4}{5}\right)$ and $\left(0, \dfrac{4}{5}\right)$.

$$
m = \frac{y_2 - y_1}{x_2 - x_1} = \frac{\frac{4}{5} - \frac{4}{5}}{-6 - 0} = \frac{0}{-6} = 0
$$

31. $y = 1.3x - 5$ is in the form $y = mx + b$ with $m = 1.3$, so the slope is 1.3.

33. The graph of $x = -2$ is a vertical line, so the slope is not defined.

35. $f(x) = -\dfrac{1}{2}x + 3$ is in the form $y = mx + b$ with $m = -\dfrac{1}{2}$, so the slope is $-\dfrac{1}{2}$.

37. $y = 9 - x$ can be written as $y = -x + 9$, or $y = -1 \cdot x + 9$. Now we have an equation in the form $y = mx + b$ with $m = -1$, so the slope is -1.

39. The graph of $y = 0.7$ is a horizontal line, so the slope is 0. (We also see this if we write the equation in the form $y = 0x + 0.7$).

41. We have the points $(2013, 8.4)$ and $(2020, 10.8)$. We find the average rate of change, or slope.

$$
m = \frac{10.8 - 8.4}{2020 - 2013} = \frac{2.4}{7} \approx 0.343
$$

The average rate of change in sales of electric bicycles from 2013 to 2020 is expected to be about $0.343 billion per year, or $343 million per year.

43. We have the data points $(2000, 478, 403)$ and $(2012, 390, 928)$. We find the average rate of change, or slope.

$$
m = \frac{390,928 - 478,403}{2012 - 2000} = \frac{-87,475}{12} \approx -7290
$$

The average rate of change in the population of Cleveland, Ohio, over the 12-year period was about -7290 people per year.

45. We have the data points $(2003, 550, 000)$ and $(2012, 810, 000)$. We find the average rate of change, or slope.

$$m = \frac{810,000 - 550,000}{2012 - 2003} = \frac{260,000}{9} \approx 28,889$$

The average rate of change in the number of acres used for growing almonds in California from 2003 to 2012 was about 28,889 acres per year.

47. We have the data points (1970, 25.3) and (2011, 5.5). We find the average rate of change, or slope.

$$m = \frac{5.5 - 25.3}{2011 - 1970} = \frac{-19.8}{41} \approx -0.5$$

The average rate of change in the per capita consumption of whole milk from 1970 to 2011 was about -0.5 gallons per year.

49. $y = \frac{3}{5}x - 7$

The equation is in the form $y = mx + b$ where $m = \frac{3}{5}$ and $b = -7$. Thus, the slope is $\frac{3}{5}$, and the y-intercept is $(0, -7)$.

51. $x = -\frac{2}{5}$

This is the equation of a vertical line $\frac{2}{5}$ unit to the left of the y-axis. The slope is not defined, and there is no y-intercept.

53. $f(x) = 5 - \frac{1}{2}x$, or $f(x) = -\frac{1}{2}x + 5$

The second equation is in the form $y = mx + b$ where $m = -\frac{1}{2}$ and $b = 5$. Thus, the slope is $-\frac{1}{2}$ and the y-intercept is $(0, 5)$.

55. Solve the equation for y.

$$3x + 2y = 10$$
$$2y = -3x + 10$$
$$y = -\frac{3}{2}x + 5$$

Slope: $-\frac{3}{2}$; y-intercept: $(0, 5)$

57. $y = -6 = 0 \cdot x - 6$

Slope: 0; y-intercept: $(0, -6)$

59. Solve the equation for y.

$$5y - 4x = 8$$
$$5y = 4x + 8$$
$$y = \frac{4}{5}x + \frac{8}{5}$$

Slope: $\frac{4}{5}$; y-intercept: $\left(0, \frac{8}{5}\right)$

61. Solve the equation for y.

$$4y - x + 2 = 0$$
$$4y = x - 2$$
$$y = \frac{1}{4}x - \frac{1}{2}$$

Slope: $\frac{1}{4}$; y-intercept: $\left(0, -\frac{1}{2}\right)$

63. Graph $y = -\frac{1}{2}x - 3$.

Plot the y-intercept, $(0, -3)$. We can think of the slope as $\frac{-1}{2}$. Start at $(0, -3)$ and find another point by moving down 1 unit and right 2 units. We have the point $(2, -4)$. We could also think of the slope as $\frac{1}{-2}$. Then we can start at $(0, -3)$ and get another point by moving up 1 unit and left 2 units. We have the point $(-2, -2)$. Connect the three points to draw the graph.

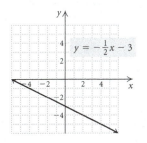

65. Graph $f(x) = 3x - 1$.

Plot the y-intercept, $(0, -1)$. We can think of the slope as $\frac{3}{1}$. Start at $(0, -1)$ and find another point by moving up 3 units and right 1 unit. We have the point $(1, 2)$. We can move from the point $(1, 2)$ in a similar manner to get a third point, $(2, 5)$. Connect the three points to draw the graph.

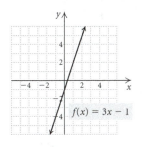

67. First solve the equation for y.

$$3x - 4y = 20$$
$$-4y = -3x + 20$$
$$y = \frac{3}{4}x - 5$$

Plot the y-intercept, $(0, -5)$. Then using the slope, $\frac{3}{4}$, start at $(0, -5)$ and find another point by moving up 3 units and right 4 units. We have the point $(4, -2)$. We can move from the point $(4, -2)$ in a similar manner to get a third point, $(8, 1)$. Connect the three points to draw the graph.

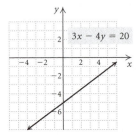

69. First solve the equation for y.

$$x + 3y = 18$$
$$3y = -x + 18$$
$$y = -\frac{1}{3}x + 6$$

Plot the y-intercept, $(0, 6)$. We can think of the slope as $\frac{-1}{3}$. Start at $(0, 6)$ and find another point by moving down 1 unit and right 3 units. We have the point $(3, 5)$. We can move from the point $(3, 5)$ in a similar manner to get a third point, $(6, 4)$. Connect the three points and draw the graph.

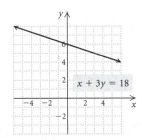

71. $P(0) = \frac{1}{33} \cdot 0 + 1 = 1$ atm

$$P(33) = \frac{1}{33} \cdot 33 + 1 = 2 \text{ atm}$$

$$P(1000) = \frac{1}{33} \cdot 1000 + 1 = 31\frac{10}{33} \text{ atm}$$

$$P(5000) = \frac{1}{33} \cdot 5000 + 1 = 152\frac{17}{33} \text{ atm}$$

$$P(7000) = \frac{1}{33} \cdot 7000 + 1 = 213\frac{4}{33} \text{ atm}$$

73. a) $D(r) = \frac{11}{10}r + \frac{1}{2}$

The slope is $\frac{11}{10}$.

For each mph faster the car travels, it takes $\frac{11}{10}$ ft longer to stop.

b) $D(5) = \frac{11}{10} \cdot 5 + \frac{1}{2} = \frac{11}{2} + \frac{1}{2} = \frac{12}{2} = 6$ ft

$$D(10) = \frac{11}{10} \cdot 10 + \frac{1}{2} = 11 + \frac{1}{2} = 11\frac{1}{2}, \text{ or } 11.5 \text{ ft}$$

$$D(20) = \frac{11}{10} \cdot 20 + \frac{1}{2} = 22 + \frac{1}{2} = 22\frac{1}{2}, \text{ or } 22.5 \text{ ft}$$

$$D(50) = \frac{11}{10} \cdot 50 + \frac{1}{2} = 55 + \frac{1}{2} = 55\frac{1}{2}, \text{ or } 55.5 \text{ ft}$$

$$D(65) = \frac{11}{10} \cdot 65 + \frac{1}{2} = \frac{143}{2} + \frac{1}{2} = \frac{144}{2} = 72 \text{ ft}$$

c) The speed cannot be negative. $D(0) = \frac{1}{2}$ which says that a stopped car travels $\frac{1}{2}$ ft before stopping. Thus, 0 is not in the domain. The speed can be positive, so the domain is $\{r | r > 0\}$, or $(0, \infty)$.

75. $C(t) = 2250 + 3380t$
$C(20) = 2250 + 3380 \cdot 20 = \$69,850$

77. $C(x) = 750 + 15x$
$C(32) = 750 + 15 \cdot 32 = \$1230$

79. $f(x) = x^2 - 3x$

$$f\left(\frac{1}{2}\right) = \left(\frac{1}{2}\right)^2 - 3 \cdot \frac{1}{2} = \frac{1}{4} - \frac{3}{2} = -\frac{5}{4}$$

81. $f(x) = x^2 - 3x$
$f(-5) = (-5)^2 - 3(-5) = 25 + 15 = 40$

83. $f(x) = x^2 - 3x$
$f(a + h) = (a + h)^2 - 3(a + h) = a^2 + 2ah + h^2 - 3a - 3h$

85. $m = \dfrac{y_2 - y_1}{x_2 - x_1} = \dfrac{(a+h)^2 - a^2}{a + h - a} = \dfrac{a^2 + 2ah + h^2 - a^2}{h} =$

$\dfrac{2ah + h^2}{h} = \dfrac{h(2a + h)}{h} = 2a + h$

87. False. For example, let $f(x) = x + 1$. Then $f(c - d) = c - d + 1$, but $f(c) - f(d) = c + 1 - (d + 1) = c - d$.

89.
$$f(x) = mx + b$$
$$f(x + 2) = f(x) + 2$$
$$m(x + 2) + b = mx + b + 2$$
$$mx + 2m + b = mx + b + 2$$
$$2m = 2$$
$$m = 1$$

Thus, $f(x) = 1 \cdot x + b$, or $f(x) = x + b$.

Chapter 1 Mid-Chapter Mixed Review

1. The statement is false. The x-intercept of a line that passes through the origin is $(0, 0)$.

3. The statement is false. The line parallel to the y-axis that passes through $(-5, 25)$ is $x = -5$.

5. Distance:
$$d = \sqrt{(-8 - 3)^2 + (-15 - 7)^2}$$
$$= \sqrt{(-11)^2 + (-22)^2}$$
$$= \sqrt{121 + 484}$$
$$= \sqrt{605} \approx 24.6$$

Midpoint: $\left(\dfrac{-8 + 3}{2}, \dfrac{-15 + 7}{2}\right) = \left(\dfrac{-5}{2}, \dfrac{-8}{2}\right) =$

$\left(-\dfrac{5}{2}, -4\right)$

7.
$$(x - h)^2 + (y - k)^2 = r^2$$
$$(x - (-5))^2 + (y - 2)^2 = 13^2$$
$$(x + 5)^2 + (y - 2)^2 = 169$$

9. Graph $3x - 6y = 6$.

We will find the intercepts along with a third point on the graph. Make a table of values, plot the points, and draw the graph.

x	y	(x, y)
2	0	$(2, 0)$
0	-1	$(0, -1)$
4	1	$(4, 1)$

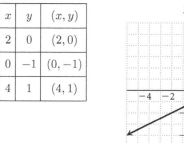

11. Graph $y = 2 - x^2$.

We choose some values for x and find the corresponding y-values. We list these points in a table, plot them, and draw the graph.

x	y	(x, y)
-2	-2	$(-2, -2)$
-1	1	$(-1, 1)$
0	2	$(0, 2)$
1	1	$(1, 1)$
2	-2	$(2, -2)$

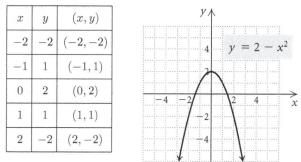

13. $f(x) = x - 2x^2$

$$f(-4) = -4 - 2(-4)^2 = -4 - 2 \cdot 16 = -4 - 32 = -36$$
$$f(0) = 0 - 2 \cdot 0^2 = 0 - 0 = 0$$
$$f(1) = 1 - 2 \cdot 1^2 = 1 - 2 \cdot 1 = 1 - 2 = -1$$

15. We can substitute any real number for x. Thus, the domain is the set of all real numbers, or $(-\infty, \infty)$.

17. We find the inputs for which the denominator is 0.
$$x^2 + 2x - 3 = 0$$
$$(x + 3)(x - 1) = 0$$
$$x + 3 = 0 \quad or \quad x - 1 = 0$$
$$x = -3 \quad or \quad x = 1$$
The domain is $\{x | x \neq -3 \text{ and } x \neq 1\}$, or $(-\infty, -3) \cup (-3, 1) \cup (1, \infty)$.

19. Graph $g(x) = x^2 - 1$.

Make a table of values, plot the points in the table, and draw the graph.

x	$g(x)$	$(x, g(x))$
-2	3	$(-2, 3)$
-1	0	$(-1, 0)$
0	-1	$(0, -1)$
1	0	$(1, 0)$
2	3	$(2, 3)$

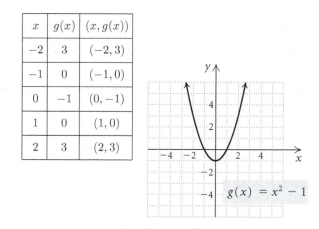

21. $m = \dfrac{y_2 - y_1}{x_2 - x_1} = \dfrac{-5 - 13}{-2 - (-2)} = \dfrac{-18}{0}$

Since division by 0 is not defined, the slope is not defined.

23. $m = \dfrac{y_2 - y_1}{x_2 - x_1} = \dfrac{\frac{1}{3} - \frac{1}{3}}{\frac{2}{7} - \frac{5}{7}} = \dfrac{0}{-\frac{3}{7}} = 0$

25. We can write $y = -6$ as $y = 0x - 6$, so the slope is 0 and the y-intercept is $(0, -6)$.

27. $3x - 16y + 1 = 0$
$$3x + 1 = 16y$$
$$\frac{3}{16}x + \frac{1}{16} = y$$
Slope: $\dfrac{3}{16}$; y-intercept: $\left(0, \dfrac{1}{16}\right)$

29. A vertical line $(x = a)$ crosses the graph more than once.

31. Let $A = (a, b)$ and $B = (c, d)$. The coordinates of a point C one-half of the way from A to B are $\left(\dfrac{a + c}{2}, \dfrac{b + d}{2}\right)$. A point D that is one-half of the way from C to B is $\dfrac{1}{2} + \dfrac{1}{2} \cdot \dfrac{1}{2}$, or $\dfrac{3}{4}$ of the way from A to B. Its coordinates are $\left(\dfrac{\frac{a+c}{2} + c}{2}, \dfrac{\frac{b+d}{2} + d}{2}\right)$, or $\left(\dfrac{a + 3c}{4}, \dfrac{b + 3d}{4}\right)$. Then a point E that is one-half of the way from D to B is $\dfrac{3}{4} + \dfrac{1}{2} \cdot \dfrac{1}{4}$, or $\dfrac{7}{8}$ of the way from A to B. Its coordinates are $\left(\dfrac{\frac{a+3c}{4} + c}{2}, \dfrac{\frac{b+3d}{4} + d}{2}\right)$, or $\left(\dfrac{a + 7c}{8}, \dfrac{b + 7d}{8}\right)$.

Exercise Set 1.4

1. We see that the y-intercept is $(0, -2)$. Another point on the graph is $(1, 2)$. Use these points to find the slope.
$$m = \frac{y_2 - y_1}{x_2 - x_1} = \frac{2 - (-2)}{1 - 0} = \frac{4}{1} = 4$$
We have $m = 4$ and $b = -2$, so the equation is $y = 4x - 2$.

3. We see that the y-intercept is $(0, 0)$. Another point on the graph is $(3, -3)$. Use these points to find the slope.

$m = \dfrac{y_2 - y_1}{x_2 - x_1} = \dfrac{-3 - 0}{3 - 0} = \dfrac{-3}{3} = -1$

We have $m = -1$ and $b = 0$, so the equation is
$y = -1 \cdot x + 0$, or $y = -x$.

5. We see that the y-intercept is $(0, -3)$. This is a horizontal line, so the slope is 0. We have $m = 0$ and $b = -3$, so the equation is $y = 0 \cdot x - 3$, or $y = -3$.

7. We substitute $\dfrac{2}{9}$ for m and 4 for b in the slope-intercept equation.

$y = mx + b$

$y = \dfrac{2}{9}x + 4$

9. We substitute -4 for m and -7 for b in the slope-intercept equation.

$y = mx + b$

$y = -4x - 7$

11. We substitute -4.2 for m and $\dfrac{3}{4}$ for b in the slope-intercept equation.

$y = mx + b$

$y = -4.2x + \dfrac{3}{4}$

13. Using the point-slope equation:

$y - y_1 = m(x - x_1)$

$y - 7 = \dfrac{2}{9}(x - 3)$ Substituting

$y - 7 = \dfrac{2}{9}x - \dfrac{2}{3}$

$y = \dfrac{2}{9}x + \dfrac{19}{3}$ Slope-intercept equation

Using the slope-intercept equation:

Substitute $\dfrac{2}{9}$ for m, 3 for x, and 7 for y in the slope-intercept equation and solve for b.

$y = mx + b$

$7 = \dfrac{2}{9} \cdot 3 + b$

$7 = \dfrac{2}{3} + b$

$\dfrac{19}{3} = b$

Now substitute $\dfrac{2}{9}$ for m and $\dfrac{19}{3}$ for b in $y = mx + b$.

$y = \dfrac{2}{9}x + \dfrac{19}{3}$

15. The slope is 0 and the second coordinate of the given point is 8, so we have a horizontal line 8 units above the x-axis. Thus, the equation is $y = 8$.

We could also use the point-slope equation or the slope-intercept equation to find the equation of the line.

Using the point-slope equation:

$y - y_1 = m(x - x_1)$

$y - 8 = 0(x - (-2))$ Substituting

$y - 8 = 0$

$y = 8$

Using the slope-intercept equation:

$y = mx + b$

$y = 0(-2) + 8$

$y = 8$

17. Using the point-slope equation:

$y - y_1 = m(x - x_1)$

$y - (-1) = -\dfrac{3}{5}(x - (-4))$

$y + 1 = -\dfrac{3}{5}(x + 4)$

$y + 1 = -\dfrac{3}{5}x - \dfrac{12}{5}$

$y = -\dfrac{3}{5}x - \dfrac{17}{5}$ Slope-intercept equation

Using the slope-intercept equation:

$y = mx + b$

$-1 = -\dfrac{3}{5}(-4) + b$

$-1 = \dfrac{12}{5} + b$

$-\dfrac{17}{5} = b$

Then we have $y = -\dfrac{3}{5}x - \dfrac{17}{5}$.

19. First we find the slope.

$m = \dfrac{-4 - 5}{2 - (-1)} = \dfrac{-9}{3} = -3$

Using the point-slope equation:

Using the point $(-1, 5)$, we get

$y - 5 = -3(x - (-1))$, or $y - 5 = -3(x + 1)$.

Using the point $(2, -4)$, we get

$y - (-4) = -3(x - 2)$, or $y + 4 = -3(x - 2)$.

In either case, the slope-intercept equation is
$y = -3x + 2$.

Using the slope-intercept equation and the point $(-1, 5)$:

$y = mx + b$

$5 = -3(-1) + b$

$5 = 3 + b$

$2 = b$

Then we have $y = -3x + 2$.

21. First we find the slope.

$m = \dfrac{4 - 0}{-1 - 7} = \dfrac{4}{-8} = -\dfrac{1}{2}$

Using the point-slope equation:

Using the point $(7, 0)$, we get

$$y - 0 = -\frac{1}{2}(x - 7).$$

Using the point $(-1, 4)$, we get

$$y - 4 = -\frac{1}{2}(x - (-1)), \text{ or}$$

$$y - 4 = -\frac{1}{2}(x + 1).$$

In either case, the slope-intercept equation is

$$y = -\frac{1}{2}x + \frac{7}{2}.$$

Using the slope-intercept equation and the point $(7, 0)$:

$$0 = -\frac{1}{2} \cdot 7 + b$$

$$\frac{7}{2} = b$$

Then we have $y = -\frac{1}{2}x + \frac{7}{2}$.

23. First we find the slope.

$$m = \frac{-4 - (-6)}{3 - 0} = \frac{2}{3}$$

We know the y-intercept is $(0, -6)$, so we substitute in the slope-intercept equation.

$$y = mx + b$$

$$y = \frac{2}{3}x - 6$$

25. First we find the slope.

$$m = \frac{7.3 - 7.3}{-4 - 0} = \frac{0}{-4} = 0$$

We know the y-intercept is $(0, \ 7.3)$, so we substitute in the slope-intercept equation.

$$y = mx + b$$

$$y = 0 \cdot x + 7.3$$

$$y = 7.3$$

27. The equation of the horizontal line through $(0, -3)$ is of the form $y = b$ where b is -3. We have $y = -3$.

The equation of the vertical line through $(0, -3)$ is of the form $x = a$ where a is 0. We have $x = 0$.

29. The equation of the horizontal line through $\left(\frac{2}{11}, -1\right)$ is of the form $y = b$ where b is -1. We have $y = -1$.

The equation of the vertical line through $\left(\frac{2}{11}, -1\right)$ is of the form $x = a$ where a is $\frac{2}{11}$. We have $x = \frac{2}{11}$.

31. We have the points $(1, 4)$ and $(-2, 13)$. First we find the slope.

$$m = \frac{13 - 4}{-2 - 1} = \frac{9}{-3} = -3$$

We will use the point-slope equation, choosing $(1, 4)$ for the given point.

$$y - 4 = -3(x - 1)$$

$$y - 4 = -3x + 3$$

$$y = -3x + 7, \text{ or}$$

$$h(x) = -3x + 7$$

Then $h(2) = -3 \cdot 2 + 7 = -6 + 7 = 1$.

33. We have the points $(5, 1)$ and $(-5, -3)$. First we find the slope.

$$m = \frac{-3 - 1}{-5 - 5} = \frac{-4}{-10} = \frac{2}{5}$$

We will use the slope-intercept equation, choosing $(5, 1)$ for the given point.

$$y = mx + b$$

$$1 = \frac{2}{5} \cdot 5 + b$$

$$1 = 2 + b$$

$$-1 = b$$

Then we have $f(x) = \frac{2}{5}x - 1$.

Now we find $f(0)$.

$$f(0) = \frac{2}{5} \cdot 0 - 1 = -1.$$

35. The slopes are $\frac{26}{3}$ and $-\frac{3}{26}$. Their product is -1, so the lines are perpendicular.

37. The slopes are $\frac{2}{5}$ and $-\frac{2}{5}$. The slopes are not the same and their product is not -1, so the lines are neither parallel nor perpendicular.

39. We solve each equation for y.

$$\begin{array}{ll} x + 2y = 5 & 2x + 4y = 8 \\[4pt] y = -\frac{1}{2}x + \frac{5}{2} & y = -\frac{1}{2}x + 2 \end{array}$$

We see that $m_1 = -\frac{1}{2}$ and $m_2 = -\frac{1}{2}$. Since the slopes are the same and the y-intercepts, $\frac{5}{2}$ and 2, are different, the lines are parallel.

41. We solve each equation for y.

$$\begin{array}{ll} y = 4x - 5 & 4y = 8 - x \\[4pt] & y = -\frac{1}{4}x + 2 \end{array}$$

We see that $m_1 = 4$ and $m_2 = -\frac{1}{4}$. Since $m_1 m_2 = 4\left(-\frac{1}{4}\right) = -1$, the lines are perpendicular.

43. $y = \frac{2}{7}x + 1$; $m = \frac{2}{7}$

The line parallel to the given line will have slope $\frac{2}{7}$. We use the point-slope equation for a line with slope $\frac{2}{7}$ and containing the point $(3, 5)$:

$$y - y_1 = m(x - x_1)$$

$$y - 5 = \frac{2}{7}(x - 3)$$

$$y - 5 = \frac{2}{7}x - \frac{6}{7}$$

$$y = \frac{2}{7}x + \frac{29}{7} \quad \text{Slope-intercept form}$$

The slope of the line perpendicular to the given line is the opposite of the reciprocal of $\frac{2}{7}$, or $-\frac{7}{2}$. We use the point-slope equation for a line with slope $-\frac{7}{2}$ and containing the point $(3, 5)$:

$$y - y_1 = m(x - x_1)$$

$$y - 5 = -\frac{7}{2}(x - 3)$$

$$y - 5 = -\frac{7}{2}x + \frac{21}{2}$$

$$y = -\frac{7}{2}x + \frac{31}{2} \quad \text{Slope-intercept form}$$

45. $y = -0.3x + 4.3; \ m = -0.3$

The line parallel to the given line will have slope -0.3. We use the point-slope equation for a line with slope -0.3 and containing the point $(-7, 0)$:

$$y - y_1 = m(x - x_1)$$

$$y - 0 = -0.3(x - (-7))$$

$$y = -0.3x - 2.1 \quad \text{Slope-intercept form}$$

The slope of the line perpendicular to the given line is the opposite of the reciprocal of -0.3, or $\frac{1}{0.3} = \frac{10}{3}$.

We use the point-slope equation for a line with slope $\frac{10}{3}$ and containing the point $(-7, 0)$:

$$y - y_1 = m(x - x_1)$$

$$y - 0 = \frac{10}{3}(x - (-7))$$

$$y = \frac{10}{3}x + \frac{70}{3} \quad \text{Slope-intercept form}$$

47. $3x + 4y = 5$

$$4y = -3x + 5$$

$$y = -\frac{3}{4}x + \frac{5}{4}; \ m = -\frac{3}{4}$$

The line parallel to the given line will have slope $-\frac{3}{4}$. We use the point-slope equation for a line with slope $-\frac{3}{4}$ and containing the point $(3, -2)$:

$$y - y_1 = m(x - x_1)$$

$$y - (-2) = -\frac{3}{4}(x - 3)$$

$$y + 2 = -\frac{3}{4}x + \frac{9}{4}$$

$$y = -\frac{3}{4}x + \frac{1}{4} \quad \text{Slope-intercept form}$$

The slope of the line perpendicular to the given line is the opposite of the reciprocal of $-\frac{3}{4}$, or $\frac{4}{3}$. We use the point-slope equation for a line with slope $\frac{4}{3}$ and containing the point $(3, -2)$:

$$y - y_1 = m(x - x_1)$$

$$y - (-2) = \frac{4}{3}(x - 3)$$

$$y + 2 = \frac{4}{3}x - 4$$

$$y = \frac{4}{3}x - 6 \quad \text{Slope-intercept form}$$

49. $x = -1$ is the equation of a vertical line. The line parallel to the given line is a vertical line containing the point $(3, -3)$, or $x = 3$.

The line perpendicular to the given line is a horizontal line containing the point $(3, -3)$, or $y = -3$.

51. $x = -3$ is a vertical line and $y = 5$ is a horizontal line, so it is true that the lines are perpendicular.

53. The lines have the same slope, $\frac{2}{5}$, and different y-intercepts, $(0, 4)$ and $(0, -4)$, so it is true that the lines are parallel.

55. $x = -1$ and $x = 1$ are both vertical lines, so it is false that they are perpendicular.

57. No. The data points fall faster from 0 to 2 than after 2 (that is, the rate of change is not constant), so they cannot be modeled by a linear function.

59. Yes. The rate of change seems to be constant, so the data points might be modeled by a linear function.

61. a) Answers may vary depending on the data points used. We will use $(1, \ 1.319)$ and $(7, \ 2.749)$.

$$m = \frac{2.749 - 1.319}{7 - 1} = \frac{1.43}{6} \approx 0.238$$

We will use the point-slope equation, letting $(x_1, y_1) = (1, \ 1.319)$.

$$y - 1.319 = 0.238(x - 1)$$

$$y - 1.319 = 0.238x - 0.238$$

$$y = 0.238x + 1.081,$$

where x is the number of years after 2006 and y is in billions.

b) In 2017, $x = 2017 - 2006 = 11$.

$$y = 0.238(11) + 1.081 = 3.699 \text{ billion Internet users}$$

In 2020, $x = 2020 - 2006 = 14$.

$$y = 0.238(14) + 1.081 = 4.413 \text{ billion users}$$

63. Answers may vary depending on the data points used. We will use $(0, \ 11,504)$ and $(3, \ 10,819)$.

$$m = \frac{10,819 - 11,504}{3 - 0} = \frac{-685}{3} \approx -228$$

We see that the y-intercept is $(0, \ 11,504)$, so using the slope-intercept equation, we have $y = -228x + 11,504$, where x is the number of years after 2010 and y is in kilowatt-hours.

In 2019, $x = 2019 - 2010 = 9$.

$y = -228(9) + 11,504 = 9452$ kilowatt-hours

65. Answers may vary depending on the data points used. We will use $(1, 28.3)$ and $(3, 30.8)$.

$$m = \frac{30.8 - 28.3}{3 - 1} = \frac{2.5}{2} = 1.25$$

We will use the point-slope equation, letting $(x_1, y_1) = (1, 28.3)$

$$y - 28.3 = 1.25(x - 1)$$
$$y - 28.3 = 1.25x - 1.25$$
$$y = 1.25x + 27.05,$$

where x is the number of years after 2009 and y is in gallons.

In 2017, $x = 2017 - 2009 = 8$.

$y = 1.25(8) + 27.05 \approx 37.1$ gallons

67. $m = \dfrac{-7 - 7}{5 - 5} = \dfrac{-14}{0}$

The slope is not defined.

69. $r = \dfrac{d}{2} = \dfrac{5}{2}$

$$(x - 0)^2 + (y - 3)^2 = \left(\frac{5}{2}\right)^2$$
$$x^2 + (y - 3)^2 = \frac{25}{4}, \text{ or}$$
$$x^2 + (y - 3)^2 = 6.25$$

71. The slope of the line containing $(-3, k)$ and $(4, 8)$ is

$$\frac{8 - k}{4 - (-3)} = \frac{8 - k}{7}.$$

The slope of the line containing $(5, 3)$ and $(1, -6)$ is

$$\frac{-6 - 3}{1 - 5} = \frac{-9}{-4} = \frac{9}{4}.$$

The slopes must be equal in order for the lines to be parallel:

$$\frac{8 - k}{7} = \frac{9}{4}$$
$$32 - 4k = 63 \quad \text{Multiplying by 28}$$
$$-4k = 31$$
$$k = -\frac{31}{4}, \text{ or } -7.75$$

73. $m = \dfrac{920.58}{13,740} = 0.067$

The road grade is 6.7%.

We find an equation of the line with slope 0.067 and containing the point $(13,740, 920.58)$:

$$y - 920.58 = 0.067(x - 13,740)$$
$$y - 920.58 = 0.067x - 920.58$$
$$y = 0.067x$$

Exercise Set 1.5

1. $4x + 5 = 21$
$$4x = 16 \quad \text{Subtracting 5 on both sides}$$
$$x = 4 \quad \text{Dividing by 4 on both sides}$$
The solution is 4.

3. $23 - \dfrac{2}{5}x = -\dfrac{2}{5}x + 23$
$$23 = 23 \qquad \text{Adding } \frac{2}{5}x \text{ on both sides}$$

We get an equation that is true for any value of x, so the solution set is the set of real numbers, $\{x | x \text{ is a real number}\}$, or $(-\infty, \infty)$.

5. $4x + 3 = 0$
$$4x = -3 \quad \text{Subtracting 3 on both sides}$$
$$x = -\frac{3}{4} \quad \text{Dividing by 4 on both sides}$$
The solution is $-\dfrac{3}{4}$.

7. $3 - x = 12$
$$-x = 9 \quad \text{Subtracting 3 on both sides}$$
$$x = -9 \quad \text{Multiplying (or dividing) by } -1 \text{ on both sides}$$
The solution is -9.

9. $3 - \dfrac{1}{4}x = \dfrac{3}{2} \qquad$ The LCD is 4.
$$4\left(3 - \frac{1}{4}x\right) = 4 \cdot \frac{3}{2} \quad \begin{array}{l}\text{Multiplying by the LCD} \\ \text{to clear fractions}\end{array}$$
$$12 - x = 6$$
$$-x = -6 \quad \text{Subtracting 12 on both sides}$$
$$x = 6 \quad \begin{array}{l}\text{Multiplying (or dividing) by } -1 \\ \text{on both sides}\end{array}$$
The solution is 6.

11. $\dfrac{2}{11} - 4x = -4x + \dfrac{9}{11}$
$$\frac{2}{11} = \frac{9}{11} \qquad \text{Adding } 4x \text{ on both sides}$$
We get a false equation. Thus, the original equation has no solution.

13. $8 = 5x - 3$
$$11 = 5x \quad \text{Adding 3 on both sides}$$
$$\frac{11}{5} = x \quad \text{Dividing by 5 on both sides}$$
The solution is $\dfrac{11}{5}$.

15.
$$\frac{2}{5}y - 2 = \frac{1}{3} \qquad \text{The LCD is 15.}$$

$$15\left(\frac{2}{5}y - 2\right) = 15 \cdot \frac{1}{3} \quad \begin{array}{l}\text{Multiplying by the LCD}\\ \text{to clear fractions}\end{array}$$

$$6y - 30 = 5$$

$$6y = 35 \qquad \text{Adding 30 on both sides}$$

$$y = \frac{35}{6} \qquad \text{Dividing by 6 on both sides}$$

The solution is $\frac{35}{6}$.

17. $y + 1 = 2y - 7$

$\quad 1 = y - 7 \quad$ Subtracting y on both sides

$\quad 8 = y \qquad$ Adding 7 on both sides

The solution is 8.

19. $2x + 7 = x + 3$

$\quad x + 7 = 3 \qquad$ Subtracting x on both sides

$\quad x = -4 \qquad$ Subtracting 7 on both sides

The solution is -4.

21. $3x - 5 = 2x + 1$

$\quad x - 5 = 1 \quad$ Subtracting $2x$ on both sides

$\quad x = 6 \quad$ Adding 5 on both sides

The solution is 6.

23. $4x - 5 = 7x - 2$

$\quad -5 = 3x - 2 \quad$ Subtracting $4x$ on both sides

$\quad -3 = 3x \qquad$ Adding 2 on both sides

$\quad -1 = x \qquad$ Dividing by 3 on both sides

The solution is -1.

25. $5x - 2 + 3x = 2x + 6 - 4x$

$\quad 8x - 2 = 6 - 2x \quad$ Collecting like terms

$\quad 8x + 2x = 6 + 2 \quad \begin{array}{l}\text{Adding } 2x \text{ and 2 on}\\ \text{both sides}\end{array}$

$\quad 10x = 8 \qquad$ Collecting like terms

$\quad x = \frac{8}{10} \qquad \begin{array}{l}\text{Dividing by 10 on both}\\ \text{sides}\end{array}$

$\quad x = \frac{4}{5} \qquad$ Simplifying

The solution is $\frac{4}{5}$.

27. $7(3x + 6) = 11 - (x + 2)$

$\quad 21x + 42 = 11 - x - 2 \quad \begin{array}{l}\text{Using the distributive}\\ \text{property}\end{array}$

$\quad 21x + 42 = 9 - x \qquad$ Collecting like terms

$\quad 21x + x = 9 - 42 \qquad \begin{array}{l}\text{Adding } x \text{ and subtract-}\\ \text{ing 42 on both sides}\end{array}$

$\quad 22x = -33 \qquad$ Collecting like terms

$\quad x = -\frac{33}{22} \qquad \begin{array}{l}\text{Dividing by 22 on both}\\ \text{sides}\end{array}$

$\quad x = -\frac{3}{2} \qquad$ Simplifying

The solution is $-\frac{3}{2}$.

29. $3(x + 1) = 5 - 2(3x + 4)$

$\quad 3x + 3 = 5 - 6x - 8 \qquad$ Removing parentheses

$\quad 3x + 3 = -6x - 3 \qquad$ Collecting like terms

$\quad 9x + 3 = -3 \qquad$ Adding $6x$

$\quad 9x = -6 \qquad$ Subtracting 3

$\quad x = -\frac{2}{3} \qquad$ Dividing by 9

The solution is $-\frac{2}{3}$.

31. $2(x - 4) = 3 - 5(2x + 1)$

$\quad 2x - 8 = 3 - 10x - 5 \quad \begin{array}{l}\text{Using the distributive}\\ \text{property}\end{array}$

$\quad 2x - 8 = -10x - 2 \qquad$ Collecting like terms

$\quad 12x = 6 \qquad$ Adding $10x$ and 8 on both sides

$\quad x = \frac{1}{2} \qquad$ Dividing by 12 on both sides

The solution is $\frac{1}{2}$.

33. **Familiarize.** Let $w =$ the number of new words that appeared in the English language in the seventeenth century. Then the number of new words that appeared in the nineteenth century is $w + 46.9\%$ of w, or $w + 0.469w$, or $1.469w$.

Translate. The number of new words that appeared in the nineteenth century is 75,029, so we have

$$75,029 = 1.469w.$$

Carry out.

$$75,029 = 1.469w$$

$$51,075 \approx w \qquad \text{Dividing by 1.469}$$

Check. 46.9% of $51,075 = 0.469(51,075) \approx 23,954$, and $51,075 + 23,954 = 75,029$. This is the number of new words that appeared in the nineteenth century, so the answer checks.

State. In the seventeenth century about 51,075 new words appeared in the English language.

35. **Familiarize.** Let $P =$ the amount Kea borrowed. We will use the formula $I = Prt$ to find the interest owed. For $r = 5\%$, or 0.05, and $t = 1$, we have $I = P(0.05)(1)$, or $0.05P$.

Translate.

$$\underbrace{\text{Amount borrowed}}_{\displaystyle P} \underset{\displaystyle +}{\text{ plus }} \underset{\displaystyle 0.05P}{\text{ interest }} \underset{\displaystyle =}{\text{ is }} \underset{\displaystyle 1365}{\$1365.}$$

Carry out. We solve the equation.

$$P + 0.05P = 1365$$

$$1.05P = 1365 \qquad \text{Adding}$$

$$P = 1300 \qquad \text{Dividing by 1.05}$$

Check. The interest due on a loan of \$1300 for 1 year at a rate of 5% is $\$1300(0.05)(1)$, or \$65, and $\$1300 + \$65 = \$1365$. The answer checks.

State. Kea borrowed \$1300.

37. Familiarize. We make a drawing.

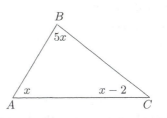

We let x = the measure of angle A. Then $5x$ = the measure of angle B, and $x - 2$ = the measure of angle C. The sum of the angle measures is 180°.

Translate.

Measure of angle A	+	Measure of angle B	+	Measure of angle C	= 180.

$$x + 5x + x - 2 = 180$$

Carry out. We solve the equation.

$$x + 5x + x - 2 = 180$$
$$7x - 2 = 180$$
$$7x = 182$$
$$x = 26$$

If $x = 26$, then $5x = 5 \cdot 26$, or 130, and $x - 2 = 26 - 2$, or 24.

Check. The measure of angle B, 130°, is five times the measure of angle A, 26°. The measure of angle C, 24°, is 2° less than the measure of angle A, 26°. The sum of the angle measures is $26° + 130° + 24°$, or 180°. The answer checks.

State. The measure of angles A, B, and C are 26°, 130°, and 24°, respectively.

39. Familiarize. Let c = the amount of clothing exports from the United States in 2012, in billions of dollars.

Translate.

Clothing imports in 2012	were 25 times	clothing exports in 2012	less	$1.459 billion.
84.916	= 25 ·	c	−	1.459.

Carry out.

$$84.916 = 25c - 1.459$$
$$86.375 = 25c$$
$$3.455 = c$$

Check. $25(3.455) - 1.459 = 86.37 - 1.459 = 84.916$. This is the amount of clothing imports, so the answer checks.

State. In 2012, clothing exports from the United States were $3.455 billion.

41. Familiarize. We make a drawing. Let t = the number of hours the passenger train travels before it overtakes the freight train. Then $t + 1$ = the number of hours the freight train travels before it is overtaken by the passenger train. Also let d = the distance the trains travel.

Freight train
60 mph $t + 1$ hr d

Passenger train
80 mph t hr d

We can also organize the information in a table.

$$d = r \cdot t$$

	Distance	Rate	Time
Freight train	d	60	$t + 1$
Passenger train	d	80	t

Translate. Using the formula $d = rt$ in each row of the table, we get two equations.

$$d = 60(t + 1) \text{ and } d = 80t.$$

Since the distances are the same, we have the equation

$$60(t + 1) = 80t.$$

Carry out. We solve the equation.

$$60(t + 1) = 80t$$
$$60t + 60 = 80t$$
$$60 = 20t$$
$$3 = t$$

When $t = 3$, then $t + 1 = 3 + 1 = 4$.

Check. In 4 hr the freight train travels $60 \cdot 4$, or 240 mi. In 3 hr the passenger train travels $80 \cdot 3$, or 240 mi. Since the distances are the same, the answer checks.

State. It will take the passenger train 3 hr to overtake the freight train.

43. Familiarize. Let p = the percentage of federal tax filers in 2000 who had zero or negative tax liability. Then the percentage of filers in 2010 who had zero or negative tax liability was $p + 15.7$.

Translate. In 2010, 40.9% of filers had zero or negative tax liability, so we have

$$40.9 = p + 15.7.$$

Carry out.

$$40.9 = p + 15.7$$
$$25.2 = p \qquad \text{Subtracting 15.7}$$

Check. $25.2 + 15.7 = 40.9$, so the answer checks.

State. In 2000, about 25.2% of federal tax filers had zero or negative tax liability.

45. Familiarize. Let a = the amount of sales for which the two choices will be equal.

Translate.

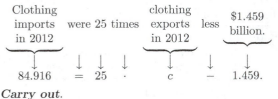

$1800	equals	$1600	plus	4%	of	amount sold.
1800	=	1600	+	0.04	·	a

Carry out.

$$1800 = 1600 + 0.04a$$
$$200 = 0.04a$$
$$5000 = a$$

Check. \$1600 + 4% of \$5000 = \$1600 + 0.04 · \$5000 = \$1600 + \$200 = \$1800, so the answer checks.

State. For sales of \$5000, the two choices will be equal.

47. *Familiarize.* Let s = the number of U.S. students who studied abroad during the 2012-2013 school year.

    Translate.

Number of U.S. students abroad	was	seven-twentieths	of	number of foreign students in U.S.
↓	↓	↓	↓	↓
s	$=$	$\dfrac{7}{20}$	$\cdot$	$820,000$

    Carry out.

    $$s = \frac{7}{20} \cdot 820,000$$
    $$s = 287,000$$

    Check. We repeat the calculation. The answer checks.

    State. About 287,000 U.S. students studied abroad during the 2012-2013 school year.

49. *Familiarize.* Let l = the length of the soccer field and $l - 35$ = the width, in yards.

    Translate. We use the formula for the perimeter of a rectangle. We substitute 330 for P and $l - 35$ for w.

    $$P = 2l + 2w$$
    $$330 = 2l + 2(l - 35)$$

    Carry out. We solve the equation.

    $$330 = 2l + 2(l - 35)$$
    $$330 = 2l + 2l - 70$$
    $$330 = 4l - 70$$
    $$400 = 4l$$
    $$100 = l$$

    If $l = 100$, then $l - 35 = 100 - 35 = 65$.

    Check. The width, 65 yd, is 35 yd less than the length, 100 yd. Also, the perimeter is

    $$2 \cdot 100 \text{ yd} + 2 \cdot 65 \text{ yd} = 200 \text{ yd} + 130 \text{ yd} = 330 \text{ yd}.$$

    The answer checks.

    State. The length of the field is 100 yd, and the width is 65 yd.

51. *Familiarize.* Using the labels on the drawing in the text, we let w = the width of the test plot and $w + 25$ = the length, in meters. Recall that for a rectangle, Perimeter = 2 · length + 2 · width.

    Translate.

Perimeter	=	2 · length	+	2 · width
↓	↓	↓	↓	↓
322	$=$	$2(w + 25)$	$+$	$2 \cdot w$

Carry out. We solve the equation.

$$322 = 2(w + 25) + 2 \cdot w$$
$$322 = 2w + 50 + 2w$$
$$322 = 4w + 50$$
$$272 = 4w$$
$$68 = w$$

When $w = 68$, then $w + 25 = 68 + 25 = 93$.

Check. The length is 25 m more than the width: $93 = 68 + 25$. The perimeter is $2 \cdot 93 + 2 \cdot 68$, or $186 + 136$, or 322 m. The answer checks.

State. The length is 93 m; the width is 68 m.

53. *Familiarize.* Let t = the number of hours it will take the plane to travel 1050 mi into the wind. The speed into the headwind is $450 - 30$, or 420 mph.

    Translate. We use the formula $d = rt$.

    $$1050 = 420 \cdot t$$

    Carry out. We solve the equation.

    $$1050 = 420 \cdot t$$
    $$2.5 = t$$

    Check. At a rate of 420 mph, in 2.5 hr the plane travels $420(2.5)$, or 1050 mi. The answer checks.

    State. It will take the plane 2.5 hr to travel 1050 mi into the wind.

55. *Familiarize.* Let x = the amount invested at 3% interest. Then $5000 - x$ = the amount invested at 4%. We organize the information in a table, keeping in mind the simple interest formula, $I = Prt$.

	Amount invested	Interest rate	Time	Amount of interest
3% invest-ment	x	3%, or 0.03	1 yr	$x(0.03)(1)$, or $0.03x$
4% invest-ment	$5000 - x$	4%, or 0.04	1 yr	$(5000 - x)(0.04)(1)$, or $0.04(5000 - x)$
Total	5000			176

    Translate.

    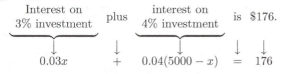

Interest on 3% investment	plus	interest on 4% investment	is	\$176.
↓	↓	↓	↓	↓
$0.03x$	$+$	$0.04(5000 - x)$	$=$	176

    Carry out. We solve the equation.

    $$0.03x + 0.04(5000 - x) = 176$$
    $$0.03x + 200 - 0.04x = 176$$
    $$-0.01x + 200 = 176$$
    $$-0.01x = -24$$
    $$x = 2400$$

    If $x = 2400$, then $5000 - x = 5000 - 2400 = 2600$.

    Check. The interest on \$2400 at 3% for 1 yr is $\$2400(0.03)(1) = \$72$. The interest on \$2600 at 4% for 1 yr is $\$2600(0.04)(1) = \$104$. Since $\$72 + \$104 = \$176$, the answer checks.

State. $2400 was invested at 3%, and $2600 was invested at 4%.

57. *Familiarize*. Let p = the number of patents Samsung received in 2013. Then $p + 2133$ = the number of patents IBM received.

Translate.

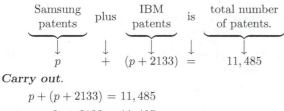

Samsung patents	plus	IBM patents	is	total number of patents.
p	$+$	$(p + 2133)$	$=$	$11,485$

Carry out.
$$p + (p + 2133) = 11,485$$
$$2p + 2133 = 11,485$$
$$2p = 9352$$
$$p = 4676$$

Check. If $p = 4676$, then $p + 2133 = 4676 + 2133 = 6809$, and $4676 + 6809 = 11,485$, the total number of patents, so the answer checks.

State. In 2013, Samsung received 4676 patents, and IBM received 6809 patents.

59. *Familiarize*. Let d = the average depth of the Atlantic Ocean, in feet. Then $\frac{4}{5}d - 272$ = the average depth of the Indian Ocean.

Translate.

Average depth of Pacific Ocean	is	Average depth of Atlantic Ocean	plus	Average depth of Indian Ocean	less 8890 ft
$14,040$	$=$	d	$+$	$\frac{4}{5}d - 272$	$-\quad 8890$

Carry out. We solve the equation.
$$14,040 = d + \frac{4}{5}d - 272 - 8890$$
$$14,040 = \frac{9}{5}d - 9162$$
$$23,202 = \frac{9}{5}d$$
$$\frac{5}{9} \cdot 23,202 = d$$
$$12,890 = d$$

If $d = 12,890$, then the average depth of the Indian Ocean is $\frac{4}{5} \cdot 12,890 - 272 = 10,040$.

Check. $12,890 + 10,040 - 8890 = 14,040$, so the answer checks.

State. The average depth of the Indian Ocean is 10,040 ft.

61. *Familiarize*. Let w = the number of pounds of Lily's body weight that is water.

Translate.

55%	of	body weight	is	water.
0.55	$\times$	135	$=$	w

Carry out. We solve the equation.
$$0.55 \times 135 = w$$
$$74.25 = w$$

Check. Since 55% of 135 is 74.25, the answer checks.

State. 74.25 lb of Lily's body weight is water.

63. *Familiarize*. Let t = the number of hours it takes the kayak to travel 36 mi upstream. The kayak travels upstream at a rate of $12 - 4$, or 8 mph.

Translate. We use the formula $d = rt$.
$$36 = 8 \cdot t$$

Carry out. We solve the equation.
$$36 = 8 \cdot t$$
$$4.5 = t$$

Check. At a rate of 8 mph, in 4.5 hr the kayak travels $8(4.5)$, or 36 mi. The answer checks.

State. It takes the kayak 4.5 hr to travel 36 mi upstream.

65. *Familiarize*. Let w = Rosalyn's regular hourly wage. She earned $40w$ for working the first 40 hr. She worked $48 - 40$, or 8 hr, of overtime. She earned $8(1.5w)$ for working 8 hr of overtime.

Translate. The total earned was $1066, so we write an equation.
$$40w + 8(1.5w) = 1066$$

Carry out. We solve the equation.
$$40w + 8(1.5)w = 1066$$
$$40w + 12w = 1066$$
$$52w = 1066$$
$$w = 20.5$$

Check. $40(\$20.50) + 8[1.5(\$20.50)] = \$820 + \$246 = \$1066$, so the answer checks.

State. Rosalyn's regular hourly wage is $20.50.

67. *Familiarize*. Let p = the percent of the world's olive oil consumed in the United States. Then the percent consumed in Italy is $3\frac{3}{4} \cdot p$, or $\frac{15}{4}p$, and the percent consumed in Spain is $\frac{2}{3} \cdot \frac{15}{4}p$, or $\frac{5}{2}p$.

Translate.

Percent of olive oil consumed in Italy, Spain, and the U.S.	is	58%.
$p + \frac{15}{4}p + \frac{5}{2}p$	$=$	58

Carry out.
$$p + \frac{15}{4}p + \frac{5}{2}p = 58$$
$$4\left(p + \frac{15}{4}p + \frac{5}{2}p\right) = 4 \cdot 58$$
$$4p + 15p + 10p = 232$$
$$29p = 232$$
$$p = 8$$

If $p = 8$, then $\frac{15}{4}p = \frac{15}{4} \cdot 8 = 30$ and $\frac{5}{2}p = \frac{5}{2} \cdot 8 = 20$.

Check. 30% is $3\frac{3}{4}$ times 8%, and 20% is $\frac{2}{3}$ of 30%. Also, $8\% + 30\% + 20\% = 58\%$, so the answer checks.

State. Italy, Spain, and the United States consume 30%, 20%, and 8% of the world's olive oil, respectively.

69.
$$x + 5 = 0 \qquad \text{Setting } f(x) = 0$$
$$x + 5 - 5 = 0 - 5 \quad \text{Subtracting 5 on both sides}$$
$$x = -5$$
The zero of the function is -5.

71.
$$-2x + 11 = 0 \qquad \text{Setting } f(x) = 0$$
$$-2x + 11 - 11 = 0 - 11 \quad \text{Subtracting 11 on both sides}$$
$$-2x = -11$$
$$x = \frac{11}{2} \quad \text{Dividing by } -2 \text{ on both sides}$$
The zero of the function is $\frac{11}{2}$.

73.
$$16 - x = 0 \qquad \text{Setting } f(x) = 0$$
$$16 - x + x = 0 + x \quad \text{Adding } x \text{ on both sides}$$
$$16 = x$$
The zero of the function is 16.

75.
$$x + 12 = 0 \qquad \text{Setting } f(x) = 0$$
$$x + 12 - 12 = 0 - 12 \quad \text{Subtracting 12 on both sides}$$
$$x = -12$$
The zero of the function is -12.

77.
$$-x + 6 = 0 \qquad \text{Setting } f(x) = 0$$
$$-x + 6 + x = 0 + x \quad \text{Adding } x \text{ on both sides}$$
$$6 = x$$
The zero of the function is 6.

79.
$$20 - x = 0 \qquad \text{Setting } f(x) = 0$$
$$20 - x + x = 0 + x \quad \text{Adding } x \text{ on both sides}$$
$$20 = x$$
The zero of the function is 20.

81.
$$\frac{2}{5}x - 10 = 0 \qquad \text{Setting } f(x) = 0$$
$$\frac{2}{5}x = 10 \qquad \text{Adding 10 on both sides}$$
$$\frac{5}{2} \cdot \frac{2}{5}x = \frac{5}{2} \cdot 10 \quad \text{Multiplying by } \frac{5}{2} \text{ on both sides}$$
$$x = 25$$
The zero of the function is 25.

83.
$$-x + 15 = 0 \quad \text{Setting } f(x) = 0$$
$$15 = x \quad \text{Adding } x \text{ on both sides}$$
The zero of the function is 15.

85. a) The graph crosses the x-axis at $(4, 0)$. This is the x-intercept.

  b) The zero of the function is the first coordinate of the x-intercept. It is 4.

87. a) The graph crosses the x-axis at $(-2, 0)$. This is the x-intercept.

  b) The zero of the function is the first coordinate of the x-intercept. It is -2.

89. a) The graph crosses the x-axis at $(-4, 0)$. This is the x-intercept.

  b) The zero of the function is the first coordinate of the x-intercept. It is -4.

91. First find the slope of the given line.
$$3x + 4y = 7$$
$$4y = -3x + 7$$
$$y = -\frac{3}{4}x + \frac{7}{4}$$

The slope is $-\frac{3}{4}$. Now write a slope-intercept equation of the line containing $(-1, 4)$ with slope $-\frac{3}{4}$.
$$y - 4 = -\frac{3}{4}[x - (-1)]$$
$$y - 4 = -\frac{3}{4}(x + 1)$$
$$y - 4 = -\frac{3}{4}x - \frac{3}{4}$$
$$y = -\frac{3}{4}x + \frac{13}{4}$$

93.
$$d = \sqrt{(x_2 - x_1)^2 + (y_2 - y_1)^2}$$
$$= \sqrt{(-10 - 2)^2 + (-3 - 2)^2}$$
$$= \sqrt{144 + 25} = \sqrt{169} = 13$$

95.
$$f(x) = \frac{x}{x - 3}$$
$$f(-3) = \frac{-3}{-3 - 3} = \frac{-3}{-6} = \frac{1}{2}$$
$$f(0) = \frac{0}{0 - 3} = \frac{0}{-3} = 0$$
$$f(3) = \frac{3}{3 - 3} = \frac{3}{0}$$
Since division by 0 is not defined, $f(3)$ does not exist.

97. $f(x) = 7 - \frac{3}{2}x = -\frac{3}{2}x + 7$

The function can be written in the form $y = mx + b$, so it is a linear function.

99. $f(x) = x^2 + 1$ cannot be written in the form $f(x) = mx + b$, so it is not a linear function.

101. $2x - \{x - [3x - (6x + 5)]\} = 4x - 1$

$\quad 2x - \{x - [3x - 6x - 5]\} = 4x - 1$

$\quad\quad 2x - \{x - [-3x - 5]\} = 4x - 1$

$\quad\quad\quad 2x - \{x + 3x + 5\} = 4x - 1$

$\quad\quad\quad\quad 2x - \{4x + 5\} = 4x - 1$

$\quad\quad\quad\quad\quad 2x - 4x - 5 = 4x - 1$

$\quad\quad\quad\quad\quad\quad -2x - 5 = 4x - 1$

$\quad\quad\quad\quad\quad\quad\quad -6x - 5 = -1$

$\quad\quad\quad\quad\quad\quad\quad\quad -6x = 4$

$\quad\quad\quad\quad\quad\quad\quad\quad\quad x = -\dfrac{2}{3}$

The solution is $-\dfrac{2}{3}$.

103. The size of the cup was reduced 8 oz $-$ 6 oz, or 2 oz, and $\dfrac{2 \text{ oz}}{8 \text{ oz}} = 0.25$, so the size was reduced 25%. The price per ounce of the 8 oz cup was $\dfrac{89\cent}{8 \text{ oz}}$, or $11.125\cent$/oz. The price per ounce of the 6 oz cup is $\dfrac{71\cent}{6 \text{ oz}}$, or $11.8\overline{3}\cent$/oz. Since the price per ounce was not reduced, it is clear that the price per ounce was not reduced by the same percent as the size of the cup. The price was increased by $11.8\overline{3} - 11.125\cent$, or $0.708\overline{3}\cent$ per ounce. This is an increase of $\dfrac{0.708\overline{3}\cent}{11.125\cent} \approx 0.064$, or about 6.4% per ounce.

105. We use a proportion to determine the number of calories c burned running for 75 minutes, or 1.25 hr.

$$\dfrac{720}{1} = \dfrac{c}{1.25}$$

$$720(1.25) = c$$

$$900 = c$$

Next we use a proportion to determine how long the person would have to walk to use 900 calories. Let t represent this time, in hours. We express 90 min as 1.5 hr.

$$\dfrac{1.5}{480} = \dfrac{t}{900}$$

$$\dfrac{900(1.5)}{480} = t$$

$$2.8125 = t$$

Then, at a rate of 4 mph, the person would have to walk $4(2.8125)$, or 11.25 mi.

Exercise Set 1.6

1. $4x - 3 > 2x + 7$

$\quad 2x - 3 > 7 \quad\quad$ Subtracting $2x$

$\quad\quad 2x > 10 \quad\quad$ Adding 3

$\quad\quad\quad x > 5 \quad\quad$ Dividing by 2

The solution set is $\{x | x > 5\}$, or $(5, \infty)$. The graph is shown below.

3. $x + 6 < 5x - 6$

$\quad 6 + 6 < 5x - x \quad$ Subtracting x and adding 6 on both sides

$\quad\quad 12 < 4x$

$\quad\quad \dfrac{12}{4} < x \quad\quad$ Dividing by 4 on both sides

$\quad\quad\quad 3 < x$

This inequality could also be solved as follows:

$\quad x + 6 < 5x - 6$

$\quad x - 5x < -6 - 6 \quad$ Subtracting $5x$ and 6 on both sides

$\quad\quad -4x < -12$

$\quad\quad x > \dfrac{-12}{-4} \quad$ Dividing by -4 on both sides and reversing the inequality symbol

$\quad\quad x > 3$

The solution set is $\{x | x > 3\}$, or $(3, \infty)$. The graph is shown below.

5. $4 - 2x \le 2x + 16$

$\quad 4 - 4x \le 16 \quad\quad$ Subtracting $2x$

$\quad\quad -4x \le 12 \quad\quad$ Subtracting 4

$\quad\quad\quad x \ge -3 \quad\quad$ Dividing by -4 and reversing the inequality symbol

The solution set is $\{x | x \ge -3\}$, or $[-3, \infty)$. The graph is shown below.

7. $14 - 5y \le 8y - 8$

$\quad 14 + 8 \le 8y + 5y$

$\quad\quad 22 \le 13y$

$\quad\quad \dfrac{22}{13} \le y$

This inequality could also be solved as follows:

$\quad 14 - 5y \le 8y - 8$

$\quad -5y - 8y \le -8 - 14$

$\quad\quad -13y \le -22$

$\quad\quad y \ge \dfrac{22}{13} \quad\quad$ Dividing by -13 on both sides and reversing the inequality symbol

The solution set is $\left\{y \left| y \ge \dfrac{22}{13}\right.\right\}$, or $\left[\dfrac{22}{13}, \infty\right)$. The graph is shown below.

9. $7x - 7 > 5x + 5$

$\quad 2x - 7 > 5 \quad\quad$ Subtracting $5x$

$\quad\quad 2x > 12 \quad\quad$ Adding 7

$\quad\quad\quad x > 6 \quad\quad$ Dividing by 2

The solution set is $\{x | x > 6\}$, or $(6, \infty)$. The graph is shown below.

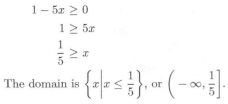

11. $3x - 3 + 2x \geq 1 - 7x - 9$

$\quad 5x - 3 \geq -7x - 8 \quad$ Collecting like terms

$\quad 5x + 7x \geq -8 + 3 \quad$ Adding $7x$ and 3

$\qquad\qquad\qquad\qquad$ on both sides

$\qquad 12x \geq -5$

$\qquad\quad x \geq -\dfrac{5}{12} \quad$ Dividing by 12 on both sides

The solution set is $\left\{x \middle| x \geq -\dfrac{5}{12}\right\}$, or $\left[-\dfrac{5}{12}, \infty\right)$. The graph is shown below.

13. $\quad -\dfrac{3}{4}x \geq -\dfrac{5}{8} + \dfrac{2}{3}x$

$\qquad \dfrac{5}{8} \geq \dfrac{3}{4}x + \dfrac{2}{3}x$

$\qquad \dfrac{5}{8} \geq \dfrac{9}{12}x + \dfrac{8}{12}x$

$\qquad \dfrac{5}{8} \geq \dfrac{17}{12}x$

$\qquad \dfrac{12}{17} \cdot \dfrac{5}{8} \geq \dfrac{12}{17} \cdot \dfrac{17}{12}x$

$\qquad \dfrac{15}{34} \geq x$

The solution set is $\left\{x \middle| x \leq \dfrac{15}{34}\right\}$, or $\left(-\infty, \dfrac{15}{34}\right]$. The graph is shown below.

15. $\quad 4x(x - 2) < 2(2x - 1)(x - 3)$

$\qquad 4x(x - 2) < 2(2x^2 - 7x + 3)$

$\qquad 4x^2 - 8x < 4x^2 - 14x + 6$

$\qquad\quad -8x < -14x + 6$

$\qquad -8x + 14x < 6$

$\qquad\qquad\quad 6x < 6$

$\qquad\qquad\quad\ x < \dfrac{6}{6}$

$\qquad\qquad\quad\ x < 1$

The solution set is $\{x | x < 1\}$, or $(-\infty, 1)$. The graph is shown below.

17. The radicand must be nonnegative, so we solve the inequality $x - 7 \geq 0$.

$\qquad\qquad\qquad x - 7 \geq 0$

$\qquad\qquad\qquad\qquad x \geq 7$

The domain is $\{x | x \geq 7\}$, or $[7, \infty)$.

19. The radicand must be nonnegative, so we solve the inequality $1 - 5x \geq 0$.

$\qquad\qquad 1 - 5x \geq 0$

$\qquad\qquad\qquad 1 \geq 5x$

$\qquad\qquad\qquad \dfrac{1}{5} \geq x$

The domain is $\left\{x \middle| x \leq \dfrac{1}{5}\right\}$, or $\left(-\infty, \dfrac{1}{5}\right]$.

21. The radicand must be positive, so we solve the inequality $4 + x > 0$.

$\qquad\qquad 4 + x > 0$

$\qquad\qquad\qquad x > -4$

The domain is $\{x | x > -4\}$, or $(-4, \infty)$.

23. $\quad -2 \leq x + 1 < 4$

$\qquad -3 \leq x < 3 \quad$ Subtracting 1

The solution set is $[-3, 3)$. The graph is shown below.

25. $\quad 5 \leq x - 3 \leq 7$

$\qquad 8 \leq x \leq 10 \quad$ Adding 3

The solution set is $[8, 10]$. The graph is shown below.

27. $\quad -3 \leq x + 4 \leq 3$

$\qquad -7 \leq x \leq -1 \quad$ Subtracting 4

The solution set is $[-7, -1]$. The graph is shown below.

29. $\quad -2 < 2x + 1 < 5$

$\qquad -3 < 2x < 4 \quad$ Adding -1

$\qquad -\dfrac{3}{2} < x < 2 \quad$ Multiplying by $\dfrac{1}{2}$

The solution set is $\left(-\dfrac{3}{2}, 2\right)$. The graph is shown below.

31. $\qquad -4 \leq 6 - 2x < 4$

$\qquad -10 \leq -2x < -2 \quad$ Adding -6

$\qquad\quad 5 \geq x > 1 \quad$ Multiplying by $-\dfrac{1}{2}$

or $\quad 1 < x \leq 5$

The solution set is $(1, 5]$. The graph is shown below.

33. $-5 < \frac{1}{2}(3x+1) < 7$

$\qquad -10 < 3x+1 < 14 \qquad$ Multiplying by 2

$\qquad -11 < 3x < 13 \qquad$ Adding -1

$\qquad -\frac{11}{3} < x < \frac{13}{3} \qquad$ Multiplying by $\frac{1}{3}$

The solution set is $\left(-\frac{11}{3}, \frac{13}{3}\right)$. The graph is shown below.

35. $3x \leq -6 \ or \ x-1 > 0$

$\qquad x \leq -2 \ or \qquad x > 1$

The solution set is $(-\infty, -2] \cup (1, \infty)$. The graph is shown below.

37. $2x+3 \leq -4 \ or \ 2x+3 \geq 4$

$\qquad 2x \leq -7 \ or \qquad 2x \geq 1$

$\qquad x \leq -\frac{7}{2} \ or \qquad x \geq \frac{1}{2}$

The solution set is $\left(-\infty, -\frac{7}{2}\right] \cup \left[\frac{1}{2}, \infty\right)$. The graph is shown below.

39. $2x-20 < -0.8 \ or \ 2x-20 > 0.8$

$\qquad 2x < 19.2 \ or \qquad 2x > 20.8$

$\qquad x < 9.6 \quad or \qquad x > 10.4$

The solution set is $(-\infty, 9.6) \cup (10.4, \infty)$. The graph is shown below.

41. $x+14 \leq -\frac{1}{4} \ or \ x+14 \geq \frac{1}{4}$

$\qquad x \leq -\frac{57}{4} \ or \qquad x \geq -\frac{55}{4}$

The solution set is $\left(-\infty, -\frac{57}{4}\right] \cup \left[-\frac{55}{4}, \infty\right)$. The graph is shown below.

43. *Familiarize and Translate.* World rice production is given by the equation $y = 9.06x + 410.81$. We want to know when production will be more than 820 million metric tons, so we have

$\qquad 9.06x + 410.81 > 820.$

Carry out. We solve the equation.

$\qquad 9.06x + 410.81 > 820$

$\qquad 9.06x > 409.19$

$\qquad x > 45 \qquad$ Rounding

Check. When $x \approx 45$, $y = 9.06(45) + 410.81 = 818.51 \approx 820$. As a partial check, we could try a value of x less than 45 and one greater than 45. When $x = 44.8$, we have $y = 9.06(44.8) + 410.81 = 816.698 < 820$; when $x = 45.2$, we have $y = 9.06(45.2) + 410.81 = 820.322 > 820$. Since $y \approx 820$ when $x = 45$ and $y > 820$ when $x > 45$, the answer is probably correct.

State. World rice production will exceed 820 million metric tons more than 45 years after 1980.

45. *Familiarize.* Let $t =$ the number of hours worked. Then Acme Movers charge $200 + 45t$ and Leo's Movers charge $65t$.

Translate.

$$\underbrace{\text{Leo's charge}}_{65t} \ \underbrace{\text{is less than}}_{<} \ \underbrace{\text{Acme's charge.}}_{200+45t}$$

Carry out. We solve the inequality.

$\qquad 65t < 200 + 45t$

$\qquad 20t < 200$

$\qquad t < 10$

Check. When $t = 10$, Leo's Movers charge $65 \cdot 10$, or \$650 and Acme Movers charge $200 + 45 \cdot 10$, or \$650, so the charges are the same. As a partial check, we find the charges for a value of $t < 10$. When $t = 9.5$, Leo's Movers charge $65(9.5) = \$617.50$ and Acme Movers charge $200 + 45(9.5) = \$627.50$. Since Leo's charge is less than Acme's, the answer is probably correct.

State. For times less than 10 hr it costs less to hire Leo's Movers.

47. *Familiarize.* Let $x =$ the amount invested at 4%. Then $7500 - x =$ the amount invested at 5%. Using the simple-interest formula, $I = Prt$, we see that in one year the 4% investment earns $0.04x$ and the 5% investment earns $0.05(7500 - x)$.

Translate.

$$\underbrace{\text{Interest at 4\%}}_{0.04x} \ \underbrace{\text{plus}}_{+} \ \underbrace{\text{interest at 5\%}}_{0.05(7500-x)} \ \underbrace{\text{is at least}}_{\geq} \ \underbrace{\$325.}_{325}$$

Carry out. We solve the inequality.

$\qquad 0.04x + 0.05(7500 - x) \geq 325$

$\qquad 0.04x + 375 - 0.05x \geq 325$

$\qquad -0.01x + 375 \geq 325$

$\qquad -0.01x \geq -50$

$\qquad x \leq 5000$

Check. When \$5000 is invested at 4%, then \$7500−\$5000, or \$2500, is invested at 5%. In one year the 4% investment earns 0.04(\$5000), or \$200, in simple interest and the 5% investment earns 0.05(\$2500), or \$125, so the total interest is $200 + $125, or $325.   As a partial check, we determine the total interest when an amount greater than $5000 is invested at 4%. Suppose $5001 is invested at 4%. Then $2499 is invested at 5%, and the total interest is $0.04($5001) + 0.05($2499)$, or $324.99. Since this amount is less than $325, the answer is probably correct.

State. The most that can be invested at 4% is $5000.

49. **Familiarize and Translate**. Let $x =$ the amount invested at 5%. Then
$\frac{1}{2}x =$ the amount invested at 3.5%, and
$1,400,000 - x - \frac{1}{2}x$, or $1,400,000 - \frac{3}{2}x =$ the
amount invested at 5.5%. The interest earned is
$0.05x + 0.035\left(\frac{1}{2}x\right) + 0.055\left(1,400,000 - \frac{3}{2}x\right)$, or
$0.05x + 0.0175x + 77,000 - 0.0825x$, or $-0.015x + 77,000$.
The foundation wants the interest to be at least $68,000, so we have

$$-0.015x + 77,000 \geq 68,000.$$

Carry out. We solve the inequality.
$$-0.015x + 77,000 \geq 68,000$$
$$-0.015x \geq -9000$$
$$x \leq 600,000$$

If $x \leq 600,000$ then $\frac{1}{2}x \leq 300,000$.

Check. If $600,000 is invested at 5% and $300,000 is invested at 3.5%, then the amount invested at 5.5% is $1,400,000 - $600,000 - $300,000 = $500,000$. The interest earned is $0.05($600,000) + 0.035($300,000) + 0.055($500,000)$, or $30,000 + $10,500 + $27,500$, or $68,000. As a partial check, we can determine if the total interest earned when more than $300,000 is invested at 3.5% is less than $68,000. This is the case, so the answer is probably correct.

State. The most than can be invested at 3.5% is $300,000.

51. **Familiarize**. Let $s =$ the monthly sales. Then the amount of sales in excess of $8000 is $s - 8000$.

Translate.

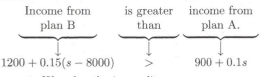

Income from plan B	is greater than	income from plan A.
$1200 + 0.15(s - 8000)$	$>$	$900 + 0.1s$

Carry out. We solve the inequality.
$$1200 + 0.15(s - 8000) > 900 + 0.1s$$
$$1200 + 0.15s - 1200 > 900 + 0.1s$$
$$0.15s > 900 + 0.1s$$
$$0.05s > 900$$
$$s > 18,000$$

Check. For sales of $18,000 the income from plan A is $900 + 0.1($18,000)$, or $2700, and the income from plan B is $1200 + 0.15(18,000 - 8000)$, or $2700 so the incomes are the same. As a partial check we can compare the incomes for an amount of sales greater than $18,000. For sales of $18,001, for example, the income from plan A is $900 +

$0.1($18,001)$, or $2700.10, and the income from plan B is $1200 + 0.15($18,001 - $8000)$, or $2700.15. Since plan B is better than plan A in this case, the answer is probably correct.

State.   Plan B is better than plan A for monthly sales greater than $18,000.

53. Function; domain; range; domain; exactly one; range

55. x-intercept

57. $2x \leq 5 - 7x < 7 + x$
$$2x \leq 5 - 7x \quad and \quad 5 - 7x < 7 + x$$
$$9x \leq 5 \quad\quad and \quad\quad -8x < 2$$
$$x \leq \frac{5}{9} \quad\quad and \quad\quad x > -\frac{1}{4}$$

The solution set is $\left(-\frac{1}{4}, \frac{5}{9}\right]$.

59. $3y < 4 - 5y < 5 + 3y$

$\quad 0 < 4 - 8y < 5$ $\quad\quad$ Subtracting $3y$

$\quad -4 < -8y < 1$ $\quad\quad$ Subtracting 4

$\quad \frac{1}{2} > y > -\frac{1}{8}$ $\quad\quad$ Dividing by -8 and reversing the inequality symbols

The solution set is $\left(-\frac{1}{8}, \frac{1}{2}\right)$.

Chapter 1 Review Exercises

1. First we solve each equation for y.
$$ax + y = c \quad\quad\quad x - by = d$$
$$y = -ax + c \quad\quad -by = -x + d$$
$$y = \frac{1}{b}x - \frac{d}{b}$$

If the lines are perpendicular, the product of their slopes is -1, so we have $-a \cdot \frac{1}{b} = -1$, or $-\frac{a}{b} = -1$, or $\frac{a}{b} = 1$. The statement is true.

3. $f(-3) = \frac{\sqrt{3 - (-3)}}{-3} = \frac{\sqrt{6}}{-3}$, so -3 is in the domain of $f(x)$. Thus, the statement is false.

5. The statement is true. See page 72 in the text.

7. For $\left(3, \frac{24}{9}\right)$:

$$\begin{array}{c|c} 2x - 9y = -18 \\ \hline 2 \cdot 3 - 9 \cdot \frac{24}{9} \ ? \ -18 \\ 6 - 24 \\ -18 \ \big| \ -18 \quad \text{TRUE} \end{array}$$

$\left(3, \frac{24}{9}\right)$ is a solution.

For $(0, -9)$:

$$\begin{array}{rcl} 2x - 9y & = & -18 \\ \hline 2(0) - 9(-9) & \overset{?}{=} & -18 \\ 0 + 81 & & \\ 81 & | & -18 \quad \text{FALSE} \end{array}$$

$(0, -9)$ is not a solution.

9. $2x - 3y = 6$

To find the x-intercept we replace y with 0 and solve for x.

$$2x - 3 \cdot 0 = 6$$
$$2x = 6$$
$$x = 3$$

The x-intercept is $(3, 0)$.

To find the y-intercept we replace x with 0 and solve for y.

$$2 \cdot 0 - 3y = 6$$
$$-3y = 6$$
$$y = -2$$

The y-intercept is $(0, -2)$.

We plot the intercepts and draw the line that contains them. We could find a third point as a check that the intercepts were found correctly.

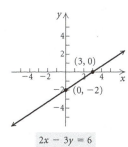

11.

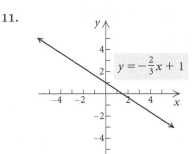

13.

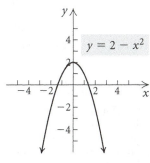

15.
$$m = \left(\frac{x_1 + x_2}{2}, \frac{y_1 + y_2}{2} \right)$$
$$= \left(\frac{3 + (-2)}{2}, \frac{7 + 4}{2} \right)$$
$$= \left(\frac{1}{2}, \frac{11}{2} \right)$$

17.
$$(x - h)^2 + (y - k)^2 = r^2$$
$$(x - 0)^2 + [y - (-4)]^2 = \left(\frac{3}{2} \right)^2 \quad \text{Substituting}$$
$$x^2 + (y + 4)^2 = \frac{9}{4}$$

19. The center is the midpoint of the diameter:
$$\left(\frac{-3 + 7}{2}, \frac{5 + 3}{2} \right) = \left(\frac{4}{2}, \frac{8}{2} \right) = (2, 4)$$

Use the center and either endpoint of the diameter to find the radius. We use the point $(7, 3)$.
$$r = \sqrt{(7 - 2)^2 + (3 - 4)^2} = \sqrt{5^2 + (-1)^2} = \sqrt{25 + 1} = \sqrt{26}$$

The equation of the circle is $(x - 2)^2 + (y - 4)^2 = (\sqrt{26})^2$, or $(x - 2)^2 + (y - 4)^2 = 26$.

21. The correspondence is a function because each member of the domain corresponds to exactly one member of the range.

23. The relation is a function, because no two ordered pairs have the same first coordinate and different second coordinates. The domain is the set of first coordinates: $\{-2, 0, 1, 2, 7\}$. The range is the set of second coordinates: $\{-7, -4, -2, 2, 7\}$.

25. $f(x) = \dfrac{x - 7}{x + 5}$

a) $f(7) = \dfrac{7 - 7}{7 + 5} = \dfrac{0}{12} = 0$

b) $f(x + 1) = \dfrac{x + 1 - 7}{x + 1 + 5} = \dfrac{x - 6}{x + 6}$

c) $f(-5) = \dfrac{-5 - 7}{-5 + 5} = \dfrac{-12}{0}$

Since division by 0 is not defined, $f(-5)$ does not exist.

d) $f\left(-\dfrac{1}{2} \right) = \dfrac{-\dfrac{1}{2} - 7}{-\dfrac{1}{2} + 5} = \dfrac{-\dfrac{15}{2}}{\dfrac{9}{2}} = -\dfrac{15}{2} \cdot \dfrac{2}{9} =$

$-\dfrac{\cancel{3} \cdot 5 \cdot \cancel{2}}{\cancel{2} \cdot \cancel{3} \cdot 3} = -\dfrac{5}{3}$

27. This is not the graph of a function, because we can find a vertical line that crosses the graph more than once.

29. This is not the graph of a function, because we can find a vertical line that crosses the graph more than once.

31. We can substitute any real number for x. Thus, the domain is the set of all real numbers, or $(-\infty, \infty)$.

33. Find the inputs that make the denominator zero:
$$x^2 - 6x + 5 = 0$$
$$(x-1)(x-5) = 0$$
$$x - 1 = 0 \quad or \quad x - 5 = 0$$
$$x = 1 \quad or \quad\quad x = 5$$

The domain is $\{x | x \neq 1 \ and \ x \neq 5\}$, or $(-\infty, 1) \cup (1, 5) \cup (5, \infty)$.

35.

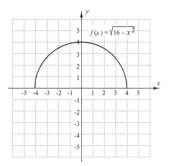

The inputs on the x axis extend from -4 to 4, so the domain is $[-4, 4]$.

The outputs on the y-axis extend from 0 to 4, so the range is $[0, 4]$.

37.

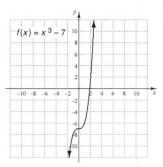

Every point on the x-axis corresponds to a point on the graph, so the domain is the set of all real numbers, or $(-\infty, \infty)$.

Each point on the y-axis also corresponds to a point on the graph, so the range is the set of all real numbers, or $(-\infty, \infty)$.

39. a) Yes. Each input is 1 more than the one that precedes it.

     b) No. The change in the output varies.

     c) No. Constant changes in inputs do not result in constant changes in outputs.

41. $m = \dfrac{y_2 - y_1}{x_2 - x_1}$

$$= \dfrac{-6 - (-11)}{5 - 2} = \dfrac{5}{3}$$

43. $m = \dfrac{y_2 - y_1}{x_2 - x_1}$

$$= \dfrac{0 - 3}{\dfrac{1}{2} - \dfrac{1}{2}} = \dfrac{-3}{0}$$

The slope is not defined.

45. $y = -\dfrac{7}{11}x - 6$

The equation is in the form $y = mx + b$. The slope is $-\dfrac{7}{11}$, and the y-intercept is $(0, -6)$.

47. Graph $y = -\dfrac{1}{4}x + 3$.

Plot the y-intercept, $(0, 3)$. We can think of the slope as $\dfrac{-1}{4}$. Start at $(0, 3)$ and find another point by moving down 1 unit and right 4 units. We have the point $(4, 2)$.

We could also think of the slope as $\dfrac{1}{-4}$. Then we can start at $(0, 3)$ and find another point by moving up 1 unit and left 4 units. We have the point $(-4, 4)$. Connect the three points and draw the graph.

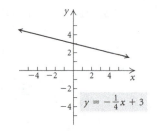

49. a) $T(d) = 10d + 20$

$$T(5) = 10(5) + 20 = 70°C$$
$$T(20) = 10(20) + 20 = 220°C$$
$$T(1000) = 10(1000) + 20 = 10,020°C$$

     b) 5600 km is the maximum depth. Domain: $[0, 5600]$.

51. $y - y_1 = m(x - x_1)$
$$y - (-1) = 3(x - (-2))$$
$$y + 1 = 3(x + 2)$$
$$y + 1 = 3x + 6$$
$$y = 3x + 5$$

53. The horizontal line that passes through $\left(-4, \dfrac{2}{5}\right)$ is $\dfrac{2}{5}$ unit above the x-axis. An equation of the line is $y = \dfrac{2}{5}$.

The vertical line that passes through $\left(-4, \dfrac{2}{5}\right)$ is 4 units to the left of the y-axis. An equation of the line is $x = -4$.

55. $\quad 3x - 2y = 8 \quad\quad\quad\quad\quad 6x - 4y = 2$

$$\quad\quad y = \dfrac{3}{2}x - 4 \quad\quad\quad\quad y = \dfrac{3}{2}x - \dfrac{1}{2}$$

The lines have the same slope, $\dfrac{3}{2}$, and different y-intercepts, $(0, -4)$ and $\left(0, -\dfrac{1}{2}\right)$, so they are parallel.

57. The slope of $y = \frac{3}{2}x + 7$ is $\frac{3}{2}$ and the slope of $y = -\frac{2}{3}x - 4$ is $-\frac{2}{3}$. Since $\frac{3}{2}\left(-\frac{2}{3}\right) = -1$, the lines are perpendicular.

59. From Exercise 58 we know that the slope of the given line is $-\frac{2}{3}$. The slope of a line perpendicular to this line is the negative reciprocal of $-\frac{2}{3}$, or $\frac{3}{2}$.

We use the slope-intercept equation to find the y-intercept.
$$y = mx + b$$
$$-1 = \frac{3}{2} \cdot 1 + b$$
$$-1 = \frac{3}{2} + b$$
$$-\frac{5}{2} = b$$

Then the equation of the desired line is $y = \frac{3}{2}x - \frac{5}{2}$.

61.
$$4y - 5 = 1$$
$$4y = 6$$
$$y = \frac{3}{2}$$
The solution is $\frac{3}{2}$.

63.
$$5(3x + 1) = 2(x - 4)$$
$$15x + 5 = 2x - 8$$
$$13x = -13$$
$$x = -1$$
The solution is -1.

65.
$$\frac{3}{5}y - 2 = \frac{3}{8} \qquad \text{The LCD is 40}$$
$$40\left(\frac{3}{5}y - 2\right) = 40 \cdot \frac{3}{8} \quad \text{Multiplying to clear fractions}$$
$$24y - 80 = 15$$
$$24y = 95$$
$$y = \frac{95}{24}$$
The solution is $\frac{95}{24}$.

67.
$$x - 13 = -13 + x$$
$$-13 = -13 \qquad \text{Subtracting } x$$
We have an equation that is true for any real number, so the solution set is the set of all real numbers, $\{x | x \text{ is a real number}\}$, or $(-\infty, \infty)$.

69. *Familiarize.* Let $a =$ the amount originally invested. Using the simple interest formula, $I = Prt$, we see that the interest earned at 5.2% interest for 1 year is $a(0.052) \cdot 1 = 0.052a$.

Translate.

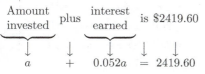

Amount invested	plus	interest earned	is	$2419.60
a	$+$	$0.052a$	$=$	2419.60

Carry out. We solve the equation.
$$a + 0.052a = 2419.60$$
$$1.052a = 2419.60$$
$$a = 2300$$

Check. 5.2% of $2300 is 0.052($2300), or $119.60, and $2300 + $119.60 = $2419.60. The answer checks.

State. $2300 was originally invested.

71.
$$6x - 18 = 0$$
$$6x = 18$$
$$x = 3$$
The zero of the function is 3.

73.
$$2 - 10x = 0$$
$$-10x = -2$$
$$x = \frac{1}{5}, \text{ or } 0.2$$
The zero of the function is $\frac{1}{5}$, or 0.2.

75.
$$2x - 5 < x + 7$$
$$x < 12$$
The solution set is $\{x | x < 12\}$, or $(-\infty, 12)$.

77.
$$-3 \leq 3x + 1 \leq 5$$
$$-4 \leq 3x \leq 4$$
$$-\frac{4}{3} \leq x \leq \frac{4}{3}$$
$$\left[-\frac{4}{3}, \frac{4}{3}\right]$$

79.
$$2x < -1 \quad or \quad x - 3 > 0$$
$$x < -\frac{1}{2} \quad or \qquad x > 3$$
The solution set is $\left\{x \,\middle|\, x < -\frac{1}{2} \text{ or } x > 3\right\}$, or $\left(-\infty, -\frac{1}{2}\right) \cup (3, \infty)$.

81. *Familiarize and Translate.* The number of home-schooled children in the U.S., in millions, is estimated by the equation $y = 0.073x + 0.848$, where x is the number of years after 1999. We want to know for what year this number will exceed 2.3 million, so we have
$$0.073x + 0.848 > 2.3.$$

Carry out. We solve the inequality.

$$0.073x + 0.848 > 2.3$$
$$0.073x > 1.452$$
$$x > 20 \qquad \text{Rounding}$$

Check. When $x = 20$, $y = 0.073(20) + 0.848 = 2.308 \approx 2.3$. As a partial check, we could try a value less than 20 and a value greater than 20. When $x = 19$, we have $y = 0.073(19) + 0.848 = 2.235 < 2.3$; when $x = 21$, we have $y = 0.073(21) + 0.848 = 2.381 > 2.3$. Since $y \approx 2.3$ when $x = 20$ and $y > 2.3$ when $x = 21 > 20$, the answer is probably correct.

State. In years more than about 20 years after 1999, or in years after 2019, the number of homeschooled children will exceed 2.3 million.

83. $f(x) = \dfrac{x + 3}{8 - 4x}$

When $x = 2$, the denominator is 0, so 2 is not in the domain of the function. Thus, the domain is $(-\infty, 2) \cup (2, \infty)$ and answer B is correct.

85. The graph of $f(x) = -\dfrac{1}{2}x - 2$ has slope $-\dfrac{1}{2}$, so it slants down from left to right. The y-intercept is $(0, -2)$. Thus, graph C is the graph of this function.

87. $f(x) = \dfrac{\sqrt{1 - x}}{x - |x|}$

We cannot find the square root of a negative number, so $x \leq 1$. Division by zero is undefined, so $x < 0$.

Domain of f is $\{x | x < 0\}$, or $(-\infty, 0)$.

89. Think of the slopes as $\dfrac{-3/5}{1}$ and $\dfrac{1/2}{1}$. The graph of $f(x)$ changes $\dfrac{3}{5}$ unit vertically for each unit of horizontal change while the graph of $g(x)$ changes $\dfrac{1}{2}$ unit vertically for each unit of horizontal change. Since $\dfrac{3}{5} > \dfrac{1}{2}$, the graph of $f(x) = -\dfrac{3}{5}x + 4$ is steeper than the graph of $g(x) = \dfrac{1}{2}x - 6$.

91. The solution set of a disjunction is a union of sets, so it is not possible for a disjunction to have no solution.

93. By definition, the notation $3 < x < 4$ indicates that $3 < x$ *and* $x < 4$. The disjunction $x < 3$ *or* $x > 4$ cannot be written $3 > x > 4$, or $4 < x < 3$, because it is not possible for x to be greater than 4 *and* less than 3.

Chapter 1 Test

1.

$$\dfrac{5y - 4 = x}{}$$

$$5 \cdot \dfrac{9}{10} - 4 \; ? \; \dfrac{1}{2}$$

$$\dfrac{9}{2} - 4 \; \Big|$$

$$\dfrac{1}{2} \; \Big| \; \dfrac{1}{2} \quad \text{TRUE}$$

$\left(\dfrac{1}{2}, \dfrac{9}{10}\right)$ is a solution.

2. $5x - 2y = -10$

To find the x-intercept we replace y with 0 and solve for x.

$$5x - 2 \cdot 0 = -10$$
$$5x = -10$$
$$x = -2$$

The x-intercept is $(-2, 0)$.

To find the y-intercept we replace x with 0 and solve for y.

$$5 \cdot 0 - 2y = -10$$
$$-2y = -10$$
$$y = 5$$

The y-intercept is $(0, 5)$.

We plot the intercepts and draw the line that contains them. We could find a third point as a check that the intercepts were found correctly.

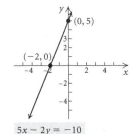

3. $d = \sqrt{(5 - (-1))^2 + (8 - 5)^2} = \sqrt{6^2 + 3^2} = \sqrt{36 + 9} = \sqrt{45} \approx 6.708$

4. $m = \left(\dfrac{-2 + (-4)}{2}, \dfrac{6 + 3}{2}\right) = \left(\dfrac{-6}{2}, \dfrac{9}{2}\right) = \left(-3, \dfrac{9}{2}\right)$

5. $(x + 4)^2 + (y - 5)^2 = 36$
$[x - (-4)]^2 + (y - 5)^2 = 6^2$
Center: $(-4, 5)$; radius: 6

6. $[x - (-1)]^2 + (y - 2)^2 = (\sqrt{5})^2$
$(x + 1)^2 + (y - 2)^2 = 5$

7. a) The relation is a function, because no two ordered pairs have the same first coordinate and different second coordinates.

b) The domain is the set of first coordinates: $\{-4, 0, 1, 3\}$.

c) The range is the set of second coordinates: $\{0, 5, 7\}$.

8. $f(x) = 2x^2 - x + 5$

a) $f(-1) = 2(-1)^2 - (-1) + 5 = 2 + 1 + 5 = 8$

b) $f(a + 2) = 2(a + 2)^2 - (a + 2) + 5$
$$= 2(a^2 + 4a + 4) - (a + 2) + 5$$
$$= 2a^2 + 8a + 8 - a - 2 + 5$$
$$= 2a^2 + 7a + 11$$

9. $f(x) = \dfrac{1-x}{x}$

a) $f(0) = \dfrac{1-0}{0} = \dfrac{1}{0}$

Since the division by 0 is not defined, $f(0)$ does not exist.

b) $f(1) = \dfrac{1-1}{1} = \dfrac{0}{1} = 0$

10. From the graph we see that when the input is -3, the output is 0, so $f(-3) = 0$.

11. a) This is not the graph of a function, because we can find a vertical line that crosses the graph more than once.

b) This is the graph of a function, because there is no vertical line that crosses the graph more than once.

12. The input 4 results in a denominator of 0. Thus the domain is $\{x|x \neq 4\}$, or $(-\infty, 4) \cup (4, \infty)$.

13. We can substitute any real number for x. Thus the domain is the set of all real numbers, or $(-\infty, \infty)$.

14. We cannot find the square root of a negative number. Thus $25 - x^2 \geq 0$ and the domain is $\{x|-5 \leq x \leq 5\}$, or $[-5, 5]$.

15. a)

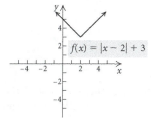

$f(x) = |x - 2| + 3$

b) Each point on the x-axis corresponds to a point on the graph, so the domain is the set of all real numbers, or $(-\infty, \infty)$.

c) The number 3 is the smallest output on the y-axis and every number greater than 3 is also an output, so the range is $[3, \infty)$.

16. $m = \dfrac{5 - \frac{2}{3}}{-2 - (-2)} = \dfrac{\frac{13}{3}}{0}$

The slope is not defined.

17. $m = \dfrac{12 - (-10)}{-8 - 4} = \dfrac{22}{-12} = -\dfrac{11}{6}$

18. $m = \dfrac{6 - 6}{\frac{3}{4} - (-5)} = \dfrac{0}{\frac{23}{4}} = 0$

19. We have the points $(1995, 21.6)$ and $(2012, 9.3)$.

$m = \dfrac{9.3 - 21.6}{2012 - 1995} = \dfrac{-12.3}{17} \approx -0.7$

The average rate of change in the percent of 12th graders who smoke daily decreased about 0.7% per year from 1995 to 2012.

20. $-3x + 2y = 5$

$2y = 3x + 5$

$y = \dfrac{3}{2}x + \dfrac{5}{2}$

Slope: $\dfrac{3}{2}$; y-intercept: $\left(0, \dfrac{5}{2}\right)$

21. $C(t) = 65 + 48t$

$C(2.25) = 65 + 48(2.25) = \$173$

22. $y = mx + b$

$y = -\dfrac{5}{8}x - 5$

23. First we find the slope:

$m = \dfrac{-2 - 4}{3 - (-5)} = \dfrac{-6}{8} = -\dfrac{3}{4}$

Use the point-slope equation.

Using $(-5, 4)$: $y - 4 = -\dfrac{3}{4}(x - (-5))$, or

$y - 4 = -\dfrac{3}{4}(x + 5)$

Using $(3, -2)$: $y - (-2) = -\dfrac{3}{4}(x - 3)$, or

$y + 2 = -\dfrac{3}{4}(x - 3)$

In either case, we have $y = -\dfrac{3}{4}x + \dfrac{1}{4}$.

24. The vertical line that passes through $\left(-\dfrac{3}{8}, 11\right)$ is $\dfrac{3}{8}$ unit to the left of the y-axis. An equation of the line is $x = -\dfrac{3}{8}$.

25. $2x + 3y = -12 \qquad 2y - 3x = 8$

$y = -\dfrac{2}{3}x - 4 \qquad y = \dfrac{3}{2}x + 4$

$m_1 = -\dfrac{2}{3},\ m_2 = \dfrac{3}{2};\ m_1 m_2 = -1.$

The lines are perpendicular.

26. First find the slope of the given line.

$x + 2y = -6$

$2y = -x - 6$

$y = -\dfrac{1}{2}x - 3;\ m = -\dfrac{1}{2}$

A line parallel to the given line has slope $-\dfrac{1}{2}$. We use the point-slope equation.

$y - 3 = -\dfrac{1}{2}(x - (-1))$

$y - 3 = -\dfrac{1}{2}(x + 1)$

$y - 3 = -\dfrac{1}{2}x - \dfrac{1}{2}$

$y = -\dfrac{1}{2}x + \dfrac{5}{2}$

27. First we find the slope of the given line.

$$x + 2y = -6$$
$$2y = -x - 6$$
$$y = -\frac{1}{2}x - 3, \; m = -\frac{1}{2}$$

The slope of a line perpendicular to this line is the negative reciprocal of $-\frac{1}{2}$, or 2. Now we find an equation of the line with slope 2 and containing $(-1, 3)$.

Using the slope-intercept equation:

$$y = mx + b$$
$$3 = 2(-1) + b$$
$$3 = -2 + b$$
$$5 = b$$

The equation is $y = 2x + 5$.

Using the point-slope equation.

$$y - y_1 = m(x - x_1)$$
$$y - 3 = 2(x - (-1))$$
$$y - 3 = 2(x + 1)$$
$$y - 3 = 2x + 2$$
$$y = 2x + 5$$

28. Answers may vary depending on the data points used. We will use $(2, 507.03)$ and $(12, 666.99)$.

$$m = \frac{666.99 - 507.03}{12 - 2} = \frac{159.96}{10} = 15.996$$

We will use the point-slope equation with the point $(2, 507.03)$.

$$y - 507.03 = 15.996(x - 2)$$
$$y - 507.03 = 15.996x - 31.992$$
$$y = 15.996x + 475.038,$$

where x is the number of years after 2000.

For 2016: $y = 15.996(16) + 475.038 \approx \$730.97$

For 2020: $y = 15.996(20) + 475.038 \approx \$794.96$

29.
$$6x + 7 = 1$$
$$6x = -6$$
$$x = -1$$

The solution is -1.

30.
$$2.5 - x = -x + 2.5$$
$$2.5 = 2.5 \qquad \text{True equation}$$

The solution set is $\{x | x \text{ is a real number}\}$, or $(-\infty, \infty)$.

31.
$$\frac{3}{2}y - 4 = \frac{5}{3}y + 6 \qquad \text{The LCD is 6.}$$

$$6\left(\frac{3}{2}y - 4\right) = 6\left(\frac{5}{3}y + 6\right)$$
$$9y - 24 = 10y + 36$$
$$-24 = y + 36$$
$$-60 = y$$

The solution is -60.

32.
$$2(4x + 1) = 8 - 3(x - 5)$$
$$8x + 2 = 8 - 3x + 15$$
$$8x + 2 = 23 - 3x$$
$$11x + 2 = 23$$
$$11x = 21$$
$$x = \frac{21}{11}$$

The solution is $\frac{21}{11}$.

33. *Familiarize.* Let $l =$ the length, in meters. Then $\frac{3}{4}l =$ the width. Recall that the formula for the perimeter P of a rectangle with length l and width w is $P = 2l + 2w$.

Translate.

$$\underbrace{\text{The perimeter}}_{} \quad \text{is} \quad \underbrace{\text{210 m.}}_{}$$
$$\downarrow \qquad\qquad \downarrow \qquad \downarrow$$
$$2l + 2 \cdot \frac{3}{4}l \quad = \quad 210$$

Carry out. We solve the equation.

$$2l + 2 \cdot \frac{3}{4}l = 210$$
$$2l + \frac{3}{2}l = 210$$
$$\frac{7}{2}l = 210$$
$$l = 60$$

If $l = 60$, then $\frac{3}{4}l = \frac{3}{4} \cdot 60 = 45$.

Check. The width, 45 m, is three-fourths of the length, 60 m. Also, $2 \cdot 60 \text{ m} + 2 \cdot 45 \text{ m} = 210 \text{ m}$, so the answer checks.

State. The length is 60 m and the width is 45 m.

34. *Familiarize.* Let $p =$ the wholesale price of the juice.

Translate. We express 25¢ as $0.25.

Wholesale price	plus	50% of wholesale price	plus	$0.25	is	$2.95.
$\downarrow$	$\downarrow$	$\downarrow$	$\downarrow$	$\downarrow$	$\downarrow$	$\downarrow$
p	$+$	$0.5p$	$+$	0.25	$=$	2.95

Carry out. We solve the equation.

$$p + 0.5p + 0.25 = 2.95$$
$$1.5p + 0.25 = 2.95$$
$$1.5p = 2.7$$
$$p = 1.8$$

Check. 50% of $1.80 is $0.90 and $1.80 + \$0.90 + \$0.25 = \$2.95$, so the answer checks.

State. The wholesale price of a bottle of juice is $1.80.

35.
$$3x + 9 = 0 \qquad \text{Setting } f(x) = 0$$
$$3x = -9$$
$$x = -3$$

The zero of the function is -3.

36.
$$5 - x \geq 4x + 20$$
$$5 - 5x \geq 20$$
$$-5x \geq 15$$
$$x \leq -3 \quad \text{Dividing by } -5 \text{ and reversing}$$
$$\text{the inequality symbol}$$

The solution set is $\{x | x \leq -3\}$, or $(-\infty, -3]$.

37.
$$-7 < 2x + 3 < 9$$
$$-10 < 2x < 6 \qquad \text{Subtracting 3}$$
$$-5 < x < 3 \qquad \text{Dividing by 2}$$

The solution set is $(-5, 3)$.

38.
$$2x - 1 \leq 3 \quad or \quad 5x + 6 \geq 26$$
$$2x \leq 4 \quad or \qquad 5x \geq 20$$
$$x \leq 2 \quad or \qquad x \geq 4$$

The solution set is $(-\infty, 2] \cup [4, \infty)$.

39. Familiarize. Let $t =$ the number of hours a move requires. Then Morgan Movers charges $200 + 40t$ to make a move and McKinley Movers charges $75t$.

Translate.

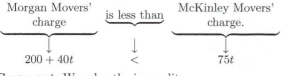

Carry out. We solve the inequality.
$$200 + 40t < 75t$$
$$200 < 35t$$
$$5.7 < t \qquad \text{Rounding}$$

Check. For $t = 5.7$, Morgan Movers charge $200 + 40(5.7)$, or \$428, and McKinley Movers charge $75(5.7)$, or \$427.5 $\approx$ 428. (Remember that we rounded the answer.) So the charge is the same for 5.7 hours. As a partial check, we can find the charges for a value of t greater than 5.7. For instance, for 6 hr Morgan Movers charge $200 + 40 \cdot 6$, or \$440, and McKinley Movers charge $75 \cdot 6$, or \$450. Since Morgan Movers cost less for a value of t greater than 5.7, the answer is probably correct.

State. It costs less to hire Morgan Movers when a move takes more than 5.7 hr.

40. The slope is $-\frac{1}{2}$, so the graph slants down from left to right. The y-intercept is $(0, 1)$. Thus, graph B is the graph of $g(x) = 1 - \frac{1}{2}x$.

41. First we find the value of x for which $x + 2 = -2$:
$$x + 2 = -2$$
$$x = -4$$

Now we find $h(-4 + 2)$, or $h(-2)$.
$$h(-4 + 2) = \frac{1}{2}(-4) = -2$$

Chapter 2

More on Functions

Exercise Set 2.1

1. a) For x-values from -5 to 1, the y-values increase from -3 to 3. Thus the function is increasing on the interval $(-5, 1)$.

   b) For x-values from 3 to 5, the y-values decrease from 3 to 1. Thus the function is decreasing on the interval $(3, 5)$.

   c) For x-values from 1 to 3, y is 3. Thus the function is constant on $(1, 3)$.

3. a) For x-values from -3 to -1, the y-values increase from -4 to 4. Also, for x-values from 3 to 5, the y-values increase from 2 to 6. Thus the function is increasing on $(-3, -1)$ and on $(3, 5)$.

   b) For x-values from 1 to 3, the y-values decrease from 3 to 2. Thus the function is decreasing on the interval $(1, 3)$.

   c) For x-values from -5 to -3, y is 1. Thus the function is constant on $(-5, -3)$.

5. a) For x-values from $-\infty$ to -8, the y-values increase from $-\infty$ to 2. Also, for x-values from -3 to -2, the y-values increase from -2 to 3. Thus the function is increasing on $(-\infty, -8)$ and on $(-3, -2)$.

   b) For x-values from -8 to -6, the y-values decrease from 2 to -2. Thus the function is decreasing on the interval $(-8, -6)$.

   c) For x-values from -6 to -3, y is -2. Also, for x-values from -2 to ∞, y is 3. Thus the function is constant on $(-6, -3)$ and on $(-2, \infty)$.

7. The x-values extend from -5 to 5, so the domain is $[-5, 5]$.

   The y-values extend from -3 to 3, so the range is $[-3, 3]$.

9. The x-values extend from -5 to -1 and from 1 to 5, so the domain is $[-5, -1] \cup [1, 5]$.

   The y-values extend from -4 to 6, so the range is $[-4, 6]$.

11. The x-values extend from $-\infty$ to ∞, so the domain is $(-\infty, \infty)$.

    The y-values extend from $-\infty$ to 3, so the range is $(-\infty, 3]$.

13. From the graph we see that a relative maximum value of the function is 3.25. It occurs at $x = 2.5$. There is no relative minimum value.

    The graph starts rising, or increasing, from the left and stops increasing at the relative maximum. From this point, the graph decreases. Thus the function is increasing on $(-\infty, 2.5)$ and is decreasing on $(2.5, \infty)$.

15. From the graph we see that a relative maximum value of the function is 2.370. It occurs at $x = -0.667$. We also see that a relative minimum value of 0 occurs at $x = 2$.

    The graph starts rising, or increasing, from the left and stops increasing at the relative maximum. From this point it decreases to the relative minimum and then increases again. Thus the function is increasing on $(-\infty, -0.667)$ and on $(2, \infty)$. It is decreasing on $(-0.667, 2)$.

17.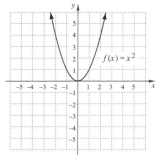

    The function is increasing on $(0, \infty)$ and decreasing on $(-\infty, 0)$. We estimate that the minimum is 0 at $x = 0$. There are no maxima.

19.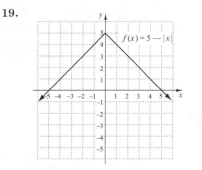

    The function is increasing on $(-\infty, 0)$ and decreasing on $(0, \infty)$. We estimate that the maximum is 5 at $x = 0$. There are no minima.

21.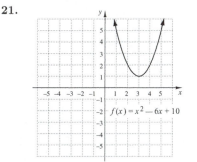

    The function is decreasing on $(-\infty, 3)$ and increasing on $(3, \infty)$. We estimate that the minimum is 1 at $x = 3$. There are no maxima.

23. If x = the length of the rectangle, in meters, then the width is $\dfrac{480 - 2x}{2}$, or $240 - x$. We use the formula Area = length × width:

$$A(x) = x(240 - x), \text{ or}$$
$$A(x) = 240x - x^2$$

25. We use the Pythagorean theorem.

$$[h(d)]^2 + 3500^2 = d^2$$
$$[h(d)]^2 = d^2 - 3500^2$$
$$h(d) = \sqrt{d^2 - 3500^2}$$

We considered only the positive square root since distance must be nonnegative.

27. Let w = the width of the rectangle. Then the length $= \dfrac{40 - 2w}{2}$, or $20 - w$. Divide the rectangle into quadrants as shown below.

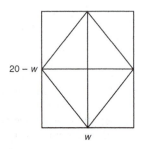

In each quadrant there are two congruent triangles. One triangle is part of the rhombus and both are part of the rectangle. Thus, in each quadrant the area of the rhombus is one-half the area of the rectangle. Then, in total, the area of the rhombus is one-half the area of the rectangle.

$$A(w) = \frac{1}{2}(20 - w)(w)$$
$$A(w) = 10w - \frac{w^2}{2}$$

29. We will use similar triangles, expressing all distances in feet. $\left(6 \text{ in.} = \dfrac{1}{2} \text{ ft}, s \text{ in.} = \dfrac{s}{12} \text{ ft, and } d \text{ yd} = 3d \text{ ft}\right)$ We have

$$\frac{3d}{7} = \frac{\frac{1}{2}}{\frac{s}{12}}$$
$$\frac{s}{12} \cdot 3d = 7 \cdot \frac{1}{2}$$
$$\frac{sd}{4} = \frac{7}{2}$$
$$d = \frac{4}{s} \cdot \frac{7}{2}, \text{ so}$$
$$d(s) = \frac{14}{s}.$$

31. a) After 4 pieces of float line, each of length x ft, are used for the sides perpendicular to the beach, there remains $(240 - 4x)$ ft of float line for the side parallel to the beach. Thus we have a rectangle with length $240 - 4x$ and width x. Then the total area of the three swimming areas is

$$A(x) = (240 - 4x)x, \text{ or } 240x - 4x^2.$$

b) The length of the sides labeled x must be positive and their total length must be less than 240 ft, so $4x < 240$, or $x < 60$. Thus the domain is $\{x | 0 < x < 60\}$, or $(0, 60)$.

c) We see from the graph that the maximum value of the area function on the interval $(0, 60)$ appears to be 3600 when $x = 30$. Thus the dimensions that yield the maximum area are 30 ft by $240 - 4 \cdot 30$, or $240 - 120$, or 120 ft.

33. a) When a square with sides of length x is cut from each corner, the length of each of the remaining sides of the piece of cardboard is $12 - 2x$. Then the dimensions of the box are x by $12 - 2x$ by $12 - 2x$. We use the formula Volume = length × width × height to find the volume of the box:

$$V(x) = (12 - 2x)(12 - 2x)(x)$$
$$V(x) = (144 - 48x + 4x^2)(x)$$
$$V(x) = 144x - 48x^2 + 4x^3$$

This can also be expressed as $V(x) = 4x(x - 6)^2$, or $V(x) = 4x(6 - x)^2$.

b) The length of the sides of the square corners that are cut out must be positive and less than half the length of a side of the piece of cardboard. Thus, the domain of the function is $\{x | 0 < x < 6\}$, or $(0, 6)$.

c) We see from the graph that the maximum value of the area function on the interval $(0, 6)$ appears to be 128 when $x = 2$. When $x = 2$, then $12 - 2x = 12 - 2 \cdot 2 = 8$, so the dimensions that yield the maximum volume are 8 cm by 8 cm by 2 cm.

35. $g(x) = \begin{cases} x + 4, & \text{for } x \le 1, \\ 8 - x, & \text{for } x > 1 \end{cases}$

Since $-4 \le 1$, $g(-4) = -4 + 4 = 0$.

Since $0 \le 1$, $g(0) = 0 + 4 = 4$.

Since $1 \le 1$, $g(1) = 1 + 4 = 5$.

Since $3 > 1$, $g(3) = 8 - 3 = 5$.

37. $h(x) = \begin{cases} -3x - 18, & \text{for } x < -5, \\ 1, & \text{for } -5 \le x < 1, \\ x + 2, & \text{for } x \ge 1 \end{cases}$

Since -5 is in the interval $[-5, 1)$, $h(-5) = 1$.

Since 0 is in the interval $[-5, 1)$, $h(0) = 1$.

Since $1 \ge 1$, $h(1) = 1 + 2 = 3$.

Since $4 \ge 1$, $h(4) = 4 + 2 = 6$.

39. $f(x) = \begin{cases} \dfrac{1}{2}x, & \text{for } x < 0, \\ \\ x + 3, & \text{for } x \geq 0 \end{cases}$

We create the graph in two parts. Graph $f(x) = \dfrac{1}{2}x$ for inputs x less than 0. Then graph $f(x) = x + 3$ for inputs x greater than or equal to 0.

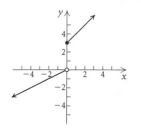

41. $f(x) = \begin{cases} -\dfrac{3}{4}x + 2, & \text{for } x < 4, \\ \\ -1, & \text{for } x \geq 4 \end{cases}$

We create the graph in two parts. Graph $f(x) = -\dfrac{3}{4}x + 2$ for inputs x less than 4. Then graph $f(x) = -1$ for inputs x greater than or equal to 4.

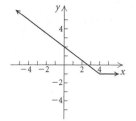

43. $f(x) = \begin{cases} x + 1, & \text{for } x \leq -3, \\ -1, & \text{for } -3 < x < 4 \\ \dfrac{1}{2}x, & \text{for } x \geq 4 \end{cases}$

We create the graph in three parts. Graph $f(x) = x + 1$ for inputs x less than or equal to -3. Graph $f(x) = -1$ for inputs greater than -3 and less than 4. Then graph $f(x) = \dfrac{1}{2}x$ for inputs greater than or equal to 4.

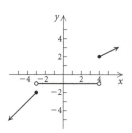

45. $g(x) = \begin{cases} \dfrac{1}{2}x - 1, & \text{for } x < 0, \\ \\ 3, & \text{for } 0 \leq x \leq 1 \\ \\ -2x, & \text{for } x > 1 \end{cases}$

We create the graph in three parts. Graph $g(x) = \dfrac{1}{2}x - 1$ for inputs less than 0. Graph $g(x) = 3$ for inputs greater than or equal to 0 and less than or equal to 1. Then graph $g(x) = -2x$ for inputs greater than 1.

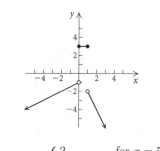

47. $f(x) = \begin{cases} 2, & \text{for } x = 5, \\ \\ \dfrac{x^2 - 25}{x - 5}, & \text{for } x \neq 5 \end{cases}$

When $x \neq 5$, the denominator of $(x^2 - 25)/(x - 5)$ is nonzero so we can simplify:

$$\frac{x^2 - 25}{x - 5} = \frac{(x + 5)(x - 5)}{x - 5} = x + 5.$$

Thus, $f(x) = x + 5$, for $x \neq 5$.

The graph of this part of the function consists of a line with a "hole" at the point $(5, 10)$, indicated by an open dot. At $x = 5$, we have $f(5) = 2$, so the point $(5, 2)$ is plotted below the open dot.

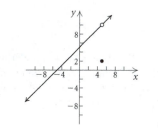

49. $f(x) = [[x]]$

See Example 9.

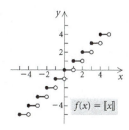

51. $f(x) = 1 + [[x]]$

This function can be defined by a piecewise function with an infinite number of statements:

$$f(x) = \begin{cases} \vdots \\ -1, & \text{for } -2 \le x < -1, \\ 0, & \text{for } -1 \le x < 0, \\ 1, & \text{for } 0 \le x < 1, \\ 2, & \text{for } 1 \le x < 2, \\ \vdots \end{cases}$$

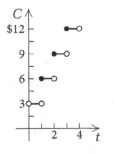

$g(x) = 1 + [[x]]$

53. From the graph we see that the domain is $(-\infty, \infty)$ and the range is $(-\infty, 0) \cup [3, \infty)$.

55. From the graph we see that the domain is $(-\infty, \infty)$ and the range is $[-1, \infty)$.

57. From the graph we see that the domain is $(-\infty, \infty)$ and the range is $\{y | y \le -2 \text{ or } y = -1 \text{ or } y \ge 2\}$.

59. From the graph we see that the domain is $(-\infty, \infty)$ and the range is $\{-5, -2, 4\}$. An equation for the function is:

$$f(x) = \begin{cases} -2, & \text{for } x < 2, \\ -5, & \text{for } x = 2, \\ 4, & \text{for } x > 2 \end{cases}$$

61. From the graph we see that the domain is $(-\infty, \infty)$ and the range is $(-\infty, -1] \cup [2, \infty)$. Finding the slope of each segment and using the slope-intercept or point-slope formula, we find that an equation for the function is:

$$g(x) = \begin{cases} x, & \text{for } x \le -1, \\ 2, & \text{for } -1 < x \le 2, \\ x, & \text{for } x > 2 \end{cases}$$

This can also be expressed as follows:

$$g(x) = \begin{cases} x, & \text{for } x \le -1, \\ 2, & \text{for } -1 < x < 2, \\ x, & \text{for } x \ge 2 \end{cases}$$

63. From the graph we see that the domain is $[-5, 3]$ and the range is $(-3, 5)$. Finding the slope of each segment and using the slope-intercept or point-slope formula, we find that an equation for the function is:

$$h(x) = \begin{cases} x + 8, & \text{for } -5 \le x < -3, \\ 3, & \text{for } -3 \le x \le 1, \\ 3x - 6, & \text{for } 1 < x \le 3 \end{cases}$$

65. $f(x) = 5x^2 - 7$

a) $f(-3) = 5(-3)^2 - 7 = 5 \cdot 9 - 7 = 45 - 7 = 38$

b) $f(3) = 5 \cdot 3^2 - 7 = 5 \cdot 9 - 7 = 45 - 7 = 38$

c) $f(a) = 5a^2 - 7$

d) $f(-a) = 5(-a)^2 - 7 = 5a^2 - 7$

67. First find the slope of the given line.

$$8x - y = 10$$
$$8x = y + 10$$
$$8x - 10 = y$$

The slope of the given line is 8. The slope of a line perpendicular to this line is the opposite of the reciprocal of 8, or $-\dfrac{1}{8}$.

$$y - y_1 = m(x - x_1)$$
$$y - 1 = -\frac{1}{8}[x - (-1)]$$
$$y - 1 = -\frac{1}{8}(x + 1)$$
$$y - 1 = -\frac{1}{8}x - \frac{1}{8}$$
$$y = -\frac{1}{8}x + \frac{7}{8}$$

69. a) The function $C(t)$ can be defined piecewise.

$$C(t) = \begin{cases} 3, & \text{for } 0 < t < 1, \\ 6, & \text{for } 1 \le t < 2, \\ 9, & \text{for } 2 \le t < 3, \\ \vdots \end{cases}$$

We graph this function.

b) From the definition of the function in part (a), we see that it can be written as

$$C(t) = 3[[t]] + 1, \ t > 0.$$

71. If $[[x]]^2 = 25$, then $[[x]] = -5$ or $[[x]] = 5$. For $-5 \le x < -4$, $[[x]] = -5$. For $5 \le x < 6$, $[[x]] = 5$. Thus, the possible inputs for x are $\{x | -5 \le x < -4 \text{ or } 5 \le x < 6\}$.

73. a) We add labels to the drawing in the text.

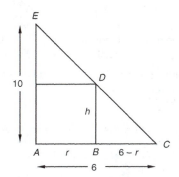

We write a proportion involving the lengths of the sides of the similar triangles BCD and ACE. Then we solve it for h.

$$\frac{h}{6-r} = \frac{10}{6}$$

$$h = \frac{10}{6}(6-r) = \frac{5}{3}(6-r)$$

$$h = \frac{30 - 5r}{3}$$

Thus, $h(r) = \dfrac{30 - 5r}{3}$.

b) $\quad V = \pi r^2 h$

$$V(r) = \pi r^2 \left(\frac{30 - 5r}{3}\right) \quad \text{Substituting for } h$$

c) We first express r in terms of h.

$$h = \frac{30 - 5r}{3}$$

$$3h = 30 - 5r$$

$$5r = 30 - 3h$$

$$r = \frac{30 - 3h}{5}$$

$$V = \pi r^2 h$$

$$V(h) = \pi \left(\frac{30 - 3h}{5}\right)^2 h$$

$$\text{Substituting for } r$$

We can also write $V(h) = \pi h \left(\dfrac{30 - 3h}{5}\right)^2$.

Exercise Set 2.2

1. $(f + g)(5) = f(5) + g(5)$
$$= (5^2 - 3) + (2 \cdot 5 + 1)$$
$$= 25 - 3 + 10 + 1$$
$$= 33$$

3. $(f - g)(-1) = f(-1) - g(-1)$
$$= ((-1)^2 - 3) - (2(-1) + 1)$$
$$= -2 - (-1) = -2 + 1$$
$$= -1$$

5. $(f/g)\left(-\dfrac{1}{2}\right) = \dfrac{f\left(-\dfrac{1}{2}\right)}{g\left(-\dfrac{1}{2}\right)}$

$$= \frac{\left(-\dfrac{1}{2}\right)^2 - 3}{2\left(-\dfrac{1}{2}\right) + 1}$$

$$= \frac{\dfrac{1}{4} - 3}{-1 + 1}$$

$$= \frac{-\dfrac{11}{4}}{0}$$

Since division by 0 is not defined, $(f/g)\left(-\dfrac{1}{2}\right)$ does not exist.

7. $(fg)\left(-\dfrac{1}{2}\right) = f\left(-\dfrac{1}{2}\right) \cdot g\left(-\dfrac{1}{2}\right)$

$$= \left[\left(-\frac{1}{2}\right)^2 - 3\right]\left[2\left(-\frac{1}{2}\right) + 1\right]$$

$$= -\frac{11}{4} \cdot 0 = 0$$

9. $(g - f)(-1) = g(-1) - f(-1)$
$$= [2(-1) + 1] - [(-1)^2 - 3]$$
$$= (-2 + 1) - (1 - 3)$$
$$= -1 - (-2)$$
$$= -1 + 2$$
$$= 1$$

11. $(h - g)(-4) = h(-4) - g(-4)$
$$= (-4 + 4) - \sqrt{-4 - 1}$$
$$= 0 - \sqrt{-5}$$

Since $\sqrt{-5}$ is not a real number, $(h - g)(-4)$ does not exist.

13. $(g/h)(1) = \dfrac{g(1)}{h(1)}$

$$= \frac{\sqrt{1 - 1}}{1 + 4}$$

$$= \frac{\sqrt{0}}{5}$$

$$= \frac{0}{5} = 0$$

15. $(g + h)(1) = g(1) + h(1)$
$$= \sqrt{1 - 1} + (1 + 4)$$
$$= \sqrt{0} + 5$$
$$= 0 + 5 = 5$$

17. $f(x) = 2x + 3$, $g(x) = 3 - 5x$

a) The domain of f and of g is the set of all real numbers, or $(-\infty, \infty)$. Then the domain of $f + g$, $f - g$, ff, and fg is also $(-\infty, \infty)$. For f/g we must exclude $\dfrac{3}{5}$ since $g\left(\dfrac{3}{5}\right) = 0$. Then the domain of f/g is $\left(-\infty, \dfrac{3}{5}\right) \cup \left(\dfrac{3}{5}, \infty\right)$. For g/f we must exclude $-\dfrac{3}{2}$ since $f\left(-\dfrac{3}{2}\right) = 0$. The domain of g/f is $\left(-\infty, -\dfrac{3}{2}\right) \cup \left(-\dfrac{3}{2}, \infty\right)$.

b) $(f + g)(x) = f(x) + g(x) = (2x + 3) + (3 - 5x) = -3x + 6$

$(f - g)(x) = f(x) - g(x) = (2x + 3) - (3 - 5x) = 2x + 3 - 3 + 5x = 7x$

$(fg)(x) = f(x) \cdot g(x) = (2x + 3)(3 - 5x) = 6x - 10x^2 + 9 - 15x = -10x^2 - 9x + 9$

$(ff)(x) = f(x) \cdot f(x) = (2x + 3)(2x + 3) = 4x^2 + 12x + 9$

$(f/g)(x) = \dfrac{f(x)}{g(x)} = \dfrac{2x + 3}{3 - 5x}$

$(g/f)(x) = \dfrac{g(x)}{f(x)} = \dfrac{3 - 5x}{2x + 3}$

19. $f(x) = x - 3$, $g(x) = \sqrt{x + 4}$

a) Any number can be an input in f, so the domain of f is the set of all real numbers, or $(-\infty, \infty)$.

The domain of g consists of all values of x for which $x + 4$ is nonnegative, so we have $x + 4 \geq 0$, or $x \geq -4$. Thus, the domain of g is $[-4, \infty)$.

The domain of $f + g$, $f - g$, and fg is the set of all numbers in the domains of both f and g. This is $[-4, \infty)$.

The domain of ff is the domain of f, or $(-\infty, \infty)$.

The domain of f/g is the set of all numbers in the domains of f and g, excluding those for which $g(x) = 0$. Since $g(-4) = 0$, the domain of f/g is $(-4, \infty)$.

The domain of g/f is the set of all numbers in the domains of g and f, excluding those for which $f(x) = 0$. Since $f(3) = 0$, the domain of g/f is $[-4, 3) \cup (3, \infty)$.

b) $(f + g)(x) = f(x) + g(x) = x - 3 + \sqrt{x + 4}$

$(f - g)(x) = f(x) - g(x) = x - 3 - \sqrt{x + 4}$

$(fg)(x) = f(x) \cdot g(x) = (x - 3)\sqrt{x + 4}$

$(ff)(x) = \left[f(x)\right]^2 = (x - 3)^2 = x^2 - 6x + 9$

$(f/g)(x) = \dfrac{f(x)}{g(x)} = \dfrac{x - 3}{\sqrt{x + 4}}$

$(g/f)(x) = \dfrac{g(x)}{f(x)} = \dfrac{\sqrt{x + 4}}{x - 3}$

21. $f(x) = 2x - 1$, $g(x) = -2x^2$

a) The domain of f and of g is $(-\infty, \infty)$. Then the domain of $f + g$, $f - g$, fg, and ff is $(-\infty, \infty)$. For f/g, we must exclude 0 since $g(0) = 0$. The domain of f/g is $(-\infty, 0) \cup (0, \infty)$. For g/f, we must exclude $\dfrac{1}{2}$ since $f\left(\dfrac{1}{2}\right) = 0$. The domain of g/f is $\left(-\infty, \dfrac{1}{2}\right) \cup \left(\dfrac{1}{2}, \infty\right)$.

b) $(f + g)(x) = f(x) + g(x) = (2x - 1) + (-2x^2) = -2x^2 + 2x - 1$

$(f - g)(x) = f(x) - g(x) = (2x - 1) - (-2x^2) = 2x^2 + 2x - 1$

$(fg)(x) = f(x) \cdot g(x) = (2x - 1)(-2x^2) = -4x^3 + 2x^2$

$(ff)(x) = f(x) \cdot f(x) = (2x - 1)(2x - 1) = 4x^2 - 4x + 1$

$(f/g)(x) = \dfrac{f(x)}{g(x)} = \dfrac{2x - 1}{-2x^2}$

$(g/f)(x) = \dfrac{g(x)}{f(x)} = \dfrac{-2x^2}{2x - 1}$

23. $f(x) = \sqrt{x - 3}$, $g(x) = \sqrt{x + 3}$

a) Since $f(x)$ is nonnegative for values of x in $[3, \infty)$, this is the domain of f. Since $g(x)$ is nonnegative for values of x in $[-3, \infty)$, this is the domain of g. The domain of $f + g$, $f - g$, and fg is the intersection of the domains of f and g, or $[3, \infty)$. The domain of ff is the same as the domain of f, or $[3, \infty)$. For f/g, we must exclude -3 since $g(-3) = 0$. This is not in $[3, \infty)$, so the domain of f/g is $[3, \infty)$. For g/f, we must exclude 3 since $f(3) = 0$. The domain of g/f is $(3, \infty)$.

b) $(f + g)(x) = f(x) + g(x) = \sqrt{x - 3} + \sqrt{x + 3}$

$(f - g)(x) = f(x) - g(x) = \sqrt{x - 3} - \sqrt{x + 3}$

$(fg)(x) = f(x) \cdot g(x) = \sqrt{x - 3} \cdot \sqrt{x + 3} = \sqrt{x^2 - 9}$

$(ff)(x) = f(x) \cdot f(x) = \sqrt{x - 3} \cdot \sqrt{x - 3} = |x - 3|$

$(f/g)(x) = \dfrac{\sqrt{x - 3}}{\sqrt{x + 3}}$

$(g/f)(x) = \dfrac{\sqrt{x + 3}}{\sqrt{x - 3}}$

25. $f(x) = x + 1$, $g(x) = |x|$

a) The domain of f and of g is $(-\infty, \infty)$. Then the domain of $f + g$, $f - g$, fg, and ff is $(-\infty, \infty)$. For f/g, we must exclude 0 since $g(0) = 0$. The domain of f/g is $(-\infty, 0) \cup (0, \infty)$. For g/f, we must exclude -1 since $f(-1) = 0$. The domain of g/f is $(-\infty, -1) \cup (-1, \infty)$.

b) $(f + g)(x) = f(x) + g(x) = x + 1 + |x|$

$(f - g)(x) = f(x) - g(x) = x + 1 - |x|$

$(fg)(x) = f(x) \cdot g(x) = (x + 1)|x|$

$(ff)(x) = f(x) \cdot f(x) = (x + 1)(x + 1) = x^2 + 2x + 1$

$$(f/g)(x) = \frac{x+1}{|x|}$$

$$(g/f)(x) = \frac{|x|}{x+1}$$

27. $f(x) = x^3$, $g(x) = 2x^2 + 5x - 3$

a) Since any number can be an input for either f or g, the domain of f, g, $f+g$, $f-g$, fg, and ff is the set of all real numbers, or $(-\infty, \infty)$.

Since $g(-3) = 0$ and $g\left(\frac{1}{2}\right) = 0$, the domain of f/g

is $(-\infty, -3) \cup \left(-3, \frac{1}{2}\right) \cup \left(\frac{1}{2}, \infty\right)$.

Since $f(0) = 0$, the domain of g/f is $(-\infty, 0) \cup (0, \infty)$.

b) $(f+g)(x) = f(x) + g(x) = x^3 + 2x^2 + 5x - 3$

$(f-g)(x) = f(x) - g(x) = x^3 - (2x^2 + 5x - 3) =$
$x^3 - 2x^2 - 5x + 3$

$(fg)(x) = f(x) \cdot g(x) = x^3(2x^2 + 5x - 3) =$
$2x^5 + 5x^4 - 3x^3$

$(ff)(x) = f(x) \cdot f(x) = x^3 \cdot x^3 = x^6$

$(f/g)(x) = \frac{f(x)}{g(x)} = \frac{x^3}{2x^2 + 5x - 3}$

$(g/f)(x) = \frac{g(x)}{f(x)} = \frac{2x^2 + 5x - 3}{x^3}$

29. $f(x) = \frac{4}{x+1}$, $g(x) = \frac{1}{6-x}$

a) Since $x + 1 = 0$ when $x = -1$, we must exclude -1 from the domain of f. It is $(-\infty, -1) \cup (-1, \infty)$. Since $6 - x = 0$ when $x = 6$, we must exclude 6 from the domain of g. It is $(-\infty, 6) \cup (6, \infty)$. The domain of $f + g$, $f - g$, and fg is the intersection of the domains of f and g, or $(-\infty, -1) \cup (-1, 6) \cup (6, \infty)$. The domain of ff is the same as the domain of f, or $(-\infty, -1) \cup (-1, \infty)$. Since there are no values of x for which $g(x) = 0$ or $f(x) = 0$, the domain of f/g and g/f is $(-\infty, -1) \cup (-1, 6) \cup (6, \infty)$.

b) $(f+g)(x) = f(x) + g(x) = \frac{4}{x+1} + \frac{1}{6-x}$

$(f-g)(x) = f(x) - g(x) = \frac{4}{x+1} - \frac{1}{6-x}$

$(fg)(x) = f(x) \cdot g(x) = \frac{4}{x+1} \cdot \frac{1}{6-x} = \frac{4}{(x+1)(6-x)}$

$(ff)(x) = f(x) \cdot f(x) = \frac{4}{x+1} \cdot \frac{4}{x+1} = \frac{16}{(x+1)^2}$, or
$\frac{16}{x^2 + 2x + 1}$

$(f/g)(x) = \frac{\frac{4}{x+1}}{\frac{1}{6-x}} = \frac{4}{x+1} \cdot \frac{6-x}{1} = \frac{4(6-x)}{x+1}$

$(g/f)(x) = \frac{\frac{1}{6-x}}{\frac{4}{x+1}} = \frac{1}{6-x} \cdot \frac{x+1}{4} = \frac{x+1}{4(6-x)}$

31. $f(x) = \frac{1}{x}$, $g(x) = x - 3$

a) Since $f(0)$ is not defined, the domain of f is $(-\infty, 0) \cup (0, \infty)$. The domain of g is $(-\infty, \infty)$. Then the domain of $f + g$, $f - g$, fg, and ff is $(-\infty, 0) \cup (0, \infty)$. Since $g(3) = 0$, the domain of f/g is $(-\infty, 0) \cup (0, 3) \cup (3, \infty)$. There are no values of x for which $f(x) = 0$, so the domain of g/f is $(-\infty, 0) \cup (0, \infty)$.

b) $(f+g)(x) = f(x) + g(x) = \frac{1}{x} + x - 3$

$(f-g)(x) = f(x) - g(x) = \frac{1}{x} - (x-3) = \frac{1}{x} - x + 3$

$(fg)(x) = f(x) \cdot g(x) = \frac{1}{x} \cdot (x-3) = \frac{x-3}{x}$, or $1 - \frac{3}{x}$

$(ff)(x) = f(x) \cdot f(x) = \frac{1}{x} \cdot \frac{1}{x} = \frac{1}{x^2}$

$(f/g)(x) = \frac{f(x)}{g(x)} = \frac{\frac{1}{x}}{x-3} = \frac{1}{x} \cdot \frac{1}{x-3} = \frac{1}{x(x-3)}$

$(g/f)(x) = \frac{g(x)}{f(x)} = \frac{x-3}{\frac{1}{x}} = (x-3) \cdot \frac{x}{1} = x(x-3)$, or
$x^2 - 3x$

33. $f(x) = \frac{3}{x-2}$, $g(x) = \sqrt{x-1}$

a) Since $f(2)$ is not defined, the domain of f is $(-\infty, 2) \cup (2, \infty)$. Since $g(x)$ is nonnegative for values of x in $[1, \infty)$, this is the domain of g. The domain of $f + g$, $f - g$, and fg is the intersection of the domains of f and g, or $[1, 2) \cup (2, \infty)$. The domain of ff is the same as the domain of f, or $(-\infty, 2) \cup (2, \infty)$. For f/g, we must exclude 1 since $g(1) = 0$, so the domain of f/g is $(1, 2) \cup (2, \infty)$. There are no values of x for which $f(x) = 0$, so the domain of g/f is $[1, 2) \cup (2, \infty)$.

b) $(f+g)(x) = f(x) + g(x) = \frac{3}{x-2} + \sqrt{x-1}$

$(f-g)(x) = f(x) - g(x) = \frac{3}{x-2} - \sqrt{x-1}$

$(fg)(x) = f(x) \cdot g(x) = \frac{3}{x-2}(\sqrt{x-1})$, or $\frac{3\sqrt{x-1}}{x-2}$

$(ff)(x) = f(x) \cdot f(x) = \frac{3}{x-2} \cdot \frac{3}{x-2} \cdot \frac{9}{(x-2)^2}$

$(f/g)(x) = \frac{f(x)}{g(x)} = \frac{\frac{3}{x-2}}{\sqrt{x-1}} = \frac{3}{(x-2)\sqrt{x-1}}$

$(g/f)(x) = \frac{g(x)}{f(x)} = \frac{\sqrt{x-1}}{\frac{3}{x-2}} = \frac{(x-2)\sqrt{x-1}}{3}$

35. From the graph we see that the domain of F is $[2, 11]$ and the domain of G is $[1, 9]$. The domain of $F + G$ is the set of numbers in the domains of both F and G. This is $[2, 9]$.

37. The domain of G/F is the set of numbers in the domains of both F and G (See Exercise 35.), excluding those for which $F = 0$. Since $F(3) = 0$, the domain of G/F is $[2, 3) \cup (3, 9]$.

39.

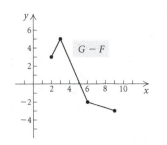

41. From the graph, we see that the domain of F is $[0,9]$ and the domain of G is $[3,10]$. The domain of $F + G$ is the set of numbers in the domains of both F and G. This is $[3,9]$.

43. The domain of G/F is the set of numbers in the domains of both F and G (See Exercise 41.), excluding those for which $F = 0$. Since $F(6) = 0$ and $F(8) = 0$, the domain of G/F is $[3,6) \cup (6,8) \cup (8,9]$.

45.

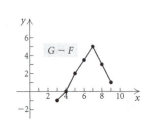

47. a) $P(x) = R(x) - C(x) = 60x - 0.4x^2 - (3x + 13) =$
$$60x - 0.4x^2 - 3x - 13 = -0.4x^2 + 57x - 13$$

   b) $R(100) = 60 \cdot 100 - 0.4(100)^2 = 6000 - 0.4(10,000) =$
$$6000 - 4000 = 2000$$
$$C(100) = 3 \cdot 100 + 13 = 300 + 13 = 313$$
$$P(100) = R(100) - C(100) = 2000 - 313 = 1687$$

49. $f(x) = 3x - 5$
$$f(x + h) = 3(x + h) - 5 = 3x + 3h - 5$$
$$\frac{f(x + h) - f(x)}{h} = \frac{3x + 3h - 5 - (3x - 5)}{h}$$
$$= \frac{3x + 3h - 5 - 3x + 5}{h}$$
$$= \frac{3h}{h} = 3$$

51. $f(x) = 6x + 2$
$$f(x + h) = 6(x + h) + 2 = 6x + 6h + 2$$
$$\frac{f(x + h) - f(x)}{h} = \frac{6x + 6h + 2 - (6x + 2)}{h}$$
$$= \frac{6x + 6h + 2 - 6x - 2}{h}$$
$$= \frac{6h}{h} = 6$$

53. $f(x) = \frac{1}{3}x + 1$
$$f(x + h) = \frac{1}{3}(x + h) + 1 = \frac{1}{3}x + \frac{1}{3}h + 1$$
$$\frac{f(x + h) - f(x)}{h} = \frac{\frac{1}{3}x + \frac{1}{3}h + 1 - \left(\frac{1}{3}x + 1\right)}{h}$$
$$= \frac{\frac{1}{3}x + \frac{1}{3}h + 1 - \frac{1}{3}x - 1}{h}$$
$$= \frac{\frac{1}{3}h}{h} = \frac{1}{3}$$

55. $f(x) = \frac{1}{3x}$
$$f(x + h) = \frac{1}{3(x + h)}$$
$$\frac{f(x + h) - f(x)}{h} = \frac{\frac{1}{3(x + h)} - \frac{1}{3x}}{h}$$
$$= \frac{\frac{1}{3(x + h)} \cdot \frac{x}{x} - \frac{1}{3x} \cdot \frac{x + h}{x + h}}{h}$$
$$= \frac{\frac{x}{3x(x + h)} - \frac{x + h}{3x(x + h)}}{h}$$
$$= \frac{\frac{x - (x + h)}{3x(x + h)}}{h} = \frac{\frac{x - x - h}{3x(x + h)}}{h}$$
$$= \frac{\frac{-h}{3x(x + h)}}{h} = \frac{-h}{3x(x + h)} \cdot \frac{1}{h}$$
$$= \frac{-h}{3x(x + h) \cdot h} = \frac{-1 \cdot h}{3x(x + h) \cdot h}$$
$$= \frac{-1}{3x(x + h)}, \text{ or } -\frac{1}{3x(x + h)}$$

57. $f(x) = -\frac{1}{4x}$
$$f(x + h) = -\frac{1}{4(x + h)}$$
$$\frac{f(x + h) - f(x)}{h} = \frac{-\frac{1}{4(x + h)} - \left(-\frac{1}{4x}\right)}{h}$$
$$= \frac{-\frac{1}{4(x + h)} \cdot \frac{x}{x} - \left(-\frac{1}{4x}\right) \cdot \frac{x + h}{x + h}}{h}$$
$$= \frac{-\frac{x}{4x(x + h)} + \frac{x + h}{4x(x + h)}}{h}$$
$$= \frac{\frac{-x + x + h}{4x(x + h)}}{h} = \frac{\frac{h}{4x(x + h)}}{h}$$
$$= \frac{h}{4x(x + h)} \cdot \frac{1}{h} = \frac{h \cdot 1}{4x(x + h) \cdot h} = \frac{1}{4x(x + h)}$$

59. $f(x) = x^2 + 1$

$f(x+h) = (x+h)^2 + 1 = x^2 + 2xh + h^2 + 1$

$\dfrac{f(x+h) - f(x)}{h} = \dfrac{x^2 + 2xh + h^2 + 1 - (x^2 + 1)}{h}$

$\phantom{\dfrac{f(x+h) - f(x)}{h}} = \dfrac{x^2 + 2xh + h^2 + 1 - x^2 - 1}{h}$

$\phantom{\dfrac{f(x+h) - f(x)}{h}} = \dfrac{2xh + h^2}{h}$

$\phantom{\dfrac{f(x+h) - f(x)}{h}} = \dfrac{h(2x + h)}{h}$

$\phantom{\dfrac{f(x+h) - f(x)}{h}} = \dfrac{h}{h} \cdot \dfrac{2x + h}{1}$

$\phantom{\dfrac{f(x+h) - f(x)}{h}} = 2x + h$

61. $f(x) = 4 - x^2$

$f(x+h) = 4 - (x+h)^2 = 4 - (x^2 + 2xh + h^2) =$
$4 - x^2 - 2xh - h^2$

$\dfrac{f(x+h) - f(x)}{h} = \dfrac{4 - x^2 - 2xh - h^2 - (4 - x^2)}{h}$

$\phantom{\dfrac{f(x+h) - f(x)}{h}} = \dfrac{4 - x^2 - 2xh - h^2 - 4 + x^2}{h}$

$\phantom{\dfrac{f(x+h) - f(x)}{h}} = \dfrac{-2xh - h^2}{h} = \dfrac{\cancel{h}(-2x - h)}{\cancel{h}}$

$\phantom{\dfrac{f(x+h) - f(x)}{h}} = -2x - h$

63. $f(x) = 3x^2 - 2x + 1$

$f(x+h) = 3(x+h)^2 - 2(x+h) + 1 =$
$3(x^2 + 2xh + h^2) - 2(x+h) + 1 =$
$3x^2 + 6xh + 3h^2 - 2x - 2h + 1$

$f(x) = 3x^2 - 2x + 1$

$\dfrac{f(x+h) - f(x)}{h} =$

$\dfrac{(3x^2 + 6xh + 3h^2 - 2x - 2h + 1) - (3x^2 - 2x + 1)}{h} =$

$\dfrac{3x^2 + 6xh + 3h^2 - 2x - 2h + 1 - 3x^2 + 2x - 1}{h} =$

$\dfrac{6xh + 3h^2 - 2h}{h} = \dfrac{h(6x + 3h - 2)}{h \cdot 1} =$

$\dfrac{h}{h} \cdot \dfrac{6x + 3h - 2}{1} = 6x + 3h - 2$

65. $f(x) = 4 + 5|x|$

$f(x+h) = 4 + 5|x+h|$

$\dfrac{f(x+h) - f(x)}{h} = \dfrac{4 + 5|x+h| - (4 + 5|x|)}{h}$

$\phantom{\dfrac{f(x+h) - f(x)}{h}} = \dfrac{4 + 5|x+h| - 4 - 5|x|}{h}$

$\phantom{\dfrac{f(x+h) - f(x)}{h}} = \dfrac{5|x+h| - 5|x|}{h}$

67. $f(x) = x^3$

$f(x+h) = (x+h)^3 = x^3 + 3x^2h + 3xh^2 + h^3$

$f(x) = x^3$

$\dfrac{f(x+h) - f(x)}{h} = \dfrac{x^3 + 3x^2h + 3xh^2 + h^3 - x^3}{h} =$

$\dfrac{3x^2h + 3xh^2 + h^3}{h} = \dfrac{h(3x^2 + 3xh + h^2)}{h \cdot 1} =$

$\dfrac{h}{h} \cdot \dfrac{3x^2 + 3xh + h^2}{1} = 3x^2 + 3xh + h^2$

69. $f(x) = \dfrac{x - 4}{x + 3}$

$\dfrac{f(x+h) - f(x)}{h} = \dfrac{\dfrac{x+h-4}{x+h+3} - \dfrac{x-4}{x+3}}{h} =$

$\dfrac{\dfrac{x+h-4}{x+h+3} - \dfrac{x-4}{x+3}}{h} \cdot \dfrac{(x+h+3)(x+3)}{(x+h+3)(x+3)} =$

$\dfrac{(x+h-4)(x+3) - (x-4)(x+h+3)}{h(x+h+3)(x+3)} =$

$\dfrac{x^2 + hx - 4x + 3x + 3h - 12 - (x^2 + hx + 3x - 4x - 4h - 12)}{h(x+h+3)(x+3)} =$

$\dfrac{x^2 + hx - x + 3h - 12 - x^2 - hx + x + 4h + 12}{h(x+h+3)(x+3)} =$

$\dfrac{7h}{h(x+h+3)(x+3)} = \dfrac{h}{h} \cdot \dfrac{7}{(x+h+3)(x+3)} =$

$\dfrac{7}{(x+h+3)(x+3)}$

71. Graph $y = 3x - 1$.

We find some ordered pairs that are solutions of the equation, plot these points, and draw the graph.

When $x = -1$, $y = 3(-1) - 1 = -3 - 1 = -4$.

When $x = 0$, $y = 3 \cdot 0 - 1 = 0 - 1 = -1$.

When $x = 2$, $y = 3 \cdot 2 - 1 = 6 - 1 = 5$.

x	y
-1	-4
0	-1
2	5

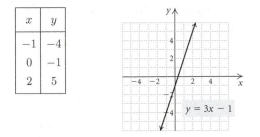

73. Graph $x - 3y = 3$.

First we find the x- and y-intercepts.

$x - 3 \cdot 0 = 3$

$x = 3$

The x-intercept is $(3, 0)$.

$0 - 3y = 3$

$-3y = 3$

$y = -1$

The y-intercept is $(0, -1)$.

We find a third point as a check. We let $x = -3$ and solve for y.

$$-3 - 3y = 3$$
$$-3y = 6$$
$$y = -2$$

Another point on the graph is $(-3, -2)$. We plot the points and draw the graph.

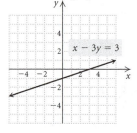

75. Answers may vary; $f(x) = \dfrac{1}{x+7}$, $g(x) = \dfrac{1}{x-3}$

77. The domain of $h(x)$ is $\left\{x \middle| x \neq \dfrac{7}{3}\right\}$, and the domain of $g(x)$ is $\{x | x \neq 3\}$, so $\dfrac{7}{3}$ and 3 are not in the domain of $(h/g)(x)$. We must also exclude the value of x for which $g(x) = 0$.

$$\frac{x^4 - 1}{5x - 15} = 0$$
$$x^4 - 1 = 0 \quad \text{Multiplying by } 5x - 15$$
$$x^4 = 1$$
$$x = \pm 1$$

Then the domain of $(h/g)(x)$ is
$\left\{x \middle| x \neq \dfrac{7}{3} \text{ and } x \neq 3 \text{ and } x \neq -1 \text{ and } x \neq 1\right\}$, or
$(-\infty, -1) \cup (-1, 1) \cup \left(1, \dfrac{7}{3}\right) \cup \left(\dfrac{7}{3}, 3\right) \cup (3, \infty)$.

Exercise Set 2.3

1. $(f \circ g)(-1) = f(g(-1)) = f((-1)^2 - 2(-1) - 6) =$
$f(1 + 2 - 6) = f(-3) = 3(-3) + 1 = -9 + 1 = -8$

3. $(h \circ f)(1) = h(f(1)) = h(3 \cdot 1 + 1) = h(3 + 1) =$
$h(4) = 4^3 = 64$

5. $(g \circ f)(5) = g(f(5)) = g(3 \cdot 5 + 1) = g(15 + 1) =$
$g(16) = 16^2 - 2 \cdot 16 - 6 = 218$

7. $(f \circ h)(-3) = f(h(-3)) = f((-3)^3) = f(-27) =$
$3(-27) + 1 = -81 + 1 = -80$

9. $(g \circ g)(-2) = g(g(-2)) = g((-2)^2 - 2(-2) - 6) =$
$g(4 + 4 - 6) = g(2) = 2^2 - 2 \cdot 2 - 6 = 4 - 4 - 6 = -6$

11. $(h \circ h)(2) = h(h(2)) = h(2^3) = h(8) = 8^3 = 512$

13. $(f \circ f)(-4) = f(f(-4)) = f(3(-4) + 1) = f(-12 + 1) =$
$f(-11) = 3(-11) + 1 = -33 + 1 = -32$

15. $(h \circ h)(x) = h(h(x)) = h(x^3) = (x^3)^3 = x^9$

17. $(f \circ g)(x) = f(g(x)) = f(x - 3) = x - 3 + 3 = x$
$(g \circ f)(x) = g(f(x)) = g(x + 3) = x + 3 - 3 = x$

The domain of f and of g is $(-\infty, \infty)$, so the domain of $f \circ g$ and of $g \circ f$ is $(-\infty, \infty)$.

19. $(f \circ g)(x) = f(g(x)) = f(3x^2 - 2x - 1) = 3x^2 - 2x - 1 + 1 = 3x^2 - 2x$
$(g \circ f)(x) = g(f(x)) = g(x + 1) = 3(x + 1)^2 - 2(x + 1) - 1 = 3(x^2 + 2x + 1) - 2(x + 1) - 1 = 3x^2 + 6x + 3 - 2x - 2 - 1 = 3x^2 + 4x$

The domain of f and of g is $(-\infty, \infty)$, so the domain of $f \circ g$ and of $g \circ f$ is $(-\infty, \infty)$.

21. $(f \circ g)(x) = f(g(x)) = f(4x - 3) = (4x - 3)^2 - 3 = 16x^2 - 24x + 9 - 3 = 16x^2 - 24x + 6$
$(g \circ f)(x) = g(f(x)) = g(x^2 - 3) = 4(x^2 - 3) - 3 = 4x^2 - 12 - 3 = 4x^2 - 15$

The domain of f and of g is $(-\infty, \infty)$, so the domain of $f \circ g$ and of $g \circ f$ is $(-\infty, \infty)$.

23. $(f \circ g)(x) = f(g(x)) = f\left(\dfrac{1}{x}\right) = \dfrac{4}{1 - 5 \cdot \dfrac{1}{x}} = \dfrac{4}{1 - \dfrac{5}{x}} =$
$\dfrac{4}{\dfrac{x - 5}{x}} = 4 \cdot \dfrac{x}{x - 5} = \dfrac{4x}{x - 5}$
$(g \circ f)(x) = g(f(x)) = g\left(\dfrac{4}{1 - 5x}\right) = \dfrac{1}{\dfrac{4}{1 - 5x}} =$
$1 \cdot \dfrac{1 - 5x}{4} = \dfrac{1 - 5x}{4}$

The domain of f is $\left\{x \middle| x \neq \dfrac{1}{5}\right\}$ and the domain of g is $\{x | x \neq 0\}$. Consider the domain of $f \circ g$. Since 0 is not in the domain of g, 0 is not in the domain of $f \circ g$. Since $\dfrac{1}{5}$ is not in the domain of f, we know that $g(x)$ cannot be $\dfrac{1}{5}$. We find the value(s) of x for which $g(x) = \dfrac{1}{5}$.

$$\frac{1}{x} = \frac{1}{5}$$
$$5 = x \quad \text{Multiplying by } 5x$$

Thus 5 is also not in the domain of $f \circ g$. Then the domain of $f \circ g$ is $\{x | x \neq 0 \text{ and } x \neq 5\}$, or $(-\infty, 0) \cup (0, 5) \cup (5, \infty)$.

Now consider the domain of $g \circ f$. Recall that $\dfrac{1}{5}$ is not in the domain of f, so it is not in the domain of $g \circ f$. Now 0 is not in the domain of g but $f(x)$ is never 0, so the domain of $g \circ f$ is $\left\{x \middle| x \neq \dfrac{1}{5}\right\}$, or $\left(-\infty, \dfrac{1}{5}\right) \cup \left(\dfrac{1}{5}, \infty\right)$.

25. $(f \circ g)(x) = f(g(x)) = f\left(\dfrac{x + 7}{3}\right) =$
$$3\left(\frac{x + 7}{3}\right) - 7 = x + 7 - 7 = x$$
$(g \circ f)(x) = g(f(x)) = g(3x - 7) = \dfrac{(3x - 7) + 7}{3} =$
$$\frac{3x}{3} = x$$

The domain of f and of g is $(-\infty, \infty)$, so the domain of $f \circ g$ and of $g \circ f$ is $(-\infty, \infty)$.

27. $(f \circ g)(x) = f(g(x)) = f(\sqrt{x}) = 2\sqrt{x} + 1$
$(g \circ f)(x) = g(f(x)) = g(2x + 1) = \sqrt{2x + 1}$

The domain of f is $(-\infty, \infty)$ and the domain of g is $\{x | x \geq 0\}$. Thus the domain of $f \circ g$ is $\{x | x \geq 0\}$, or $[0, \infty)$.

Now consider the domain of $g \circ f$. There are no restrictions on the domain of f, but the domain of g is $\{x | x \geq 0\}$. Since $f(x) \geq 0$ for $x \geq -\dfrac{1}{2}$, the domain of $g \circ f$ is $\left\{x \middle| x \geq -\dfrac{1}{2}\right\}$, or $\left[-\dfrac{1}{2}, \infty\right)$.

29. $(f \circ g)(x) = f(g(x)) = f(0.05) = 20$
$(g \circ f)(x) = g(f(x)) = g(20) = 0.05$

The domain of f and of g is $(-\infty, \infty)$, so the domain of $f \circ g$ and of $g \circ f$ is $(-\infty, \infty)$.

31. $(f \circ g)(x) = f(g(x)) = f(x^2 - 5) =$
$\sqrt{x^2 - 5 + 5} = \sqrt{x^2} = |x|$
$(g \circ f)(x) = g(f(x)) = g(\sqrt{x + 5}) =$
$(\sqrt{x + 5})^2 - 5 = x + 5 - 5 = x$

The domain of f is $\{x | x \geq -5\}$ and the domain of g is $(-\infty, \infty)$. Since $x^2 \geq 0$ for all values of x, then $x^2 - 5 \geq -5$ for all values of x and the domain of $g \circ f$ is $(-\infty, \infty)$.

Now consider the domain of $f \circ g$. There are no restrictions on the domain of g, so the domain of $f \circ g$ is the same as the domain of f, $\{x | x \geq -5\}$, or $[-5, \infty)$.

33. $(f \circ g)(x) = f(g(x)) = f(\sqrt{3 - x}) = (\sqrt{3 - x})^2 + 2 =$
$3 - x + 2 = 5 - x$
$(g \circ f)(x) = g(f(x)) = g(x^2 + 2) = \sqrt{3 - (x^2 + 2)} =$
$\sqrt{3 - x^2 - 2} = \sqrt{1 - x^2}$

The domain of f is $(-\infty, \infty)$ and the domain of g is $\{x | x \leq 3\}$, so the domain of $f \circ g$ is $\{x | x \leq 3\}$, or $(-\infty, 3]$.

Now consider the domain of $g \circ f$. There are no restrictions on the domain of f and the domain of g is $\{x | x \leq 3\}$, so we find the values of x for which $f(x) \leq 3$. We see that $x^2 + 2 \leq 3$ for $-1 \leq x \leq 1$, so the domain of $g \circ f$ is $\{x | -1 \leq x \leq 1\}$, or $[-1, 1]$.

35. $(f \circ g)(x) = f(g(x)) = f\left(\dfrac{1}{1 + x}\right) =$

$\dfrac{1 - \left(\dfrac{1}{1 + x}\right)}{\dfrac{1}{1 + x}} = \dfrac{\dfrac{1 + x - 1}{1 + x}}{\dfrac{1}{1 + x}} =$

$\dfrac{x}{1 + x} \cdot \dfrac{1 + x}{1} = x$

$(g \circ f)(x) = g(f(x)) = g\left(\dfrac{1 - x}{x}\right) =$

$\dfrac{1}{1 + \left(\dfrac{1 - x}{x}\right)} = \dfrac{1}{\dfrac{x + 1 - x}{x}} =$

$\dfrac{1}{\dfrac{1}{x}} = 1 \cdot \dfrac{x}{1} = x$

The domain of f is $\{x | x \neq 0\}$ and the domain of g is $\{x | x \neq -1\}$, so we know that -1 is not in the domain of $f \circ g$. Since 0 is not in the domain of f, values of x for which $g(x) = 0$ are not in the domain of $f \circ g$. But $g(x)$ is never 0, so the domain of $f \circ g$ is $\{x | x \neq -1\}$, or $(-\infty, -1) \cup (-1, \infty)$.

Now consider the domain of $g \circ f$. Recall that 0 is not in the domain of f. Since -1 is not in the domain of g, we know that $g(x)$ cannot be -1. We find the value(s) of x for which $f(x) = -1$.

$\dfrac{1 - x}{x} = -1$
$1 - x = -x$ Multiplying by x
$1 = 0$ False equation

We see that there are no values of x for which $f(x) = -1$, so the domain of $g \circ f$ is $\{x | x \neq 0\}$, or $(-\infty, 0) \cup (0, \infty)$.

37. $(f \circ g)(x) = f(g(x)) = f(x + 1) =$
$(x + 1)^3 - 5(x + 1)^2 + 3(x + 1) + 7 =$
$x^3 + 3x^2 + 3x + 1 - 5x^2 - 10x - 5 + 3x + 3 + 7 =$
$x^3 - 2x^2 - 4x + 6$

$(g \circ f)(x) = g(f(x)) = g(x^3 - 5x^2 + 3x + 7) =$
$x^3 - 5x^2 + 3x + 7 + 1 = x^3 - 5x^2 + 3x + 8$

The domain of f and of g is $(-\infty, \infty)$, so the domain of $f \circ g$ and of $g \circ f$ is $(-\infty, \infty)$.

39. $h(x) = (4 + 3x)^5$

This is $4 + 3x$ to the 5th power. The most obvious answer is $f(x) = x^5$ and $g(x) = 4 + 3x$.

41. $h(x) = \dfrac{1}{(x - 2)^4}$

This is 1 divided by $(x - 2)$ to the 4th power. One obvious answer is $f(x) = \dfrac{1}{x^4}$ and $g(x) = x - 2$. Another possibility is $f(x) = \dfrac{1}{x}$ and $g(x) = (x - 2)^4$.

43. $f(x) = \dfrac{x - 1}{x + 1}$, $g(x) = x^3$

45. $f(x) = x^6$, $g(x) = \dfrac{2 + x^3}{2 - x^3}$

47. $f(x) = \sqrt{x}$, $g(x) = \dfrac{x - 5}{x + 2}$

49. $f(x) = x^3 - 5x^2 + 3x - 1$, $g(x) = x + 2$

51. a) Use the distance formula, distance = rate $\times$ time. Substitute 3 for the rate and t for time.

$r(t) = 3t$

b) Use the formula for the area of a circle.

$A(r) = \pi r^2$

c) $(A \circ r)(t) = A(r(t)) = A(3t) = \pi(3t)^2 = 9\pi t^2$

This function gives the area of the ripple in terms of time t.

53. $f(x) = (t \circ s)(x) = t(s(x)) = t(x-3) = x-3+4 = x+1$

We have $f(x) = x + 1$.

55. Equations $(a) - (f)$ are in the form $y = mx + b$, so we can read the y-intercepts directly from the equations. Equations (g) and (h) can be written in this form as $y = \dfrac{2}{3}x - 2$ and $y = -2x + 3$, respectively. We see that only equation (c) has y-intercept $(0, 1)$.

57. If a line slopes down from left to right, its slope is negative. The equations $y = mx + b$ for which m is negative are (b), (d), (f), and (h). (See Exercise 55.)

59. The only equation that has $(0, 0)$ as a solution is (a).

61. Only equations (c) and (g) have the same slope and different y-intercepts. They represent parallel lines.

63. Only the composition $(c \circ p)(a)$ makes sense. It represents the cost of the grass seed required to seed a lawn with area a.

Chapter 2 Mid-Chapter Mixed Review

1. The statement is true. See page 100 in the text.

3. The statement is true. See Example 2 in Section 2.3 in the text, for instance.

5. From the graph we see that a relative maximum value of 6.30 occurs at $x = -1.29$. We also see that a relative minimum value of -2.30 occurs at $x = 1.29$.

The graph starts rising, or increasing, from the left and stops increasing at the relative maximum. From this point it decreases to the relative minimum and then increases again. Thus the function is increasing on $(-\infty, -1.29)$ and on $(1.29, \infty)$. It is decreasing on $(-1.29, 1.29)$.

7. $A(h) = \dfrac{1}{2}(h+4)h$

$A(h) = \dfrac{1}{2}h^2 + 2h$, or $\dfrac{h^2}{2} + 2h$

9. $g(x) = \begin{cases} x+2, & \text{for } x < -4, \\ -x, & \text{for } x \geq -4 \end{cases}$

We create the graph in two parts. Graph $g(x) = x + 2$ for inputs less than -4. Then graph $g(x) = -x$ for inputs greater than or equal to -4.

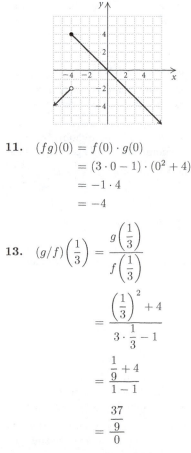

11. $(fg)(0) = f(0) \cdot g(0)$

$= (3 \cdot 0 - 1) \cdot (0^2 + 4)$

$= -1 \cdot 4$

$= -4$

13. $(g/f)\left(\dfrac{1}{3}\right) = \dfrac{g\left(\dfrac{1}{3}\right)}{f\left(\dfrac{1}{3}\right)}$

$= \dfrac{\left(\dfrac{1}{3}\right)^2 + 4}{3 \cdot \dfrac{1}{3} - 1}$

$= \dfrac{\dfrac{1}{9} + 4}{1 - 1}$

$= \dfrac{\dfrac{37}{9}}{0}$

Since division by 0 is not defined, $(g/f)\left(\dfrac{1}{3}\right)$ does not exist.

15. $f(x) = x - 1$, $g(x) = \sqrt{x+2}$

a) Any number can be an input for f, so the domain of f is the set of all real numbers, or $(-\infty, \infty)$.

The domain of g consists of all values for which $x+2$ is nonnegative, so we have $x + 2 \geq 0$, or $x \geq -2$, or $[-2, \infty)$. Then the domain of $f + g$, $f - g$, and fg is $[-2, \infty)$.

The domain of ff is $(-\infty, \infty)$.

Since $g(-2) = 0$, the domain of f/g is $(-2, \infty)$.

Since $f(1) = 0$, the domain of g/f is $[-2, 1) \cup (1, \infty)$.

b) $(f+g)(x) = f(x) + g(x) = x - 1 + \sqrt{x+2}$

$(f-g)(x) = f(x) - g(x) = x - 1 - \sqrt{x+2}$

$(fg)(x) = f(x) \cdot g(x) = (x-1)\sqrt{x+2}$

$(ff)(x) = f(x) \cdot f(x) = (x-1)(x-1) = x^2 - x - x + 1 = x^2 - 2x + 1$

$(f/g)(x) = \dfrac{f(x)}{g(x)} = \dfrac{x-1}{\sqrt{x+2}}$

$(g/f)(x) = \dfrac{g(x)}{f(x)} = \dfrac{\sqrt{x+2}}{x-1}$

17. $f(x) = 6 - x^2$

$$\frac{f(x+h) - f(x)}{h} = \frac{6 - (x+h)^2 - (6 - x^2)}{h} =$$

$$\frac{6 - (x^2 + 2xh + h^2) - 6 + x^2}{h} = \frac{6 - x^2 - 2xh - h^2 - 6 + x^2}{h} =$$

$$\frac{-2xh - h^2}{h} = \frac{\cancel{h}(-2x - h)}{\cancel{h} \cdot 1} = -2x - h$$

19. $(g \circ h)(2) = g(h(2)) = g(2^2 - 2 \cdot 2 + 3) = g(4 - 4 + 3) =$
$g(3) = 3^3 + 1 = 27 + 1 = 28$

21. $(h \circ f)(-1) = h(f(-1)) = h(5(-1) - 4) = h(-5 - 4) =$
$h(-9) = (-9)^2 - 2(-9) + 3 = 81 + 18 + 3 = 102$

23. $(f \circ g)(x) = f(g(x)) = f(\sqrt{x}) = 3\sqrt{x} + 2$
$(g \circ f)(x) = g(f(x)) = g(3x + 2) = \sqrt{3x + 2}$

The domain of f is $(-\infty, \infty)$ and the domain of g is $[0, \infty)$.
Consider the domain of $f \circ g$. Since any number can be an input for f, the domain of $f \circ g$ is the same as the domain of g, $[0, \infty)$.

Now consider the domain of $g \circ f$. Since the inputs of g must be nonnegative, we must have $3x + 2 \geq 0$, or $x \geq -\frac{2}{3}$.
Thus the domain of $g \circ f$ is $\left[-\frac{2}{3}, \infty \right)$.

25. Under the given conditions, $(f + g)(x)$ and $(f/g)(x)$ have different domains if $g(x) = 0$ for one or more real numbers x.

27. This approach is not valid. Consider Exercise 23 in Exercise Set 2.3 in the text, for example. Since $(f \circ g)(x) = \frac{4x}{x - 5}$, an examination of only this composed function would lead to the incorrect conclusion that the domain of $f \circ g$ is $(-\infty, 5) \cup (5, \infty)$. However, we must also exclude from the domain of $f \circ g$ those values of x that are not in the domain of g. Thus, the domain of $f \circ g$ is $(-\infty, 0) \cup (0, 5) \cup (5, \infty)$.

Exercise Set 2.4

1. If the graph were folded on the x-axis, the parts above and below the x-axis would not coincide, so the graph is not symmetric with respect to the x-axis.

If the graph were folded on the y-axis, the parts to the left and right of the y-axis would coincide, so the graph is symmetric with respect to the y-axis.

If the graph were rotated $180°$, the resulting graph would not coincide with the original graph, so it is not symmetric with respect to the origin.

3. If the graph were folded on the x-axis, the parts above and below the x-axis would coincide, so the graph is symmetric with respect to the x-axis.

If the graph were folded on the y-axis, the parts to the left and right of the y-axis would not coincide, so the graph is not symmetric with respect to the y-axis.

If the graph were rotated $180°$, the resulting graph would not coincide with the original graph, so it is not symmetric with respect to the origin.

5. If the graph were folded on the x-axis, the parts above and below the x-axis would not coincide, so the graph is not symmetric with respect to the x-axis.

If the graph were folded on the y-axis, the parts to the left and right of the y-axis would not coincide, so the graph is not symmetric with respect to the y-axis.

If the graph were rotated $180°$, the resulting graph would coincide with the original graph, so it is symmetric with respect to the origin.

7.

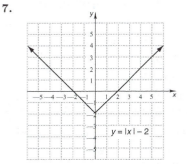

The graph is symmetric with respect to the y-axis. It is not symmetric with respect to the x-axis or the origin.

Test algebraically for symmetry with respect to the x-axis:

$\begin{aligned} y &= |x| - 2 &&\text{Original equation} \\ -y &= |x| - 2 &&\text{Replacing } y \text{ by } -y \\ y &= -|x| + 2 &&\text{Simplifying} \end{aligned}$

The last equation is not equivalent to the original equation, so the graph is not symmetric with respect to the x-axis.

Test algebraically for symmetry with respect to the y-axis:

$\begin{aligned} y &= |x| - 2 &&\text{Original equation} \\ y &= |-x| - 2 &&\text{Replacing } x \text{ by } -x \\ y &= |x| - 2 &&\text{Simplifying} \end{aligned}$

The last equation is equivalent to the original equation, so the graph is symmetric with respect to the y-axis.

Test algebraically for symmetry with respect to the origin:

$\begin{aligned} y &= |x| - 2 &&\text{Original equation} \\ -y &= |-x| - 2 &&\text{Replacing } x \text{ by } -x \text{ and} \\ &&&\quad y \text{ by } -y \\ -y &= |x| - 2 &&\text{Simplifying} \\ y &= -|x| + 2 \end{aligned}$

The last equation is not equivalent to the original equation, so the graph is not symmetric with respect to the origin.

9.

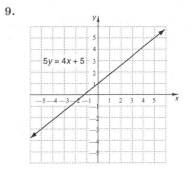

The graph is not symmetric with respect to the x-axis, the y-axis, or the origin.

Test algebraically for symmetry with respect to the x-axis:

$$5y = 4x + 5 \qquad \text{Original equation}$$
$$5(-y) = 4x + 5 \qquad \text{Replacing } y \text{ by } -y$$
$$-5y = 4x + 5 \qquad \text{Simplifying}$$
$$5y = -4x - 5$$

The last equation is not equivalent to the original equation, so the graph is not symmetric with respect to the x-axis.

Test algebraically for symmetry with respect to the y-axis:

$$5y = 4x + 5 \qquad \text{Original equation}$$
$$5y = 4(-x) + 5 \qquad \text{Replacing } x \text{ by } -x$$
$$5y = -4x + 5 \qquad \text{Simplifying}$$

The last equation is not equivalent to the original equation, so the graph is not symmetric with respect to the y-axis.

Test algebraically for symmetry with respect to the origin:

$$5y = 4x + 5 \qquad \text{Original equation}$$
$$5(-y) = 4(-x) + 5 \qquad \text{Replacing } x \text{ by } -x \text{ and } y \text{ by } -y$$
$$-5y = -4x + 5 \qquad \text{Simplifying}$$
$$5y = 4x - 5$$

The last equation is not equivalent to the original equation, so the graph is not symmetric with respect to the origin.

11.

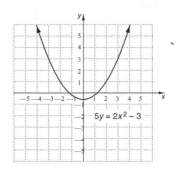

The graph is symmetric with respect to the y-axis. It is not symmetric with respect to the x-axis or the origin.

Test algebraically for symmetry with respect to the x-axis:

$$5y = 2x^2 - 3 \qquad \text{Original equation}$$
$$5(-y) = 2x^2 - 3 \qquad \text{Replacing } y \text{ by } -y$$
$$-5y = 2x^2 - 3 \qquad \text{Simplifying}$$
$$5y = -2x^2 + 3$$

The last equation is not equivalent to the original equation, so the graph is not symmetric with respect to the x-axis.

Test algebraically for symmetry with respect to the y-axis:

$$5y = 2x^2 - 3 \qquad \text{Original equation}$$
$$5y = 2(-x)^2 - 3 \qquad \text{Replacing } x \text{ by } -x$$
$$5y = 2x^2 - 3$$

The last equation is equivalent to the original equation, so the graph is symmetric with respect to the y-axis.

Test algebraically for symmetry with respect to the origin:

$$5y = 2x^2 - 3 \qquad \text{Original equation}$$
$$5(-y) = 2(-x)^2 - 3 \qquad \text{Replacing } x \text{ by } -x \text{ and } y \text{ by } -y$$
$$-5y = 2x^2 - 3 \qquad \text{Simplifying}$$
$$5y = -2x^2 + 3$$

The last equation is not equivalent to the original equation, so the graph is not symmetric with respect to the origin.

13.

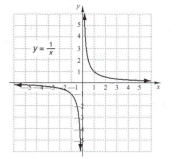

The graph is not symmetric with respect to the x-axis or the y-axis. It is symmetric with respect to the origin.

Test algebraically for symmetry with respect to the x-axis:

$$y = \frac{1}{x} \qquad \text{Original equation}$$
$$-y = \frac{1}{x} \qquad \text{Replacing } y \text{ by } -y$$
$$y = -\frac{1}{x} \qquad \text{Simplifying}$$

The last equation is not equivalent to the original equation, so the graph is not symmetric with respect to the x-axis.

Test algebraically for symmetry with respect to the y-axis:

$$y = \frac{1}{x} \qquad \text{Original equation}$$
$$y = \frac{1}{-x} \qquad \text{Replacing } x \text{ by } -x$$
$$y = -\frac{1}{x} \qquad \text{Simplifying}$$

The last equation is not equivalent to the original equation, so the graph is not symmetric with respect to the y-axis.

Test algebraically for symmetry with respect to the origin:

$$y = \frac{1}{x} \qquad \text{Original equation}$$
$$-y = \frac{1}{-x} \qquad \text{Replacing } x \text{ by } -x \text{ and } y \text{ by } -y$$
$$y = \frac{1}{x} \qquad \text{Simplifying}$$

The last equation is equivalent to the original equation, so the graph is symmetric with respect to the origin.

15. Test for symmetry with respect to the x-axis:

$$5x - 5y = 0 \qquad \text{Original equation}$$
$$5x - 5(-y) = 0 \qquad \text{Replacing } y \text{ by } -y$$
$$5x + 5y = 0 \qquad \text{Simplifying}$$

The last equation is not equivalent to the original equation, so the graph is not symmetric with respect to the x-axis.

Test for symmetry with respect to the y-axis:

$$5x - 5y = 0 \quad \text{Original equation}$$
$$5(-x) - 5y = 0 \quad \text{Replacing } x \text{ by } -x$$
$$-5x - 5y = 0 \quad \text{Simplifying}$$
$$5x + 5y = 0$$

The last equation is not equivalent to the original equation, so the graph is not symmetric with respect to the y-axis.

Test for symmetry with respect to the origin:

$$5x - 5y = 0 \quad \text{Original equation}$$
$$5(-x) - 5(-y) = 0 \quad \text{Replacing } x \text{ by } -x \text{ and } y \text{ by } -y$$
$$-5x + 5y = 0 \quad \text{Simplifying}$$
$$5x - 5y = 0$$

The last equation is equivalent to the original equation, so the graph is symmetric with respect to the origin.

17. Test for symmetry with respect to the x-axis:

$$3x^2 - 2y^2 = 3 \quad \text{Original equation}$$
$$3x^2 - 2(-y)^2 = 3 \quad \text{Replacing } y \text{ by } -y$$
$$3x^2 - 2y^2 = 3 \quad \text{Simplifying}$$

The last equation is equivalent to the original equation, so the graph is symmetric with respect to the x-axis.

Test for symmetry with respect to the y-axis:

$$3x^2 - 2y^2 = 3 \quad \text{Original equation}$$
$$3(-x)^2 - 2y^2 = 3 \quad \text{Replacing } x \text{ by } -x$$
$$3x^2 - 2y^2 = 3 \quad \text{Simplifying}$$

The last equation is equivalent to the original equation, so the graph is symmetric with respect to the y-axis.

Test for symmetry with respect to the origin:

$$3x^2 - 2y^2 = 3 \quad \text{Original equation}$$
$$3(-x)^2 - 2(-y)^2 = 3 \quad \text{Replacing } x \text{ by } -x \text{ and } y \text{ by } -y$$
$$3x^2 - 2y^2 = 3 \quad \text{Simplifying}$$

The last equation is equivalent to the original equation, so the graph is symmetric with respect to the origin.

19. Test for symmetry with respect to the x-axis:

$$y = |2x| \quad \text{Original equation}$$
$$-y = |2x| \quad \text{Replacing } y \text{ by } -y$$
$$y = -|2x| \quad \text{Simplifying}$$

The last equation is not equivalent to the original equation, so the graph is not symmetric with respect to the x-axis.

Test for symmetry with respect to the y-axis:

$$y = |2x| \quad \text{Original equation}$$
$$y = |2(-x)| \quad \text{Replacing } x \text{ by } -x$$
$$y = |-2x| \quad \text{Simplifying}$$
$$y = |2x|$$

The last equation is equivalent to the original equation, so the graph is symmetric with respect to the y-axis.

Test for symmetry with respect to the origin:

$$y = |2x| \quad \text{Original equation}$$
$$-y = |2(-x)| \quad \text{Replacing } x \text{ by } -x \text{ and } y \text{ by } -y$$
$$-y = |-2x| \quad \text{Simplifying}$$
$$-y = |2x|$$
$$y = -|2x|$$

The last equation is not equivalent to the original equation, so the graph is not symmetric with respect to the origin.

21. Test for symmetry with respect to the x-axis:

$$2x^4 + 3 = y^2 \quad \text{Original equation}$$
$$2x^4 + 3 = (-y)^2 \quad \text{Replacing } y \text{ by } -y$$
$$2x^4 + 3 = y^2 \quad \text{Simplifying}$$

The last equation is equivalent to the original equation, so the graph is symmetric with respect to the x-axis.

Test for symmetry with respect to the y-axis:

$$2x^4 + 3 = y^2 \quad \text{Original equation}$$
$$2(-x)^4 + 3 = y^2 \quad \text{Replacing } x \text{ by } -x$$
$$2x^4 + 3 = y^2 \quad \text{Simplifying}$$

The last equation is equivalent to the original equation, so the graph is symmetric with respect to the y-axis.

Test for symmetry with respect to the origin:

$$2x^4 + 3 = y^2 \quad \text{Original equation}$$
$$2(-x)^4 + 3 = (-y)^2 \quad \text{Replacing } x \text{ by } -x \text{ and } y \text{ by } -y$$
$$2x^4 + 3 = y^2 \quad \text{Simplifying}$$

The last equation is equivalent to the original equation, so the graph is symmetric with respect to the origin.

23. Test for symmetry with respect to the x-axis:

$$3y^3 = 4x^3 + 2 \quad \text{Original equation}$$
$$3(-y)^3 = 4x^3 + 2 \quad \text{Replacing } y \text{ by } -y$$
$$-3y^3 = 4x^3 + 2 \quad \text{Simplifying}$$
$$3y^3 = -4x^3 - 2$$

The last equation is not equivalent to the original equation, so the graph is not symmetric with respect to the x-axis.

Test for symmetry with respect to the y-axis:

$$3y^3 = 4x^3 + 2 \quad \text{Original equation}$$
$$3y^3 = 4(-x)^3 + 2 \quad \text{Replacing } x \text{ by } -x$$
$$3y^3 = -4x^3 + 2 \quad \text{Simplifying}$$

The last equation is not equivalent to the original equation, so the graph is not symmetric with respect to the y-axis.

Test for symmetry with respect to the origin:

$$3y^3 = 4x^3 + 2 \quad \text{Original equation}$$
$$3(-y)^3 = 4(-x)^3 + 2 \quad \text{Replacing } x \text{ by } -x \text{ and } y \text{ by } -y$$
$$-3y^3 = -4x^3 + 2 \quad \text{Simplifying}$$
$$3y^3 = 4x^3 - 2$$

The last equation is not equivalent to the original equation, so the graph is not symmetric with respect to the origin.

25. Test for symmetry with respect to the x-axis:

$$xy = 12 \qquad \text{Original equation}$$
$$x(-y) = 12 \qquad \text{Replacing } y \text{ by } -y$$
$$-xy = 12 \qquad \text{Simplifying}$$
$$xy = -12$$

The last equation is not equivalent to the original equation, so the graph is not symmetric with respect to the x-axis.

Test for symmetry with respect to the y-axis:

$$xy = 12 \qquad \text{Original equation}$$
$$-xy = 12 \qquad \text{Replacing } x \text{ by } -x$$
$$xy = -12 \qquad \text{Simplifying}$$

The last equation is not equivalent to the original equation, so the graph is not symmetric with respect to the y-axis.

Test for symmetry with respect to the origin:

$$xy = 12 \qquad \text{Original equation}$$
$$-x(-y) = 12 \qquad \text{Replacing } x \text{ with } -x \text{ and } y \text{ by } -y$$
$$xy = 12 \qquad \text{Simplifying}$$

The last equation is equivalent to the original equation, so the graph is symmetric with respect to the origin.

27. x-axis: Replace y with $-y$; $(-5, -6)$

y-axis: Replace x with $-x$; $(5, 6)$

Origin: Replace x with $-x$ and y with $-y$; $(5, -6)$

29. x-axis: Replace y with $-y$; $(-10, 7)$

y-axis: Replace x with $-x$; $(10, -7)$

Origin: Replace x with $-x$ and y with $-y$; $(10, 7)$

31. x-axis: Replace y with $-y$; $(0, 4)$

y-axis: Replace x with $-x$; $(0, -4)$

Origin: Replace x with $-x$ and y with $-y$; $(0, 4)$

33. The graph is symmetric with respect to the y-axis, so the function is even.

35. The graph is symmetric with respect to the origin, so the function is odd.

37. The graph is not symmetric with respect to either the y-axis or the origin, so the function is neither even nor odd.

39.
$$f(x) = -3x^3 + 2x$$
$$f(-x) = -3(-x)^3 + 2(-x) = 3x^3 - 2x$$
$$-f(x) = -(-3x^3 + 2x) = 3x^3 - 2x$$
$$f(-x) = -f(x), \text{ so } f \text{ is odd.}$$

41.
$$f(x) = 5x^2 + 2x^4 - 1$$
$$f(-x) = 5(-x)^2 + 2(-x)^4 - 1 = 5x^2 + 2x^4 - 1$$
$$f(x) = f(-x), \text{ so } f \text{ is even.}$$

43.
$$f(x) = x^{17}$$
$$f(-x) = (-x)^{17} = -x^{17}$$
$$-f(x) = -x^{17}$$
$$f(-x) = -f(x), \text{ so } f \text{ is odd.}$$

45.
$$f(x) = x - |x|$$
$$f(-x) = (-x) - |(-x)| = -x - |x|$$
$$-f(x) = -(x - |x|) = -x + |x|$$
$$f(x) \neq f(-x), \text{ so } f \text{ is not even.}$$
$$f(-x) \neq -f(x), \text{ so } f \text{ is not odd.}$$

Thus, $f(x) = x - |x|$ is neither even nor odd.

47.
$$f(x) = 8$$
$$f(-x) = 8$$
$$f(x) = f(-x), \text{ so } f \text{ is even.}$$

49.

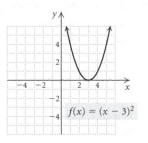

51.
$$f(x) = x\sqrt{10 - x^2}$$
$$f(-x) = -x\sqrt{10 - (-x)^2} = -x\sqrt{10 - x^2}$$
$$-f(x) = -x\sqrt{10 - x^2}$$

Since $f(-x) = -f(x)$, f is odd.

53. If the graph were folded on the x-axis, the parts above and below the x-axis would coincide, so the graph is symmetric with respect to the x-axis.

If the graph were folded on the y-axis, the parts to the left and right of the y-axis would not coincide, so the graph is not symmetric with respect to the y-axis.

If the graph were rotated $180°$, the resulting graph would not coincide with the original graph, so it is not symmetric with respect to the origin.

55. See the answer section in the text.

57. a), b) See the answer section in the text.

59. Let $f(x)$ and $g(x)$ be even functions. Then by definition, $f(x) = f(-x)$ and $g(x) = g(-x)$. Thus, $(f + g)(x) = f(x) + g(x) = f(-x) + g(-x) = (f + g)(-x)$ and $f + g$ is even. The statement is true.

Exercise Set 2.5

1. Shift the graph of $f(x) = x^2$ right 3 units.

$$f(x) = (x - 3)^2$$

3. Shift the graph of $g(x) = x$ down 3 units.

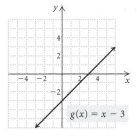

$g(x) = x - 3$

5. Reflect the graph of $h(x) = \sqrt{x}$ across the x-axis.

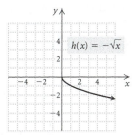

$h(x) = -\sqrt{x}$

7. Shift the graph of $h(x) = \dfrac{1}{x}$ up 4 units.

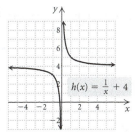

$h(x) = \frac{1}{x} + 4$

9. First stretch the graph of $h(x) = x$ vertically by multiplying each y-coordinate by 3. Then reflect it across the x-axis and shift it up 3 units.

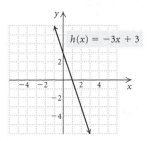

$h(x) = -3x + 3$

11. First shrink the graph of $h(x) = |x|$ vertically by multiplying each y-coordinate by $\dfrac{1}{2}$. Then shift it down 2 units.

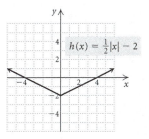

$h(x) = \frac{1}{2}|x| - 2$

13. Shift the graph of $g(x) = x^3$ right 2 units and reflect it across the x-axis.

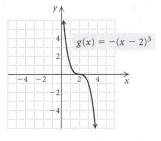

$g(x) = -(x - 2)^3$

15. Shift the graph of $g(x) = x^2$ left 1 unit and down 1 unit.

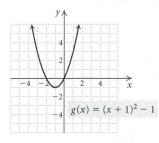

$g(x) = (x + 1)^2 - 1$

17. First shrink the graph of $g(x) = x^3$ vertically by multiplying each y-coordinate by $\dfrac{1}{3}$. Then shift it up 2 units.

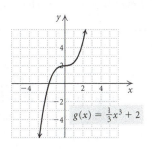

$g(x) = \frac{1}{3}x^3 + 2$

19. Shift the graph of $f(x) = \sqrt{x}$ left 2 units.

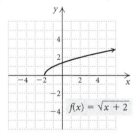

21. Shift the graph of $f(x) = \sqrt[3]{x}$ down 2 units.

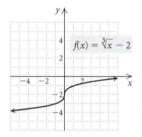

23. Think of the graph of $f(x) = |x|$. Since $g(x) = f(3x)$, the graph of $g(x) = |3x|$ is the graph of $f(x) = |x|$ shrunk horizontally by dividing each x-coordinate by 3 $\left(\text{or multiplying each } x\text{-coordinate by } \dfrac{1}{3}\right)$.

25. Think of the graph of $f(x) = \dfrac{1}{x}$. Since $h(x) = 2f(x)$, the graph of $h(x) = \dfrac{2}{x}$ is the graph of $f(x) = \dfrac{1}{x}$ stretched vertically by multiplying each y-coordinate by 2.

27. Think of the graph of $g(x) = \sqrt{x}$. Since $f(x) = 3g(x) - 5$, the graph of $f(x) = 3\sqrt{x} - 5$ is the graph of $g(x) = \sqrt{x}$ stretched vertically by multiplying each y-coordinate by 3 and then shifted down 5 units.

29. Think of the graph of $f(x) = |x|$. Since $g(x) = f\left(\dfrac{1}{3}x\right) - 4$, the graph of $g(x) = \left|\dfrac{1}{3}x\right| - 4$ is the graph of $f(x) = |x|$ stretched horizontally by multiplying each x-coordinate by 3 and then shifted down 4 units.

31. Think of the graph of $g(x) = x^2$. Since $f(x) = -\dfrac{1}{4}g(x-5)$, the graph of $f(x) = -\dfrac{1}{4}(x-5)^2$ is the graph of $g(x) = x^2$ shifted right 5 units, shrunk vertically by multiplying each y-coordinate by $\dfrac{1}{4}$, and reflected across the x-axis.

33. Think of the graph of $g(x) = \dfrac{1}{x}$. Since $f(x) = g(x+3) + 2$, the graph of $f(x) = \dfrac{1}{x+3} + 2$ is the graph of $g(x) = \dfrac{1}{x}$ shifted left 3 units and up 2 units.

35. Think of the graph of $f(x) = x^2$. Since $h(x) = -f(x-3) + 5$, the graph of $h(x) = -(x-3)^2 + 5$ is the graph of $f(x) = x^2$ shifted right 3 units, reflected across the x-axis, and shifted up 5 units.

37. The graph of $y = g(x)$ is the graph of $y = f(x)$ shrunk vertically by a factor of $\dfrac{1}{2}$. Multiply the y-coordinate by $\dfrac{1}{2}$: $(-12, 2)$.

39. The graph of $y = g(x)$ is the graph of $y = f(x)$ reflected across the y-axis, so we reflect the point across the y-axis: $(12, 4)$.

41. The graph of $y = g(x)$ is the graph of $y = f(x)$ shifted down 2 units. Subtract 2 from the y-coordinate: $(-12, 2)$.

43. The graph of $y = g(x)$ is the graph of $y = f(x)$ stretched vertically by a factor of 4. Multiply the y-coordinate by 4: $(-12, 16)$.

45. $g(x) = x^2 + 4$ is the function $f(x) = x^2 + 3$ shifted up 1 unit, so $g(x) = f(x) + 1$. Answer B is correct.

47. If we substitute $x - 2$ for x in f, we get $(x-2)^3 + 3$, so $g(x) = f(x-2)$. Answer A is correct.

49. Shape: $h(x) = x^2$

Turn $h(x)$ upside-down (that is, reflect it across the x-axis): $g(x) = -h(x) = -x^2$

Shift $g(x)$ right 8 units: $f(x) = g(x-8) = -(x-8)^2$

51. Shape: $h(x) = |x|$

Shift $h(x)$ left 7 units: $g(x) = h(x+7) = |x+7|$

Shift $g(x)$ up 2 units: $f(x) = g(x) + 2 = |x+7| + 2$

53. Shape: $h(x) = \dfrac{1}{x}$

Shrink $h(x)$ vertically by a factor of $\dfrac{1}{2}$ $\left(\text{that is,}\right.$ multiply each function value by $\left.\dfrac{1}{2}\right)$:

$g(x) = \dfrac{1}{2}h(x) = \dfrac{1}{2} \cdot \dfrac{1}{x}$, or $\dfrac{1}{2x}$

Shift $g(x)$ down 3 units: $f(x) = g(x) - 3 = \dfrac{1}{2x} - 3$

55. Shape: $m(x) = x^2$

Turn $m(x)$ upside-down (that is, reflect it across the x-axis): $h(x) = -m(x) = -x^2$

Shift $h(x)$ right 3 units: $g(x) = h(x-3) = -(x-3)^2$

Shift $g(x)$ up 4 units: $f(x) = g(x) + 4 = -(x-3)^2 + 4$

57. Shape: $m(x) = \sqrt{x}$

Reflect $m(x)$ across the y-axis: $h(x) = m(-x) = \sqrt{-x}$

Shift $h(x)$ left 2 units: $g(x) = h(x+2) = \sqrt{-(x+2)}$

Shift $g(x)$ down 1 unit: $f(x) = g(x) - 1 = \sqrt{-(x+2)} - 1$

59. Each y-coordinate is multiplied by -2. We plot and connect $(-4, 0)$, $(-3, 4)$, $(-1, 4)$, $(2, -6)$, and $(5, 0)$.

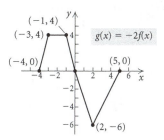

61. The graph is reflected across the y-axis and stretched horizontally by a factor of 2. That is, each x-coordinate is multiplied by -2 $\left(\text{or divided by } -\dfrac{1}{2}\right)$. We plot and connect $(8, 0)$, $(6, -2)$, $(2, -2)$, $(-4, 3)$, and $(-10, 0)$.

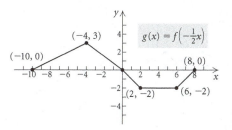

63. The graph is shifted right 1 unit so each x-coordinate is increased by 1. The graph is also reflected across the x-axis, shrunk vertically by a factor of 2, and shifted up 3 units. Thus, each y-coordinate is multiplied by $-\dfrac{1}{2}$ and then increased by 3. We plot and connect $(-3, 3)$, $(-2, 4)$, $(0, 4)$, $(3, 1.5)$, and $(6, 3)$.

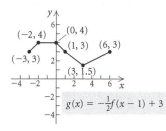

65. The graph is reflected across the y-axis so each x-coordinate is replaced by its opposite.

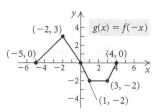

67. The graph is shifted left 2 units so each x-coordinate is decreased by 2. It is also reflected across the x-axis so each y-coordinate is replaced with its opposite. In addition, the graph is shifted up 1 unit, so each y-coordinate is then increased by 1.

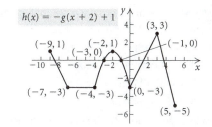

69. The graph is shrunk horizontally. The x-coordinates of $y = h(x)$ are one-half the corresponding x-coordinates of $y = g(x)$.

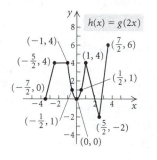

71. $g(x) = f(-x) + 3$

The graph of $g(x)$ is the graph of $f(x)$ reflected across the y-axis and shifted up 3 units. This is graph (f).

73. $g(x) = -f(x) + 3$

The graph of $g(x)$ is the graph of $f(x)$ reflected across the x-axis and shifted up 3 units. This is graph (f).

75. $g(x) = \dfrac{1}{3} f(x - 2)$

The graph of $g(x)$ is the graph of $f(x)$ shrunk vertically by a factor of 3 $\left(\text{that is, each } y\text{-coordinate is multiplied by } \dfrac{1}{3}\right)$ and then shifted right 2 units. This is graph (d).

77. $g(x) = \dfrac{1}{3} f(x + 2)$

The graph of $g(x)$ is the graph of $f(x)$ shrunk vertically by a factor of 3 $\left(\text{that is, each } y\text{-coordinate is multiplied by } \dfrac{1}{3}\right)$ and then shifted left 2 units. This is graph (c).

79. $f(-x) = 2(-x)^4 - 35(-x)^3 + 3(-x) - 5 =$
$2x^4 + 35x^3 - 3x - 5 = g(x)$

81. The graph of $f(x) = x^3 - 3x^2$ is shifted up 2 units. A formula for the transformed function is $g(x) = f(x) + 2$, or $g(x) = x^3 - 3x^2 + 2$.

83. The graph of $f(x) = x^3 - 3x^2$ is shifted left 1 unit. A formula for the transformed function is $k(x) = f(x + 1)$, or $k(x) = (x + 1)^3 - 3(x + 1)^2$.

85. Test for symmetry with respect to the x-axis.

$\quad y = 3x^4 - 3 \qquad$ Original equation

$-y = 3x^4 - 3 \qquad$ Replacing y by $-y$

$\quad y = -3x^4 + 3 \quad$ Simplifying

The last equation is not equivalent to the original equation, so the graph is not symmetric with respect to the x-axis.

Test for symmetry with respect to the y-axis.

$$y = 3x^4 - 3 \qquad \text{Original equation}$$
$$y = 3(-x)^4 - 3 \quad \text{Replacing } x \text{ by } -x$$
$$y = 3x^4 - 3 \qquad \text{Simplifying}$$

The last equation is equivalent to the original equation, so the graph is symmetric with respect to the y-axis.

Test for symmetry with respect to the origin:

$$y = 3x^4 - 3$$
$$-y = 3(-x)^4 - 3 \quad \text{Replacing } x \text{ by } -x \text{ and}$$
$$\qquad\qquad\qquad\quad y \text{ by } -y$$
$$-y = 3x^4 - 3$$
$$y = -3x^4 + 3 \qquad \text{Simplifying}$$

The last equation is not equivalent to the original equation, so the graph is not symmetric with respect to the origin.

87. Test for symmetry with respect to the x-axis:

$$2x - 5y = 0 \qquad \text{Original equation}$$
$$2x - 5(-y) = 0 \quad \text{Replacing } y \text{ by } -y$$
$$2x + 5y = 0 \qquad \text{Simplifying}$$

The last equation is not equivalent to the original equation, so the graph is not symmetric with respect to the x-axis.

Test for symmetry with respect to the y-axis:

$$2x - 5y = 0 \qquad \text{Original equation}$$
$$2(-x) - 5y = 0 \quad \text{Replacing } x \text{ by } -x$$
$$-2x - 5y = 0 \quad \text{Simplifying}$$

The last equation is not equivalent to the original equation, so the graph is not symmetric with respect to the y-axis.

Test for symmetry with respect to the origin:

$$2x - 5y = 0 \qquad \text{Original equation}$$
$$2(-x) - 5(-y) = 0 \quad \text{Replacing } x \text{ by } -x \text{ and}$$
$$\qquad\qquad\qquad\qquad y \text{ by } -y$$
$$-2x + 5y = 0$$
$$2x - 5y = 0 \quad \text{Simplifying}$$

The last equation is equivalent to the original equation, so the graph is symmetric with respect to the origin.

89. *Familiarize.* Let $n =$ the number of guns that were found with airline travelers in 2010.

Translate.

Number of guns found in 2013 was twice number of guns found in 2010 less 418.

$$\underbrace{1828}_{} \;\; \underset{=}{\downarrow} \;\; \underset{2 \,\cdot}{\downarrow} \;\; \underset{n}{\downarrow} \;\; \underset{-}{\downarrow}\,\underset{418}{\downarrow}$$

Solve.

$$1828 = 2n - 418$$
$$2246 = 2n$$
$$1123 = n$$

Check. $2 \cdot 1123 = 2246$ and $2246 - 418 = 1828$. This is the number of guns found in 2013, so the answer checks.

State. In 2010, 1123 guns were found with airline travelers.

91. Each point for which $f(x) < 0$ is reflected across the x-axis.

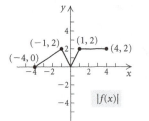

93. The graph of $y = g(|x|)$ consists of the points of $y = g(x)$ for which $x \geq 0$ along with their reflections across the y-axis.

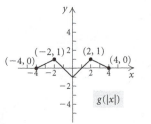

95. $f(2 - 3) = f(-1) = 5$, so $b = 5$.

(The graph of $y = f(x - 3)$ is the graph of $y = f(x)$ shifted right 3 units, so the point $(-1, 5)$ on $y = f(x)$ is transformed to the point $(-1 + 3, 5)$, or $(2, 5)$ on $y = f(x - 3)$.)

Exercise Set 2.6

1.
$$y = kx$$
$$54 = k \cdot 12$$
$$\frac{54}{12} = k, \text{ or } k = \frac{9}{2}$$

The variation constant is $\frac{9}{2}$, or 4.5. The equation of variation is $y = \frac{9}{2}x$, or $y = 4.5x$.

3.
$$y = \frac{k}{x}$$
$$3 = \frac{k}{12}$$
$$36 = k$$

The variation constant is 36. The equation of variation is $y = \frac{36}{x}$.

5.
$$y = kx$$
$$1 = k \cdot \frac{1}{4}$$
$$4 = k$$

The variation constant is 4. The equation of variation is $y = 4x$.

7.
$$y = \frac{k}{x}$$
$$32 = \frac{k}{\frac{1}{8}}$$
$$\frac{1}{8} \cdot 32 = k$$
$$4 = k$$

The variation constant is 4. The equation of variation is $y = \frac{4}{x}$.

9.
$$y = kx$$
$$\frac{3}{4} = k \cdot 2$$
$$\frac{1}{2} \cdot \frac{3}{4} = k$$
$$\frac{3}{8} = k$$

The variation constant is $\frac{3}{8}$. The equation of variation is $y = \frac{3}{8}x$.

11.
$$y = \frac{k}{x}$$
$$1.8 = \frac{k}{0.3}$$
$$0.54 = k$$

The variation constant is 0.54. The equation of variation is $y = \frac{0.54}{x}$.

13. Let W = the weekly allowance and a = the child's age.
$$W = ka$$
$$5.50 = k \cdot 6$$
$$\frac{11}{12} = k$$

$$W = \frac{11}{12}x$$
$$W = \frac{11}{12} \cdot 9$$
$$W = \$8.25$$

15.
$$t = \frac{k}{r}$$
$$5 = \frac{k}{80}$$
$$400 = r$$
$$t = \frac{400}{r}$$
$$t = \frac{400}{70}$$
$$t = \frac{40}{7}, \text{ or } 5\frac{5}{7} \text{ hr}$$

17. Let F = the number of grams of fat and w = the weight.

$$F = kw \qquad F \text{ varies directly as } w.$$
$$60 = k \cdot 120 \quad \text{Substituting}$$
$$\frac{60}{120} = k, \text{ or } \quad \text{Solving for } k$$
$$\frac{1}{2} = k \qquad \text{Variation constant}$$
$$F = \frac{1}{2}w \qquad \text{Equation of variation}$$
$$F = \frac{1}{2} \cdot 180 \quad \text{Substituting}$$
$$F = 90$$

The maximum daily fat intake for a person weighing 180 lb is 90 g.

19.
$$T = \frac{k}{P} \qquad T \text{ varies inversely as } P.$$
$$5 = \frac{k}{7} \qquad \text{Substituting}$$
$$35 = k \qquad \text{Variation constant}$$
$$T = \frac{35}{P} \qquad \text{Equation of variation}$$
$$T = \frac{35}{10} \qquad \text{Substituting}$$
$$T = 3.5$$

It will take 10 bricklayers 3.5 hr to complete the job.

21.
$$d = km \quad d \text{ varies directly as } m.$$
$$40 = k \cdot 3 \quad \text{Substituting}$$
$$\frac{40}{3} = k \qquad \text{Variation constant}$$
$$d = \frac{40}{3}m \qquad \text{Equation of variation}$$
$$d = \frac{40}{3} \cdot 5 = \frac{200}{3} \qquad \text{Substituting}$$
$$d = 66\frac{2}{3}$$

A 5-kg mass will stretch the spring $66\frac{2}{3}$ cm.

23.
$$P = \frac{k}{W} \qquad P \text{ varies inversely as } W.$$
$$330 = \frac{k}{3.2} \qquad \text{Substituting}$$
$$1056 = k \qquad \text{Variation constant}$$
$$P = \frac{1056}{W} \qquad \text{Equation of variation}$$
$$550 = \frac{1056}{W} \qquad \text{Substituting}$$
$$550W = 1056 \qquad \text{Multiplying by } W$$
$$W = \frac{1056}{550} \qquad \text{Dividing by 550}$$
$$W = 1.92 \qquad \text{Simplifying}$$

A tone with a pitch of 550 vibrations per second has a wavelength of 1.92 ft.

25.
$$y = \frac{k}{x^2}$$

$$0.15 = \frac{k}{(0.1)^2} \quad \text{Substituting}$$

$$0.15 = \frac{k}{0.01}$$

$$0.15(0.01) = k$$

$$0.0015 = k$$

The equation of variation is $y = \frac{0.0015}{x^2}$.

27.
$$y = kx^2$$

$$0.15 = k(0.1)^2 \quad \text{Substituting}$$

$$0.15 = 0.01k$$

$$\frac{0.15}{0.01} = k$$

$$15 = k$$

The equation of variation is $y = 15x^2$.

29.
$$y = kxz$$

$$56 = k \cdot 7 \cdot 8 \quad \text{Substituting}$$

$$56 = 56k$$

$$1 = k$$

The equation of variation is $y = xz$.

31.
$$y = kxz^2$$

$$105 = k \cdot 14 \cdot 5^2 \quad \text{Substituting}$$

$$105 = 350k$$

$$\frac{105}{350} = k$$

$$\frac{3}{10} = k$$

The equation of variation is $y = \frac{3}{10}xz^2$.

33.
$$y = k\frac{xz}{wp}$$

$$\frac{3}{28} = k\frac{3 \cdot 10}{7 \cdot 8} \quad \text{Substituting}$$

$$\frac{3}{28} = k \cdot \frac{30}{56}$$

$$\frac{3}{28} \cdot \frac{56}{30} = k$$

$$\frac{1}{5} = k$$

The equation of variation is $y = \frac{1}{5}\frac{xz}{wp}$, or $\frac{xz}{5wp}$.

35.
$$I = \frac{k}{d^2}$$

$$90 = \frac{k}{5^2} \quad \text{Substituting}$$

$$90 = \frac{k}{25}$$

$$2250 = k$$

The equation of variation is $I = \frac{2250}{d^2}$.

Substitute 40 for I and find d.

$$40 = \frac{2250}{d^2}$$

$$40d^2 = 2250$$

$$d^2 = 56.25$$

$$d = 7.5$$

The distance from 5 m to 7.5 m is $7.5 - 5$, or 2.5 m, so it is 2.5 m further to a point where the intensity is 40 W/m².

37.
$$d = kr^2$$

$$200 = k \cdot 60^2 \quad \text{Substituting}$$

$$200 = 3600k$$

$$\frac{200}{3600} = k$$

$$\frac{1}{18} = k$$

The equation of variation is $d = \frac{1}{18}r^2$.

Substitute 72 for d and find r.

$$72 = \frac{1}{18}r^2$$

$$1296 = r^2$$

$$36 = r$$

A car can travel 36 mph and still stop in 72 ft.

39.
$$E = \frac{kR}{I}$$

We first find k.

$$3.75 = \frac{k \cdot 89}{213.1} \quad \text{Substituting}$$

$$3.75\left(\frac{213.1}{89}\right) = k \qquad \text{Multiplying by } \frac{213.1}{89}$$

$$9 \approx k$$

The equation of variation is $E = \frac{9R}{I}$.

Substitute 3.75 for E and 235 for I and solve for R.

$$3.75 = \frac{9R}{235}$$

$$3.75\left(\frac{235}{9}\right) = R \qquad \text{Multiplying by } \frac{235}{9}$$

$$98 \approx R$$

John Lester would have given up about 98 earned runs if he had pitched 235 innings.

41. parallel

43. relative minimum

45. inverse variation

47. Let V represent the volume and p represent the price of a jar of peanut butter.

$$V = kp \qquad V \text{ varies directly as } p.$$

$$\pi\left(\frac{3}{2}\right)^2(5) = k(2.89) \quad \text{Substituting}$$

$$3.89\pi \approx k \qquad \text{Variation constant}$$

$$V = 3.89\pi p \qquad \text{Equation of variation}$$

$$\pi(1.625)^2(5.25) = 3.89\pi p \quad \text{Substituting}$$

$$3.56 \approx p$$

If cost is directly proportional to volume, the larger jar should cost $3.56.

Now let W represent the weight and p represent the price of a jar of peanut butter.

$$W = kp$$

$$18 = k(2.89) \quad \text{Substituting}$$

$$6.23 \approx k \qquad \text{Variation constant}$$

$$W = 6.23p \qquad \text{Equation of variation}$$

$$22 = 6.23p \qquad \text{Substituting}$$

$$3.53 = p$$

If cost is directly proportional to weight, the larger jar should cost $3.53.

Chapter 2 Review Exercises

1. This statement is true by the definition of the greatest integer function.

3. The graph of $y = f(x - d)$ is the graph of $y = f(x)$ shifted right d units, so the statement is true.

5. a) For x-values from -4 to -2, the y-values increase from 1 to 4. Thus the function is increasing on the interval $(-4, -2)$.

   b) For x-values from 2 to 5, the y-values decrease from 4 to 3. Thus the function is decreasing on the interval $(2, 5)$.

   c) For x-values from -2 to 2, y is 4. Thus the function is constant on the interval $(-2, 2)$.

7.

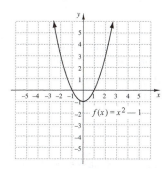

The function is increasing on $(0, \infty)$ and decreasing on $(-\infty, 0)$. We estimate that the minimum value is -1 at $x = 0$. There are no maxima.

9. If two sides of the patio are each x feet, then the remaining side will be $(48 - 2x)$ ft. We use the formula Area = length $\times$ width.

$$A(x) = x(48 - 2x), \text{ or } 48x - 2x^2$$

11. a) Let $h =$ the height of the box. Since the volume is 108 in^3, we have:

$$108 = x \cdot x \cdot h$$

$$108 = x^2 h$$

$$\frac{108}{x^2} = h$$

Now find the surface area.

$$S = x^2 + 4 \cdot x \cdot h$$

$$S(x) = x^2 + 4 \cdot x \cdot \frac{108}{x^2}$$

$$S(x) = x^2 + \frac{432}{x}$$

   b) x must be positive, so the domain is $(0, \infty)$.

   c) From the graph, we see that the minimum value of the function occurs when $x = 6$ in. For this value of x,

$$h = \frac{108}{x^2} = \frac{108}{6^2} = \frac{108}{36} = 3 \text{ in.}$$

13. $f(x) = \begin{cases} x^3, & \text{for } x < -2, \\ |x|, & \text{for } -2 \leq x \leq 2, \\ \sqrt{x-1}, & \text{for } x > 2 \end{cases}$

We create the graph in three parts. Graph $f(x) = x^3$ for inputs less than -2. Then graph $f(x) = |x|$ for inputs greater than or equal to -2 and less than or equal to 2. Finally graph $f(x) = \sqrt{x-1}$ for inputs greater than 2.

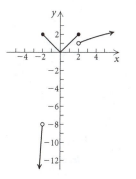

15. $f(x) = [[x]]$. See Example 9 in Section 2.1 of the text.

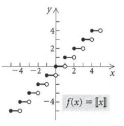

17. $f(x) = \begin{cases} x^3, & \text{for } x < -2, \\ |x|, & \text{for } -2 \leq x \leq 2, \\ \sqrt{x-1}, & \text{for } x > 2 \end{cases}$

Since -1 is in the interval $[-2, 2]$, $f(-1) = |-1| = 1$.

Since $5 > 2$, $f(5) = \sqrt{5-1} = \sqrt{4} = 2$.

Since -2 is in the interval $[-2, 2]$, $f(-2) = |-2| = 2$.

Since $-3 < -2$, $f(-3) = (-3)^3 = -27$.

19. $(f - g)(6) = f(6) - g(6)$
$$= \sqrt{6-2} - (6^2 - 1)$$
$$= \sqrt{4} - (36 - 1)$$
$$= 2 - 35$$
$$= -33$$

21. $(f + g)(-1) = f(-1) + g(-1)$
$$= \sqrt{-1-2} + ((-1)^2 - 1)$$
$$= \sqrt{-3} + (1 - 1)$$

Since $\sqrt{-3}$ is not a real number, $(f+g)(-1)$ does not exist.

23. $f(x) = 3x^2 + 4x$, $g(x) = 2x - 1$

a) The domain of f, g, $f + g$, $f - g$, and fg is all real numbers, or $(-\infty, \infty)$. Since $g\left(\frac{1}{2}\right) = 0$, the domain of f/g is $\left\{x \middle| x \neq \frac{1}{2}\right\}$, or $\left(-\infty, \frac{1}{2}\right) \cup \left(\frac{1}{2}, \infty\right)$.

b) $(f + g)(x) = (3x^2 + 4x) + (2x - 1) = 3x^2 + 6x - 1$

$(f - g)(x) = (3x^2 + 4x) - (2x - 1) = 3x^2 + 2x + 1$

$(fg)(x) = (3x^2 + 4x)(2x - 1) = 6x^3 + 5x^2 - 4x$

$(f/g)(x) = \dfrac{3x^2 + 4x}{2x - 1}$

25. $f(x) = 2x + 7$

$\dfrac{f(x + h) - f(x)}{h} = \dfrac{2(x + h) + 7 - (2x + 7)}{h} =$

$\dfrac{2x + 2h + 7 - 2x - 7}{h} = \dfrac{2h}{h} = 2$

27. $f(x) = \dfrac{4}{x}$

$\dfrac{f(x + h) - f(x)}{h} = \dfrac{\dfrac{4}{x+h} - \dfrac{4}{x}}{h} = \dfrac{\dfrac{4}{x+h} \cdot \dfrac{x}{x} - \dfrac{4}{x} \cdot \dfrac{x+h}{x+h}}{h} =$

$\dfrac{\dfrac{4x - 4(x+h)}{x(x+h)}}{h} = \dfrac{\dfrac{4x - 4x - 4h}{x(x+h)}}{h} = \dfrac{\dfrac{-4h}{x(x+h)}}{h} =$

$\dfrac{-4h}{x(x+h)} \cdot \dfrac{1}{h} = \dfrac{-4 \cdot h}{x(x+h) \cdot h} = \dfrac{-4}{x(x+h)}$, or $-\dfrac{4}{x(x+h)}$

29. $(g \circ f)(1) = g(f(1)) = g(2 \cdot 1 - 1) = g(2 - 1) = g(1) =$
$1^2 + 4 = 1 + 4 = 5$

31. $(g \circ h)(3) = g(h(3)) = g(3 - 3^3) = g(3 - 27) =$
$g(-24) = (-24)^2 + 4 = 576 + 4 = 580$

33. $(h \circ g)(2) = h(g(2)) = h(2^2 + 4) = h(4 + 4) =$
$h(8) = 3 - 8^3 = 3 - 512 = -509$

35. $(h \circ h)(x) = h(h(x)) = h(3 - x^3) = 3 - (3 - x^3)^3 =$
$3 - (27 - 27x^3 + 9x^6 - x^9) = 3 - 27 + 27x^3 - 9x^6 + x^9 =$
$-24 + 27x^3 - 9x^6 + x^9$

37. a) $f \circ g(x) = f(2x - 1)$
$$= 3(2x - 1)^2 + 4(2x - 1)$$
$$= 3(4x^2 - 4x + 1) + 4(2x - 1)$$
$$= 12x^2 - 12x + 3 + 8x - 4$$
$$= 12x^2 - 4x - 1$$

$(g \circ f)(x) = g(3x^2 + 4x)$
$$= 2(3x^2 + 4x) - 1$$
$$= 6x^2 + 8x - 1$$

b) Domain of f = domain of g = all real numbers, so domain of $f \circ g$ = domain of $g \circ f$ = all real numbers, or $(-\infty, \infty)$.

39. $f(x) = 4x^2 + 9$, $g(x) = 5x - 1$. Answers may vary.

41. $y^2 = x^2 + 3$

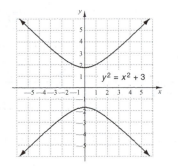

The graph is symmetric with respect to the x-axis, the y-axis, and the origin.

Replace y with $-y$ to test algebraically for symmetry with respect to the x-axis.

$$(-y)^2 = x^2 + 3$$
$$y^2 = x^2 + 3$$

The resulting equation is equivalent to the original equation, so the graph is symmetric with respect to the x-axis.

Replace x with $-x$ to test algebraically for symmetry with respect to the y-axis.

$$y^2 = (-x)^2 + 3$$
$$y^2 = x^2 + 3$$

The resulting equation is equivalent to the original equation, so the graph is symmetric with respect to the y-axis.

Replace x and $-x$ and y with $-y$ to test for symmetry with respect to the origin.

$$(-y)^2 = (-x)^2 + 3$$
$$y^2 = x^2 + 3$$

The resulting equation is equivalent to the original equation, so the graph is symmetric with respect to the origin.

43. $y = x^2$

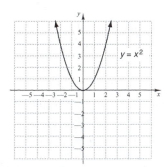

The graph is symmetric with respect to the y-axis. It is not symmetric with respect to the x-axis or the origin.

Replace y with $-y$ to test algebraically for symmetry with respect to the x-axis.

$$-y = x^2$$
$$y = -x^2$$

The resulting equation is not equivalent to the original equation, so the graph is not symmetric with respect to the x-axis.

Replace x with $-x$ to test algebraically for symmetry with respect to the y-axis.

$$y = (-x)^2$$
$$y = x^2$$

The resulting equation is equivalent to the original equation, so the graph is symmetric with respect to the y-axis.

Replace x and $-x$ and y with $-y$ to test for symmetry with respect to the origin.

$$-y = (-x)^2$$
$$-y = x^2$$
$$y = -x^2$$

The resulting equation is not equivalent to the original equation, so the graph is not symmetric with respect to the origin.

45. $y = x^4 - x^2$

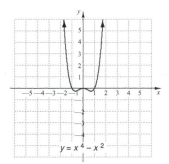
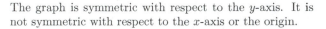

The graph is symmetric with respect to the y-axis. It is not symmetric with respect to the x-axis or the origin.

Replace y with $-y$ to test algebraically for symmetry with respect to the x-axis.

$$-y = x^4 - x^2$$
$$y = -x^4 + x^2$$

The resulting equation is not equivalent to the original equation, so the graph is not symmetric with respect to the x-axis.

Replace x with $-x$ to test algebraically for symmetry with respect to the y-axis.

$$y = (-x)^4 - (-x)^2$$
$$y = x^4 - x^2$$

The resulting equation is equivalent to the original equation, so the graph is symmetric with respect to the y-axis.

Replace x and $-x$ and y with $-y$ to test for symmetry with respect to the origin.

$$-y = (-x)^4 - (-x)^2$$
$$-y = x^4 - x^2$$
$$y = -x^4 + x^2$$

The resulting equation is not equivalent to the original equation, so the graph is not symmetric with respect to the origin.

47. The graph is symmetric with respect to the y-axis, so the function is even.

49. The graph is symmetric with respect to the y-axis, so the function is even.

51. $f(x) = x^3 - 2x + 4$

$f(-x) = (-x)^3 - 2(-x) + 4 = -x^3 + 2x + 4$

$f(x) \neq f(-x)$, so f is not even.

$-f(x) = -(x^3 - 2x + 4) = -x^3 + 2x - 4$

$f(-x) \neq -f(x)$, so f is not odd.

Thus, $f(x) = x^3 - 2x + 4$ is neither even or odd.

53. $f(x) = |x|$

$f(-x) = |-x| = |x|$

$f(x) = f(-x)$, so f is even.

55. $f(x) = \dfrac{10x}{x^2 + 1}$

$f(-x) = \dfrac{10(-x)}{(-x)^2 + 1} = -\dfrac{10x}{x^2 + 1}$

$f(x) \neq f(-x)$, so $f(x)$ is not even.

$-f(x) = -\dfrac{10x}{x^2 + 1}$

$f(-x) = -f(x)$, so f is odd.

57. Shape: $t(x) = \sqrt{x}$

Turn $t(x)$ upside down (that is, reflect it across the x-axis): $h(x) = -t(x) = -\sqrt{x}$.

Shift $h(x)$ right 3 units: $g(x) = h(x - 3) = -\sqrt{x - 3}$.

Shift $g(x)$ up 4 units: $f(x) = g(x) + 4 = -\sqrt{x - 3} + 4$.

59. The graph is shifted right 1 unit so each x-coordinate is increased by 1. We plot and connect $(-4, 3)$, $(-2, 0)$, $(1, 1)$ and $(5, -2)$.

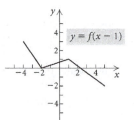

61. Each y-coordinate is multiplied by -2. We plot and connect $(-5, -6)$, $(-3, 0)$, $(0, -2)$ and $(4, 4)$.

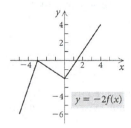

63.
$$y = kx$$
$$100 = 25x$$
$$4 = x$$

Equation of variation: $y = 4x$

65.
$$y = \frac{k}{x}$$
$$100 = \frac{k}{25}$$
$$2500 = k$$

Equation of variation: $y = \dfrac{2500}{x}$

67.
$$y = \frac{k}{x^2}$$
$$12 = \frac{k}{2^2}$$
$$48 = k$$
$$y = \frac{48}{x^2}$$

69.
$$t = \frac{k}{r}$$
$$35 = \frac{k}{800}$$
$$28,000 = k$$
$$t = \frac{28,000}{r}$$
$$t = \frac{28,000}{1400}$$
$$t = 20 \text{ min}$$

71.
$$P = kC^2$$
$$180 = k \cdot 6^2$$
$$5 = k \qquad \text{Variation constant}$$
$$P = 5C^2 \qquad \text{Variation equation}$$
$$P = 5 \cdot 10^2$$
$$P = 500 \text{ watts}$$

73. For $b > 0$, the graph of $y = f(x) + b$ is the graph of $y = f(x)$ shifted up b units. Answer C is correct.

75. Let $f(x)$ and $g(x)$ be odd functions. Then by definition, $f(-x) = -f(x)$, or $f(x) = -f(-x)$, and $g(-x) = -g(-x)$, or $g(x) = -g(-x)$. Thus $(f + g)(x) = f(x) + g(x) = -f(-x) + [-g(-x)] = -[f(-x) + g(-x)] = -(f + g)(-x)$ and $f + g$ is odd.

77. $f(x) = 4x^3 - 2x + 7$

a) $f(x) + 2 = 4x^3 - 2x + 7 + 2 = 4x^3 - 2x + 9$

b) $f(x + 2) = 4(x + 2)^3 - 2(x + 2) + 7$
$$= 4(x^3 + 6x^2 + 12x + 8) - 2(x + 2) + 7$$
$$= 4x^3 + 24x^2 + 48x + 32 - 2x - 4 + 7$$
$$= 4x^3 + 24x^2 + 46x + 35$$

c) $f(x) + f(2) = 4x^3 - 2x + 7 + 4 \cdot 2^3 - 2 \cdot 2 + 7$
$$= 4x^3 - 2x + 7 + 32 - 4 + 7$$
$$= 4x^3 - 2x + 42$$

$f(x) + 2$ adds 2 to each function value; $f(x + 2)$ adds 2 to each input before the function value is found; $f(x) + f(2)$ adds the output for 2 to the output for x.

79. The graph of $f(x) = 0$ is symmetric with respect to the x-axis, the y-axis, and the origin. This function is both even and odd.

81. Let $y(x) = kx^2$. Then $y(2x) = k(2x)^2 = k \cdot 4x^2 = 4 \cdot kx^2 = 4 \cdot y(x)$. Thus, doubling x causes y to be quadrupled.

Chapter 2 Test

1. a) For x-values from -5 to -2, the y-values increase from -4 to 3. Thus the function is increasing on the interval $(-5, -2)$.

b) For x-values from 2 to 5, the y-values decrease from 2 to -1. Thus the function is decreasing on the interval $(2, 5)$.

c) For x-values from -2 to 2, y is 2. Thus the function is constant on the interval $(-2, 2)$.

2.

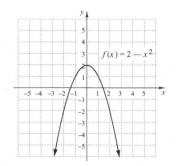

The function is increasing on $(-\infty, 0)$ and decreasing on $(0, \infty)$. The relative maximum is 2 at $x = 0$. There are no minima.

3. If b = the length of the base, in inches, then the height = $4b - 6$. We use the formula for the area of a triangle, $A = \frac{1}{2}bh$.

$$A(b) = \frac{1}{2}b(4b - 6), \text{ or}$$
$$A(b) = 2b^2 - 3b$$

4. $f(x) = \begin{cases} x^2, & \text{for } x < -1, \\ |x|, & \text{for } -1 \le x \le 1, \\ \sqrt{x - 1}, & \text{for } x > 1 \end{cases}$

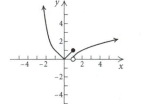

5. Since $-1 \le -\frac{7}{8} \le 1$, $f\left(-\frac{7}{8}\right) = \left|-\frac{7}{8}\right| = \frac{7}{8}$.

Since $5 > 1$, $f(5) = \sqrt{5 - 1} = \sqrt{4} = 2$.

Since $-4 < -1$, $f(-4) = (-4)^2 = 16$.

6. $(f + g)(-6) = f(-6) + g(-6) =$
$(-6)^2 - 4(-6) + 3 + \sqrt{3 - (-6)} =$
$36 + 24 + 3 + \sqrt{3 + 6} = 63 + \sqrt{9} = 63 + 3 = 66$

7. $(f - g)(-1) = f(-1) - g(-1) =$
$(-1)^2 - 4(-1) + 3 - \sqrt{3 - (-1)} =$
$1 + 4 + 3 - \sqrt{3 + 1} = 8 - \sqrt{4} = 8 - 2 = 6$

8. $(fg)(2) = f(2) \cdot g(2) = (2^2 - 4 \cdot 2 + 3)(\sqrt{3 - 2}) =$
$(4 - 8 + 3)(\sqrt{1}) = -1 \cdot 1 = -1$

9. $(f/g)(1) = \frac{f(1)}{g(1)} = \frac{1^2 - 4 \cdot 1 + 3}{\sqrt{3 - 1}} = \frac{1 - 4 + 3}{\sqrt{2}} = \frac{0}{\sqrt{2}} = 0$

10. Any real number can be an input for $f(x) = x^2$, so the domain is the set of real numbers, or $(-\infty, \infty)$.

11. The domain of $g(x) = \sqrt{x - 3}$ is the set of real numbers for which $x - 3 \ge 0$, or $x \ge 3$. Thus the domain is $\{x | x \ge 3\}$, or $[3, \infty)$.

12. The domain of $f + g$ is the intersection of the domains of f and g. This is $\{x | x \ge 3\}$, or $[3, \infty)$.

13. The domain of $f - g$ is the intersection of the domains of f and g. This is $\{x | x \ge 3\}$, or $[3, \infty)$.

14. The domain of fg is the intersection of the domains of f and g. This is $\{x | x \ge 3\}$, or $[3, \infty)$.

15. The domain of f/g is the intersection of the domains of f and g, excluding those x-values for which $g(x) = 0$. Since $x - 3 = 0$ when $x = 3$, the domain is $(3, \infty)$.

16. $(f + g)(x) = f(x) + g(x) = x^2 + \sqrt{x - 3}$

17. $(f - g)(x) = f(x) - g(x) = x^2 - \sqrt{x - 3}$

18. $(fg)(x) = f(x) \cdot g(x) = x^2\sqrt{x - 3}$

19. $(f/g)(x) = \frac{f(x)}{g(x)} = \frac{x^2}{\sqrt{x - 3}}$

20. $f(x) = \frac{1}{2}x + 4$

$$f(x + h) = \frac{1}{2}(x + h) + 4 = \frac{1}{2}x + \frac{1}{2}h + 4$$

$$\frac{f(x + h) - f(x)}{h} = \frac{\frac{1}{2}x + \frac{1}{2}h + 4 - \left(\frac{1}{2}x + 4\right)}{h}$$

$$= \frac{\frac{1}{2}x + \frac{1}{2}h + 4 - \frac{1}{2}x - 4}{h}$$

$$= \frac{\frac{1}{2}h}{h} = \frac{1}{2}h \cdot \frac{1}{h} = \frac{1}{2} \cdot \frac{h}{h} = \frac{1}{2}$$

21. $f(x) = 2x^2 - x + 3$
$f(x+h) = 2(x+h)^2 - (x+h) + 3 = 2(x^2 + 2xh + h^2) - x - h + 3 = 2x^2 + 4xh + 2h^2 - x - h + 3$

$$\frac{f(x+h) - f(x)}{h} = \frac{2x^2 + 4xh + 2h^2 - x - h + 3 - (2x^2 - x + 3)}{h}$$

$$= \frac{2x^2 + 4xh + 2h^2 - x - h + 3 - 2x^2 + x - 3}{h}$$

$$= \frac{4xh + 2h^2 - h}{h}$$

$$= \frac{\cancel{h}(4x + 2h - 1)}{\cancel{h}}$$

$$= 4x + 2h - 1$$

22. $(g \circ h)(2) = g(h(2)) = g(3 \cdot 2^2 + 2 \cdot 2 + 4) =$
$g(3 \cdot 4 + 4 + 4) = g(12 + 4 + 4) = g(20) = 4 \cdot 20 + 3 =$
$80 + 3 = 83$

23. $(f \circ g)(-1) = f(g(-1)) = f(4(-1) + 3) = f(-4 + 3) =$
$f(-1) = (-1)^2 - 1 = 1 - 1 = 0$

24. $(h \circ f)(1) = h(f(1)) = h(1^2 - 1) = h(1 - 1) = h(0) =$
$3 \cdot 0^2 + 2 \cdot 0 + 4 = 0 + 0 + 4 = 4$

25. $(g \circ g)(x) = g(g(x)) = g(4x + 3) = 4(4x + 3) + 3 =$
$16x + 12 + 3 = 16x + 15$

26. $(f \circ g)(x) = f(g(x)) = f(x^2 + 1) = \sqrt{x^2 + 1 - 5} = \sqrt{x^2 - 4}$

$(g \circ f)(x) = g(f(x)) = g(\sqrt{x - 5}) = (\sqrt{x - 5})^2 + 1 = x - 5 + 1 = x - 4$

27. The inputs for $f(x)$ must be such that $x - 5 \geq 0$, or $x \geq 5$. Then for $(f \circ g)(x)$ we must have $g(x) \geq 5$, or $x^2 + 1 \geq 5$, or $x^2 \geq 4$. Then the domain of $(f \circ g)(x)$ is $(-\infty, -2] \cup [2, \infty)$.

Since we can substitute any real number for x in g, the domain of $(g \circ f)(x)$ is the same as the domain of $f(x)$, $[5, \infty)$.

28. Answers may vary. $f(x) = x^4$, $g(x) = 2x - 7$

29. $y = x^4 - 2x^2$

Replace y with $-y$ to test for symmetry with respect to the x-axis.
$$-y = x^4 - 2x^2$$
$$y = -x^4 + 2x^2$$

The resulting equation is not equivalent to the original equation, so the graph is not symmetric with respect to the x-axis.

Replace x with $-x$ to test for symmetry with respect to the y-axis.
$$y = (-x)^4 - 2(-x)^2$$
$$y = x^4 - 2x^2$$

The resulting equation is equivalent to the original equation, so the graph is symmetric with respect to the y-axis.

Replace x with $-x$ and y with $-y$ to test for symmetry with respect to the origin.
$$-y = (-x)^4 - 2(-x)^2$$
$$-y = x^4 - 2x^2$$
$$y = -x^4 + 2x^2$$

The resulting equation is not equivalent to the original equation, so the graph is not symmetric with respect to the origin.

30. $f(x) = \dfrac{2x}{x^2 + 1}$

$f(-x) = \dfrac{2(-x)}{(-x)^2 + 1} = -\dfrac{2x}{x^2 + 1}$

$f(x) \neq f(-x)$, so f is not even.

$-f(x) = -\dfrac{2x}{x^2 + 1}$

$f(-x) = -f(x)$, so f is odd.

31. Shape: $h(x) = x^2$

Shift $h(x)$ right 2 units: $g(x) = h(x - 2) = (x - 2)^2$

Shift $g(x)$ down 1 unit: $f(x) = (x - 2)^2 - 1$

32. Shape: $h(x) = x^2$

Shift $h(x)$ left 2 units: $g(x) = h(x + 2) = (x + 2)^2$

Shift $g(x)$ down 3 units: $f(x) = (x + 2)^2 - 3$

33. Each y-coordinate is multiplied by $-\dfrac{1}{2}$. We plot and connect $(-5, 1)$, $(-3, -2)$, $(1, 2)$ and $(4, -1)$.

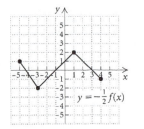

34. $y = \dfrac{k}{x}$

$5 = \dfrac{k}{6}$

$30 = k$ Variation constant

Equation of variation: $y = \dfrac{30}{x}$

35. $y = kx$

$60 = k \cdot 12$

$5 = k$ Variation constant

Equation of variation: $y = 5x$

36. $y = \dfrac{kxz^2}{w}$

$100 = \dfrac{k(0.1)(10)^2}{5}$

$100 = 2k$

$50 = k$ Variation constant

$y = \dfrac{50xz^2}{w}$ Equation of variation

37. $d = kr^2$

$200 = k \cdot 60^2$

$\dfrac{1}{18} = k$ Variation constant

$d = \dfrac{1}{18}r^2$ Equation of variation

$d = \dfrac{1}{18} \cdot 30^2$

$d = 50$ ft

38. The graph of $g(x) = 2f(x) - 1$ is the graph of $y = f(x)$ stretched vertically by a factor of 2 and shifted down 1 unit. The correct graph is C.

39. Each x-coordinate on the graph of $y = f(x)$ is divided by 3 on the graph of $y = f(3x)$. Thus the point $\left(\dfrac{-3}{3}, 1\right)$, or $(-1, 1)$ is on the graph of $f(3x)$.

Chapter 3

Quadratic Functions and Equations; Inequalities

Exercise Set 3.1

1. $\sqrt{-3} = \sqrt{-1 \cdot 3} = \sqrt{-1} \cdot \sqrt{3} = i\sqrt{3}$, or $\sqrt{3}i$

3. $\sqrt{-25} = \sqrt{-1 \cdot 25} = \sqrt{-1} \cdot \sqrt{25} = i \cdot 5 = 5i$

5. $-\sqrt{-33} = -\sqrt{-1 \cdot 33} = -\sqrt{-1} \cdot \sqrt{33} = -i\sqrt{33}$, or $-\sqrt{33}i$

7. $-\sqrt{-81} = -\sqrt{-1 \cdot 81} = -\sqrt{-1} \cdot \sqrt{81} = -i \cdot 9 = -9i$

9. $\sqrt{-98} = \sqrt{-1 \cdot 98} = \sqrt{-1} \cdot \sqrt{98} = i\sqrt{49 \cdot 2} =$
$i \cdot 7\sqrt{2} = 7i\sqrt{2}$, or $7\sqrt{2}i$

11. $\quad (-5 + 3i) + (7 + 8i)$
$= (-5 + 7) + (3i + 8i)$ Collecting the real parts
                        and the imaginary parts

$= 2 + (3 + 8)i$
$= 2 + 11i$

13. $\quad (4 - 9i) + (1 - 3i)$
$= (4 + 1) + (-9i - 3i)$ Collecting the real parts
                        and the imaginary parts

$= 5 + (-9 - 3)i$
$= 5 - 12i$

15. $\quad (12 + 3i) + (-8 + 5i)$
$= (12 - 8) + (3i + 5i)$
$= 4 + 8i$

17. $\quad (-1 - i) + (-3 - i)$
$= (-1 - 3) + (-i - i)$
$= -4 - 2i$

19. $(3 + \sqrt{-16}) + (2 + \sqrt{-25}) = (3 + 4i) + (2 + 5i)$
$= (3 + 2) + (4i + 5i)$
$= 5 + 9i$

21. $\quad (10 + 7i) - (5 + 3i)$
$= (10 - 5) + (7i - 3i)$ The 5 and the $3i$ are
                        both being subtracted.
$= 5 + 4i$

23. $\quad (13 + 9i) - (8 + 2i)$
$= (13 - 8) + (9i - 2i)$ The 8 and the $2i$ are
                        both being subtracted.
$= 5 + 7i$

25. $\quad (6 - 4i) - (-5 + i)$
$= [6 - (-5)] + (-4i - i)$
$= (6 + 5) + (-4i - i)$
$= 11 - 5i$

27. $\quad (-5 + 2i) - (-4 - 3i)$
$= [-5 - (-4)] + [2i - (-3i)]$
$= (-5 + 4) + (2i + 3i)$
$= -1 + 5i$

29. $\quad (4 - 9i) - (2 + 3i)$
$= (4 - 2) + (-9i - 3i)$
$= 2 - 12i$

31. $\sqrt{-4} \cdot \sqrt{-36} = 2i \cdot 6i = 12i^2 = 12(-1) = -12$

33. $\sqrt{-81} \cdot \sqrt{-25} = 9i \cdot 5i = 45i^2 = 45(-1) = -45$

35. $\quad 7i(2 - 5i)$
$= 14i - 35i^2$ Using the distributive law
$= 14i + 35$ $i^2 = -1$
$= 35 + 14i$ Writing in the form $a + bi$

37. $\quad -2i(-8 + 3i)$
$= 16i - 6i^2$ Using the distributive law
$= 16i + 6$ $i^2 = -1$
$= 6 + 16i$ Writing in the form $a + bi$

39. $\quad (1 + 3i)(1 - 4i)$
$= 1 - 4i + 3i - 12i^2$ Using FOIL
$= 1 - 4i + 3i - 12(-1)$ $i^2 = -1$
$= 1 - i + 12$
$= 13 - i$

41. $\quad (2 + 3i)(2 + 5i)$
$= 4 + 10i + 6i + 15i^2$ Using FOIL
$= 4 + 10i + 6i - 15$ $i^2 = -1$
$= -11 + 16i$

43. $\quad (-4 + i)(3 - 2i)$
$= -12 + 8i + 3i - 2i^2$ Using FOIL
$= -12 + 8i + 3i + 2$ $i^2 = -1$
$= -10 + 11i$

45. $\quad (8 - 3i)(-2 - 5i)$
$= -16 - 40i + 6i + 15i^2$
$= -16 - 40i + 6i - 15$ $i^2 = -1$
$= -31 - 34i$

47. $(3 + \sqrt{-16})(2 + \sqrt{-25})$
$= (3 + 4i)(2 + 5i)$
$= 6 + 15i + 8i + 20i^2$
$= 6 + 15i + 8i - 20 \qquad i^2 = -1$
$= -14 + 23i$

49. $(5 - 4i)(5 + 4i) = 5^2 - (4i)^2$
$= 25 - 16i^2$
$= 25 + 16 \qquad i^2 = -1$
$= 41$

51. $(3 + 2i)(3 - 2i)$
$= 9 - 6i + 6i - 4i^2$
$= 9 - 6i + 6i + 4 \qquad i^2 = -1$
$= 13$

53. $(7 - 5i)(7 + 5i)$
$= 49 + 35i - 35i - 25i^2$
$= 49 + 35i - 35i + 25 \qquad i^2 = -1$
$= 74$

55. $(4 + 2i)^2$
$= 16 + 2 \cdot 4 \cdot 2i + (2i)^2 \quad$ Recall $(A + B)^2 =$
$\qquad\qquad\qquad\qquad\qquad A^2 + 2AB + B^2$
$= 16 + 16i + 4i^2$
$= 16 + 16i - 4 \qquad\qquad i^2 = -1$
$= 12 + 16i$

57. $(-2 + 7i)^2$
$= (-2)^2 + 2(-2)(7i) + (7i)^2 \quad$ Recall $(A+B)^2 =$
$\qquad\qquad\qquad\qquad\qquad\qquad A^2 + 2AB + B^2$
$= 4 - 28i + 49i^2$
$= 4 - 28i - 49 \qquad\qquad i^2 = -1$
$= -45 - 28i$

59. $(1 - 3i)^2$
$= 1^2 - 2 \cdot 1 \cdot (3i) + (3i)^2$
$= 1 - 6i + 9i^2$
$= 1 - 6i - 9 \qquad i^2 = -1$
$= -8 - 6i$

61. $(-1 - i)^2$
$= (-1)^2 - 2(-1)(i) + i^2$
$= 1 + 2i + i^2$
$= 1 + 2i - 1 \qquad i^2 = -1$
$= 2i$

63. $(3 + 4i)^2$
$= 9 + 2 \cdot 3 \cdot 4i + (4i)^2$
$= 9 + 24i + 16i^2$
$= 9 + 24i - 16 \qquad i^2 = -1$
$= -7 + 24i$

65. $\dfrac{3}{5 - 11i}$
$= \dfrac{3}{5 - 11i} \cdot \dfrac{5 + 11i}{5 + 11i} \qquad$ 5 − 11i is the conjugate
$\qquad\qquad\qquad\qquad\qquad$ of 5 + 11i.
$= \dfrac{3(5 + 11i)}{(5 - 11i)(5 + 11i)}$
$= \dfrac{15 + 33i}{25 - 121i^2}$
$= \dfrac{15 + 33i}{25 + 121} \qquad i^2 = -1$
$= \dfrac{15 + 33i}{146}$
$= \dfrac{15}{146} + \dfrac{33}{146}i \qquad$ Writing in the form $a + bi$

67. $\dfrac{5}{2 + 3i}$
$= \dfrac{5}{2 + 3i} \cdot \dfrac{2 - 3i}{2 - 3i} \qquad$ 2 − 3i is the conjugate
$\qquad\qquad\qquad\qquad\qquad$ of 2 + 3i.
$= \dfrac{5(2 - 3i)}{(2 + 3i)(2 - 3i)}$
$= \dfrac{10 - 15i}{4 - 9i^2}$
$= \dfrac{10 - 15i}{4 + 9} \qquad i^2 = -1$
$= \dfrac{10 - 15i}{13}$
$= \dfrac{10}{13} - \dfrac{15}{13}i \qquad$ Writing in the form $a + bi$

69. $\dfrac{4 + i}{-3 - 2i}$
$= \dfrac{4 + i}{-3 - 2i} \cdot \dfrac{-3 + 2i}{-3 + 2i} \qquad$ −3 + 2i is the conjugate
$\qquad\qquad\qquad\qquad\qquad\qquad$ of the divisor.
$= \dfrac{(4 + i)(-3 + 2i)}{(-3 - 2i)(-3 + 2i)}$
$= \dfrac{-12 + 5i + 2i^2}{9 - 4i^2}$
$= \dfrac{-12 + 5i - 2}{9 + 4} \qquad i^2 = -1$
$= \dfrac{-14 + 5i}{13}$
$= -\dfrac{14}{13} + \dfrac{5}{13}i \qquad$ Writing in the form $a + bi$

71. $\dfrac{5-3i}{4+3i}$

$= \dfrac{5-3i}{4+3i} \cdot \dfrac{4-3i}{4-3i}$ $\quad$ 4 − 3i is the conjugate of 4 + 3i.

$= \dfrac{(5-3i)(4-3i)}{(4+3i)(4-3i)}$

$= \dfrac{20-27i+9i^2}{16-9i^2}$

$= \dfrac{20-27i-9}{16+9}$ $\quad i^2 = -1$

$= \dfrac{11-27i}{25}$

$= \dfrac{11}{25} - \dfrac{27}{25}i$ $\quad$ Writing in the form $a+bi$

73. $\dfrac{2+\sqrt{3}i}{5-4i}$

$= \dfrac{2+\sqrt{3}i}{5-4i} \cdot \dfrac{5+4i}{5+4i}$ $\quad$ 5 + 4i is the conjugate of the divisor.

$= \dfrac{(2+\sqrt{3}i)(5+4i)}{(5-4i)(5+4i)}$

$= \dfrac{10+8i+5\sqrt{3}i+4\sqrt{3}i^2}{25-16i^2}$

$= \dfrac{10+8i+5\sqrt{3}i-4\sqrt{3}}{25+16}$ $\quad i^2 = -1$

$= \dfrac{10-4\sqrt{3}+(8+5\sqrt{3})i}{41}$

$= \dfrac{10-4\sqrt{3}}{41} + \dfrac{8+5\sqrt{3}}{41}i$ $\quad$ Writing in the form $a+bi$

75. $\dfrac{1+i}{(1-i)^2}$

$= \dfrac{1+i}{1-2i+i^2}$

$= \dfrac{1+i}{1-2i-1}$ $\quad i^2 = -1$

$= \dfrac{1+i}{-2i}$

$= \dfrac{1+i}{-2i} \cdot \dfrac{2i}{2i}$ $\quad$ 2i is the conjugate of −2i.

$= \dfrac{(1+i)(2i)}{(-2i)(2i)}$

$= \dfrac{2i+2i^2}{-4i^2}$

$= \dfrac{2i-2}{4}$ $\quad i^2 = -1$

$= -\dfrac{2}{4} + \dfrac{2}{4}i$

$= -\dfrac{1}{2} + \dfrac{1}{2}i$

77. $\dfrac{4-2i}{1+i} + \dfrac{2-5i}{1+i}$

$= \dfrac{6-7i}{1+i}$ $\quad$ Adding

$= \dfrac{6-7i}{1+i} \cdot \dfrac{1-i}{1-i}$ $\quad$ 1 − i is the conjugate of 1 + i.

$= \dfrac{(6-7i)(1-i)}{(1+i)(1-i)}$

$= \dfrac{6-13i+7i^2}{1-i^2}$

$= \dfrac{6-13i-7}{1+1}$ $\quad i^2 = -1$

$= \dfrac{-1-13i}{2}$

$= -\dfrac{1}{2} - \dfrac{13}{2}i$

79. $i^{11} = i^{10} \cdot i = (i^2)^5 \cdot i = (-1)^5 \cdot i = -1 \cdot i = -i$

81. $i^{35} = i^{34} \cdot i = (i^2)^{17} \cdot i = (-1)^{17} \cdot i = -1 \cdot i = -i$

83. $i^{64} = (i^2)^{32} = (-1)^{32} = 1$

85. $(-i)^{71} = (-1 \cdot i)^{71} = (-1)^{71} \cdot i^{71} = -i^{70} \cdot i =$
$-(i^2)^{35} \cdot i = -(-1)^{35} \cdot i = -(-1)i = i$

87. $(5i)^4 = 5^4 \cdot i^4 = 625(i^2)^2 = 625(-1)^2 = 625 \cdot 1 = 625$

89. First find the slope of the given line.

$$3x - 6y = 7$$
$$-6y = -3x + 7$$
$$y = \frac{1}{2}x - \frac{7}{6}$$

The slope is $\dfrac{1}{2}$. The slope of the desired line is the opposite of the reciprocal of $\dfrac{1}{2}$, or −2. Write a slope-intercept equation of the line containing $(3, -5)$ with slope −2.

$$y - (-5) = -2(x - 3)$$
$$y + 5 = -2x + 6$$
$$y = -2x + 1$$

91. The domain of f is the set of all real numbers as is the domain of g. When $x = -\dfrac{5}{3}$, $g(x) = 0$, so the domain of f/g is $\left(-\infty, -\dfrac{5}{3}\right) \cup \left(-\dfrac{5}{3}, \infty\right)$.

93. $(f/g)(2) = \dfrac{f(2)}{g(2)} = \dfrac{2^2+4}{3 \cdot 2 + 5} = \dfrac{4+4}{6+5} = \dfrac{8}{11}$

95. $(a+bi)+(a-bi) = 2a$, a real number. Thus, the statement is true.

97. $(a+bi)(c+di) = (ac-bd) + (ad+bc)i$. The conjugate of the product is $(ac-bd) - (ad+bc)i =$
$(a-bi)(c-di)$, the product of the conjugates of the individual complex numbers. Thus, the statement is true.

99. $z\bar{z} = (a+bi)(a-bi) = a^2 - b^2 i^2 = a^2 + b^2$

101. $[x - (3 + 4i)][x - (3 - 4i)]$

$= [x - 3 - 4i][x - 3 + 4i]$

$= [(x - 3) - 4i][(x - 3) + 4i]$

$= (x - 3)^2 - (4i)^2$

$= x^2 - 6x + 9 - 16i^2$

$= x^2 - 6x + 9 + 16 \qquad i^2 = -1$

$= x^2 - 6x + 25$

Exercise Set 3.2

1. $(2x - 3)(3x - 2) = 0$

$2x - 3 = 0 \quad or \quad 3x - 2 = 0$ Using the principle
of zero products

$\qquad 2x = 3 \quad or \qquad 3x = 2$

$\qquad x = \dfrac{3}{2} \quad or \qquad x = \dfrac{2}{3}$

The solutions are $\dfrac{3}{2}$ and $\dfrac{2}{3}$.

3. $x^2 - 8x - 20 = 0$

$(x - 10)(x + 2) = 0$ Factoring

$x - 10 = 0 \quad or \quad x + 2 = 0$ Using the principle
of zero products

$\qquad x = 10 \quad or \qquad x = -2$

The solutions are 10 and -2.

5. $3x^2 + x - 2 = 0$

$(3x - 2)(x + 1) = 0$ Factoring

$3x - 2 = 0 \quad or \quad x + 1 = 0$ Using the principle
of zero products

$\qquad x = \dfrac{2}{3} \quad or \qquad x = -1$

The solutions are $\dfrac{2}{3}$ and -1.

7. $4x^2 - 12 = 0$

$\qquad 4x^2 = 12$

$\qquad x^2 = 3$

$\qquad x = \sqrt{3} \ or \ x = -\sqrt{3}$ Using the principle
of square roots

The solutions are $\sqrt{3}$ and $-\sqrt{3}$.

9. $3x^2 = 21$

$\qquad x^2 = 7$

$\qquad x = \sqrt{7} \ or \ x = -\sqrt{7}$ Using the principle
of square roots

The solutions are $\sqrt{7}$ and $-\sqrt{7}$.

11. $5x^2 + 10 = 0$

$\qquad 5x^2 = -10$

$\qquad x^2 = -2$

$\qquad x = \sqrt{2}i \ or \ x = -\sqrt{2}i$

The solutions are $\sqrt{2}i$ and $-\sqrt{2}i$.

13. $x^2 + 16 = 0$

$\qquad x^2 = -16$

$\qquad x = \sqrt{-16} \ or \ x = -\sqrt{-16}$

$\qquad x = 4i \qquad or \ x = -4i$

The solutions are $4i$ and $-4i$.

15. $\qquad 2x^2 = 6x$

$2x^2 - 6x = 0$ Subtracting $6x$ on both sides

$2x(x - 3) = 0$

$2x = 0 \quad or \quad x - 3 = 0$

$\ x = 0 \quad or \qquad x = 3$

The solutions are 0 and 3.

17. $3y^3 - 5y^2 - 2y = 0$

$y(3y^2 - 5y - 2) = 0$

$y(3y + 1)(y - 2) = 0$

$y = 0 \ or \ 3y + 1 = 0 \quad or \ y - 2 = 0$

$y = 0 \ or \qquad y = -\dfrac{1}{3} \ or \qquad y = 2$

The solutions are $-\dfrac{1}{3}$, 0 and 2.

19. $7x^3 + x^2 - 7x - 1 = 0$

$x^2(7x + 1) - (7x + 1) = 0$

$(x^2 - 1)(7x + 1) = 0$

$(x + 1)(x - 1)(7x + 1) = 0$

$x + 1 = 0 \quad or \ x - 1 = 0 \ or \ 7x + 1 = 0$

$x = -1 \ or \qquad x = 1 \ or \qquad x = -\dfrac{1}{7}$

The solutions are -1, $-\dfrac{1}{7}$, and 1.

21. a) The graph crosses the x-axis at $(-4, 0)$ and at $(2, 0)$.
These are the x-intercepts.

b) The zeros of the function are the first coordinates of
the x-intercepts of the graph. They are -4 and 2.

23. a) The graph crosses the x-axis at $(-1, 0)$ and at $(3, 0)$.
These are the x-intercepts.

b) The zeros of the function are the first coordinates of
the x-intercepts of the graph. They are -1 and 3.

25. a) The graph crosses the x-axis at $(-2, 0)$ and at $(2, 0)$.
These are the x-intercepts.

b) The zeros of the function are the first coordinates of
the x-intercepts of the graph. They are -2 and 2.

27. a) The graph has only one x-intercept, $(1, 0)$.

b) The zero of the function is the first coordinate of
the x-intercept of the graph, 1.

29. $\qquad x^2 + 6x = 7$

$x^2 + 6x + 9 = 7 + 9$ Completing the square:
$\frac{1}{2} \cdot 6 = 3$ and $3^2 = 9$

$\qquad (x + 3)^2 = 16$ Factoring

$\qquad x + 3 = \pm 4$ Using the principle
of square roots

$\qquad x = -3 \pm 4$

$x = -3 - 4$ *or* $x = -3 + 4$

$x = -7$ *or* $x = 1$

The solutions are -7 and 1.

31.
$$x^2 = 8x - 9$$

$x^2 - 8x = -9$ Subtracting $8x$

$x^2 - 8x + 16 = -9 + 16$ Completing the square:
$$\tfrac{1}{2}(-8) = -4 \text{ and } (-4)^2 = 16$$

$(x - 4)^2 = 7$ Factoring

$x - 4 = \pm\sqrt{7}$ Using the principle
                of square roots

$x = 4 \pm \sqrt{7}$

The solutions are $4 - \sqrt{7}$ and $4 + \sqrt{7}$, or $4 \pm \sqrt{7}$.

33. $x^2 + 8x + 25 = 0$

$x^2 + 8x = -25$ Subtracting 25

$x^2 + 8x + 16 = -25 + 16$ Completing the
                        square:
$$\tfrac{1}{2} \cdot 8 = 4 \text{ and } 4^2 = 16$$

$(x + 4)^2 = -9$ Factoring

$x + 4 = \pm 3i$ Using the principle
                of square roots

$x = -4 \pm 3i$

The solutions are $-4 - 3i$ and $-4 + 3i$, or $-4 \pm 3i$.

35. $3x^2 + 5x - 2 = 0$

$3x^2 + 5x = 2$ Adding 2

$x^2 + \dfrac{5}{3}x = \dfrac{2}{3}$ Dividing by 3

$x^2 + \dfrac{5}{3}x + \dfrac{25}{36} = \dfrac{2}{3} + \dfrac{25}{36}$ Completing the
                                     square:
$$\tfrac{1}{2} \cdot \tfrac{5}{3} = \tfrac{5}{6} \text{ and } \left(\tfrac{5}{6}\right)^2 = \tfrac{25}{36}$$

$\left(x + \dfrac{5}{6}\right)^2 = \dfrac{49}{36}$ Factoring and
                       simplifying

$x + \dfrac{5}{6} = \pm\dfrac{7}{6}$ Using the principle
                    of square roots

$x = -\dfrac{5}{6} \pm \dfrac{7}{6}$

$x = -\dfrac{5}{6} - \dfrac{7}{6}$ *or* $x = -\dfrac{5}{6} + \dfrac{7}{6}$

$x = -\dfrac{12}{6}$ *or* $x = \dfrac{2}{6}$

$x = -2$ *or* $x = \dfrac{1}{3}$

The solutions are -2 and $\dfrac{1}{3}$.

37.
$$x^2 - 2x = 15$$

$x^2 - 2x - 15 = 0$

$(x - 5)(x + 3) = 0$ Factoring

$x - 5 = 0$ *or* $x + 3 = 0$

$x = 5$ *or* $x = -3$

The solutions are 5 and -3.

39.
$$5m^2 + 3m = 2$$

$5m^2 + 3m - 2 = 0$

$(5m - 2)(m + 1) = 0$ Factoring

$5m - 2 = 0$ *or* $m + 1 = 0$

$m = \dfrac{2}{5}$ *or* $m = -1$

The solutions are $\dfrac{2}{5}$ and -1.

41.
$$3x^2 + 6 = 10x$$

$3x^2 - 10x + 6 = 0$

We use the quadratic formula. Here $a = 3$, $b = -10$, and $c = 6$.

$$x = \frac{-b \pm \sqrt{b^2 - 4ac}}{2a}$$

$$= \frac{-(-10) \pm \sqrt{(-10)^2 - 4 \cdot 3 \cdot 6}}{2 \cdot 3} \quad \text{Substituting}$$

$$= \frac{10 \pm \sqrt{28}}{6} = \frac{10 \pm 2\sqrt{7}}{6}$$

$$= \frac{2(5 \pm \sqrt{7})}{2 \cdot 3} = \frac{5 \pm \sqrt{7}}{3}$$

The solutions are $\dfrac{5 - \sqrt{7}}{3}$ and $\dfrac{5 + \sqrt{7}}{3}$, or $\dfrac{5 \pm \sqrt{7}}{3}$.

43. $x^2 + x + 2 = 0$

We use the quadratic formula. Here $a = 1$, $b = 1$, and $c = 2$.

$$x = \frac{-b \pm \sqrt{b^2 - 4ac}}{2a}$$

$$= \frac{-1 \pm \sqrt{1^2 - 4 \cdot 1 \cdot 2}}{2 \cdot 1} \quad \text{Substituting}$$

$$= \frac{-1 \pm \sqrt{-7}}{2}$$

$$= \frac{-1 \pm \sqrt{7}i}{2} = -\frac{1}{2} \pm \frac{\sqrt{7}}{2}i$$

The solutions are $-\dfrac{1}{2} - \dfrac{\sqrt{7}}{2}i$ and $-\dfrac{1}{2} + \dfrac{\sqrt{7}}{2}i$, or $-\dfrac{1}{2} \pm \dfrac{\sqrt{7}}{2}i$.

45.
$$5t^2 - 8t = 3$$

$5t^2 - 8t - 3 = 0$

We use the quadratic formula. Here $a = 5$, $b = -8$, and $c = -3$.

$$t = \frac{-b \pm \sqrt{b^2 - 4ac}}{2a}$$

$$= \frac{-(-8) \pm \sqrt{(-8)^2 - 4 \cdot 5(-3)}}{2 \cdot 5}$$

$$= \frac{8 \pm \sqrt{124}}{10} = \frac{8 \pm 2\sqrt{31}}{10}$$

$$= \frac{2(4 \pm \sqrt{31})}{2 \cdot 5} = \frac{4 \pm \sqrt{31}}{5}$$

The solutions are $\dfrac{4 - \sqrt{31}}{5}$ and $\dfrac{4 + \sqrt{31}}{5}$, or

$$\frac{4 \pm \sqrt{31}}{5}.$$

47.
$$3x^2 + 4 = 5x$$
$$3x^2 - 5x + 4 = 0$$

We use the quadratic formula. Here $a = 3$, $b = -5$, and $c = 4$.

$$x = \frac{-b \pm \sqrt{b^2 - 4ac}}{2a}$$

$$= \frac{-(-5) \pm \sqrt{(-5)^2 - 4 \cdot 3 \cdot 4}}{2 \cdot 3}$$

$$= \frac{5 \pm \sqrt{-23}}{6} = \frac{5 \pm \sqrt{23}i}{6}$$

$$= \frac{5}{6} \pm \frac{\sqrt{23}}{6}i$$

The solutions are $\frac{5}{6} - \frac{\sqrt{23}}{6}i$ and $\frac{5}{6} + \frac{\sqrt{23}}{6}i$, or $\frac{5}{6} \pm \frac{\sqrt{23}}{6}i$.

49. $x^2 - 8x + 5 = 0$

We use the quadratic formula. Here $a = 1$, $b = -8$, and $c = 5$.

$$x = \frac{-b \pm \sqrt{b^2 - 4ac}}{2a}$$

$$= \frac{-(-8) \pm \sqrt{(-8)^2 - 4 \cdot 1 \cdot 5}}{2 \cdot 1}$$

$$= \frac{8 \pm \sqrt{44}}{2} = \frac{8 \pm 2\sqrt{11}}{2}$$

$$= \frac{2(4 \pm \sqrt{11})}{2} = 4 \pm \sqrt{11}$$

The solutions are $4 - \sqrt{11}$ and $4 + \sqrt{11}$, or $4 \pm \sqrt{11}$.

51.
$$3x^2 + x = 5$$
$$3x^2 + x - 5 = 0$$

We use the quadratic formula. We have $a = 3$, $b = 1$, and $c = -5$.

$$x = \frac{-b \pm \sqrt{b^2 - 4ac}}{2a}$$

$$= \frac{-1 \pm \sqrt{1^2 - 4 \cdot 3 \cdot (-5)}}{2 \cdot 3}$$

$$= \frac{-1 \pm \sqrt{61}}{6}$$

The solutions are $\frac{-1 - \sqrt{61}}{6}$ and $\frac{-1 + \sqrt{61}}{6}$, or $\frac{-1 \pm \sqrt{61}}{6}$.

53.
$$2x^2 + 1 = 5x$$
$$2x^2 - 5x + 1 = 0$$

We use the quadratic formula. We have $a = 2$, $b = -5$, and $c = 1$.

$$x = \frac{-b \pm \sqrt{b^2 - 4ac}}{2a}$$

$$= \frac{-(-5) \pm \sqrt{(-5)^2 - 4 \cdot 2 \cdot 1}}{2 \cdot 2} = \frac{5 \pm \sqrt{17}}{4}$$

The solutions are $\frac{5 - \sqrt{17}}{4}$ and $\frac{5 + \sqrt{17}}{4}$, or $\frac{5 \pm \sqrt{17}}{4}$.

55.
$$5x^2 + 2x = -2$$
$$5x^2 + 2x + 2 = 0$$

We use the quadratic formula. We have $a = 5$, $b = 2$, and $c = 2$.

$$x = \frac{-b \pm \sqrt{b^2 - 4ac}}{2a}$$

$$= \frac{-2 \pm \sqrt{2^2 - 4 \cdot 5 \cdot 2}}{2 \cdot 5}$$

$$= \frac{-2 \pm \sqrt{-36}}{10} = \frac{-2 \pm 6i}{10}$$

$$= \frac{2(-1 \pm 3i)}{2 \cdot 5} = \frac{-1 \pm 3i}{5}$$

$$= -\frac{1}{5} \pm \frac{3}{5}i$$

The solutions are $-\frac{1}{5} - \frac{3}{5}i$ and $-\frac{1}{5} + \frac{3}{5}i$, or $-\frac{1}{5} \pm \frac{3}{5}i$.

57.
$$4x^2 = 8x + 5$$
$$4x^2 - 8x - 5 = 0$$
$$a = 4, b = -8, c = -5$$
$$b^2 - 4ac = (-8)^2 - 4 \cdot 4(-5) = 144$$

Since $b^2 - 4ac > 0$, there are two different real-number solutions.

59. $x^2 + 3x + 4 = 0$
$$a = 1, b = 3, c = 4$$
$$b^2 - 4ac = 3^2 - 4 \cdot 1 \cdot 4 = -7$$

Since $b^2 - 4ac < 0$, there are two different imaginary-number solutions.

61. $5t^2 - 7t = 0$
$$a = 5, b = -7, c = 0$$
$$b^2 - 4ac = (-7)^2 - 4 \cdot 5 \cdot 0 = 49$$

Since $b^2 - 4ac > 0$, there are two different real-number solutions.

63.
$$x^2 + 6x + 5 = 0 \quad \text{Setting } f(x) = 0$$
$$(x + 5)(x + 1) = 0 \quad \text{Factoring}$$
$$x + 5 = 0 \quad or \quad x + 1 = 0$$
$$x = -5 \quad or \quad x = -1$$

The zeros of the function are -5 and -1.

65. $x^2 - 3x - 3 = 0$
$$a = 1, b = -3, c = -3$$

$$x = \frac{-b \pm \sqrt{b^2 - 4ac}}{2a}$$

$$= \frac{-(-3) \pm \sqrt{(-3)^2 - 4 \cdot 1 \cdot (-3)}}{2 \cdot 1}$$

$$= \frac{3 \pm \sqrt{9 + 12}}{2}$$

$$= \frac{3 \pm \sqrt{21}}{2}$$

The zeros of the function are $\dfrac{3 - \sqrt{21}}{2}$ and $\dfrac{3 + \sqrt{21}}{2}$, or $\dfrac{3 \pm \sqrt{21}}{2}$.

We use a calculator to find decimal approximations for the zeros:

$$\frac{3 + \sqrt{21}}{2} \approx 3.791 \text{ and } \frac{3 - \sqrt{21}}{2} \approx -0.791.$$

67. $x^2 - 5x + 1 = 0$

$a = 1,\ b = -5,\ c = 1$

$$x = \frac{-b \pm \sqrt{b^2 - 4ac}}{2a}$$

$$= \frac{-(-5) \pm \sqrt{(-5)^2 - 4 \cdot 1 \cdot 1}}{2 \cdot 1}$$

$$= \frac{5 \pm \sqrt{25 - 4}}{2}$$

$$= \frac{5 \pm \sqrt{21}}{2}$$

The zeros of the function are $\dfrac{5 - \sqrt{21}}{2}$ and $\dfrac{5 + \sqrt{21}}{2}$, or $\dfrac{5 \pm \sqrt{21}}{2}$.

We use a calculator to find decimal approximations for the zeros:

$$\frac{5 + \sqrt{21}}{2} \approx 4.791 \text{ and } \frac{5 - \sqrt{21}}{2} \approx 0.209.$$

69. $x^2 + 2x - 5 = 0$

$a = 1,\ b = 2,\ c = -5$

$$x = \frac{-b \pm \sqrt{b^2 - 4ac}}{2a}$$

$$= \frac{-2 \pm \sqrt{2^2 - 4 \cdot 1 \cdot (-5)}}{2 \cdot 1}$$

$$= \frac{-2 \pm \sqrt{4 + 20}}{2} = \frac{-2 \pm \sqrt{24}}{2}$$

$$= \frac{-2 \pm 2\sqrt{6}}{2} = -1 \pm \sqrt{6}$$

The zeros of the function are $-1 + \sqrt{6}$ and $-1 - \sqrt{6}$, or $-1 \pm \sqrt{6}$.

We use a calculator to find decimal approximations for the zeros:

$$-1 + \sqrt{6} \approx 1.449 \text{ and } -1 - \sqrt{6} \approx -3.449$$

71. $2x^2 - x + 4 = 0$

$a = 2,\ b = -1,\ c = 4$

$$x = \frac{-b \pm \sqrt{b^2 - 4ac}}{2a}$$

$$= \frac{-(-1) \pm \sqrt{(-1)^2 - 4 \cdot 2 \cdot 4}}{2 \cdot 2}$$

$$= \frac{1 \pm \sqrt{-31}}{4} = \frac{1 \pm \sqrt{31}i}{4}$$

$$= \frac{1}{4} \pm \frac{\sqrt{31}}{4}i$$

The zeros of the function are $\dfrac{1}{4} - \dfrac{\sqrt{31}}{4}i$ and $\dfrac{1}{4} + \dfrac{\sqrt{31}}{4}i$, or $\dfrac{1}{4} \pm \dfrac{\sqrt{31}}{4}i$.

73. $3x^2 - x - 1 = 0$

$a = 3,\ b = -1,\ c = -1$

$$x = \frac{-b \pm \sqrt{b^2 - 4ac}}{2a}$$

$$= \frac{-(-1) \pm \sqrt{(-1)^2 - 4 \cdot 3 \cdot (-1)}}{2 \cdot 3}$$

$$= \frac{1 \pm \sqrt{13}}{6}$$

The zeros of the function are $\dfrac{1 - \sqrt{13}}{6}$ and $\dfrac{1 + \sqrt{13}}{6}$, or $\dfrac{1 \pm \sqrt{13}}{6}$.

We use a calculator to find decimal approximations for the zeros:

$$\frac{1 + \sqrt{13}}{6} \approx 0.768 \text{ and } \frac{1 - \sqrt{13}}{6} \approx -0.434.$$

75. $5x^2 - 2x - 1 = 0$

$a = 5,\ b = -2,\ c = -1$

$$x = \frac{-b \pm \sqrt{b^2 - 4ac}}{2a}$$

$$= \frac{-(-2) \pm \sqrt{(-2)^2 - 4 \cdot 5 \cdot (-1)}}{2 \cdot 5}$$

$$= \frac{2 \pm \sqrt{24}}{10} = \frac{2 \pm 2\sqrt{6}}{10}$$

$$= \frac{2(1 \pm \sqrt{6})}{2 \cdot 5} = \frac{1 \pm \sqrt{6}}{5}$$

The zeros of the function are $\dfrac{1 - \sqrt{6}}{5}$ and $\dfrac{1 + \sqrt{6}}{5}$, or $\dfrac{1 \pm \sqrt{6}}{5}$.

We use a calculator to find decimal approximations for the zeros:

$$\frac{1 + \sqrt{6}}{5} \approx 0.690 \text{ and } \frac{1 - \sqrt{6}}{5} \approx -0.290.$$

77. $4x^2 + 3x - 3 = 0$

$a = 4,\ b = 3,\ c = -3$

$$x = \frac{-b \pm \sqrt{b^2 - 4ac}}{2a}$$

$$= \frac{-3 \pm \sqrt{3^2 - 4 \cdot 4 \cdot (-3)}}{2 \cdot 4}$$

$$= \frac{-3 \pm \sqrt{57}}{8}$$

The zeros of the function are $\dfrac{-3 - \sqrt{57}}{8}$ and $\dfrac{-3 + \sqrt{57}}{8}$, or $\dfrac{-3 \pm \sqrt{57}}{8}$.

We use a calculator to find decimal approximations for the zeros:

$$\frac{-3+\sqrt{57}}{8} \approx 0.569 \text{ and } \frac{-3-\sqrt{57}}{8} \approx -1.319.$$

79. $x^4 - 3x^2 + 2 = 0$

Let $u = x^2$.

$u^2 - 3u + 2 = 0$ Substituting u for x^2

$(u-1)(u-2) = 0$

$u - 1 = 0$ or $u - 2 = 0$

$u = 1$ or $u = 2$

Now substitute x^2 for u and solve for x.

$x^2 = 1$ or $x^2 = 2$

$x = \pm 1$ or $x = \pm\sqrt{2}$

The solutions are -1, 1, $-\sqrt{2}$, and $\sqrt{2}$.

81. $x^4 + 3x^2 = 10$

$x^4 + 3x^2 - 10 = 0$

Let $u = x^2$.

$u^2 + 3u - 10 = 0$ Substituting u for x^2

$(u+5)(u-2) = 0$

$u + 5 = 0$ or $u - 2 = 0$

$u = -5$ or $u = 2$

Now substitute x^2 for u and solve for x.

$x^2 = -5$ or $x^2 = 2$

$x = \pm\sqrt{5}i$ or $x = \pm\sqrt{2}$

The solutions are $-\sqrt{5}i$, $\sqrt{5}i$, $-\sqrt{2}$, and $\sqrt{2}$.

83. $y^4 + 4y^2 - 5 = 0$

Let $u = y^2$.

$u^2 + 4u - 5 = 0$ Substituting u for y^2

$(u+5)(u-1) = 0$

$u + 5 = 0$ or $u - 1 = 0$

$u = -5$ or $u = 1$

Now substitute y^2 for u and solve for y.

$y^2 = -5$ or $y^2 = 1$

$y = \pm\sqrt{5}i$ or $y = \pm 1$

The solutions are $-\sqrt{5}i$, $\sqrt{5}i$, -1, and 1.

85. $x - 3\sqrt{x} - 4 = 0$

Let $u = \sqrt{x}$.

$u^2 - 3u - 4 = 0$ Substituting u for $\sqrt{x}$

$(u+1)(u-4) = 0$

$u + 1 = 0$ or $u - 4 = 0$

$u = -1$ or $u = 4$

Now substitute $\sqrt{x}$ for u and solve for x.

$\sqrt{x} = -1$ or $\sqrt{x} = 4$

No solution $x = 16$

Note that $\sqrt{x}$ must be nonnegative, so $\sqrt{x} = -1$ has no solution. The number 16 checks and is the solution. The solution is 16.

87. $m^{2/3} - 2m^{1/3} - 8 = 0$

Let $u = m^{1/3}$.

$u^2 - 2u - 8 = 0$ Substituting u for $m^{1/3}$

$(u+2)(u-4) = 0$

$u + 2 = 0$ or $u - 4 = 0$

$u = -2$ or $u = 4$

Now substitute $m^{1/3}$ for u and solve for m.

$m^{1/3} = -2$ or $m^{1/3} = 4$

$(m^{1/3})^3 = (-2)^3$ or $(m^{1/3})^3 = 4^3$ Using the
$\qquad\qquad\qquad\qquad\qquad\qquad$ principle of powers

$m = -8$ or $m = 64$

The solutions are -8 and 64.

89. $x^{1/2} - 3x^{1/4} + 2 = 0$

Let $u = x^{1/4}$.

$u^2 - 3u + 2 = 0$ Substituting u for $x^{1/4}$

$(u-1)(u-2) = 0$

$u - 1 = 0$ or $u - 2 = 0$

$u = 1$ or $u = 2$

Now substitute $x^{1/4}$ for u and solve for x.

$x^{1/4} = 1$ or $x^{1/4} = 2$

$(x^{1/4})^4 = 1^4$ or $(x^{1/4})^4 = 2^4$

$x = 1$ or $x = 16$

The solutions are 1 and 16.

91. $(2x-3)^2 - 5(2x-3) + 6 = 0$

Let $u = 2x - 3$.

$u^2 - 5u + 6 = 0$ Substituting u for $2x - 3$

$(u-2)(u-3) = 0$

$u - 2 = 0$ or $u - 3 = 0$

$u = 2$ or $u = 3$

Now substitute $2x - 3$ for u and solve for x.

$2x - 3 = 2$ or $2x - 3 = 3$

$2x = 5$ or $2x = 6$

$x = \dfrac{5}{2}$ or $x = 3$

The solutions are $\dfrac{5}{2}$ and 3.

93. $(2t^2 + t)^2 - 4(2t^2 + t) + 3 = 0$

Let $u = 2t^2 + t$.

$u^2 - 4u + 3 = 0$ Substituting u for $2t^2 + t$

$(u-1)(u-3) = 0$

$u - 1 = 0$ or $u - 3 = 0$

$u = 1$ or $u = 3$

Now substitute $2t^2 + t$ for u and solve for t.

$2t^2 + t = 1$ or $2t^2 + t = 3$

$2t^2 + t - 1 = 0$ or $2t^2 + t - 3 = 0$

$(2t-1)(t+1) = 0$ or $(2t+3)(t-1) = 0$

$2t-1=0$ or $t+1=0$ or $2t+3=0$ or $t-1=0$

$t=\dfrac{1}{2}$ or $t=-1$ or $t=-\dfrac{3}{2}$ or $t=1$

The solutions are $\dfrac{1}{2}$, -1, $-\dfrac{3}{2}$ and 1.

95. Substitute 10.5 for $f(x)$ and solve for x.

$10.5 = -1.321x^2 + 5.156x + 5.517$

$0 = -1.321x^2 + 5.156x - 4.983$

$a = -1.321,\ b = 5.156,\ c = -4.983$

$x = \dfrac{-b \pm \sqrt{b^2 - 4ac}}{2a}$

$= \dfrac{-5.156 \pm \sqrt{(5.156)^2 - 4(-1.321)(-4.983)}}{2(-1.321)}$

$= \dfrac{-5.156 \pm \sqrt{0.254164}}{-2.641}$

$x \approx 1.76$ or $x \approx 2.14$

In either case, $x \approx 2$, so U.S. funding for Afghan security forces was \$10.5 billion about 2 years after 2009, or in 2011.

97. Substitute 40 for $h(x)$ and solve for x.

$40 = 0.012x^2 - 0.583x + 35.727$

$0 = 0.012x^2 - 0.583x - 4.273$

$a = 0.012,\ b = -0.583,\ c = -4.273$

$x = \dfrac{-b \pm \sqrt{b^2 - 4ac}}{2a}$

$= \dfrac{-(-0.583) \pm \sqrt{(-0.583)^2 - 4(0.012)(-4.273)}}{2(0.012)}$

$= \dfrac{0.583 \pm \sqrt{0.544993}}{0.024}$

$x \approx -6.5$ or $x \approx 55.0$

Since we are looking for a year after 1940, we use the positive solution. There were 40 million multigenerational households about 55 yr after 1940, or in 1995.

99. Familiarize and Translate. We will use the formula $s = 16t^2$, substituting 1670 for s.

$1670 = 16t^2$

Carry out. We solve the equation.

$1670 = 16t^2$

$104.375 = t^2$ Dividing by 16 on both sides

$10.216 \approx t$ Taking the square root on both sides

Check. When $t = 10.216$, $s = 16(10.216)^2 \approx 1670$. The answer checks.

State. It would take an object about 10.216 sec to reach the ground.

101. Familiarize. Let w = the width of the rug. Then $w + 1$ = the length.

Translate. We use the Pythagorean equation.

$w^2 + (w + 1)^2 = 5^2$

Carry out. We solve the equation.

$w^2 + (w + 1)^2 = 5^2$

$w^2 + w^2 + 2w + 1 = 25$

$2w^2 + 2w + 1 = 25$

$2w^2 + 2w - 24 = 0$

$2(w + 4)(w - 3) = 0$

$w + 4 = 0$ or $w - 3 = 0$

$w = -4$ or $w = 3$

Since the width cannot be negative, we consider only 3. When $w = 3$, $w + 1 = 3 + 1 = 4$.

Check. The length, 4 ft, is 1 ft more than the width, 3 ft. The length of a diagonal of a rectangle with width 3 ft and length 4 ft is $\sqrt{3^2 + 4^2} = \sqrt{9 + 16} = \sqrt{25} = 5$. The answer checks.

State. The length is 4 ft, and the width is 3 ft.

103. Familiarize. Let n = the smaller number. Then $n + 5$ = the larger number.

Translate.

$\underbrace{\text{The product of the numbers}}$ is 36.

$\qquad\qquad \downarrow \qquad\qquad\quad \downarrow\ \ \downarrow$

$\qquad\quad n(n + 5) \qquad\quad = \ \ 36$

Carry out.

$n(n + 5) = 36$

$n^2 + 5n = 36$

$n^2 + 5n - 36 = 0$

$(n + 9)(n - 4) = 0$

$n + 9 = 0$ or $n - 4 = 0$

$n = -9$ or $n = 4$

If $n = -9$, then $n + 5 = -9 + 5 = -4$. If $n = 4$, then $n + 5 = 4 + 5 = 9$.

Check. The number -4 is 5 more than -9 and $(-4)(-9) = 36$, so the pair -9 and -4 check. The number 9 is 5 more than 4 and $9 \cdot 4 = 36$, so the pair 4 and 9 also check.

State. The numbers are -9 and -4 or 4 and 9.

105. Familiarize. We add labels to the drawing in the text.

We let x represent the length of a side of the square in each corner. Then the length and width of the resulting base are represented by $20 - 2x$ and $10 - 2x$, respectively. Recall that for a rectangle, Area = length × width.

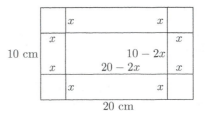

Translate.

$\underbrace{\text{The area of the base}}$ is $\underbrace{96\ \text{cm}^2}$.

$(20 - 2x)(10 - 2x) \quad = \qquad 96$

Carry out. We solve the equation.

$$200 - 60x + 4x^2 = 96$$
$$4x^2 - 60x + 104 = 0$$
$$x^2 - 15x + 26 = 0$$
$$(x - 13)(x - 2) = 0$$
$$x - 13 = 0 \quad or \quad x - 2 = 0$$
$$x = 13 \quad or \quad x = 2$$

Check. When $x = 13$, both $20 - 2x$ and $10 - 2x$ are negative numbers, so we only consider $x = 2$. When $x = 2$, then $20 - 2x = 20 - 2 \cdot 2 = 16$ and $10 - 2x = 10 - 2 \cdot 2 = 6$, and the area of the base is $16 \cdot 6$, or 96 cm^2. The answer checks.

State. The length of the sides of the squares is 2 cm.

107. *Familiarize*. We have $P = 2l + 2w$, or $28 = 2l + 2w$. Solving for w, we have

$$28 = 2l + 2w$$
$$14 = l + w \quad \text{Dividing by 2}$$
$$14 - l = w.$$

Then we have $l =$ the length of the rug and $14 - l =$ the width, in feet. Recall that the area of a rectangle is the product of the length and the width.

Translate.

$$\underbrace{\text{The area}}_{l(14-l)} \;\; \underbrace{\text{is}}_{=} \;\; \underbrace{48 \text{ ft}^2.}_{48}$$

Carry out. We solve the equation.

$$l(14 - l) = 48$$
$$14l - l^2 = 48$$
$$0 = l^2 - 14l + 48$$
$$0 = (l - 6)(l - 8)$$
$$l - 6 = 0 \quad or \quad l - 8 = 0$$
$$l = 6 \quad or \quad l = 8$$

If $l = 6$, then $14 - l = 14 - 6 = 8$.

If $l = 8$, then $14 - l = 14 - 8 = 6$.

In either case, the dimensions are 8 ft by 6 ft. Since we usually consider the length to be greater than the width, we let 8 ft = the length and 6 ft = the width.

Check. The perimeter is $2 \cdot 8$ ft $+ 2 \cdot 6$ ft $= 16$ ft $+ 12$ ft $= 28$ ft. The answer checks.

State. The length of the rug is 8 ft, and the width is 6 ft.

109. $f(x) = 4 - 5x = -5x + 4$

The function can be written in the form $y = mx + b$, so it is a linear function.

111. $f(x) = 7x^2$

The function is in the form $f(x) = ax^2 + bx + c$, $a \neq 0$, so it is a quadratic function.

113. $f(x) = 1.2x - (3.6)^2$

The function is in the form $f(x) = mx + b$, so it is a linear function.

115. In 2014, $x = 2014 - 2004 = 10$.

$$C(10) = 0.17(10) + 2.25 = 1.7 + 2.25 = 3.95$$

We estimate that the cost of a 30-sec Super Bowl ad was $3.95 million in 2014.

117. Test for symmetry with respect to the x-axis:

$$3x^2 + 4y^2 = 5 \quad \text{Original equation}$$
$$3x^2 + 4(-y)^2 = 5 \quad \text{Replacing } y \text{ by } -y$$
$$3x^2 + 4y^2 = 5 \quad \text{Simplifying}$$

The last equation is equivalent to the original equation, so the graph is symmetric with respect to the x-axis.

Test for symmetry with respect to the y-axis:

$$3x^2 + 4y^2 = 5 \quad \text{Original equation}$$
$$3(-x)^2 + 4y^2 = 5 \quad \text{Replacing } x \text{ by } -x$$
$$3x^2 + 4y^2 = 5 \quad \text{Simplifying}$$

The last equation is equivalent to the original equation, so the equation is symmetric with respect to the y-axis.

Test for symmetry with respect to the origin:

$$3x^2 + 4y^2 = 5 \quad \text{Original equation}$$
$$3(-x)^2 + 4(-y)^2 = 5 \quad \text{Replacing } x \text{ by } -x$$
$$\text{and } y \text{ by } -y$$
$$3x^2 + 4y^2 = 5 \quad \text{Simplifying}$$

The last equation is equivalent to the original equation, so the equation is symmetric with respect to the origin.

119. $f(x) = 2x^3 - x$

$$f(-x) = 2(-x)^3 - (-x) = -2x^3 + x$$
$$-f(x) = -2x^3 + x$$
$$f(x) \neq f(-x) \text{ so } f \text{ is not even}$$
$$f(-x) = -f(x), \text{ so } f \text{ is odd.}$$

121. a) $\quad kx^2 - 17x + 33 = 0$

$$k(3)^2 - 17(3) + 33 = 0 \quad \text{Substituting 3 for } x$$
$$9k - 51 + 33 = 0$$
$$9k = 18$$
$$k = 2$$

b) $\quad 2x^2 - 17x + 33 = 0 \quad \text{Substituting 2 for } k$

$$(2x - 11)(x - 3) = 0$$
$$2x - 11 = 0 \quad or \quad x - 3 = 0$$
$$x = \frac{11}{2} \quad or \quad x = 3$$

The other solution is $\frac{11}{2}$.

123. $x^2 - kx + 2 = 0$

a) $\quad (1 + i)^2 - k(1 + i) + 2 = 0 \quad \text{Substituting}$
$$1 + i \text{ for } x$$
$$1 + 2i - 1 - k - ki + 2 = 0$$
$$2 + 2i = k + ki$$
$$2(1 + i) = k(1 + i)$$
$$2 = k$$

b) $x^2 - 2x + 2 = 0$ Substituting 2 for k

$$x = \frac{-(-2) \pm \sqrt{(-2)^2 - 4 \cdot 1 \cdot 2}}{2 \cdot 1}$$

$$= \frac{2 \pm \sqrt{-4}}{2}$$

$$= \frac{2 \pm 2i}{2} = 1 \pm i$$

The other solution is $1 - i$.

125. $$(x-2)^3 = x^3 - 2$$

$$x^3 - 6x^2 + 12x - 8 = x^3 - 2$$

$$0 = 6x^2 - 12x + 6$$

$$0 = 6(x^2 - 2x + 1)$$

$$0 = 6(x - 1)(x - 1)$$

$x - 1 = 0$ or $x - 1 = 0$

$x = 1$ or $x = 1$

The solution is 1.

127. $(6x^3 + 7x^2 - 3x)(x^2 - 7) = 0$

$$x(6x^2 + 7x - 3)(x^2 - 7) = 0$$

$$x(3x - 1)(2x + 3)(x^2 - 7) = 0$$

$x = 0$ or $3x - 1 = 0$ or $2x + 3 = 0$ or $x^2 - 7 = 0$

$x = 0$ or $x = \dfrac{1}{3}$ or $x = -\dfrac{3}{2}$ or $x = \sqrt{7}$ or

$$x = -\sqrt{7}$$

The exact solutions are $-\sqrt{7}$, $-\dfrac{3}{2}$, 0, $\dfrac{1}{3}$, and $\sqrt{7}$.

129. $x^2 + x - \sqrt{2} = 0$

$$x = \frac{-b \pm \sqrt{b^2 - 4ac}}{2a}$$

$$= \frac{-1 \pm \sqrt{1^2 - 4 \cdot 1 (-\sqrt{2})}}{2 \cdot 1} = \frac{-1 \pm \sqrt{1 + 4\sqrt{2}}}{2}$$

The solutions are $\dfrac{-1 \pm \sqrt{1 + 4\sqrt{2}}}{2}$.

131. $2t^2 + (t-4)^2 = 5t(t-4) + 24$

$$2t^2 + t^2 - 8t + 16 = 5t^2 - 20t + 24$$

$$0 = 2t^2 - 12t + 8$$

$$0 = t^2 - 6t + 4 \quad \text{Dividing by 2}$$

Use the quadratic formula.

$$t = \frac{-b \pm \sqrt{b^2 - 4ac}}{2a}$$

$$= \frac{-(-6) \pm \sqrt{(-6)^2 - 4 \cdot 1 \cdot 4}}{2 \cdot 1}$$

$$= \frac{6 \pm \sqrt{20}}{2} = \frac{6 \pm 2\sqrt{5}}{2}$$

$$= \frac{2(3 \pm \sqrt{5})}{2} = 3 \pm \sqrt{5}$$

The solutions are $3 \pm \sqrt{5}$.

133. $\sqrt{x-3} - \sqrt[4]{x-3} = 2$

Substitute u for $\sqrt[4]{x-3}$.

$$u^2 - u - 2 = 0$$

$$(u - 2)(u + 1) = 0$$

$u - 2 = 0$ or $u + 1 = 0$

$u = 2$ or $u = -1$

Substitute $\sqrt[4]{x-3}$ for u and solve for x.

$\sqrt[4]{x-3} = 2$ or $\sqrt[4]{x-3} = 1$

$x - 3 = 16$ No solution

$x = 19$

The value checks. The solution is 19.

135. $$\left(y + \frac{2}{y}\right)^2 + 3y + \frac{6}{y} = 4$$

$$\left(y + \frac{2}{y}\right)^2 + 3\left(y + \frac{2}{y}\right) - 4 = 0$$

Substitute u for $y + \dfrac{2}{y}$.

$$u^2 + 3u - 4 = 0$$

$$(u + 4)(u - 1) = 0$$

$u = -4$ or $u = 1$

Substitute $y + \dfrac{2}{y}$ for u and solve for y.

$y + \dfrac{2}{y} = -4$ or $y + \dfrac{2}{y} = 1$

$y^2 + 2 = -4y$ or $y^2 + 2 = y$

$y^2 + 4y + 2 = 0$ or $y^2 - y + 2 = 0$

$y = \dfrac{-4 \pm \sqrt{4^2 - 4 \cdot 1 \cdot 2}}{2 \cdot 1}$ or

$$y = \frac{-(-1) \pm \sqrt{(-1)^2 - 4 \cdot 1 \cdot 2}}{2 \cdot 1}$$

$y = \dfrac{-4 \pm \sqrt{8}}{2}$ or $y = \dfrac{1 \pm \sqrt{-7}}{2}$

$y = \dfrac{-4 \pm 2\sqrt{2}}{2}$ or $y = \dfrac{1 \pm \sqrt{7}i}{2}$

$y = -2 \pm \sqrt{2}$ or $y = \dfrac{1}{2} \pm \dfrac{\sqrt{7}}{2}i$

The solutions are $-2 \pm \sqrt{2}$ and $\dfrac{1}{2} \pm \dfrac{\sqrt{7}}{2}i$.

Exercise Set 3.3

1. a) The minimum function value occurs at the vertex, so the vertex is $\left(-\dfrac{1}{2}, -\dfrac{9}{4}\right)$.

b) The axis of symmetry is a vertical line through the vertex. It is $x = -\dfrac{1}{2}$.

c) The minimum value of the function is $-\dfrac{9}{4}$.

3. $f(x) = x^2 - 8x + 12$ 16 completes the square for $x^2 - 8x$.

$= x^2 - 8x + 16 - 16 + 12$ Adding $16 - 16$ on the right side

$= (x^2 - 8x + 16) - 16 + 12$

$= (x - 4)^2 - 4$ Factoring and simplifying

$= (x - 4)^2 + (-4)$ Writing in the form $f(x) = a(x - h)^2 + k$

a) Vertex: $(4, -4)$

b) Axis of symmetry: $x = 4$

c) Minimum value: -4

d) We plot the vertex and find several points on either side of it. Then we plot these points and connect them with a smooth curve.

x	$f(x)$
4	-4
2	0
1	5
5	-3
6	0

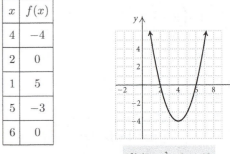

$f(x) = x^2 - 8x + 12$

5. $f(x) = x^2 - 7x + 12$ $\dfrac{49}{4}$ completes the square for $x^2 - 7x$.

$= x^2 - 7x + \dfrac{49}{4} - \dfrac{49}{4} + 12$ Adding

$\dfrac{49}{4} - \dfrac{49}{4}$ on the right side

$= \left(x^2 - 7x + \dfrac{49}{4}\right) - \dfrac{49}{4} + 12$

$= \left(x - \dfrac{7}{2}\right)^2 - \dfrac{1}{4}$ Factoring and simplifying

$= \left(x - \dfrac{7}{2}\right)^2 + \left(-\dfrac{1}{4}\right)$ Writing in the form $f(x) = a(x - h)^2 + k$

a) Vertex: $\left(\dfrac{7}{2}, -\dfrac{1}{4}\right)$

b) Axis of symmetry: $x = \dfrac{7}{2}$

c) Minimum value: $-\dfrac{1}{4}$

d) We plot the vertex and find several points on either side of it. Then we plot these points and connect them with a smooth curve.

x	$f(x)$
$\dfrac{7}{2}$	$-\dfrac{1}{4}$
4	0
5	2
3	0
1	6

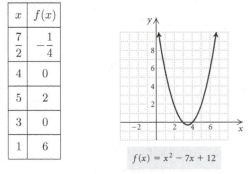

$f(x) = x^2 - 7x + 12$

7. $f(x) = x^2 + 4x + 5$ 4 completes the square for $x^2 + 4x$

$= x^2 + 4x + 4 - 4 + 5$ Adding $4 - 4$ on the right side

$= (x + 2)^2 + 1$ Factoring and simplifying

$= [x - (-2)]^2 + 1$ Writing in the form $f(x) = a(x - h)^2 + k$

a) Vertex: $(-2, 1)$

b) Axis of symmetry: $x = -2$

c) Minimum value: 1

d) We plot the vertex and find several points on either side of it. Then we plot these points and connect them with a smooth curve.

x	$f(x)$
-2	1
-1	2
0	5
-3	2
-4	5

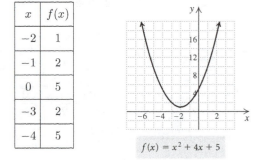

$f(x) = x^2 + 4x + 5$

9. $g(x) = \dfrac{x^2}{2} + 4x + 6$

$= \dfrac{1}{2}(x^2 + 8x) + 6$ Factoring $\dfrac{1}{2}$ out of the first two terms

$= \dfrac{1}{2}(x^2 + 8x + 16 - 16) + 6$ Adding $16 - 16$ inside the parentheses

$= \dfrac{1}{2}(x^2 + 8x + 16) - \dfrac{1}{2} \cdot 16 + 6$ Removing -16 from within the parentheses

$= \dfrac{1}{2}(x + 4)^2 - 2$ Factoring and simplifying

$= \dfrac{1}{2}[x - (-4)]^2 + (-2)$

a) Vertex: $(-4, -2)$

b) Axis of symmetry: $x = -4$

c) Minimum value: -2

d) We plot the vertex and find several points on either side of it. Then we plot these points and connect them with a smooth curve.

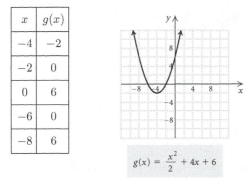

x	$g(x)$
-4	-2
-2	0
0	6
-6	0
-8	6

$g(x) = \dfrac{x^2}{2} + 4x + 6$

11. $g(x) = 2x^2 + 6x + 8$

$\qquad = 2(x^2 + 3x) + 8 \qquad$ Factoring 2 out of the first two terms

$\qquad = 2\left(x^2 + 3x + \dfrac{9}{4} - \dfrac{9}{4}\right) + 8 \qquad$ Adding $\dfrac{9}{4} - \dfrac{9}{4}$ inside the parentheses

$\qquad = 2\left(x^2 + 3x + \dfrac{9}{4}\right) - 2 \cdot \dfrac{9}{4} + 8$ Removing $-\dfrac{9}{4}$ from within the parentheses

$\qquad = 2\left(x + \dfrac{3}{2}\right)^2 + \dfrac{7}{2} \qquad$ Factoring and simplifying

$\qquad = 2\left[x - \left(-\dfrac{3}{2}\right)\right]^2 + \dfrac{7}{2}$

a) Vertex: $\left(-\dfrac{3}{2}, \dfrac{7}{2}\right)$

b) Axis of symmetry: $x = -\dfrac{3}{2}$

c) Minimum value: $\dfrac{7}{2}$

d) We plot the vertex and find several points on either side of it. Then we plot these points and connect them with a smooth curve.

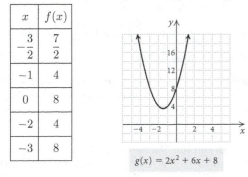

x	$f(x)$
$-\dfrac{3}{2}$	$\dfrac{7}{2}$
-1	4
0	8
-2	4
-3	8

$g(x) = 2x^2 + 6x + 8$

13. $f(x) = -x^2 - 6x + 3$

$\qquad = -(x^2 + 6x) + 3 \qquad$ 9 completes the square for $x^2 + 6x$.

$\qquad = -(x^2 + 6x + 9 - 9) + 3$

$\qquad = -(x + 3)^2 - (-9) + 3 \qquad$ Removing -9 from the parentheses

$\qquad = -(x + 3)^2 + 9 + 3$

$\qquad = -[x - (-3)]^2 + 12$

a) Vertex: $(-3, 12)$

b) Axis of symmetry: $x = -3$

c) Maximum value: 12

d) We plot the vertex and find several points on either side of it. Then we plot these points and connect them with a smooth curve.

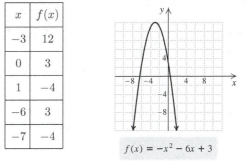

x	$f(x)$
-3	12
0	3
1	-4
-6	3
-7	-4

$f(x) = -x^2 - 6x + 3$

15. $g(x) = -2x^2 + 2x + 1$

$\qquad = -2(x^2 - x) + 1 \qquad$ Factoring -2 out of the first two terms

$\qquad = -2\left(x^2 - x + \dfrac{1}{4} - \dfrac{1}{4}\right) + 1 \qquad$ Adding $\dfrac{1}{4} - \dfrac{1}{4}$ inside the parentheses

$\qquad = -2\left(x^2 - x + \dfrac{1}{4}\right) - 2\left(-\dfrac{1}{4}\right) + 1$

Removing $-\dfrac{1}{4}$ from within the parentheses

$\qquad = -2\left(x - \dfrac{1}{2}\right)^2 + \dfrac{3}{2}$

a) Vertex: $\left(\dfrac{1}{2}, \dfrac{3}{2}\right)$

b) Axis of symmetry: $x = \dfrac{1}{2}$

c) Maximum value: $\dfrac{3}{2}$

d) We plot the vertex and find several points on either side of it. Then we plot these points and connect them with a smooth curve.

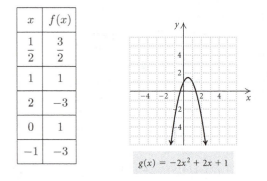

x	$f(x)$
$\frac{1}{2}$	$\frac{3}{2}$
1	1
2	-3
0	1
-1	-3

$g(x) = -2x^2 + 2x + 1$

17. The graph of $y = (x+3)^2$ has vertex $(-3, 0)$ and opens up. It is graph (f).

19. The graph of $y = 2(x-4)^2 - 1$ has vertex $(4, -1)$ and opens up. It is graph (b).

21. The graph of $y = -\frac{1}{2}(x+3)^2 + 4$ has vertex $(-3, 4)$ and opens down. It is graph (h).

23. The graph of $y = -(x+3)^2 + 4$ has vertex $(-3, 4)$ and opens down. It is graph (c).

25. The function $f(x) = -3x^2 + 2x + 5$ is of the form $f(x) = ax^2 + bx + c$ with $a < 0$, so it is true that it has a maximum value.

27. The statement is false. The graph of $h(x) = (x+2)^2$ can be obtained by translating the graph of $h(x) = x^2$ two units to the *left*.

29. The function $f(x) = -(x+2)^2 - 4$ can be written as $f(x) = -[x-(-2)]^2 - 4$, so it is true that the axis of symmetry is $x = -2$.

31. $f(x) = x^2 - 6x + 5$

a) The x-coordinate of the vertex is
$$-\frac{b}{2a} = -\frac{-6}{2 \cdot 1} = 3.$$
Since $f(3) = 3^2 - 6 \cdot 3 + 5 = -4$, the vertex is $(3, -4)$.

b) Since $a = 1 > 0$, the graph opens up so the second coordinate of the vertex, -4, is the minimum value of the function.

c) The range is $[-4, \infty)$.

d) Since the graph opens up, function values decrease to the left of the vertex and increase to the right of the vertex. Thus, $f(x)$ is increasing on $(3, \infty)$ and decreasing on $(-\infty, 3)$.

33. $f(x) = 2x^2 + 4x - 16$

a) The x-coordinate of the vertex is
$$-\frac{b}{2a} = -\frac{4}{2 \cdot 2} = -1.$$
Since $f(-1) = 2(-1)^2 + 4(-1) - 16 = -18$, the vertex is $(-1, -18)$.

b) Since $a = 2 > 0$, the graph opens up so the second coordinate of the vertex, -18, is the minimum value of the function.

c) The range is $[-18, \infty)$.

d) Since the graph opens up, function values decrease to the left of the vertex and increase to the right of the vertex. Thus, $f(x)$ is increasing on $(-1, \infty)$ and decreasing on $(-\infty, -1)$.

35. $f(x) = -\frac{1}{2}x^2 + 5x - 8$

a) The x-coordinate of the vertex is
$$-\frac{b}{2a} = -\frac{5}{2\left(-\frac{1}{2}\right)} = 5.$$
Since $f(5) = -\frac{1}{2} \cdot 5^2 + 5 \cdot 5 - 8 = \frac{9}{2}$, the vertex is $\left(5, \frac{9}{2}\right)$.

b) Since $a = -\frac{1}{2} < 0$, the graph opens down so the second coordinate of the vertex, $\frac{9}{2}$, is the maximum value of the function.

c) The range is $\left(-\infty, \frac{9}{2}\right]$.

d) Since the graph opens down, function values increase to the left of the vertex and decrease to the right of the vertex. Thus, $f(x)$ is increasing on $(-\infty, 5)$ and decreasing on $(5, \infty)$.

37. $f(x) = 3x^2 + 6x + 5$

a) The x-coordinate of the vertex is
$$-\frac{b}{2a} = -\frac{6}{2 \cdot 3} = -1.$$
Since $f(-1) = 3(-1)^2 + 6(-1) + 5 = 2$, the vertex is $(-1, 2)$.

b) Since $a = 3 > 0$, the graph opens up so the second coordinate of the vertex, 2, is the minimum value of the function.

c) The range is $[2, \infty)$.

d) Since the graph opens up, function values decrease to the left of the vertex and increase to the right of the vertex. Thus, $f(x)$ is increasing on $(-1, \infty)$ and decreasing on $(-\infty, -1)$.

39. $g(x) = -4x^2 - 12x + 9$

a) The x-coordinate of the vertex is
$$-\frac{b}{2a} = -\frac{-12}{2(-4)} = -\frac{3}{2}.$$
Since $g\left(-\frac{3}{2}\right) = -4\left(-\frac{3}{2}\right)^2 - 12\left(-\frac{3}{2}\right) + 9 = 18$, the vertex is $\left(-\frac{3}{2}, 18\right)$.

b) Since $a = -4 < 0$, the graph opens down so the second coordinate of the vertex, 18, is the maximum value of the function.

c) The range is $(-\infty, 18]$.

d) Since the graph opens down, function values increase to the left of the vertex and decrease to the right of the vertex. Thus, $g(x)$ is increasing on $\left(-\infty, -\frac{3}{2}\right)$ and decreasing on $\left(-\frac{3}{2}, \infty\right)$.

41. Familiarize and Translate. The function
$s(t) = -16t^2 + 20t + 6$ is given in the statement of the problem.

Carry out. The function $s(t)$ is quadratic and the coefficient of t^2 is negative, so $s(t)$ has a maximum value. It occurs at the vertex of the graph of the function. We find the first coordinate of the vertex. This is the time at which the ball reaches its maximum height.
$$t = -\frac{b}{2a} = -\frac{20}{2(-16)} = 0.625$$
The second coordinate of the vertex gives the maximum height.
$$s(0.625) = -16(0.625)^2 + 20(0.625) + 6 = 12.25$$

Check. Completing the square, we write the function in the form $s(t) = -16(t - 0.625)^2 + 12.25$. We see that the coordinates of the vertex are $(0.625, 12.25)$, so the answer checks.

State. The ball reaches its maximum height after 0.625 seconds. The maximum height is 12.25 ft.

43. Familiarize and Translate. The function
$s(t) = -16t^2 + 120t + 80$ is given in the statement of the problem.

Carry out. The function $s(t)$ is quadratic and the coefficient of t^2 is negative, so $s(t)$ has a maximum value. It occurs at the vertex of the graph of the function. We find the first coordinate of the vertex. This is the time at which the rocket reaches its maximum height.
$$t = -\frac{b}{2a} = -\frac{120}{2(-16)} = 3.75$$
The second coordinate of the vertex gives the maximum height.
$$s(3.75) = -16(3.75)^2 + 120(3.75) + 80 = 305$$

Check. Completing the square, we write the function in the form $s(t) = -16(t - 3.75)^2 + 305$. We see that the coordinates of the vertex are $(3.75, 305)$, so the answer checks.

State. The rocket reaches its maximum height after 3.75 seconds. The maximum height is 305 ft.

45. Familiarize. Using the label in the text, we let $x =$ the height of the file. Then the length $= 10$ and the width $= 18 - 2x$.

Translate. Since the volume of a rectangular solid is length × width × height we have
$$V(x) = 10(18 - 2x)x, \text{ or } -20x^2 + 180x.$$

Carry out. Since $V(x)$ is a quadratic function with $a = -20 < 0$, the maximum function value occurs at the vertex of the graph of the function. The first coordinate of the vertex is
$$-\frac{b}{2a} = -\frac{180}{2(-20)} = 4.5.$$

Check. When $x = 4.5$, then $18 - 2x = 9$ and $V(x) = 10 \cdot 9(4.5)$, or 405. As a partial check, we can find $V(x)$ for a value of x less than 4.5 and for a value of x greater than 4.5. For instance, $V(4.4) = 404.8$ and $V(4.6) = 404.8$.

Since both of these values are less than 405, our result appears to be correct.

State. The file should be 4.5 in. tall in order to maximize the volume.

47. Familiarize. Let $b =$ the length of the base of the triangle. Then the height $= 20 - b$.

Translate. Since the area of a triangle is $\frac{1}{2} \times$ base × height, we have
$$A(b) = \frac{1}{2}b(20 - b), \text{ or } -\frac{1}{2}b^2 + 10b.$$

Carry out. Since $A(b)$ is a quadratic function with $a = -\frac{1}{2} < 0$, the maximum function value occurs at the vertex of the graph of the function. The first coordinate of the vertex is
$$-\frac{b}{2a} = -\frac{10}{2\left(-\frac{1}{2}\right)} = 10.$$
When $b = 10$, then $20 - b = 20 - 10 = 10$, and the area is $\frac{1}{2} \cdot 10 \cdot 10 = 50 \text{ cm}^2$.

Check. As a partial check, we can find $A(b)$ for a value of b less than 10 and for a value of b greater than 10. For instance, $V(9.9) = 49.995$ and $V(10.1) = 49.995$. Since both of these values are less than 50, our result appears to be correct.

State. The area is a maximum when the base and the height are both 10 cm.

49. $C(x) = 0.1x^2 - 0.7x + 1.625$

Since $C(x)$ is a quadratic function with $a = 0.1 > 0$, a minimum function value occurs at the vertex of the graph of $C(x)$. The first coordinate of the vertex is
$$-\frac{b}{2a} = -\frac{-0.7}{2(0.1)} = 3.5.$$
Thus, 3.5 hundred, or 350 doghouses should be built to minimize the average cost per doghouse.

51. $P(x) = R(x) - C(x)$
$P(x) = (50x - 0.5x^2) - (10x + 3)$
$P(x) = -0.5x^2 + 40x - 3$

Since $P(x)$ is a quadratic function with $a = -0.5 < 0$, a maximum function value occurs at the vertex of the graph of the function. The first coordinate of the vertex is
$$-\frac{b}{2a} = -\frac{40}{2(-0.5)} = 40.$$
$$P(40) = -0.5(40)^2 + 40 \cdot 40 - 3 = 797$$
Thus, the maximum profit is \$797. It occurs when 40 units are sold.

53. Familiarize. Using the labels on the drawing in the text, we let $x =$ the width of each field and $240 - 3x =$ the total length of the fields.

Translate. Since the area of a rectangle is length × width, we have
$$A(x) = (240 - 3x)x = -3x^2 + 240x.$$

Carry out. Since $A(x)$ is a quadratic function with $a = -3 < 0$, the maximum function value occurs at the vertex of the graph of $A(x)$. The first coordinate of the vertex is

$$-\frac{b}{2a} = -\frac{240}{2(-3)} = 40.$$

$$A(40) = -3(40)^2 + 240(40) = 4800$$

Check. As a partial check we can find $A(x)$ for a value of x less than 40 and for a value of x greater than 40. For instance, $A(39.9) = 4799.97$ and $A(40.1) = 4799.97$. Since both of these values are less than 4800, our result appears to be correct.

State. The largest total area that can be enclosed is 4800 yd².

55. ***Familiarize.*** We let s = the depth of the well, t_1 = the time it takes the chlorine tablet to reach the bottom of the well, and t_2 = the time it takes the sound to reach the top of the well.

Translate. We know that $t_1 + t_2 = 2$. Using the information in Example 7 we also know that

$$s = 16t_1^2, \quad or \quad t_1 = \frac{\sqrt{s}}{4} \text{ and}$$

$$s = 1100t_2, \quad or \quad t_2 = \frac{s}{1100}.$$

Then $\dfrac{\sqrt{s}}{4} + \dfrac{s}{1100} = 2.$

Carry out. We solve the last equation above.

$$\frac{\sqrt{s}}{4} + \frac{s}{1100} = 2$$

$$275\sqrt{s} + s = 2200 \quad \text{Multiplying by 1100}$$

$$s + 275\sqrt{s} - 2200 = 0$$

Let $u = \sqrt{s}$ and substitute.

$$u^2 + 275u - 2200 = 0$$

$$u = \frac{-b + \sqrt{b^2 - 4ac}}{2a} \qquad \begin{array}{l}\text{We only want the}\\ \text{positive solution.}\end{array}$$

$$= \frac{-275 + \sqrt{275^2 - 4 \cdot 1 (-2200)}}{2 \cdot 1}$$

$$= \frac{-275 + \sqrt{84,425}}{2} \approx 7.78$$

Since $u \approx 7.78$, we have $\sqrt{s} = 7.78$, so $s \approx 60.5$.

Check. If $s \approx 60.5$, then $t_1 = \dfrac{\sqrt{s}}{4} = \dfrac{\sqrt{60.5}}{4} \approx$ 1.94 and $t_2 = \dfrac{s}{1100} = \dfrac{60.5}{1100} \approx 0.06$, so $t_1 + t_2 = 1.94 + 0.06 = 2$. The result checks.

State. The well is about 60.5 ft deep.

57. $f(x) = 3x - 7$

$$\frac{f(x+h) - f(x)}{h} = \frac{3(x+h) - 7 - (3x - 7)}{h}$$

$$= \frac{3x + 3h - 7 - 3x + 7}{h}$$

$$= \frac{3h}{h} = 3$$

59. The graph of $f(x)$ is stretched vertically and reflected across the x-axis.

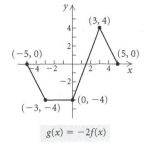

$$g(x) = -2f(x)$$

61. $f(x) = -0.2x^2 - 3x + c$

The x-coordinate of the vertex of $f(x)$ is $-\dfrac{b}{2a} = -\dfrac{-3}{2(-0.2)} = -7.5$. Now we find c such that $f(-7.5) = -225$.

$$-0.2(-7.5)^2 - 3(-7.5) + c = -225$$

$$-11.25 + 22.5 + c = -225$$

$$c = -236.25$$

63.

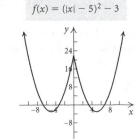

$$f(x) = (|x| - 5)^2 - 3$$

65. First we find the radius r of a circle with circumference x:

$$2\pi r = x$$

$$r = \frac{x}{2\pi}$$

Then we find the length s of a side of a square with perimeter $24 - x$:

$$4s = 24 - x$$

$$s = \frac{24 - x}{4}$$

Then S = area of circle + area of square

$$S = \pi r^2 + s^2$$

$$S(x) = \pi \left(\frac{x}{2\pi}\right)^2 + \left(\frac{24 - x}{4}\right)^2$$

$$S(x) = \left(\frac{1}{4\pi} + \frac{1}{16}\right)x^2 - 3x + 36$$

Since $S(x)$ is a quadratic function with $a = \dfrac{1}{4\pi} + \dfrac{1}{16} > 0$, the minimum function value occurs at the vertex of the graph of $S(x)$. The first coordinate of the vertex is

$$-\frac{b}{2a} = -\frac{-3}{2\left(\dfrac{1}{4\pi} + \dfrac{1}{16}\right)} = \frac{24\pi}{4 + \pi}.$$

Then the string should be cut so that one piece is $\dfrac{24\pi}{4+\pi}$ in., or about 10.56 in. The other piece will be $24 - \dfrac{24\pi}{4+\pi}$, or $\dfrac{96}{4+\pi}$ in., or about 13.44 in.

Chapter 3 Mid-Chapter Mixed Review

1. The statement is true. See page 171 in the text.

3. The statement is true. See page 181 in the text.

5. $\sqrt{-36} = \sqrt{-1 \cdot 36} = \sqrt{-1} \cdot \sqrt{36} = i \cdot 6 = 6i$

7. $-\sqrt{-16} = -\sqrt{-1 \cdot 16} = -\sqrt{-1} \cdot \sqrt{16} = -i \cdot 4 = -4i$

9. $(3 - 2i) + (-4 + 3i) = (3 - 4) + (-2i + 3i) = -1 + i$

11. $(2 + 3i)(4 - 5i) = 8 - 10i + 12i - 15i^2$
$$= 8 + 2i + 15$$
$$= 23 + 2i$$

13. $i^{13} = i^{12} \cdot i = (i^2)^6 \cdot i = (-1)^6 \cdot i = i$

15. $(-i)^5 = (-1 \cdot i)^5 = (-1)^5 i^5 = -i^4 \cdot i = -(i^2)^2 \cdot i =$
$-(-1)^2 \cdot i = -i$

17. $x^2 + 3x - 4 = 0$
$$(x + 4)(x - 1) = 0$$
$$x + 4 = 0 \quad or \quad x - 1 = 0$$
$$x = -4 \quad or \quad x = 1$$
The solutions are -4 and 1.

19. $4x^2 = 24$
$$x^2 = 6$$
$$x = \sqrt{6} \ or \ x = -\sqrt{6}$$
The solutions are $\sqrt{6}$ and $-\sqrt{6}$, or $\pm\sqrt{6}$.

21. $4x^2 - 8x - 3 = 0$
$$4x^2 - 8x = 3$$
$$x^2 - 2x = \frac{3}{4}$$
$$x^2 - 2x + 1 = \frac{3}{4} + 1 \quad \text{Completing the square:}$$
$$\frac{1}{2}(-2) = -1 \text{ and } (-1)^2 = 1$$
$$(x - 1)^2 = \frac{7}{4}$$
$$x - 1 = \pm\frac{\sqrt{7}}{2}$$
$$x = 1 + \frac{\sqrt{7}}{2}$$
$$x = \frac{2 \pm \sqrt{7}}{2}$$
The zeros are $\dfrac{2 + \sqrt{7}}{2}$ and $\dfrac{2 - \sqrt{7}}{4}$, or $\dfrac{2 \pm \sqrt{7}}{2}$.

23. $4x^2 - 12x + 9 = 0$
   a) $b^2 - 4ac = (-12)^2 - 4 \cdot 4 \cdot 9 = 144 - 144 = 0$
   There is one real-number solution.
   b) $4x^2 - 12x + 9 = 0$
$$(2x - 3)^2 = 0$$
$$2x - 3 = 0$$
$$2x = 3$$
$$x = \frac{3}{2}$$
   The solution is $\dfrac{3}{2}$.

25. $x^4 + 5x^2 - 6 = 0$
   Let $u = x^2$.
$$u^2 + 5u - 6 = 0 \quad \text{Substituting}$$
$$(u + 6)(u - 1) = 0$$
$$u + 6 = 0 \qquad or \quad u - 1 = 0$$
$$u = -6 \quad or \qquad u = 1$$
$$x^2 = -6 \quad or \qquad x^2 = 1$$
$$x = \pm\sqrt{6}i \ or \qquad x = \pm 1$$
The solutions are $\pm\sqrt{6}i$ and ± 1.

27. **Familiarize.** Let $x =$ the smaller number. Then $x + 2 =$ the larger number.

Translate.

$$\underbrace{\text{The product of the numbers}}_{\downarrow} \ \text{is} \ 35.$$
$$x(x + 2) \qquad\qquad = \ 35$$

Carry out.
$$x(x + 2) = 35$$
$$x^2 + 2x = 35$$
$$x^2 + 2x - 35 = 0$$
$$(x + 7)(x - 5) = 0$$
$$x + 7 = 0 \quad or \quad x - 5 = 0$$
$$x = -7 \quad or \qquad x = 5$$
If $x = -7$, then $x + 2 = -7 + 2 = -5$; if $x = 5$, then $x + 2 = 5 + 2 = 7$.

Check. -5 is 2 more than -7, and $(-7)(-5) = 35$. Also, 7 is 2 more than 5, and $5 \cdot 7 = 35$. The numbers check.

State. The numbers are 5 and 7 or -7 and -5.

29. $f(x) = -2x^2 - 4x - 5$
$$= -2(x^2 + 2x) - 5$$
$$= -2(x^2 + 2x + 1 - 1) - 5$$
$$= -2(x^2 + 2x + 1) - 2(-1) - 5$$
$$= -2(x + 1)^2 - 3$$
$$= -2[x - (-1)]^2 + (-3)$$
   a) Vertex: $(-1, -3)$
   b) Axis of symmetry: $x = -1$
   c) Maximum value: -3

d) Range: $(-\infty, -3]$

e) Increasing: $(-\infty, -3)$; decreasing: $(-3, \infty)$

f)

x	$f(x)$
-1	-3
-3	-11
-2	-5
0	-5
1	-11

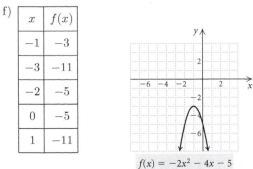

$f(x) = -2x^2 - 4x - 5$

31. The sum of two imaginary numbers is not always an imaginary number. For example, $(2 + i) + (3 - i) = 5$, a real number.

33. Completing the square was used in Section 3.2 to solve quadratic equations. It was used again in this section to write quadratic functions in the form $f(x) = a(x - h)^2 + k$.

Exercise Set 3.4

1. $\qquad \dfrac{1}{4} + \dfrac{1}{5} = \dfrac{1}{t}$, LCD is $20t$

$$20t\left(\dfrac{1}{4} + \dfrac{1}{5}\right) = 20t \cdot \dfrac{1}{t}$$

$$20t \cdot \dfrac{1}{4} + 20t \cdot \dfrac{1}{5} = 20t \cdot \dfrac{1}{t}$$

$$5t + 4t = 20$$

$$9t = 20$$

$$t = \dfrac{20}{9}$$

Check:

$$\dfrac{\dfrac{1}{4} + \dfrac{1}{5} = \dfrac{1}{t}}{}$$

$$\dfrac{1}{4} + \dfrac{1}{5} \ ? \ \dfrac{1}{\frac{20}{9}}$$

$$\dfrac{5}{20} + \dfrac{4}{20} \ \bigg| \ 1 \cdot \dfrac{9}{20}$$

$$\dfrac{9}{20} \ \bigg| \ \dfrac{9}{20} \qquad \text{TRUE}$$

The solution is $\dfrac{20}{9}$.

3. $\qquad \dfrac{x + 2}{4} - \dfrac{x - 1}{5} = 15$, LCD is 20

$$20\left(\dfrac{x + 2}{4} - \dfrac{x - 1}{5}\right) = 20 \cdot 15$$

$$5(x + 2) - 4(x - 1) = 300$$

$$5x + 10 - 4x + 4 = 300$$

$$x + 14 = 300$$

$$x = 286$$

The solution is 286.

5. $\qquad \dfrac{1}{2} + \dfrac{2}{x} = \dfrac{1}{3} + \dfrac{3}{x}$, LCD is $6x$

$$6x\left(\dfrac{1}{2} + \dfrac{2}{x}\right) = 6x\left(\dfrac{1}{3} + \dfrac{3}{x}\right)$$

$$3x + 12 = 2x + 18$$

$$3x - 2x = 18 - 12$$

$$x = 6$$

Check:

$$\dfrac{\dfrac{1}{2} + \dfrac{2}{x} = \dfrac{1}{3} + \dfrac{3}{x}}{}$$

$$\dfrac{1}{2} + \dfrac{2}{6} \ ? \ \dfrac{1}{3} + \dfrac{3}{6}$$

$$\dfrac{1}{2} + \dfrac{1}{3} \ \bigg| \ \dfrac{1}{3} + \dfrac{1}{2} \qquad \text{TRUE}$$

The solution is 6.

7. $\qquad \dfrac{5}{3x + 2} = \dfrac{3}{2x}$, LCD is $2x(3x + 2)$

$$2x(3x + 2) \cdot \dfrac{5}{3x + 2} = 2x(3x + 2) \cdot \dfrac{3}{2x}$$

$$2x \cdot 5 = 3(3x + 2)$$

$$10x = 9x + 6$$

$$x = 6$$

6 checks, so the solution is 6.

9. $\qquad \dfrac{y^2}{y + 4} = \dfrac{16}{y + 4}$, LCD is $y + 4$

$$(y + 4) \cdot \dfrac{y^2}{y + 4} = (y + 4) \cdot \dfrac{16}{y + 4}$$

$$y^2 = 16$$

$$y = 4 \ \ or \ \ y = -4$$

Only 4 checks. It is the solution.

11. $\qquad x + \dfrac{6}{x} = 5$, LCD is x

$$x\left(x + \dfrac{6}{x}\right) = x \cdot 5$$

$$x^2 + 6 = 5x$$

$$x^2 - 5x + 6 = 0$$

$$(x - 2)(x - 3) = 0$$

$$x - 2 = 0 \ \ or \ \ x - 3 = 0$$

$$x = 2 \ \ or \qquad x = 3$$

Both numbers check. The solutions are 2 and 3.

13.
$$\frac{6}{y+3} + \frac{2}{y} = \frac{5y-3}{y^2-9}$$

$$\frac{6}{y+3} + \frac{2}{y} = \frac{5y-3}{(y+3)(y-3)},$$
$$\text{LCD is } y(y+3)(y-3)$$

$$y(y+3)(y-3)\left(\frac{6}{y+3} + \frac{2}{y}\right) = y(y+3)(y-3)\cdot\frac{5y-3}{(y+3)(y-3)}$$

$$6y(y-3) + 2(y+3)(y-3) = y(5y-3)$$
$$6y^2 - 18y + 2(y^2-9) = 5y^2 - 3y$$
$$6y^2 - 18y + 2y^2 - 18 = 5y^2 - 3y$$
$$8y^2 - 18y - 18 = 5y^2 - 3y$$
$$3y^2 - 15y - 18 = 0$$
$$y^2 - 5y - 6 = 0$$
$$(y-6)(y+1) = 0$$
$$y - 6 = 0 \quad or \quad y + 1 = 0$$
$$y = 6 \quad or \qquad y = -1$$

Both numbers check. The solutions are 6 and -1.

15.
$$\frac{2x}{x-1} = \frac{5}{x-3}, \text{ LCD is } (x-1)(x-3)$$

$$(x-1)(x-3)\cdot\frac{2x}{x-1} = (x-1)(x-3)\cdot\frac{5}{x-3}$$
$$2x(x-3) = 5(x-1)$$
$$2x^2 - 6x = 5x - 5$$
$$2x^2 - 11x + 5 = 0$$
$$(2x-1)(x-5) = 0$$
$$2x - 1 = 0 \quad or \quad x - 5 = 0$$
$$2x = 1 \quad or \qquad x = 5$$
$$x = \frac{1}{2} \quad or \qquad x = 5$$

Both numbers check. The solutions are $\frac{1}{2}$ and 5.

17.
$$\frac{2}{x+5} + \frac{1}{x-5} = \frac{16}{x^2-25}$$

$$\frac{2}{x+5} + \frac{1}{x-5} = \frac{16}{(x+5)(x-5)},$$
$$\text{LCD is } (x+5)(x-5)$$

$$(x+5)(x-5)\left(\frac{2}{x+5} + \frac{1}{x-5}\right) = (x+5)(x-5)\cdot\frac{16}{(x+5)(x-5)}$$
$$2(x-5) + x + 5 = 16$$
$$2x - 10 + x + 5 = 16$$
$$3x - 5 = 16$$
$$3x = 21$$
$$x = 7$$

7 checks, so the solution is 7.

19.
$$\frac{3x}{x+2} + \frac{6}{x} = \frac{12}{x^2+2x}$$

$$\frac{3x}{x+2} + \frac{6}{x} = \frac{12}{x(x+2)}, \text{ LCD is } x(x+2)$$

$$x(x+2)\left(\frac{3x}{x+2} + \frac{6}{x}\right) = x(x+2)\cdot\frac{12}{x(x+2)}$$
$$3x\cdot x + 6(x+2) = 12$$
$$3x^2 + 6x + 12 = 12$$
$$3x^2 + 6x = 0$$
$$3x(x+2) = 0$$
$$3x = 0 \quad or \quad x + 2 = 0$$
$$x = 0 \quad or \qquad x = -2$$

Neither 0 nor -2 checks, so the equation has no solution.

21.
$$\frac{1}{5x+20} - \frac{1}{x^2-16} = \frac{3}{x-4}$$

$$\frac{1}{5(x+4)} - \frac{1}{(x+4)(x-4)} = \frac{3}{x-4},$$
$$\text{LCD is } 5(x+4)(x-4)$$

$$5(x+4)(x-4)\left(\frac{1}{5(x+4)} - \frac{1}{(x+4)(x-4)}\right) = 5(x+4)(x-4)\cdot\frac{3}{x-4}$$
$$x - 4 - 5 = 15(x+4)$$
$$x - 9 = 15x + 60$$
$$-14x - 9 = 60$$
$$-14x = 69$$
$$x = -\frac{69}{14}$$

$-\frac{69}{14}$ checks, so the solution is $-\frac{69}{14}$.

23.
$$\frac{2}{5x+5} - \frac{3}{x^2-1} = \frac{4}{x-1}$$

$$\frac{2}{5(x+1)} - \frac{3}{(x+1)(x-1)} = \frac{4}{x-1},$$
$$\text{LCD is } 5(x+1)(x-1)$$

$$5(x+1)(x-1)\left(\frac{2}{5(x+1)} - \frac{3}{(x+1)(x-1)}\right) = 5(x+1)(x-1)\cdot\frac{4}{x-1}$$
$$2(x-1) - 5\cdot 3 = 20(x+1)$$
$$2x - 2 - 15 = 20x + 20$$
$$2x - 17 = 20x + 20$$
$$-18x - 17 = 20$$
$$-18x = 37$$
$$x = -\frac{37}{18}$$

$-\frac{37}{18}$ checks, so the solution is $-\frac{37}{18}$.

25.
$$\frac{8}{x^2 - 2x + 4} = \frac{x}{x + 2} + \frac{24}{x^3 + 8},$$
$$\text{LCD is } (x + 2)(x^2 - 2x + 4)$$

$$(x+2)(x^2-2x+4) \cdot \frac{8}{x^2-2x+4} =$$
$$(x+2)(x^2-2x+4)\left(\frac{x}{x+2} + \frac{24}{(x+2)(x^2-2x+4)}\right)$$
$$8(x + 2) = x(x^2-2x+4)+24$$
$$8x + 16 = x^3-2x^2+4x+24$$
$$0 = x^3-2x^2-4x+8$$
$$0 = x^2(x-2) - 4(x-2)$$
$$0 = (x-2)(x^2-4)$$
$$0 = (x-2)(x+2)(x-2)$$

$$x - 2 = 0 \ \text{ or } \ x + 2 = 0 \ \ \text{ or } \ x - 2 = 0$$
$$x = 2 \ \text{ or } \quad\quad x = -2 \ \text{ or } \quad\quad x = 2$$

Only 2 checks. The solution is 2.

27.
$$\frac{x}{x - 4} - \frac{4}{x + 4} = \frac{32}{x^2 - 16}$$
$$\frac{x}{x - 4} - \frac{4}{x + 4} = \frac{32}{(x+4)(x-4)},$$
$$\text{LCD is } (x+4)(x-4)$$
$$(x+4)(x-4)\left(\frac{x}{x-4} - \frac{4}{x+4}\right) = (x+4)(x-4)\cdot\frac{32}{(x+4)(x-4)}$$
$$x(x + 4) - 4(x - 4) = 32$$
$$x^2 + 4x - 4x + 16 = 32$$
$$x^2 + 16 = 32$$
$$x^2 = 16$$
$$x = \pm 4$$

Neither 4 nor -4 checks, so the equation has no solution.

29.
$$\frac{1}{x - 6} - \frac{1}{x} = \frac{6}{x^2 - 6x}$$
$$\frac{1}{x - 6} - \frac{1}{x} = \frac{6}{x(x-6)}, \text{ LCD is } x(x-6)$$
$$x(x-6)\left(\frac{1}{x-6} - \frac{1}{x}\right) = x(x-6)\cdot\frac{6}{x(x-6)}$$
$$x - (x - 6) = 6$$
$$x - x + 6 = 6$$
$$6 = 6$$

We get an equation that is true for all real numbers. Note, however, that when $x = 6$ or $x = 0$, division by 0 occurs in the original equation. Thus, the solution set is $\{x | x \text{ is a real number } and \ x \neq 6 \ and \ x \neq 0\}$, or $(-\infty, 0) \cup (0, 6) \cup (6, \infty)$.

31.
$$\sqrt{3x - 4} = 1$$
$$(\sqrt{3x - 4})^2 = 1^2$$
$$3x - 4 = 1$$
$$3x = 5$$
$$x = \frac{5}{3}$$

Check:
$$\frac{\sqrt{3x - 4} = 1}{}$$
$$\sqrt{3 \cdot \frac{5}{3} - 4} \ ? \ 1$$
$$\sqrt{5 - 4} \ \Big|$$
$$\sqrt{1} \ \Big|$$
$$1 \ \Big| \ 1 \quad \text{TRUE}$$

The solution is $\frac{5}{3}$.

33.
$$\sqrt{2x - 5} = 2$$
$$(\sqrt{2x - 5})^2 = 2^2$$
$$2x - 5 = 4$$
$$2x = 9$$
$$x = \frac{9}{2}$$

Check:
$$\frac{\sqrt{2x - 5} = 2}{}$$
$$\sqrt{2 \cdot \frac{9}{2} - 5} \ ? \ 2$$
$$\sqrt{9 - 5} \ \Big|$$
$$\sqrt{4} \ \Big|$$
$$2 \ \Big| \ 2 \quad \text{TRUE}$$

The solution is $\frac{9}{2}$.

35.
$$\sqrt{7 - x} = 2$$
$$(\sqrt{7 - x})^2 = 2^2$$
$$7 - x = 4$$
$$-x = -3$$
$$x = 3$$

Check:
$$\frac{\sqrt{7 - x} = 2}{}$$
$$\sqrt{7 - 3} \ ? \ 2$$
$$\sqrt{4} \ \Big|$$
$$2 \ \Big| \ 2 \quad \text{TRUE}$$

The solution is 3.

37.
$$\sqrt{1 - 2x} = 3$$
$$(\sqrt{1 - 2x})^2 = 3^2$$
$$1 - 2x = 9$$
$$-2x = 8$$
$$x = -4$$

Check:
$$\frac{\sqrt{1 - 2x} = 3}{}$$
$$\sqrt{1 - 2(-4)} \ ? \ 3$$
$$\sqrt{1 + 8} \ \Big|$$
$$\sqrt{9} \ \Big|$$
$$3 \ \Big| \ 3 \quad \text{TRUE}$$

The solution is -4.

39. $\sqrt[3]{5x-2} = -3$

$(\sqrt[3]{5x-2})^3 = (-3)^3$

$5x - 2 = -27$

$5x = -25$

$x = -5$

Check:

$\sqrt[3]{5x-2} = -3$

$\sqrt[3]{5(-5)-2}$? -3

$\sqrt[3]{-25-2}$

$\sqrt[3]{-27}$

-3 | -3 TRUE

The solution is -5.

41. $\sqrt[4]{x^2-1} = 1$

$(\sqrt[4]{x^2-1})^4 = 1^4$

$x^2 - 1 = 1$

$x^2 = 2$

$x = \pm\sqrt{2}$

Check:

$\sqrt[4]{x^2-1} = 1$

$\sqrt[4]{(\pm\sqrt{2})^2-1}$? 1

$\sqrt[4]{2-1}$

$\sqrt[4]{1}$

1 | 1 TRUE

The solutions are $\pm\sqrt{2}$.

43. $\sqrt{y-1} + 4 = 0$

$\sqrt{y-1} = -4$

The principal square root is never negative. Thus, there is no solution.

If we do not observe the above fact, we can continue and reach the same answer.

$(\sqrt{y-1})^2 = (-4)^2$

$y - 1 = 16$

$y = 17$

Check:

$\sqrt{y-1} + 4 = 0$

$\sqrt{17-1} + 4$? 0

$\sqrt{16} + 4$

$4 + 4$

8 | 0 FALSE

Since 17 does not check, there is no solution.

45. $\sqrt{b+3} - 2 = 1$

$\sqrt{b+3} = 3$

$(\sqrt{b+3})^2 = 3^2$

$b + 3 = 9$

$b = 6$

Check:

$\sqrt{b+3} - 2 = 1$

$\sqrt{6+3} - 2$? 1

$\sqrt{9} - 2$

$3 - 2$

1 | 1 TRUE

The solution is 6.

47. $\sqrt{z+2} + 3 = 4$

$\sqrt{z+2} = 1$

$(\sqrt{z+2})^2 = 1^2$

$z + 2 = 1$

$z = -1$

Check:

$\sqrt{z+2} + 3 = 4$

$\sqrt{-1+2} + 3$? 4

$\sqrt{1} + 3$

$1 + 3$

4 | 4 TRUE

The solution is -1.

49. $\sqrt{2x+1} - 3 = 3$

$\sqrt{2x+1} = 6$

$(\sqrt{2x+1})^2 = 6^2$

$2x + 1 = 36$

$2x = 35$

$x = \dfrac{35}{2}$

Check:

$\sqrt{2x+1} - 3 = 3$

$\sqrt{2 \cdot \dfrac{35}{2} + 1} - 3$? 3

$\sqrt{35+1} - 3$

$\sqrt{36} - 3$

$6 - 3$

3 | 3 TRUE

The solution is $\dfrac{35}{2}$.

51. $\sqrt{2-x} - 4 = 6$

$\sqrt{2-x} = 10$

$(\sqrt{2-x})^2 = 10^2$

$2 - x = 100$

$-x = 98$

$x = -98$

Check:
$$
\begin{array}{c|c}
\sqrt{2-x} - 4 = 6 \\
\hline
\sqrt{2-(-98)} - 4 \ ? \ 6 \\
\sqrt{100} - 4 \\
10 - 4 \\
6 & 6 \quad \text{TRUE}
\end{array}
$$
The solution is -98.

53. $\sqrt[3]{6x+9} + 8 = 5$

$\sqrt[3]{6x+9} = -3$

$(\sqrt[3]{6x+9})^3 = (-3)^3$

$6x + 9 = -27$

$6x = -36$

$x = -6$

Check:
$$
\begin{array}{c|c}
\sqrt[3]{6x+9} + 8 = 5 \\
\hline
\sqrt[3]{6(-6)+9} + 8 \ ? \ 5 \\
\sqrt[3]{-27} + 8 \\
-3 + 8 \\
5 & 5 \quad \text{TRUE}
\end{array}
$$
The solution is -6.

55. $\sqrt{x+4} + 2 = x$

$\sqrt{x+4} = x - 2$

$(\sqrt{x+4})^2 = (x-2)^2$

$x + 4 = x^2 - 4x + 4$

$0 = x^2 - 5x$

$0 = x(x-5)$

$x = 0 \ \text{ or } \ x - 5 = 0$

$x = 0 \ \text{ or } \quad\quad x = 5$

Check:

For 0:
$$
\begin{array}{c|c}
\sqrt{x+4} + 2 = x \\
\hline
\sqrt{0+4} + 2 \ ? \ 0 \\
2 + 2 \\
4 & 0 \quad \text{FALSE}
\end{array}
$$

For 5:
$$
\begin{array}{c|c}
\sqrt{x+4} + 2 = x \\
\hline
\sqrt{5+4} + 2 \ ? \ 5 \\
\sqrt{9} + 2 \\
3 + 2 \\
5 & 5 \quad \text{TRUE}
\end{array}
$$
The number 5 checks but 0 does not. The solution is 5.

57. $\sqrt{x-3} + 5 = x$

$\sqrt{x-3} = x - 5$

$(\sqrt{x-3})^2 = (x-5)^2$

$x - 3 = x^2 - 10x + 25$

$0 = x^2 - 11x + 28$

$0 = (x-4)(x-7)$

$x - 4 = 0 \ \text{ or } \ x - 7 = 0$

$x = 4 \ \text{ or } \quad\quad x = 7$

Check:

For 4:
$$
\begin{array}{c|c}
\sqrt{x-3} + 5 = x \\
\hline
\sqrt{4-3} + 5 \ ? \ 4 \\
\sqrt{1} + 5 \\
1 + 5 \\
6 & 4 \quad \text{FALSE}
\end{array}
$$

For 7:
$$
\begin{array}{c|c}
\sqrt{x-3} + 5 = x \\
\hline
\sqrt{7-3} + 5 \ ? \ 7 \\
\sqrt{4} + 5 \\
2 + 5 \\
7 & 7 \quad \text{TRUE}
\end{array}
$$
The number 7 checks but 4 does not. The solution is 7.

59. $\sqrt{x+7} = x + 1$

$(\sqrt{x+7})^2 = (x+1)^2$

$x + 7 = x^2 + 2x + 1$

$0 = x^2 + x - 6$

$0 = (x+3)(x-2)$

$x + 3 = 0 \ \text{ or } \ x - 2 = 0$

$x = -3 \ \text{ or } \quad\quad x = 2$

Check:

For -3:
$$
\begin{array}{c|c}
\sqrt{x+7} = x + 1 \\
\hline
\sqrt{-3+7} \ ? \ -3 + 1 \\
\sqrt{4} & -2 \\
2 & -2 \quad \text{FALSE}
\end{array}
$$

For 2:
$$
\begin{array}{c|c}
\sqrt{x+7} = x + 1 \\
\hline
\sqrt{2+7} \ ? \ 2 + 1 \\
\sqrt{9} & 3 \\
3 & 3 \quad \text{TRUE}
\end{array}
$$
The number 2 checks but -3 does not. The solution is 2.

61. $\sqrt{3x+3} = x + 1$

$(\sqrt{3x+3})^2 = (x+1)^2$

$3x + 3 = x^2 + 2x + 1$

$0 = x^2 - x - 2$

$0 = (x-2)(x+1)$

$x - 2 = 0 \ \text{ or } \ x + 1 = 0$

$x = 2 \ \text{ or } \quad\quad x = -1$

Check:

For 2:

$$\frac{\sqrt{3x+3} = x+1}{\sqrt{3\cdot 2 + 3} \ ? \ 2+1}$$

$$\begin{array}{c|c} \sqrt{9} & 3 \\ 3 & 3 \end{array} \quad \text{TRUE}$$

For -1:

$$\frac{\sqrt{3x+3} = x+1}{\sqrt{3(-1)+3} \ ? \ -1+1}$$

$$\begin{array}{c|c} \sqrt{0} & 0 \\ 0 & 0 \end{array} \quad \text{TRUE}$$

Both numbers check. The solutions are 2 and -1.

63. $\sqrt{5x+1} = x-1$

$$(\sqrt{5x+1})^2 = (x-1)^2$$
$$5x+1 = x^2 - 2x + 1$$
$$0 = x^2 - 7x$$
$$0 = x(x-7)$$

$x = 0 \ \ or \ \ x - 7 = 0$

$x = 0 \ \ or \qquad x = 7$

Check:

For 0:

$$\frac{\sqrt{5x+1} = x-1}{\sqrt{5\cdot 0 + 1} \ ? \ 0 - 1}$$

$$\begin{array}{c|c} \sqrt{1} & -1 \\ 1 & -1 \end{array} \quad \text{FALSE}$$

For 7:

$$\frac{\sqrt{5x+1} = x-1}{\sqrt{5\cdot 7 + 1} \ ? \ 7 - 1}$$

$$\begin{array}{c|c} \sqrt{36} & 6 \\ 6 & 6 \end{array} \quad \text{TRUE}$$

The number 7 checks but 0 does not. The solution is 7.

65. $\sqrt{x-3} + \sqrt{x+2} = 5$

$$\sqrt{x+2} = 5 - \sqrt{x-3}$$
$$(\sqrt{x+2})^2 = (5 - \sqrt{x-3})^2$$
$$x+2 = 25 - 10\sqrt{x-3} + (x-3)$$
$$x+2 = 22 - 10\sqrt{x-3} + x$$
$$10\sqrt{x-3} = 20$$
$$\sqrt{x-3} = 2$$
$$(\sqrt{x-3})^2 = 2^2$$
$$x-3 = 4$$
$$x = 7$$

Check:

$$\frac{\sqrt{x-3} + \sqrt{x+2} = 5}{\sqrt{7-3} + \sqrt{7+2} \ ? \ 5}$$

$$\begin{array}{c|c} \sqrt{4} + \sqrt{9} & \\ 2 + 3 & \\ 5 & 5 \end{array} \quad \text{TRUE}$$

The solution is 7.

67. $\sqrt{3x-5} + \sqrt{2x+3} + 1 = 0$

$$\sqrt{3x-5} + \sqrt{2x+3} = -1$$

The principal square root is never negative. Thus the sum of two principal square roots cannot equal -1. There is no solution.

69. $\sqrt{x} - \sqrt{3x-3} = 1$

$$\sqrt{x} = \sqrt{3x-3} + 1$$
$$(\sqrt{x})^2 = (\sqrt{3x-3} + 1)^2$$
$$x = (3x-3) + 2\sqrt{3x-3} + 1$$
$$2 - 2x = 2\sqrt{3x-3}$$
$$1 - x = \sqrt{3x-3}$$
$$(1-x)^2 = (\sqrt{3x-3})^2$$
$$1 - 2x + x^2 = 3x - 3$$
$$x^2 - 5x + 4 = 0$$
$$(x-4)(x-1) = 0$$

$x = 4 \ \ or \ \ x = 1$

The number 4 does not check, but 1 does. The solution is 1.

71. $\sqrt{2y-5} - \sqrt{y-3} = 1$

$$\sqrt{2y-5} = \sqrt{y-3} + 1$$
$$(\sqrt{2y-5})^2 = (\sqrt{y-3} + 1)^2$$
$$2y-5 = (y-3) + 2\sqrt{y-3} + 1$$
$$y-3 = 2\sqrt{y-3}$$
$$(y-3)^2 = (2\sqrt{y-3})^2$$
$$y^2 - 6y + 9 = 4(y-3)$$
$$y^2 - 6y + 9 = 4y - 12$$
$$y^2 - 10y + 21 = 0$$
$$(y-7)(y-3) = 0$$

$y = 7 \ \ or \ \ y = 3$

Both numbers check. The solutions are 7 and 3.

73. $\sqrt{y+4} - \sqrt{y-1} = 1$

$$\sqrt{y+4} = \sqrt{y-1} + 1$$
$$(\sqrt{y+4})^2 = (\sqrt{y-1} + 1)^2$$
$$y+4 = y-1 + 2\sqrt{y-1} + 1$$
$$4 = 2\sqrt{y-1}$$
$$2 = \sqrt{y-1} \qquad \text{Dividing by 2}$$
$$2^2 = (\sqrt{y-1})^2$$
$$4 = y-1$$
$$5 = y$$

The answer checks. The solution is 5.

75.
$$\sqrt{x+5} + \sqrt{x+2} = 3$$
$$\sqrt{x+5} = 3 - \sqrt{x+2}$$
$$(\sqrt{x+5})^2 = (3 - \sqrt{x+2})^2$$
$$x+5 = 9 - 6\sqrt{x+2} + x + 2$$
$$-6 = -6\sqrt{x+2}$$
$$1 = \sqrt{x+2} \qquad \text{Dividing by } -6$$
$$1^2 = (\sqrt{x+2})^2$$
$$1 = x+2$$
$$-1 = x$$

The answer checks. The solution is -1.

77.
$$x^{1/3} = -2$$
$$(x^{1/3})^3 = (-2)^3 \qquad (x^{1/3} = \sqrt[3]{x})$$
$$x = -8$$

The value checks. The solution is -8.

79.
$$t^{1/4} = 3$$
$$(t^{1/4})^4 = 3^4 \qquad (t^{1/4} = \sqrt[4]{t})$$
$$t = 81$$

The value checks. The solution is 81.

81.
$$\frac{P_1 V_1}{T_1} = \frac{P_2 V_2}{T_2}$$
$$P_1 V_1 T_2 = P_2 V_2 T_1 \qquad \text{Multiplying by } T_1 T_2 \text{ on both sides}$$
$$\frac{P_1 V_1 T_2}{P_2 V_2} = T_1 \qquad \text{Dividing by } P_2 V_2 \text{ on both sides}$$

83.
$$W = \sqrt{\frac{1}{LC}}$$
$$W^2 = \left(\sqrt{\frac{1}{LC}}\right)^2 \qquad \text{Squaring both sides}$$
$$W^2 = \frac{1}{LC}$$
$$CW^2 = \frac{1}{L} \qquad \text{Multiplying by } C$$
$$C = \frac{1}{LW^2} \qquad \text{Dividing by } W^2$$

85.
$$\frac{1}{R} = \frac{1}{R_1} + \frac{1}{R_2}$$
$$RR_1 R_2 \cdot \frac{1}{R} = RR_1 R_2 \left(\frac{1}{R_1} + \frac{1}{R_2}\right)$$
$$\text{Multiplying by } RR_1 R_2 \text{ on both sides}$$
$$R_1 R_2 = RR_2 + RR_1$$
$$R_1 R_2 - RR_2 = RR_1 \qquad \text{Subtracting } RR_2 \text{ on both sides}$$
$$R_2 (R_1 - R) = RR_1 \qquad \text{Factoring}$$
$$R_2 = \frac{RR_1}{R_1 - R} \qquad \text{Dividing by } R_1 - R \text{ on both sides}$$

87.
$$I = \sqrt{\frac{A}{P}} - 1$$
$$I + 1 = \sqrt{\frac{A}{P}} \qquad \text{Adding 1}$$
$$(I+1)^2 = \left(\sqrt{\frac{A}{P}}\right)^2$$
$$I^2 + 2I + 1 = \frac{A}{P}$$
$$P(I^2 + 2I + 1) = A \qquad \text{Multplying by } P$$
$$P = \frac{A}{I^2 + 2I + 1} \qquad \text{Dividing by } I^2 + 2I + 1$$

We could also express this result as $P = \dfrac{A}{(I+1)^2}$.

89.
$$\frac{1}{F} = \frac{1}{m} + \frac{1}{p}$$
$$Fmp \cdot \frac{1}{F} = Fmp\left(\frac{1}{m} + \frac{1}{p}\right) \qquad \text{Multiplying by } Fmp \text{ on both sides}$$
$$mp = Fp + Fm$$
$$mp - Fp = Fm \qquad \text{Subtracting } Fp \text{ on both sides}$$
$$p(m - F) = Fm \qquad \text{Factoring}$$
$$p = \frac{Fm}{m - F} \qquad \text{Dividing by } m - F \text{ on both sides}$$

91.
$$15 - 2x = 0 \qquad \text{Setting } f(x) = 0$$
$$15 = 2x$$
$$\frac{15}{2} = x, \text{ or}$$
$$7.5 = x$$

The zero of the function is $\dfrac{15}{2}$, or 7.5.

93. Familiarize. Let p = the number of metric tons of pork produced in the United States in 2013. Then $5p + 1,260,000$ = the number of metric tons of pork produced in China.

Translate.

China pork production	plus	U.S. pork production	was	total pork production.
$5p + 1,260,000$	$+$	p	$=$	$64,308,000$

Carry out.
$$5p + 1,260,000 + p = 64,308,000$$
$$6p + 1,260,000 = 64,308,000$$
$$6p = 63,048,000$$
$$p = 10,508,000$$

Then $5p + 1,260,000 = 5(10,508,000) + 1,260,000 = 53,800,000$.

Check. $53,800,000$ is $1,260,000$ more than 5 times $10,508,000$, and $53,800,000 + 10,508,000 = 64,308,000$, so the answer checks.

State. In 2013, China produced 53,800,000 metric tons of pork, and the United States produced 10,508,000 metric tons.

95.
$$(x-3)^{2/3} = 2$$
$$[(x-3)^{2/3}]^3 = 2^3$$
$$(x-3)^2 = 8$$
$$x^2 - 6x + 9 = 8$$
$$x^2 - 6x + 1 = 0$$
$$a = 1, \ b = -6, \ c = 1$$
$$x = \frac{-b \pm \sqrt{b^2 - 4ac}}{2a}$$
$$= \frac{-(-6) \pm \sqrt{(-6)^2 - 4 \cdot 1 \cdot 1}}{2 \cdot 1}$$
$$= \frac{6 \pm \sqrt{32}}{2} = \frac{6 \pm 4\sqrt{2}}{2}$$
$$= \frac{2(3 \pm 2\sqrt{2})}{2} = 3 \pm 2\sqrt{2}$$

Both values check. The solutions are $3 \pm 2\sqrt{2}$.

97.
$$\sqrt{x+5} + 1 = \frac{6}{\sqrt{x+5}}, \quad \text{LCD is } \sqrt{x+5}$$
$$x + 5 + \sqrt{x+5} = 6 \qquad \text{Multiplying by } \sqrt{x+5}$$
$$\sqrt{x+5} = 1 - x$$
$$x + 5 = 1 - 2x + x^2$$
$$0 = x^2 - 3x - 4$$
$$0 = (x-4)(x+1)$$
$$x = 4 \ \text{ or } \ x = -1$$

Only -1 checks. The solution set is -1.

99.
$$x^{2/3} = x$$
$$(x^{2/3})^3 = x^3$$
$$x^2 = x^3$$
$$0 = x^3 - x^2$$
$$0 = x^2(x-1)$$
$$x^2 = 0 \ \text{ or } \ x - 1 = 0$$
$$x = 0 \ \text{ or } \qquad x = 1$$

Both numbers check. The solutions are 0 and 1.

Exercise Set 3.5

1. $|x| = 7$

The solutions are those numbers whose distance from 0 on a number line is 7. They are -7 and 7. That is,
$$x = -7 \ \text{ or } \ x = 7.$$
The solutions are -7 and 7.

3. $|x| = 0$

The distance of 0 from 0 on a number line is 0. That is,
$$x = 0.$$
The solution is 0.

5. $|x| = \dfrac{5}{6}$
$$x = -\frac{5}{6} \ \text{ or } \ x = \frac{5}{6}$$
The solutions are $-\dfrac{5}{6}$ and $\dfrac{5}{6}$.

7. $|x| = -10.7$

The absolute value of a number is nonnegative. Thus, the equation has no solution.

9. $|3x| = 1$
$$3x = -1 \ \text{ or } \ 3x = 1$$
$$x = -\frac{1}{3} \ \text{ or } \quad x = \frac{1}{3}$$
The solutions are $-\dfrac{1}{3}$ and $\dfrac{1}{3}$.

11. $|8x| = 24$
$$8x = -24 \ \text{ or } \ 8x = 24$$
$$x = -3 \ \text{ or } \quad x = 3$$
The solutions are -3 and 3.

13. $|x - 1| = 4$
$$x - 1 = -4 \ \text{ or } \ x - 1 = 4$$
$$x = -3 \ \text{ or } \qquad x = 5$$
The solutions are -3 and 5.

15. $|x + 2| = 6$
$$x + 2 = -6 \ \text{ or } \ x + 2 = 6$$
$$x = -8 \ \text{ or } \qquad x = 4$$
The solutions are -8 and 4.

17. $|3x + 2| = 1$
$$3x + 2 = -1 \ \text{ or } \ 3x + 2 = 1$$
$$3x = -3 \ \text{ or } \qquad 3x = -1$$
$$x = -1 \ \text{ or } \qquad x = -\frac{1}{3}$$
The solutions are -1 and $-\dfrac{1}{3}$.

19. $\left|\dfrac{1}{2}x - 5\right| = 17$
$$\frac{1}{2}x - 5 = -17 \ \text{ or } \ \frac{1}{2}x - 5 = 17$$
$$\frac{1}{2}x = -12 \ \text{ or } \qquad \frac{1}{2}x = 22$$
$$x = -24 \ \text{ or } \qquad x = 44$$
The solutions are -24 and 44.

21. $|x - 1| + 3 = 6$
$$|x - 1| = 3$$
$$x - 1 = -3 \ \text{ or } \ x - 1 = 3$$
$$x = -2 \ \text{ or } \qquad x = 4$$
The solutions are -2 and 4.

23. $|x + 3| - 2 = 8$

$|x + 3| = 10$

$x + 3 = -10 \quad or \quad x + 3 = 10$

$x = -13 \quad or \qquad x = 7$

The solutions are -13 and 7.

25. $|3x + 1| - 4 = -1$

$|3x + 1| = 3$

$3x + 1 = -3 \quad or \quad 3x + 1 = 3$

$3x = -4 \quad or \qquad 3x = 2$

$x = -\dfrac{4}{3} \quad or \qquad x = \dfrac{2}{3}$

The solutions are $-\dfrac{4}{3}$ and $\dfrac{2}{3}$.

27. $|4x - 3| + 1 = 7$

$|4x - 3| = 6$

$4x - 3 = -6 \quad or \quad 4x - 3 = 6$

$4x = -3 \quad or \qquad 4x = 9$

$x = -\dfrac{3}{4} \quad or \qquad x = \dfrac{9}{4}$

The solutions are $-\dfrac{3}{4}$ and $\dfrac{9}{4}$.

29. $12 - |x + 6| = 5$

$-|x + 6| = -7$

$|x + 6| = 7 \quad$ Multiplying by -1

$x + 6 = -7 \quad or \quad x + 6 = 7$

$x = -13 \quad or \qquad x = 1$

The solutions are -13 and 1.

31. $7 - |2x - 1| = 6$

$-|2x - 1| = -1$

$|2x - 1| = 1 \quad$ Multiplying by -1

$2x - 1 = -1 \quad or \quad 2x - 1 = 1$

$2x = 0 \quad or \qquad 2x = 2$

$x = 0 \quad or \qquad x = 1$

The solutions are 0 and 1.

33. $|x| < 7$

To solve we look for all numbers x whose distance from 0 is less than 7. These are the numbers between -7 and 7. That is, $-7 < x < 7$. The solution set is $(-7, 7)$. The graph is shown below.

35. $|x| \le 2$

$-2 \le x \le 2$

The solution set is $[-2, 2]$. The graph is shown below.

37. $|x| \ge 4.5$

To solve we look for all numbers x whose distance from 0 is greater than or equal to 4.5. That is, $x \le -4.5$ or $x \ge 4.5$. The solution set and its graph are as follows.

$\{x | x \le -4.5 \text{ or } x \ge 4.5\}$, or $(-\infty, -4.5] \cup [4.5, \infty)$

39. $|x| > 3$

$x < -3 \quad or \quad x > 3$

The solution set is $(-\infty, -3) \cup (3, \infty)$. The graph is shown below.

41. $|3x| < 1$

$-1 < 3x < 1$

$-\dfrac{1}{3} < x < \dfrac{1}{3} \quad$ Dividing by 3

The solution set is $\left(-\dfrac{1}{3}, \dfrac{1}{3}\right)$. The graph is shown below.

43. $|2x| \ge 6$

$2x \le -6 \quad or \quad 2x \ge 6$

$x \le -3 \quad or \quad x \ge 3$

The solution set is $(-\infty, -3] \cup [3, \infty)$. The graph is shown below.

45. $|x + 8| < 9$

$-9 < x + 8 < 9$

$-17 < x < 1 \qquad$ Subtracting 8

The solution set is $(-17, 1)$. The graph is shown below.

47. $|x + 8| \ge 9$

$x + 8 \le -9 \quad or \quad x + 8 \ge 9$

$x \le -17 \quad or \qquad x \ge 1 \quad$ Subtracting 8

The solution set is $(-\infty, -17] \cup [1, \infty)$. The graph is shown below.

49. $\left|x - \dfrac{1}{4}\right| < \dfrac{1}{2}$

$-\dfrac{1}{2} < x - \dfrac{1}{4} < \dfrac{1}{2}$

$-\dfrac{1}{4} < x < \dfrac{3}{4}$ \quad Adding $\dfrac{1}{4}$

The solution set is $\left(-\dfrac{1}{4}, \dfrac{3}{4}\right)$. The graph is shown below.

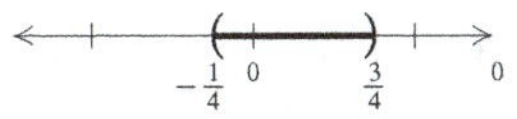

51. $|2x + 3| \le 9$

$-9 \le 2x + 3 \le 9$

$-12 \le 2x \le 6$ \quad Subtracting 3

$-6 \le x \le 3$ \quad Dividing by 2

The solution set is $[-6, 3]$. The graph is shown below.

53. $|x - 5| > 0.1$

$x - 5 < -0.1$ \; or \; $x - 5 > 0.1$

$x < 4.9$ \quad or \quad $x > 5.1$ \quad Adding 5

The solution set is $(-\infty, 4.9) \cup (5.1, \infty)$. The graph is shown below.

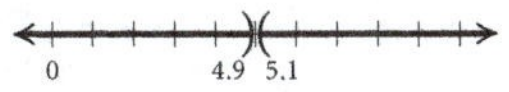

55. $|6 - 4x| \ge 8$

$6 - 4x \le -8$ \; or \; $6 - 4x \ge 8$

$-4x \le -14$ or $-4x \ge 2$ \quad Subtracting 6

$x \ge \dfrac{14}{4}$ \quad or \quad $x \le -\dfrac{2}{4}$ Dividing by -4 and

reversing the inequality symbols

$x \ge \dfrac{7}{2}$ \quad or \quad $x \le -\dfrac{1}{2}$ Simplifying

The solution set is $\left(-\infty, -\dfrac{1}{2}\right] \cup \left[\dfrac{7}{2}, \infty\right)$. The graph is shown below.

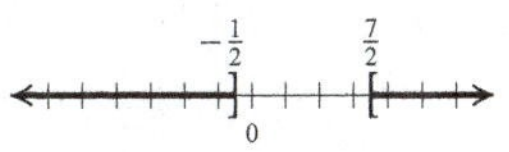

57. $\left|x + \dfrac{2}{3}\right| \le \dfrac{5}{3}$

$-\dfrac{5}{3} \le x + \dfrac{2}{3} \le \dfrac{5}{3}$

$-\dfrac{7}{3} \le x \le 1$ \quad Subtracting $\dfrac{2}{3}$

The solution set is $\left[-\dfrac{7}{3}, 1\right]$. The graph is shown below.

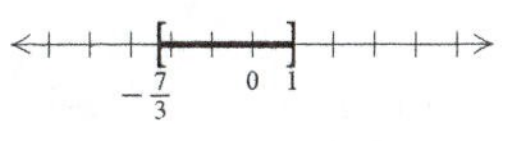

59. $\left|\dfrac{2x + 1}{3}\right| > 5$

$\dfrac{2x + 1}{3} < -5$ \; or \; $\dfrac{2x + 1}{3} > 5$

$2x + 1 < -15$ or $2x + 1 > 15$ Multiplying by 3

$2x < -16$ or \quad $2x > 14$ Subtracting 1

$x < -8$ \quad or \quad $x > 7$ Dividing by 2

The solution set is $\{x | x < -8 \text{ or } x > 7\}$, or
$(-\infty, -8) \cup (7, \infty)$. The graph is shown below.

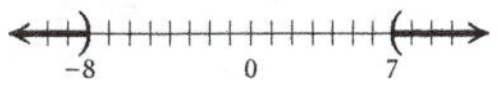

61. $|2x - 4| < -5$

Since $|2x - 4| \ge 0$ for all x, there is no x such that $|2x - 4|$ would be less than -5. There is no solution.

63. $|7 - x| \ge -4$

Since absolute value is nonnegative, for any real-number value of x we have $|7 - x| \ge 0 \ge -4$. Thus the solution set is $(-\infty, \infty)$. The graph is the entire number line.

65. y-intercept

67. relation

69. horizontal lines

71. decreasing

73. $|3x - 1| > 5x - 2$

$3x - 1 < -(5x - 2)$ \; or \; $3x - 1 > 5x - 2$

$3x - 1 < -5x + 2$ \quad or \quad $1 > 2x$

$8x < 3$ \quad or \quad $\dfrac{1}{2} > x$

$x < \dfrac{3}{8}$ \quad or \quad $\dfrac{1}{2} > x$

The solution set is $\left(-\infty, \dfrac{3}{8}\right) \cup \left(-\infty, \dfrac{1}{2}\right)$. This is equivalent to $\left(-\infty, \dfrac{1}{2}\right)$.

75. $|p - 4| + |p + 4| < 8$

If $p < -4$, then $|p - 4| = -(p - 4)$ and $|p + 4| = -(p + 4)$.
Solve: $-(p - 4) + [-(p + 4)] < 8$

$-p + 4 - p - 4 < 8$

$-2p < 8$

$p > -4$

Since this is false for all values of p in the interval $(-\infty, -4)$ there is no solution in this interval.

If $p \ge -4$, then $|p + 4| = p + 4$.
Solve: $|p - 4| + p + 4 < 8$

$|p - 4| < 4 - p$

$p - 4 > -(4 - p)$ \; and \; $p - 4 < 4 - p$

$p - 4 > p - 4$ \quad and \quad $2p < 8$

$-4 > -4$ \quad and \quad $p < 4$

Since $-4 > -4$ is false for all values of p, there is no solution in the interval $[-4, \infty)$.

Thus, $|p - 4| + |p + 4| < 8$ has no solution.

77. $|x - 3| + |2x + 5| > 6$

Divide the set of real numbers into three intervals:
$\left(-\infty, -\dfrac{5}{2}\right)$, $\left[-\dfrac{5}{2}, 3\right)$, and $[3, \infty)$.

Find the solution set of $|x - 3| + |2x + 5| > 6$ in each interval. Then find the union of the three solution sets.

If $x < -\dfrac{5}{2}$, then $|x-3| = -(x-3)$ and $|2x+5| = -(2x+5)$.

Solve: $x < -\dfrac{5}{2}$ *and* $-(x-3) + [-(2x+5)] > 6$

$\quad x < -\dfrac{5}{2}$ *and* $\quad -x + 3 - 2x - 5 > 6$

$\quad x < -\dfrac{5}{2}$ *and* $\quad\quad\quad -3x > 8$

$\quad x < -\dfrac{5}{2}$ *and* $\quad\quad\quad x < -\dfrac{8}{3}$

The solution set in this interval is $\left(-\infty, -\dfrac{8}{3}\right)$.

If $-\dfrac{5}{2} \le x < 3$, then $|x-3| = -(x-3)$ and $|2x+5| = 2x+5$.

Solve: $-\dfrac{5}{2} \le x < 3$ *and* $-(x-3) + 2x + 5 > 6$

$\quad -\dfrac{5}{2} \le x < 3$ *and* $\quad -x + 3 + 2x + 5 > 6$

$\quad -\dfrac{5}{2} \le x < 3$ *and* $\quad\quad\quad x > -2$

The solution set in this interval is $(-2, 3)$.

If $x \ge 3$, then $|x - 3| = x - 3$ and $|2x + 5| = 2x + 5$.

Solve: $x \ge 3$ *and* $x - 3 + 2x + 5 > 6$

$\quad x \ge 3$ *and* $\quad\quad 3x > 4$

$\quad x \ge 3$ *and* $\quad\quad x > \dfrac{4}{3}$

The solution set in this interval is $[3, \infty)$.

The union of the above solution sets is
$\left(-\infty, -\dfrac{8}{3}\right) \cup (-2, \infty)$. This is the solution set of $|x - 3| + |2x + 5| > 6$.

Chapter 3 Review Exercises

1. The statement is true. See page 179 in the text.

3. The statement is false. For example, $3^2 = (-3)^2$, but $3 \ne -3$.

5. $(2y + 5)(3y - 1) = 0$

$\quad 2y + 5 = 0$ *or* $3y - 1 = 0$

$\quad 2y = -5$ *or* $\quad 3y = 1$

$\quad y = -\dfrac{5}{2}$ *or* $\quad y = \dfrac{1}{3}$

The solutions are $-\dfrac{5}{2}$ and $\dfrac{1}{3}$.

7. $\quad 3x^2 + 2x = 8$

$\quad 3x^2 + 2x - 8 = 0$

$\quad (x + 2)(3x - 4) = 0$

$\quad x + 2 = 0$ *or* $3x - 4 = 0$

$\quad x = -2$ *or* $\quad 3x = 4$

$\quad x = -2$ *or* $\quad x = \dfrac{4}{3}$

The solutions are -2 and $\dfrac{4}{3}$.

9. $x^2 + 10 = 0$

$\quad x^2 = -10$

$\quad x = -\sqrt{-10}$ *or* $x = \sqrt{-10}$

$\quad x = -\sqrt{10}i$ *or* $x = \sqrt{10}i$

The solutions are $-\sqrt{10}i$ and $\sqrt{10}i$.

11. $\quad x^2 + 2x - 15 = 0$

$\quad (x + 5)(x - 3) = 0$

$\quad x + 5 = 0$ *or* $x - 3 = 0$

$\quad x = -5$ *or* $\quad x = 3$

The zeros of the function are -5 and 3.

13. $3x^2 + 2x + 3 = 0$

$\quad a = 3, \ b = 2, \ c = 3$

$\quad x = \dfrac{-b \pm \sqrt{b^2 - 4ac}}{2a}$

$\quad x = \dfrac{-2 \pm \sqrt{2^2 - 4 \cdot 3 \cdot 3}}{2 \cdot 3}$

$\quad = \dfrac{-2 \pm \sqrt{-32}}{2 \cdot 3} = \dfrac{-2 \pm \sqrt{-16 \cdot 2}}{2 \cdot 3} = \dfrac{-2 \pm 4i\sqrt{2}}{2 \cdot 3}$

$\quad = \dfrac{2(-1 \pm 2i\sqrt{2})}{2 \cdot 3} = \dfrac{-1 \pm 2i\sqrt{2}}{3}$

The zeros of the function are $\dfrac{-1 \pm 2i\sqrt{2}}{3}$.

15. $\quad\quad \dfrac{3}{8x + 1} + \dfrac{8}{2x + 5} = 1$

LCD is $(8x+1)(2x+5)$

$(8x+1)(2x+5)\left(\dfrac{3}{8x+1} + \dfrac{8}{2x+5}\right) = (8x+1)(2x+5) \cdot 1$

$3(2x + 5) + 8(8x + 1) = (8x+1)(2x+5)$

$6x + 15 + 64x + 8 = 16x^2 + 42x + 5$

$70x + 23 = 16x^2 + 42x + 5$

$0 = 16x^2 - 28x - 18$

$0 = 2(8x^2 - 14x - 9)$

$0 = 2(2x+1)(4x-9)$

$2x + 1 = 0$ *or* $4x - 9 = 0$

$\quad 2x = -1$ *or* $\quad 4x = 9$

$\quad x = -\dfrac{1}{2}$ *or* $\quad x = \dfrac{9}{4}$

Both numbers check. The solutions are $-\dfrac{1}{2}$ and $\dfrac{9}{4}$.

17. $\sqrt{x-1} - \sqrt{x-4} = 1$

$\sqrt{x-1} = \sqrt{x-4} + 1$

$(\sqrt{x-1})^2 = (\sqrt{x-4} + 1)^2$

$x - 1 = x - 4 + 2\sqrt{x-4} + 1$

$x - 1 = x - 3 + 2\sqrt{x-4}$

$2 = 2\sqrt{x-4}$

$1 = \sqrt{x-4}$ Dividing by 2

$1^2 = (\sqrt{x-4})^2$

$1 = x - 4$

$5 = x$

This number checks. The solution is 5.

19. $|2y + 7| = 9$

$2y + 7 = -9$ *or* $2y + 7 = 9$

$2y = -16$ *or* $2y = 2$

$y = -8$ *or* $y = 1$

The solutions are -8 and 1.

21. $|3x + 4| < 10$

$-10 < 3x + 4 < 10$

$-14 < 3x < 6$

$-\dfrac{14}{3} < x < 2$

The solution set is $\left(-\dfrac{14}{3}, 2 \right)$. The graph is shown below.

23. $|x + 4| \geq 2$

$x + 4 \leq -2$ *or* $x + 4 \geq 2$

$x \leq -6$ *or* $x \geq -2$

The solution is $(-\infty, -6] \cup [-2, \infty)$.

25. $-\sqrt{-40} = -\sqrt{-1} \cdot \sqrt{4} \cdot \sqrt{10} = -2\sqrt{10}i$

27. $\dfrac{\sqrt{-49}}{-\sqrt{-64}} = \dfrac{7i}{-8i} = -\dfrac{7}{8}$

29. $(3 - 5i) - (2 - i) = (3 - 2) + [-5i - (-i)]$
$= 1 - 4i$

31. $\dfrac{2 - 3i}{1 - 3i} = \dfrac{2 - 3i}{1 - 3i} \cdot \dfrac{1 + 3i}{1 + 3i}$

$= \dfrac{2 + 3i - 9i^2}{1 - 9i^2}$

$= \dfrac{2 + 3i + 9}{1 + 9}$

$= \dfrac{11 + 3i}{10}$

$= \dfrac{11}{10} + \dfrac{3}{10}i$

33. $x^2 - 3x = 18$

$x^2 - 3x + \dfrac{9}{4} = 18 + \dfrac{9}{4}$ $\left(\dfrac{1}{2}(-3) = -\dfrac{3}{2} \text{ and } \left(-\dfrac{3}{2} \right)^2 = \dfrac{9}{4} \right)$

$\left(x - \dfrac{3}{2} \right)^2 = \dfrac{81}{4}$

$x - \dfrac{3}{2} = \pm\dfrac{9}{2}$

$x = \dfrac{3}{2} \pm \dfrac{9}{2}$

$x = \dfrac{3}{2} - \dfrac{9}{2}$ *or* $x = \dfrac{3}{2} + \dfrac{9}{2}$

$x = -3$ *or* $x = 6$

The solutions are -3 and 6.

35. $3x^2 + 10x = 8$

$3x^2 + 10x - 8 = 0$

$(x + 4)(3x - 2) = 0$

$x + 4 = 0$ *or* $3x - 2 = 0$

$x = -4$ *or* $3x = 2$

$x = -4$ *or* $x = \dfrac{2}{3}$

The solutions are -4 and $\dfrac{2}{3}$.

37. $x^2 = 10 + 3x$

$x^2 - 3x - 10 = 0$

$(x + 2)(x - 5) = 0$

$x + 2 = 0$ *or* $x - 5 = 0$

$x = -2$ *or* $x = 5$

The solutions are -2 and 5.

39. $y^4 - 3y^2 + 1 = 0$

Let $u = y^2$.

$u^2 - 3u + 1 = 0$

$u = \dfrac{-(-3) \pm \sqrt{(-3)^2 - 4 \cdot 1 \cdot 1}}{2 \cdot 1} = \dfrac{3 \pm \sqrt{5}}{2}$

Substitute y^2 for u and solve for y.

$y^2 = \dfrac{3 \pm \sqrt{5}}{2}$

$y = \pm\sqrt{\dfrac{3 \pm \sqrt{5}}{2}}$

The solutions are $\pm\sqrt{\dfrac{3 \pm \sqrt{5}}{2}}$.

41. $(p - 3)(3p + 2)(p + 2) = 0$

$p - 3 = 0$ *or* $3p + 2 = 0$ *or* $p + 2 = 0$

$p = 3$ *or* $3p = -2$ *or* $p = -2$

$p = 3$ *or* $p = -\dfrac{2}{3}$ *or* $p = -2$

The solutions are -2, $-\dfrac{2}{3}$ and 3.

43. $f(x) = -4x^2 + 3x - 1$

$$= -4\left(x^2 - \frac{3}{4}x\right) - 1$$

$$= -4\left(x^2 - \frac{3}{4}x + \frac{9}{64} - \frac{9}{64}\right) - 1$$

$$= -4\left(x^2 - \frac{3}{4}x + \frac{9}{64}\right) - 4\left(-\frac{9}{64}\right) - 1$$

$$= -4\left(x^2 - \frac{3}{4}x + \frac{9}{64}\right) + \frac{9}{16} - 1$$

$$= -4\left(x - \frac{3}{8}\right)^2 - \frac{7}{16}$$

a) Vertex: $\left(\frac{3}{8}, -\frac{7}{16}\right)$

b) Axis of symmetry: $x = \frac{3}{8}$

c) Maximum value: $-\frac{7}{16}$

d) Range: $\left(-\infty, -\frac{7}{16}\right]$

e)

$f(x) = -4x^2 + 3x - 1$

45. The graph of $y = (x - 2)^2$ has vertex $(2, 0)$ and opens up. It is graph (d).

47. The graph of $y = -2(x + 3)^2 + 4$ has vertex $(-3, 4)$ and opens down. It is graph (b).

49. *Familiarize.* Using the labels in the textbook, the legs of the right triangle are represented by x and $x + 10$.

Translate. We use the Pythagorean theorem.
$$x^2 + (x + 10)^2 = 50^2$$

Carry out. We solve the equation.
$$x^2 + (x + 10)^2 = 50^2$$
$$x^2 + x^2 + 20x + 100 = 2500$$
$$2x^2 + 20x - 2400 = 0$$
$$2(x^2 + 10x - 1200) = 0$$
$$2(x + 40)(x - 30) = 0$$
$$x + 40 = 0 \quad or \quad x - 30 = 0$$
$$x = -40 \quad or \qquad x = 30$$

Check. Since the length cannot be negative, we need to check only 30. If $x = 30$, then $x + 10 = 30 + 10 = 40$. Since $30^2 + 40^2 = 900 + 1600 = 2500 = 50^2$, the answer checks.

State. The lengths of the legs are 30 ft and 40 ft.

51. *Familiarize.* Using the drawing in the textbook, let $w =$ the width of the sidewalk, in ft. Then the length of the new parking lot is $80 - 2w$, and its width is $60 - 2w$.

Translate. We use the formula for the area of a rectangle, $A = lw$.

$$\underbrace{\text{New area}} \quad \text{is} \quad \tfrac{2}{3} \text{ of } \underbrace{\text{old area}}$$

$$(80 - 2w)(60 - 2w) \;=\; \frac{2}{3} \cdot \;\; 80 \cdot 60$$

Carry out. We solve the equation.

$$(80 - 2w)(60 - 2w) = \frac{2}{3} \cdot 80 \cdot 60$$

$$4800 - 280w + 4w^2 = \frac{2}{3} \cdot 80 \cdot 3 \cdot 20$$

$$4w^2 - 280w + 4800 = 3200$$

$$4w^2 - 280w + 1600 = 0$$

$$w^2 - 70w + 400 = 0 \qquad \text{Dividing by 4}$$

We use the quadratic formula.

$$w = \frac{-b \pm \sqrt{b^2 - 4ac}}{2a}$$

$$= \frac{-(-70) \pm \sqrt{(-70)^2 - 4 \cdot 1 \cdot 400}}{2 \cdot 1}$$

$$= \frac{70 \pm \sqrt{3300}}{2}$$

$$= \frac{70 \pm \sqrt{33 \cdot 100}}{2} = \frac{70 \pm 10\sqrt{33}}{2}$$

$$= 35 \pm 5\sqrt{33}$$

$35 + 5\sqrt{33} \approx 63.7$ and $35 - 5\sqrt{33} \approx 6.3$

Check. The width of the sidewalk cannot be 63.7 ft because this width exceeds the width of the original parking lot, 60 ft. We check $35 - 5\sqrt{33} \approx 6.3$. If the width of the sidewalk in about 6.3 ft, then the length of the new parking lot is $80 - 2(6.3)$, or 67.4, and the width is $60 - 2(6.3)$, or 47.4. The area of a parking lot with these dimensions is $(67.4)(47.4) = 3194.76$. Two-thirds of the area of the original parking lot is $\frac{2}{3} \cdot 80 \cdot 60 = 3200$. Since $3194.76 \approx 3200$, this answer checks.

State. The width of the sidewalk is $35 - 5\sqrt{33}$ ft, or about 6.3 ft.

53. *Familiarize.* Using the labels in the textbook, let $x =$ the length of the sides of the squares, in cm. Then the length of the base of the box is $20 - 2x$ and the width of the base is $10 - 2x$.

Translate. We use the formula for the area of a rectangle, $A = lw$.
$$90 = (20 - 2x)(10 - 2x)$$
$$90 = 200 - 60x + 4x^2$$
$$0 = 4x^2 - 60x + 110$$
$$0 = 2x^2 - 30x + 55 \qquad \text{Dividing by 2}$$

We use the quadratic formula.

Carry out.

$$x = \frac{-b \pm \sqrt{b^2 - 4ac}}{2a}$$

$$= \frac{-(-30) \pm \sqrt{(-30)^2 - 4 \cdot 2 \cdot 55}}{2 \cdot 2}$$

$$= \frac{30 \pm \sqrt{460}}{4}$$

$$= \frac{30 \pm \sqrt{4 \cdot 115}}{4} = \frac{30 \pm 2\sqrt{115}}{4}$$

$$= \frac{15 \pm \sqrt{115}}{2}$$

$\dfrac{15 + \sqrt{115}}{2} \approx 12.9$ and $\dfrac{15 - \sqrt{115}}{2} \approx 2.1$.

Check. The length of the sides of the squares cannot be 12.9 cm because this length exceeds the width of the piece of aluminum. We check 2.1 cm. If the sides of the squares are 2.1 cm, then the length of the base of the box is $20 - 2(2.1) = 15.8$, and the width is $10 - 2(2.1) = 5.8$. The area of the base is $15.8(5.8) = 91.64 \approx 90$. This answer checks.

State. The length of the sides of the squares is $\dfrac{15 - \sqrt{115}}{2}$ cm, or about 2.1 cm.

55. $\sqrt{4x+1} + \sqrt{2x} = 1$

$$\sqrt{4x+1} = 1 - \sqrt{2x}$$

$$(\sqrt{4x+1})^2 = (1 - \sqrt{2x})^2$$

$$4x + 1 = 1 - 2\sqrt{2x} + 2x$$

$$2x = -2\sqrt{2x}$$

$$x = -\sqrt{2x}$$

$$x^2 = (-\sqrt{2x})^2$$

$$x^2 = 2x$$

$$x^2 - 2x = 0$$

$$x(x - 2) = 0$$

$$x = 0 \ \ or \ \ x = 2$$

Only 0 checks, so answer B is correct.

57. $\sqrt{\sqrt{\sqrt{x}}} = 2$

$$\left(\sqrt{\sqrt{\sqrt{x}}}\right)^2 = 2^2$$

$$\sqrt{\sqrt{x}} = 4$$

$$\left(\sqrt{\sqrt{x}}\right)^2 = 4^2$$

$$\sqrt{x} = 16$$

$$(\sqrt{x})^2 = 16^2$$

$$x = 256$$

The answer checks. The solution is 256.

59. $(x - 1)^{2/3} = 4$

$$(x - 1)^2 = 4^3$$

$$x - 1 = \pm\sqrt{64}$$

$$x - 1 = \pm 8$$

$$x - 1 = -8 \ \ or \ \ x - 1 = 8$$

$$x = -7 \ \ or \ \ \ \ \ \ x = 9$$

Both numbers check. The solutions are -7 and 9.

61. $\sqrt{x+2} + \sqrt[4]{x+2} - 2 = 0$

Let $u = \sqrt[4]{x+2}$, so $u^2 = (\sqrt[4]{x+2})^2 = \sqrt{x+2}$.

$$u^2 + u - 2 = 0$$

$$(u + 2)(u - 1) = 0$$

$$u = -2 \ \ or \ \ u = 1$$

Substitute $\sqrt[4]{x+2}$ for u and solve for x.

$$\sqrt[4]{x+2} = -2 \ \ \ \ \ or \ \sqrt[4]{x+2} = 1$$

No real solution $\ \ \ \ \ \ \ x + 2 = 1$

$$x = -1$$

This number checks. The solution is -1.

63. The maximum value occurs at the vertex. The first coordinate of the vertex is $-\dfrac{b}{2a} = -\dfrac{b}{2(-3)} = \dfrac{b}{6}$ and $f\left(\dfrac{b}{6}\right) = 2$.

$$-3\left(\frac{b}{6}\right)^2 + b\left(\frac{b}{6}\right) - 1 = 2$$

$$-\frac{b^2}{12} + \frac{b^2}{6} - 1 = 2$$

$$-b^2 + 2b^2 - 12 = 24$$

$$b^2 = 36$$

$$b = \pm 6$$

65. No; consider the quadratic formula $x = \dfrac{-b \pm \sqrt{b^2 - 4ac}}{2a}$. If $b^2 - 4ac = 0$, then $x = \dfrac{-b}{2a}$, so there is one real zero. If $b^2 - 4ac > 0$, then $\sqrt{b^2 - 4ac}$ is a real number and there are two real zeros. If $b^2 - 4ac < 0$, then $\sqrt{b^2 - 4ac}$ is an imaginary number and there are two imaginary zeros. Thus, a quadratic function cannot have one real zero and one imaginary zero.

67. When both sides of an equation are multiplied by the LCD, the resulting equation might not be equivalent to the original equation. One or more of the possible solutions of the resulting equation might make a denominator of the original equation 0.

69. Absolute value is nonnegative.

Chapter 3 Test

1. $(2x - 1)(x + 5) = 0$

$$2x - 1 = 0 \ \ or \ \ x + 5 = 0$$

$$2x = 1 \ \ or \ \ \ \ \ \ \ x = -5$$

$$x = \frac{1}{2} \ \ or \ \ \ \ \ \ \ x = -5$$

The solutions are $\dfrac{1}{2}$ and -5.

2. $6x^2 - 36 = 0$

$$6x^2 = 36$$
$$x^2 = 6$$
$$x = -\sqrt{6} \ \ or \ \ x = \sqrt{6}$$

The solutions are $-\sqrt{6}$ and $\sqrt{6}$.

3. $x^2 + 4 = 0$

$$x^2 = -4$$
$$x = \pm\sqrt{-4}$$
$$x = -2i \ \ or \ \ x = 2i$$

The solutions are $-2i$ and $2i$.

4. $x^2 - 2x - 3 = 0$

$$(x+1)(x-3) = 0$$
$$x + 1 = 0 \ \ or \ \ x - 3 = 0$$
$$x = -1 \ \ or \ \ \ \ \ x = 3$$

The solutions are -1 and 3.

5. $x^2 - 5x + 3 = 0$

$$a = 1, \ b = -5, \ c = 3$$
$$x = \frac{-b \pm \sqrt{b^2 - 4ac}}{2a}$$
$$x = \frac{-(-5) \pm \sqrt{(-5)^2 - 4 \cdot 1 \cdot 3}}{2 \cdot 1}$$
$$= \frac{5 \pm \sqrt{13}}{2}$$

The solutions are $\dfrac{5 + \sqrt{13}}{2}$ and $\dfrac{5 - \sqrt{13}}{2}$.

6. $2t^2 - 3t + 4 = 0$

$$a = 2, \ b = -3, \ c = 4$$
$$x = \frac{-b \pm \sqrt{b^2 - 4ac}}{2a}$$
$$x = \frac{-(-3) \pm \sqrt{(-3)^2 - 4 \cdot 2 \cdot 4}}{2 \cdot 2}$$
$$= \frac{3 \pm \sqrt{-23}}{4} = \frac{3 \pm i\sqrt{23}}{4}$$
$$= \frac{3}{4} \pm \frac{\sqrt{23}}{4}i$$

The solutions are $\dfrac{3}{4} + \dfrac{\sqrt{23}}{4}i$ and $\dfrac{3}{4} - \dfrac{\sqrt{23}}{4}i$.

7. $x + 5\sqrt{x} - 36 = 0$

Let $u = \sqrt{x}$.

$$u^2 + 5u - 36 = 0$$
$$(u+9)(u-4) = 0$$
$$u + 9 = 0 \ \ or \ \ u - 4 = 0$$
$$u = -9 \ \ or \ \ \ \ \ u = 4$$

Substitute $\sqrt{x}$ for u and solve for x.

$$\sqrt{x} = -9 \ \ \ \ \ or \ \ \sqrt{x} = 4$$

No solution $\ \ \ \ \ \ \ \ \ \ x = 16$

The number 16 checks. It is the solution.

8.
$$\frac{3}{3x+4} + \frac{2}{x-1} = 2, \ \text{LCD is } (3x+4)(x-1)$$
$$(3x+4)(x-1)\left(\frac{3}{3x+4} + \frac{2}{x-1}\right) = (3x+4)(x-1)(2)$$
$$3(x-1) + 2(3x+4) = 2(3x^2 + x - 4)$$
$$3x - 3 + 6x + 8 = 6x^2 + 2x - 8$$
$$9x + 5 = 6x^2 + 2x - 8$$
$$0 = 6x^2 - 7x - 13$$
$$0 = (x+1)(6x - 13)$$
$$x + 1 = 0 \ \ \ or \ \ 6x - 13 = 0$$
$$x = -1 \ or \ \ \ \ \ \ \ 6x = 13$$
$$x = -1 \ or \ \ \ \ \ \ \ \ \ x = \frac{13}{6}$$

Both numbers check. The solutions are -1 and $\dfrac{13}{6}$.

9. $\sqrt{x+4} - 2 = 1$

$$\sqrt{x+4} = 3$$
$$(\sqrt{x+4})^2 = 3^2$$
$$x + 4 = 9$$
$$x = 5$$

This number checks. The solution is 5.

10. $\sqrt{x+4} - \sqrt{x-4} = 2$

$$\sqrt{x+4} = \sqrt{x-4} + 2$$
$$(\sqrt{x+4})^2 = (\sqrt{x-4} + 2)^2$$
$$x + 4 = x - 4 + 4\sqrt{x-4} + 4$$
$$4 = 4\sqrt{x-4}$$
$$1 = \sqrt{x-4}$$
$$1^2 = (\sqrt{x-4})^2$$
$$1 = x - 4$$
$$5 = x$$

This number checks. The solution is 5.

11. $|x + 4| = 7$

$$x + 4 = -7 \ \ \ or \ \ x + 4 = 7$$
$$x = -11 \ or \ \ \ \ \ \ \ x = 3$$

The solutions are -11 and 3.

12. $|4y - 3| = 5$

$$4y - 3 = -5 \ \ \ or \ \ 4y - 3 = 5$$
$$4y = -2 \ \ \ or \ \ \ \ \ \ \ 4y = 8$$
$$y = -\frac{1}{2} \ \ or \ \ \ \ \ \ \ \ \ y = 2$$

The solutions are $-\dfrac{1}{2}$ and 2.

13. $|x + 3| \le 4$

$$-4 \le x + 3 \le 4$$
$$-7 \le x \le 1$$

The solution set is $[-7, 1]$.

14. $|2x - 1| < 5$

$$-5 < 2x - 1 < 5$$
$$-4 < 2x < 6$$
$$-2 < x < 3$$

The solution set is $(-2, 3)$.

15. $|x + 5| > 2$

$$x + 5 < -2 \quad or \quad x + 5 > 2$$
$$x < -7 \quad or \quad \quad x > -3$$

The solution set is $(-\infty, -7) \cup (-3, \infty)$.

16. $|3 - 2x| \geq 7$

$$3 - 2x \leq -7 \quad or \quad 3 - 2x \geq 7$$
$$-2x \leq -10 \quad or \quad \quad -2x \geq 4$$
$$x \geq 5 \quad \quad or \quad \quad x \leq -2$$

The solution set is $(-\infty, -2] \cup [5, \infty)$.

17.
$$\frac{1}{A} + \frac{1}{B} = \frac{1}{C}$$
$$ABC \left(\frac{1}{A} + \frac{1}{B} \right) = ABC \cdot \frac{1}{C}$$
$$BC + AC = AB$$
$$AC = AB - BC$$
$$AC = B(A - C)$$
$$\frac{AC}{A - C} = B$$

18. $R = \sqrt{3np}$

$$R^2 = (\sqrt{3np})^2$$
$$R^2 = 3np$$
$$\frac{R^2}{3p} = n$$

19. $x^2 + 4x = 1$

$$x^2 + 4x + 4 = 1 + 4 \quad \left(\frac{1}{2}(4) = 2 \text{ and } 2^2 = 4 \right)$$
$$(x + 2)^2 = 5$$
$$x + 2 = \pm\sqrt{5}$$
$$x = -2 \pm \sqrt{5}$$

The solutions are $-2 + \sqrt{5}$ and $-2 - \sqrt{5}$.

20. *Familiarize and Translate.* We will use the formula $s = 16t^2$, substituting 2063 for s.

$$2063 = 16t^2$$

Carry out. We solve the equation.

$$2063 = 16t^2$$
$$\frac{2063}{16} = t^2$$
$$11.4 \approx t$$

Check. When $t = 11.4$, $s = 16(11.4)^2 = 2079.36 \approx 2063$. The answer checks.

State. It would take an object about 11.4 sec to reach the ground.

21. $\sqrt{-43} = \sqrt{-1} \cdot \sqrt{43} = i\sqrt{43}$, or $\sqrt{43}i$

22. $-\sqrt{-25} = -\sqrt{-1} \cdot \sqrt{25} = -5i$

23. $(5 - 2i) - (2 + 3i) = (5 - 2) + (-2i - 3i)$
$$= 3 - 5i$$

24. $(3 + 4i)(2 - i) = 6 - 3i + 8i - 4i^2$
$$= 6 + 5i + 4 \quad (i^2 = -1)$$
$$= 10 + 5i$$

25.
$$\frac{1 - i}{6 + 2i} = \frac{1 - i}{6 + 2i} \cdot \frac{6 - 2i}{6 - 2i}$$
$$= \frac{6 - 2i - 6i + 2i^2}{36 - 4i^2}$$
$$= \frac{6 - 8i - 2}{36 + 4}$$
$$= \frac{4 - 8i}{40}$$
$$= \frac{4}{40} - \frac{8}{40}i$$
$$= \frac{1}{10} - \frac{1}{5}i$$

26. $i^{33} = (i^2)^{16} \cdot i = (-1)^{16} \cdot i = 1 \cdot i = i$

27.
$$4x^2 - 11x - 3 = 0$$
$$(4x + 1)(x - 3) = 0$$
$$4x + 1 = 0 \quad or \quad x - 3 = 0$$
$$4x = -1 \quad or \quad \quad x = 3$$
$$x = -\frac{1}{4} \quad or \quad \quad x = 3$$

The zeros of the functions are $-\frac{1}{4}$ and 3.

28. $2x^2 - x - 7 = 0$

$a = 2$, $b = -1$, $c = -7$

$$x = \frac{-b \pm \sqrt{b^2 - 4ac}}{2a}$$
$$x = \frac{-(-1) \pm \sqrt{(-1)^2 - 4 \cdot 2 \cdot (-7)}}{2 \cdot 2}$$
$$= \frac{1 \pm \sqrt{57}}{4}$$

The solutions are $\frac{1 + \sqrt{57}}{4}$ and $\frac{1 - \sqrt{57}}{4}$.

29. $f(x) = -x^2 + 2x + 8$
$$= -(x^2 - 2x) + 8$$
$$= -(x^2 - 2x + 1 - 1) + 8$$
$$= -(x^2 - 2x + 1) - (-1) + 8$$
$$= -(x^2 - 2x + 1) + 1 + 8$$
$$= -(x - 1)^2 + 9$$

a) Vertex: $(1, 9)$

b) Axis of symmetry: $x = 1$

c) Maximum value: 9

d) Range: $(-\infty, 9]$

e)

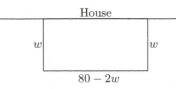

$$f(x) = -x^2 + 2x + 8$$

30. Familiarize. We make a drawing, letting $w =$ the width of the rectangle, in ft. This leaves $80 - w - w$, or $80 - 2w$ ft of fencing for the length.

House

w $\qquad$ w

$80 - 2w$

Translating. The area of a rectangle is given by length times width.
$$A(w) = (80 - 2w)w$$
$$= 80w - 2w^2, \text{ or } -2w^2 + 80w$$

Carry out. This is a quadratic function with $a < 0$, so it has a maximum value that occurs at the vertex of the graph of the function. The first coordinate of the vertex is
$$w = -\frac{b}{2a} = -\frac{80}{2(-2)} = 20.$$
If $w = 20$, then $80 - 2w = 80 - 2 \cdot 20 = 40$.

Check. The area of a rectangle with length 40 ft and width 20 ft is $40 \cdot 20$, or 800 ft^2. As a partial check, we can find $A(w)$ for a value of w less than 20 and for a value of w greater than 20. For instance, $A(19.9) = 799.98$ and $A(20.1) = 799.98$. Since both of these values are less than 800, the result appears to be correct.

State. The dimensions for which the area is a maximum are 20 ft by 40 ft.

31. $f(x) = (x - 1)^2 - 2$

The graph of this function opens up and has vertex $(1, -2)$. Thus the correct graph is C.

32. The maximum value occurs at the vertex. The first coordinate of the vertex is $-\frac{b}{2a} = -\frac{(-4)}{2a} = \frac{2}{a}$ and $f\left(\frac{2}{a}\right) = 12$.
Then we have:
$$a\left(\frac{2}{a}\right)^2 - 4\left(\frac{2}{a}\right) + 3 = 12$$
$$a \cdot \frac{4}{a^2} - \frac{8}{a} + 3 = 12$$
$$\frac{4}{a} - \frac{8}{a} + 3 = 12$$
$$-\frac{4}{a} + 3 = 12$$
$$-\frac{4}{a} = 9$$
$$-4 = 9a$$
$$-\frac{4}{9} = a$$

Chapter 4

Polynomial and Rational Functions

Exercise Set 4.1

---

1. $g(x) = \frac{1}{2}x^3 - 10x + 8$

The leading term is $\frac{1}{2}x^3$ and the leading coefficient is $\frac{1}{2}$. The degree of the polynomial is 3, so the polynomial is cubic.

3. $h(x) = 0.9x - 0.13$

The leading term is $0.9x$ and the leading coefficient is 0.9. The degree of the polynomial is 1, so the polynomial is linear.

5. $g(x) = 305x^4 + 4021$

The leading term is $305x^4$ and the leading coefficient is 305. The degree of the polynomial is 4, so the polynomial is quartic.

7. $h(x) = -5x^2 + 7x^3 + x^4 = x^4 + 7x^3 - 5x^2$

The leading term is x^4 and the leading coefficient is 1 ($x^4 = 1 \cdot x^4$). The degree of the polynomial is 4, so the polynomial is quartic.

9. $g(x) = 4x^3 - \frac{1}{2}x^2 + 8$

The leading term is $4x^3$ and the leading coefficient is 4. The degree of the polynomial is 3, so the polynomial is cubic.

11. $f(x) = -3x^3 - x + 4$

The leading term is $-3x^3$. The degree, 3, is odd and the leading coefficient, -3, is negative. Thus the end behavior of the graph is like that of (d).

13. $f(x) = -x^6 + \frac{3}{4}x^4$

The leading term is $-x^6$. The degree, 6, is even and the leading coefficient, -1, is negative. Thus the end behavior of the graph is like that of (b).

15. $f(x) = -3.5x^4 + x^6 + 0.1x^7 = 0.1x^7 + x^6 - 3.5x^4$

The leading term is $0.1x^7$. The degree, 7, is odd and the leading coefficient, 0.1, is positive. Thus the end behavior of the graph is like that of (c).

17. $f(x) = 10 + \frac{1}{10}x^4 - \frac{2}{5}x^3 = \frac{1}{10}x^4 - \frac{2}{5}x^3 + 10$

The leading term is $\frac{1}{10}x^4$. The degree, 4, is even and the leading coefficient, $\frac{1}{10}$, is positive. Thus the end behavior of the graph is like that of (a).

19. $f(x) = -x^6 + 2x^5 - 7x^2$

The leading term is $-x^6$. The degree, 6, is even and the leading coefficient, -1, is negative. Thus, (c) is the correct graph.

21. $f(x) = x^5 + \frac{1}{10}x - 3$

The leading term is x^5. The degree, 5, is odd and the leading coefficient, 1, is positive. Thus, (d) is the correct graph.

23. $f(x) = x^3 - 9x^2 + 14x + 24$

$f(4) = 4^3 - 9 \cdot 4^2 + 14 \cdot 4 + 24 = 0$

Since $f(4) = 0$, 4 is a zero of $f(x)$.

$f(5) = 5^3 - 9 \cdot 5^2 + 14 \cdot 5 + 24 = -6$

Since $f(5) \neq 0$, 5 is not a zero of $f(x)$.

$f(-2) = (-2)^3 - 9(-2)^2 + 14(-2) + 24 = -48$

Since $f(-2) \neq 0$, -2 is not a zero of $f(x)$.

25. $g(x) = x^4 - 6x^3 + 8x^2 + 6x - 9$

$g(2) = 2^4 - 6 \cdot 2^3 + 8 \cdot 2^2 + 6 \cdot 2 - 9 = 3$

Since $g(2) \neq 0$, 2 is not a zero of $g(x)$.

$g(3) = 3^4 - 6 \cdot 3^3 + 8 \cdot 3^2 + 6 \cdot 3 - 9 = 0$

Since $g(3) = 0$, 3 is a zero of $g(x)$.

$g(-1) = (-1)^4 - 6(-1)^3 + 8(-1)^2 + 6(-1) - 9 = 0$

Since $g(-1) = 0$, -1 is a zero of $g(x)$.

27. $f(x) = (x + 3)^2(x - 1) = (x + 3)(x + 3)(x - 1)$

To solve $f(x) = 0$ we use the principle of zero products, solving $x + 3 = 0$ and $x - 1 = 0$. The zeros of $f(x)$ are -3 and 1.

The factor $x + 3$ occurs twice. Thus the zero -3 has a multiplicity of two.

The factor $x - 1$ occurs only one time. Thus the zero 1 has a multiplicity of one.

29. $f(x) = -2(x - 4)(x - 4)(x - 4)(x + 6) = -2(x - 4)^3(x + 6)$

To solve $f(x) = 0$ we use the principle of zero products, solving $x - 4 = 0$ and $x + 6 = 0$. The zeros of $f(x)$ are 4 and -6.

The factor $x - 4$ occurs three times. Thus the zero 4 has a multiplicity of 3.

The factor $x + 6$ occurs only one time. Thus the zero -6 has a multiplicity of 1.

31. $f(x) = (x^2 - 9)^3 = [(x + 3)(x - 3)]^3 = (x + 3)^3(x - 3)^3$

To solve $f(x) = 0$ we use the principle of zero products, solving $x + 3 = 0$ and $x - 3 = 0$. The zeros of $f(x)$ are -3 and 3.

The factors $x + 3$ and $x - 3$ each occur three times so each zero has a multiplicity of 3.

33. $f(x) = x^3(x - 1)^2(x + 4)$

To solve $f(x) = 0$ we use the principle of zero products, solving $x = 0$, $x - 1 = 0$, and $x + 4 = 0$. The zeros of $f(x)$ are 0, 1, and -4.

The factor x occurs three times. Thus the zero 0 has a multiplicity of three.

The factor $x - 1$ occurs twice. Thus the zero 1 has a multiplicity of two.

The factor $x + 4$ occurs only one time. Thus the zero -4 has a multiplicity of one.

35. $f(x) = -8(x - 3)^2(x + 4)^3 x^4$

To solve $f(x) = 0$ we use the principle of zero products, solving $x - 3 = 0$, $x + 4 = 0$, and $x = 0$. The zeros of $f(x)$ are 3, -4, and 0.

The factor $x - 3$ occurs twice. Thus the zero 3 has a multiplicity of 2.

The factor $x + 4$ occurs three times. Thus the zero -4 has a multiplicity of 3.

The factor x occurs four times. Thus the zero 0 has a multiplicity of 4.

37. $f(x) = x^4 - 4x^2 + 3$

We factor as follows:
$$f(x) = (x^2 - 3)(x^2 - 1)$$
$$= (x - \sqrt{3})(x + \sqrt{3})(x - 1)(x + 1)$$

The zeros of the function are $\sqrt{3}$, $-\sqrt{3}$, 1, and -1. Each has a multiplicity of 1.

39. $f(x) = x^3 + 3x^2 - x - 3$

We factor by grouping:
$$f(x) = x^2(x + 3) - (x + 3)$$
$$= (x^2 - 1)(x + 3)$$
$$= (x - 1)(x + 1)(x + 3)$$

The zeros of the function are 1, -1, and -3. Each has a multiplicity of 1.

41. $f(x) = 2x^3 - x^2 - 8x + 4$
$$= x^2(2x - 1) - 4(2x - 1)$$
$$= (2x - 1)(x^2 - 4)$$
$$= (2x - 1)(x + 2)(x - 2)$$

The zeros of the function are $\dfrac{1}{2}$, -2, and 2. Each has a multiplicity of 1.

43. Graphing the function, we see that the graph touches the x-axis at $(3, 0)$ but does not cross it, so the statement is false.

45. Graphing the function, we see that the statement is true.

47. For 2008, $x = 2008 - 2001 = 7$.

$f(7) = -0.000913(7)^4 + 0.0248(7)^3 - 0.1515(7)^2 + 0.2136(7) + 1.2779 \approx 1.7$ million albums

For 2012, $x = 2012 - 2001 = 11$.

$f(11) = -0.000913(11)^4 + 0.0248(11)^3 - 0.1515(11)^2 + 0.2136(11) + 1.2779 \approx 4.9$ million albums

For 2016, $x = 2016 - 2001 = 15$.

$f(15) = -0.000913(15)^4 + 0.0248(15)^3 - 0.1515(15)^2 + 0.2136(15) + 1.2779 \approx 7.9$ million albums

49. $d(3) = 0.010255(3)^3 - 0.340119(3)^2 + 7.397499(3) + 6.618361 \approx 26$ yr

$d(12) = 0.010255(12)^3 - 0.340119(12)^2 + 7.397499(12) + 6.618361 \approx 64$ yr

$d(16) = 0.010255(16)^3 - 0.340119(16)^2 + 7.397499(16) + 6.618361 \approx 80$ yr

51. We substitute 294 for $s(t)$ and solve for t.
$$294 = 4.9t^2 + 34.3t$$
$$0 = 4.9t^2 + 34.3t - 294$$
$$t = \frac{-b \pm \sqrt{b^2 - 4ac}}{2a}$$
$$= \frac{-34.3 \pm \sqrt{(34.3)^2 - 4(4.9)(-294)}}{2(4.9)}$$
$$= \frac{-34.3 \pm \sqrt{6938.89}}{9.8}$$

$t = 5$ or $t = -12$

Only the positive number has meaning in the situation. It will take the stone 5 sec to reach the ground.

53. For 2003, $x = 2003 - 2001 = 2$.

$p(2) = 6.213(2)^4 - 432.347(2)^3 + 1922.987(2)^2 + 20,503.912(2) + 638,684.984 \approx 684,025$ admissions

For 2006, $x = 2006 - 2001 = 5$.

$p(5) = 6.213(5)^4 - 432.347(5)^3 + 1922.987(5)^2 + 20,503.912(5) + 638,684.984 \approx 739,119$ admissions

For 2011, $x = 2011 - 2001 = 10$.

$p(10) = 6.213(10)^4 - 432.347(10)^3 + 1922.987(10)^2 + 20,503.912(10) + 638,684.984 \approx 665,806$ admissions

55.
$$A = P(1 + i)^t$$
$$9039.75 = 8000(1 + i)^2 \quad \text{Substituting}$$
$$\frac{9039.75}{8000} = (1 + i)^2$$
$$\pm 1.063 \approx 1 + i \quad \text{Taking the square root}$$
$$-1 \pm 1.063 \approx i \qquad \text{on both sides}$$
$$-1 + 1.063 \approx i \quad \text{or} \quad -1 - 1.063 \approx i$$
$$0.063 \approx i \quad \text{or} \qquad -2.063 \approx i$$

Only the positive result has meaning in this application. The interest rate is about 0.063, or 6.3%.

57. $d = \sqrt{(x_2 - x_1)^2 + (y_2 - y_1)^2}$

$ = \sqrt{[-1 - (-5)]^2 + (0 - 3)^2}$

$ = \sqrt{4^2 + (-3)^2} = \sqrt{16 + 9}$

$ = \sqrt{25} = 5$

59. $\quad (x - 3)^2 + (y + 5)^2 = 49$

$\quad (x - 3)^2 + [y - (-5)]^2 = 7^2$

Center: $(3, -5)$; radius: 7

61. $2y - 3 \geq 1 - y + 5$

$\quad 2y - 3 \geq 6 - y \qquad$ Collecting like terms

$\quad 3y - 3 \geq 6 \qquad\qquad$ Adding y

$\quad 3y \geq 9 \qquad\qquad\quad$ Adding 3

$\quad y \geq 3 \qquad\qquad\quad$ Dividing by 3

The solution set is $\{y | y \geq 3\}$, or $[3, \infty)$.

63. $|x + 6| \geq 7$

$\quad x + 6 \leq -7 \quad or \quad x + 6 \geq 7$

$\quad x \leq -13 \quad or \quad\quad x \geq 1$

The solution set is $\{x | x \leq -13 \; or \; x \geq 1\}$, or
$(-\infty, -13] \cup [1, \infty)$.

65. $f(x) = (x^5 - 1)^2(x^2 + 2)^3$

The leading term of $(x^5 - 1)^2$ is $(x^5)^2$, or x^{10}. The leading term of $(x^2 + 2)^3$ is $(x^2)^3$, or x^6. Then the leading term of $f(x)$ is $x^{10} \cdot x^6$, or x^{16}, and the degree of $f(x)$ is 16.

Exercise Set 4.2

1. $f(x) = x^5 - x^2 + 6$

a) This function has degree 5, so its graph can have at most 5 real zeros.

b) This function has degree 5, so its graph can have most 5 x-intercepts.

c) This function has degree 5, so its graph can have most $5 - 1$, or 4, turning points.

3. $f(x) = x^{10} - 2x^5 + 4x - 2$

a) This function has degree 10, so its graph can have at most 10 real zeros.

b) This function has degree 10, so its graph can have at most 10 x-intercepts.

c) This function has degree 10, so its graph can have at most $10 - 1$, or 9, turning points.

5. $f(x) = -x - x^3 = -x^3 - x$

a) This function has degree 3, so its graph can have at most 3 real zeros.

b) This function has degree 3, so its graph can have at most 3 x-intercepts.

c) This function has degree 3, so its graph can have at most $3 - 1$, or 2, turning points.

7. $f(x) = \dfrac{1}{4}x^2 - 5$

The leading term is $\dfrac{1}{4}x^2$. The sign of the leading coefficient, $\dfrac{1}{4}$, is positive and the degree, 2, is even, so we would choose either graph (b) or graph (d). Note also that $f(0) = -5$, so the y-intercept is $(0, -5)$. Thus, graph (d) is the graph of this function.

9. $f(x) = x^5 - x^4 + x^2 + 4$

The leading term is x^5. The sign of the leading coefficient, 1, is positive and the degree, 5, is odd. Thus, graph (f) is the graph of this function.

11. $f(x) = x^4 - 2x^3 + 12x^2 + x - 20$

The leading term is x^4. The sign of the leading coefficient, 1, is positive and the degree, 4, is even, so we would choose either graph (b) or graph (d). Note also that $f(0) = -20$, so the y-intercept is $(0, -20)$. Thus, graph (b) is the graph of this function.

13. $f(x) = -x^3 - 2x^2$

1. The leading term is $-x^3$. The degree, 3, is odd and the leading coefficient, -1, is negative so as $x \to \infty$, $f(x) \to -\infty$ and as $x \to -\infty$, $f(x) \to \infty$.

2. We solve $f(x) = 0$.

$\quad -x^3 - 2x^2 = 0$

$\quad -x^2(x + 2) = 0$

$\quad -x^2 = 0 \quad or \quad x + 2 = 0$

$\quad\quad x^2 = 0 \quad or \quad\quad\quad x = -2$

$\quad\quad\;\; x = 0 \quad or \quad\quad\quad x = -2$

The zeros of the function are 0 and -2, so the x-intercepts of the graph are $(0, 0)$ and $(-2, 0)$.

3. The zeros divide the x-axis into 3 intervals, $(-\infty, -2)$, $(-2, 0)$, and $(0, \infty)$. We choose a value for x from each interval and find $f(x)$. This tells us the sign of $f(x)$ for all values of x in that interval.

In $(-\infty, -2)$, test -3:

$f(-3) = -(-3)^3 - 2(-3)^2 = 9 > 0$

In $(-2, 0)$, test -1:

$f(-1) = -(-1)^3 - 2(-1)^2 = -1 < 0$

In $(0, \infty)$, test 1:

$f(1) = -1^3 - 2 \cdot 1^2 = -3 < 0$

Thus the graph lies above the x-axis on $(-\infty, -2)$ and below the x-axis on $(-2, 0)$ and $(0, \infty)$. We also know the points $(-3, 9)$, $(-1, -1)$, and $(1, -3)$ are on the graph.

4. From Step 2 we see that the y-intercept is $(0, 0)$.

5. We find additional points on the graph and then draw the graph.

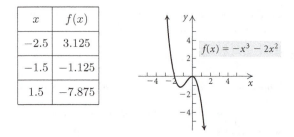

x	$f(x)$
-2.5	3.125
-1.5	-1.125
1.5	-7.875

6. Checking the graph as described on page 240 in the text, we see that it appears to be correct.

15. $h(x) = x^2 + 2x - 3$

1. The leading term is x^2. The degree, 2, is even and leading coefficient, 1, is positive so as $x \to \infty$, $h(x) \to \infty$ and as $x \to -\infty$, $h(x) \to \infty$.

2. We solve $h(x) = 0$.
$$x^2 + 2x - 3 = 0$$
$$(x + 3)(x - 1) = 0$$
$$x + 3 = 0 \quad or \quad x - 1 = 0$$
$$x = -3 \quad or \qquad x = 1$$

   The zeros of the function are -3 and 1, so the x-intercepts of the graph are $(-3, 0)$ and $(1, 0)$.

3. The zeros divide the x-axis into 3 intervals, $(-\infty, -3)$, $(-3, 1)$, and $(1, \infty)$. We choose a value for x from each interval and find $h(x)$. This tells us the sign of $h(x)$ for all values of x in that interval.
   In $(-\infty, -3)$, test -4:
   $$h(-4) = (-4)^2 + 2(-4) - 3 = 5 > 0$$
   In $(-3, 1)$, test 0:
   $$h(0) = 0^2 + 2 \cdot 0 - 3 = -3 < 0$$
   In $(1, \infty)$, test 2:
   $$h(2) = 2^2 + 2 \cdot 2 - 3 = 5 > 0$$

   Thus the graph lies above the x-axis on $(-\infty, -3)$ and on $(1, \infty)$. It lies below the x-axis on $(-3, 1)$. We also know the points $(-4, 5)$, $(0, -3)$, and $(2, 5)$ are on the graph.

4. From Step 3 we see that the y-intercept is $(0, -3)$.

5. We find additional points on the graph and then draw the graph.

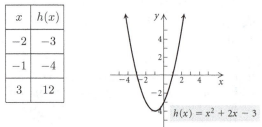

x	$h(x)$
-2	-3
-1	-4
3	12

6. Checking the graph as described on page 240 in the text, we see that it appears to be correct.

17. $h(x) = x^5 - 4x^3$

1. The leading term is x^5. The degree, 5, is odd and the leading coefficient, 1, is positive so as $x \to \infty$, $h(x) \to \infty$ and as $x \to -\infty$, $h(x) \to -\infty$.

2. We solve $h(x) = 0$.
$$x^5 - 4x^3 = 0$$
$$x^3(x^2 - 4) = 0$$
$$x^3(x + 2)(x - 2) = 0$$
$$x^3 = 0 \quad or \quad x + 2 = 0 \quad or \quad x - 2 = 0$$
$$x = 0 \quad or \qquad x = -2 \quad or \qquad x = 2$$

   The zeros of the function are 0, -2, and 2 so the x-intercepts of the graph are $(0, 0)$, $(-2, 0)$, and $(2, 0)$.

3. The zeros divide the x-axis into 4 intervals, $(-\infty, -2)$, $(-2, 0)$, $(0, 2)$, and $(2, \infty)$. We choose a value for x from each interval and find $h(x)$. This tells us the sign of $h(x)$ for all values of x in that interval.
   In $(-\infty, -2)$, test -3:
   $$h(-3) = (-3)^5 - 4(-3)^3 = -135 < 0$$
   In $(-2, 0)$, test -1:
   $$h(-1) = (-1)^5 - 4(-1)^3 = 3 > 0$$
   In $(0, 2)$, test 1:
   $$h(1) = 1^5 - 4 \cdot 1^3 = -3 < 0$$
   In $(2, \infty)$, test 3:
   $$h(3) = 3^5 - 4 \cdot 3^3 = 135 > 0$$

   Thus the graph lies below the x-axis on $(-\infty, -2)$ and on $(0, 2)$. It lies above the x-axis on $(-2, 0)$ and on $(2, \infty)$. We also know the points $(-3, -135)$, $(-1, 3)$, $(1, -3)$, and $(3, 135)$ are on the graph.

4. From Step 2 we see that the y-intercept is $(0, 0)$.

5. We find additional points on the graph and then draw the graph.

x	$h(x)$
-2.5	-35.2
-1.5	5.9
1.5	-5.9
2.5	35.2

$h(x) = x^5 - 4x^3$

6. Checking the graph as described on page 240 in the text, we see that it appears to be correct.

19. $h(x) = x(x - 4)(x + 1)(x - 2)$

1. The leading term is $x \cdot x \cdot x \cdot x$, or x^4. The degree, 4, is even and the leading coefficient, 1, is positive so as $x \to \infty$, $h(x) \to \infty$ and as $x \to -\infty$, $h(x) \to \infty$.

2. We see that the zeros of the function are 0, 4, -1, and 2 so the x-intercepts of the graph are $(0, 0)$, $(4, 0)$, $(-1, 0)$, and $(2, 0)$.

3. The zeros divide the x-axis into 5 intervals, $(-\infty, -1)$, $(-1, 0)$, $(0, 2)$, $(2, 4)$, and $(4, \infty)$. We choose a value for x from each interval and find $h(x)$. This tells us the sign of $h(x)$ for all values of x in that interval.

In $(-\infty, -1)$, test -2:

$h(-2) = -2(-2-4)(-2+1)(-2-2) = 48 > 0$

In $(-1, 0)$, test -0.5:

$h(-0.5) = (-0.5)(-0.5-4)(-0.5+1)(-0.5-2) = -2.8125 < 0$

In $(0, 2)$, test 1:

$h(1) = 1(1-4)(1+1)(1-2) = 6 > 0$

In $(2, 4)$, test 3:

$h(3) = 3(3-4)(3+1)(3-2) = -12 < 0$

In $(4, \infty)$, test 5:

$h(5) = 5(5-4)(5+1)(5-2) = 90 > 0$

Thus the graph lies above the x-axis on $(-\infty, -1)$, $(0, 2)$, and $(4, \infty)$. It lies below the x-axis on $(-1, 0)$ and on $(2, 4)$. We also know the points $(-2, 48)$, $(-0.5, -2.8125)$, $(1, 6)$, $(3, -12)$, and $(5, 90)$ are on the graph.

4. From Step 2 we see that the y-intercept is $(0, 0)$.

5. We find additional points on the graph and then draw the graph.

x	$h(x)$
-1.5	14.4
1.5	4.7
2.5	-6.6
4.5	30.9

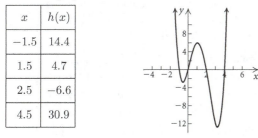

$h(x) = x(x-4)(x+1)(x-2)$

6. Checking the graph as described on page 240 in the text, we see that it appears to be correct.

21. $g(x) = -\dfrac{1}{4}x^3 - \dfrac{3}{4}x^2$

1. The leading term is $-\dfrac{1}{4}x^3$. The degree, 3, is odd and the leading coefficient, $-\dfrac{1}{4}$, is negative so as $x \to \infty$, $g(x) \to -\infty$ and as $x \to -\infty$, $g(x) \to \infty$.

2. We solve $g(x) = 0$.

$-\dfrac{1}{4}x^3 - \dfrac{3}{4}x^2 = 0$

$-\dfrac{1}{4}x^2(x+3) = 0$

$-\dfrac{1}{4}x^2 = 0 \ \ or \ \ x+3 = 0$

$x^2 = 0 \ \ or \ \ \ \ \ \ \ x = -3$

$x = 0 \ \ or \ \ \ \ \ \ \ x = -3$

The zeros of the function are 0 and -3, so the x-intercepts of the graph are $(0, 0)$ and $(-3, 0)$.

3. The zeros divide the x-axis into 3 intervals, $(-\infty, -3)$, $(-3, 0)$, and $(0, \infty)$. We choose a value for x from each interval and find $g(x)$. This tells us the sign of $g(x)$ for all values of x in that interval.

In $(-\infty, -3)$, test -4:

$g(-4) = -\dfrac{1}{4}(-4)^3 - \dfrac{3}{4}(-4)^2 = 4 > 0$

In $(-3, 0)$, test -1:

$g(-1) = -\dfrac{1}{4}(-1)^3 - \dfrac{3}{4}(-1)^2 = -\dfrac{1}{2} < 0$

In $(0, \infty)$, test 1:

$g(1) = -\dfrac{1}{4} \cdot 1^3 - \dfrac{3}{4} \cdot 1^2 = -1 < 0$

Thus the graph lies above the x-axis on $(-\infty, -3)$ and below the x-axis on $(-3, 0)$ and on $(0, \infty)$. We also know the points $(-4, 4)$, $\left(-1, -\dfrac{1}{2}\right)$, and $(1, -1)$ are on the graph.

4. From Step 2 we see that the y-intercept is $(0, 0)$.

5. We find additional points on the graph and then draw the graph.

x	$g(x)$
-2	-1
2	-5
2.5	-8.6

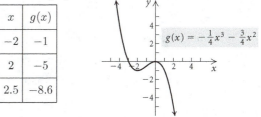

$g(x) = -\frac{1}{4}x^3 - \frac{3}{4}x^2$

6. Checking the graph as described on page 240 in the text, we see that it appears to be correct.

23. $g(x) = -x^4 - 2x^3$

1. The leading term is $-x^4$. The degree, 4, is even and the leading coefficient, -1, is negative so as $x \to \infty$, $g(x) \to -\infty$ and as $x \to -\infty$, $g(x) \to -\infty$.

2. We solve $f(x) = 0$.

$-x^4 - 2x^3 = 0$

$-x^3(x+2) = 0$

$-x^3 = 0 \ \ or \ \ x+2 = 0$

$x = 0 \ \ or \ \ \ \ \ \ \ x = -2$

The zeros of the function are 0 and -2, so the x-intercepts of the graph are $(0, 0)$ and $(-2, 0)$.

3. The zeros divide the x-axis into 3 intervals, $(-\infty, -2)$, $(-2, 0)$, and $(0, \infty)$. We choose a value for x from each interval and find $g(x)$. This tells us the sign of $g(x)$ for all values of x in that interval.

In $(-\infty, -2)$, test -3:

$g(-3) = -(-3)^4 - 2(-3)^3 = -27 < 0$

In $(-2, 0)$, test -1:

$g(-1) = -(-1)^4 - 2(-1)^3 = 1 > 0$

In $(0, \infty)$, test 1:

$g(1) = -(1)^4 - 2(1)^3 = -3 < 0$

Thus the graph lies below the x-axis on $(-\infty, -2)$ and $(0, \infty)$ and above the x-axis on $(-2, 0)$. We also know the points $(-3, -27)$, $(-1, 1)$, and $(1, -3)$ are on the graph.

4. From Step 2 we see that the y-intercept is $(0, 0)$.

5. We find additional points on the graph and then draw the graph.

x	$g(x)$
-2.5	-7.8
-1.5	1.7
0.5	-0.3
1.5	-11.8
2	-32

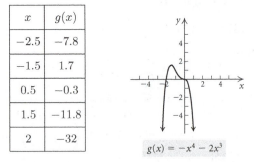

$$g(x) = -x^4 - 2x^3$$

6. Checking the graph as described on page 240 in the text, we see that it appears to be correct.

25. $f(x) = -\dfrac{1}{2}(x - 2)(x + 1)^2(x - 1)$

1. The leading term is $-\dfrac{1}{2} \cdot x \cdot x \cdot x \cdot x$, or $-\dfrac{1}{2}x^4$. The degree, 4, is even and the leading coefficient, $-\dfrac{1}{2}$, is negative so as $x \to \infty$, $f(x) \to -\infty$ and as $x \to -\infty$, $f(x) \to -\infty$.

2. We solve $f(x) = 0$.

$$-\frac{1}{2}(x - 2)(x + 1)^2(x - 1) = 0$$

$x - 2 = 0 \quad or \quad (x + 1)^2 = 0 \quad or \quad x - 1 = 0$

$\quad x = 2 \quad or \quad x + 1 = 0 \quad or \quad \quad x = 1$

$\quad x = 2 \quad or \quad \quad x = -1 \quad or \quad \quad x = 1$

The zeros of the function are 2, -1, and 1, so the x-intercepts of the graph are $(2, 0)$, $(-1, 0)$, and $(1, 0)$.

3. The zeros divide the x-axis into 4 intervals, $(-\infty, -1)$, $(-1, 1)$, $(1, 2)$, and $(2, \infty)$. We choose a value for x from each interval and find $f(x)$. This tells us the sign of $f(x)$ for all values of x in that interval.

In $(-\infty, -1)$, test -2:

$$f(-2) = -\frac{1}{2}(-2 - 2)(-2 + 1)^2(-2 - 1) = -6 < 0$$

In $(-1, 1)$, test 0:

$$f(0) = -\frac{1}{2}(0 - 2)(0 + 1)^2(0 - 1) = -1 < 0$$

In $(1, 2)$, test 1.5:

$$f(1.5) = -\frac{1}{2}(1.5 - 2)(1.5 + 1)^2(1.5 - 1) = 0.78125 > 0$$

In $(2, \infty)$, test 3:

$$f(3) = -\frac{1}{2}(3 - 2)(3 + 1)^2(3 - 1) = -16 < 0$$

Thus the graph lies below the x-axis on $(-\infty, -1)$, $(-1, 1)$, and $(2, \infty)$ and above the x-axis on $(1, 2)$. We also know the points $(-2, -6)$, $(0, -1)$, $(1.5, 0.78125)$, and $(3, -16)$ are on the graph.

4. From Step 2 we know that $f(0) = -1$ so the y-intercept is $(0, -1)$.

5. We find additional points on the graph and then draw the graph.

x	$f(x)$
-3	-40
-0.5	-0.5
0.5	-0.8
1.5	0.8

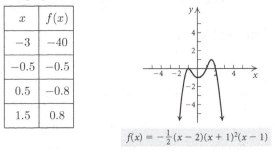

$$f(x) = -\frac{1}{2}(x - 2)(x + 1)^2(x - 1)$$

6. Checking the graph as described on page 240 in the text, we see that it appears to be correct.

27. $g(x) = -x(x - 1)^2(x + 4)^2$

1. The leading term is $-x \cdot x \cdot x \cdot x \cdot x$, or $-x^5$. The degree, 5, is odd and the leading coefficient, -1, is negative so as $x \to \infty$, $g(x) \to -\infty$ and as $x \to -\infty$, $g(x) \to \infty$.

2. We solve $g(x) = 0$.

$$-x(x - 1)^2(x + 4)^2 = 0$$

$-x = 0 \quad or \quad (x - 1)^2 = 0 \quad or \quad (x + 4)^2 = 0$

$\quad x = 0 \quad or \quad \quad x - 1 = 0 \quad or \quad \quad x + 4 = 0$

$\quad x = 0 \quad or \quad \quad \quad x = 1 \quad or \quad \quad \quad x = -4$

The zeros of the function are 0, 1, and -4, so the x-intercepts are $(0, 0)$, $(1, 0)$, and $(-4, 0)$.

3. The zeros divide the x-axis into 4 intervals, $(-\infty, -4)$, $(-4, 0)$, $(0, 1)$, and $(1, \infty)$. We choose a value for x from each interval and find $g(x)$. This tells us the sign of $g(x)$ for all values of x in that interval.

In $(-\infty, -4)$, test -5:

$$g(-5) = -(-5)(-5 - 1)^2(-5 + 4)^2 = 180 > 0$$

In $(-4, 0)$, test -1:

$$g(-1) = -(-1)(-1 - 1)^2(-1 + 4)^2 = 36 > 0$$

In $(0, 1)$, test 0.5:

$$g(0.5) = -0.5(0.5 - 1)^2(-0.5 + 4)^2 = -2.53125 < 0$$

In $(1, \infty)$, test 2:

$$g(2) = -2(2 - 1)^2(2 + 4)^2 = -72 < 0$$

Thus the graph lies above the x-axis on $(-\infty, -4)$ and on $(-4, 0)$ and below the x-axis on $(0, 1)$ and $(1, \infty)$. We also know the points $(-5, 180)$, $(-1, 36)$, $(0.5, -2.53125)$, and $(2, -72)$ are on the graph.

4. From Step 2 we see that the y-intercept is $(0, 0)$.

5. We find additional points on the graph and then draw the graph.

x	$g(x)$
-3	4.8
-2	72
1.5	-11.3

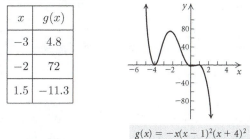

$$g(x) = -x(x-1)^2(x+4)^2$$

6. Checking the graph as described on page 240 in the text, we see that it appears to be correct.

29. $f(x) = (x-2)^2(x+1)^4$

1. The leading term is $x \cdot x \cdot x \cdot x \cdot x \cdot x$, or x^6. The degree, 6, is even and the leading coefficient, 1, is positive so as $x \to \infty$, $f(x) \to \infty$ and as $x \to -\infty$, $f(x) \to \infty$.

2. We see that the zeros of the function are 2 and -1 so the x-intercepts of the graph are $(2, 0)$ and $(-1, 0)$.

3. The zeros divide the x-axis into 3 intervals, $(-\infty, -1)$, $(-1, 2)$, and $(2, \infty)$. We choose a value for x from each interval and find $f(x)$. This tells us the sign of $f(x)$ for all values of x in that interval.
   In $(-\infty, -1)$, test -2:
   $$f(-2) = (-2-2)^2(-2+1)^4 = 16 > 0$$
   In $(-1, 2)$, test 0:
   $$f(0) = (0-2)^2(0+1)^4 = 4 > 0$$
   In $(2, \infty)$, test 3:
   $$f(3) = (3-2)^2(3+1)^4 = 256 > 0$$
   Thus the graph lies above the x-axis on all 3 intervals. We also know the points $(-2, 16)$, $(0, 4)$, and $(3, 256)$ are on the graph.

4. From Step 3 we know that $f(0) = 4$ so the y-intercept is $(0, 4)$.

5. We find additional points on the graph and then draw the graph.

x	$f(x)$
-1.5	0.8
-0.5	0.4
1	16
1.5	9.8

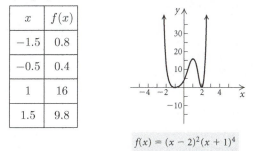

$$f(x) = (x-2)^2(x+1)^4$$

6. Checking the graph as described on page 240 in the text, we see that it appears to be correct.

31. $g(x) = -(x-1)^4$

1. The leading term is $-1 \cdot x \cdot x \cdot x \cdot x$, or $-x^4$. The degree, 4, is even and the leading coefficient, -1, is negative so as $x \to \infty$, $g(x) \to -\infty$ and as $x \to -\infty$, $g(x) \to -\infty$.

2. We see that the zero of the function is 1, so the x-intercept is $(1, 0)$.

3. The zero divides the x-axis into 2 intervals, $(-\infty, 1)$ and $(1, \infty)$. We choose a value for x from each interval and find $g(x)$. This tells us the sign of $g(x)$ for all values of x in that interval.
   In $(-\infty, 1)$, test 0:
   $$g(0) = -(0-1)^4 = -1 < 0$$
   In $(1, \infty)$, test 2:
   $$g(2) = -(2-1)^4 = -1 < 0$$
   Thus the graph lies below the x-axis on both intervals. We also know the points $(0, -1)$ and $(2, -1)$ are on the graph.

4. From Step 3 we know that $g(0) = -1$ so the y-intercept is $(0, -1)$.

5. We find additional points on the graph and then draw the graph.

x	$g(x)$
-1	-16
-0.5	-5.1
1.5	0.1
3	-16

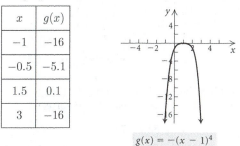

$$g(x) = -(x-1)^4$$

6. Checking the graph as described on page 240 in the text, we see that it appears to be correct.

33. $h(x) = x^3 + 3x^2 - x - 3$

1. The leading term is x^3. The degree, 3, is odd and the leading coefficient, 1, is positive so as $x \to \infty$, $h(x) \to \infty$ and as $x \to -\infty$, $h(x) \to -\infty$.

2. We solve $h(x) = 0$.
$$x^3 + 3x^2 - x - 3 = 0$$
$$x^2(x+3) - (x+3) = 0$$
$$(x+3)(x^2 - 1) = 0$$
$$(x+3)(x+1)(x-1) = 0$$
$$x+3 = 0 \quad or \quad x+1 = 0 \quad or \quad x-1 = 0$$
$$x = -3 \quad or \quad x = -1 \quad or \quad x = 1$$

The zeros of the function are -3, -1, and 1 so the x-intercepts of the graph are $(-3, 0)$, $(-1, 0)$, and $(1, 0)$.

3. The zeros divide the x-axis into 4 intervals, $(-\infty, -3)$, $(-3, -1)$, $(-1, 1)$, and $(1, \infty)$. We choose a value for x from each interval and find $h(x)$. This tells us the sign of $h(x)$ for all values of x in that interval.

In $(-\infty, -3)$, test -4:

$h(-4) = (-4)^3 + 3(-4)^2 - (-4) - 3 = -15 < 0$

In $(-3, -1)$, test -2:

$h(-2) = (-2)^3 + 3(-2)^2 - (-2) - 3 = 3 > 0$

In $(-1, 1)$, test 0:

$h(0) = 0^3 + 3 \cdot 0^2 - 0 - 3 = -3 < 0$

In $(1, \infty)$, test 2:

$h(2) = 2^3 + 3 \cdot 2^2 - 2 - 3 = 15 > 0$

Thus the graph lies below the x-axis on $(-\infty, -3)$ and on $(-1, 1)$ and above the x-axis on $(-3, -1)$ and on $(1, \infty)$. We also know the points $(-4, -15)$, $(-2, 3)$, $(0, -3)$, and $(2, 15)$ are on the graph.

4. From Step 3 we know that $h(0) = -3$ so the y-intercept is $(0, -3)$.

5. We find additional points on the graph and then draw the graph.

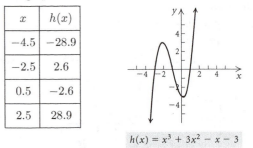

x	$h(x)$
-4.5	-28.9
-2.5	2.6
0.5	-2.6
2.5	28.9

$h(x) = x^3 + 3x^2 - x - 3$

6. Checking the graph as described on page 240 in the text, we see that it appears to be correct.

35. $f(x) = 6x^3 - 8x^2 - 54x + 72$

1. The leading term is $6x^3$. The degree, 3, is odd and the leading coefficient, 6, is positive so as $x \to \infty$, $f(x) \to \infty$ and as $x \to -\infty$, $f(x) \to -\infty$.

2. We solve $f(x) = 0$.

$6x^3 - 8x^2 - 54x + 72 = 0$

$2(3x^3 - 4x^2 - 27x + 36) = 0$

$2[x^2(3x - 4) - 9(3x - 4)] = 0$

$2(3x - 4)(x^2 - 9) = 0$

$2(3x - 4)(x + 3)(x - 3) = 0$

$3x - 4 = 0 \quad or \quad x + 3 = 0 \quad or \quad x - 3 = 0$

$x = \dfrac{4}{3} \quad or \quad\quad x = -3 \quad or \quad\quad x = 3$

The zeros of the function are $\dfrac{4}{3}$, -3, and 3, so the x-intercepts of the graph are $\left(\dfrac{4}{3}, 0\right)$, $(-3, 0)$, and $(3, 0)$.

3. The zeros divide the x-axis into 4 intervals, $(-\infty, -3)$, $\left(-3, \dfrac{4}{3}\right)$, $\left(\dfrac{4}{3}, 3\right)$, and $(3, \infty)$. We choose a value for x from each interval and find $f(x)$. This tells us the sign of $f(x)$ for all values of x in that interval.

In $(-\infty, -3)$, test -4:

$f(-4) = 6(-4)^3 - 8(-4)^2 - 54(-4) + 72 = -224 < 0$

In $\left(-3, \dfrac{4}{3}\right)$, test 0:

$f(0) = 6 \cdot 0^3 - 8 \cdot 0^2 - 54 \cdot 0 + 72 = 72 > 0$

In $\left(\dfrac{4}{3}, 3\right)$, test 2:

$f(2) = 6 \cdot 2^3 - 8 \cdot 2^2 - 54 \cdot 2 + 72 = -20 < 0$

In $(3, \infty)$, test 4:

$f(4) = 6 \cdot 4^3 - 8 \cdot 4^2 - 54 \cdot 4 + 72 = 112 > 0$

Thus the graph lies below the x-axis on $(-\infty, -3)$ and on $\left(\dfrac{4}{3}, 3\right)$ and above the x-axis on $\left(-3, \dfrac{4}{3}\right)$ and on $(3, \infty)$. We also know the points $(-4, -224)$, $(0, 72)$, $(2, -20)$, and $(4, 112)$ are on the graph.

4. From Step 3 we know that $f(0) = 72$ so the y-intercept is $(0, 72)$.

5. We find additional points on the graph and then draw the graph.

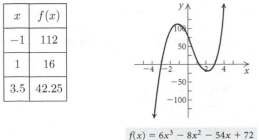

x	$f(x)$
-1	112
1	16
3.5	42.25

$f(x) = 6x^3 - 8x^2 - 54x + 72$

6. Checking the graph as described on page 240 in the text, we see that it appears to be correct.

37. We graph $g(x) = -x + 3$ for $x \le -2$, $g(x) = 4$ for $2 < x < 1$, and $g(x) = \dfrac{1}{2}x^3$ for $x \ge 1$.

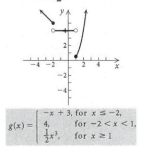

$$g(x) = \begin{cases} -x + 3, & \text{for } x \le -2, \\ 4, & \text{for } -2 < x < 1, \\ \frac{1}{2}x^3, & \text{for } x \ge 1 \end{cases}$$

39. $f(-5) = (-5)^3 + 3(-5)^2 - 9(-5) - 13 = -18$

$f(-4) = (-4)^3 + 3(-4)^2 - 9(-4) - 13 = 7$

By the intermediate value theorem, since $f(-5)$ and $f(-4)$ have opposite signs then $f(x)$ has a zero between -5 and -4.

41. $f(-3) = 3(-3)^2 - 2(-3) - 11 = 22$

$f(-2) = 3(-2)^2 - 2(-2) - 11 = 5$

Since both $f(-3)$ and $f(-2)$ are positive, we cannot use the intermediate value theorem to determine if there is a zero between -3 and -2.

43. $f(2) = 2^4 - 2 \cdot 2^2 - 6 = 2$

$f(3) = 3^4 - 2 \cdot 3^2 - 6 = 57$

Since both $f(2)$ and $f(3)$ are positive, we cannot use the intermediate value theorem to determine if there is a zero between 2 and 3.

45. $f(4) = 4^3 - 5 \cdot 4^2 + 4 = -12$

$f(5) = 5^3 - 5 \cdot 5^2 + 4 = 4$

By the intermediate value theorem, since $f(4)$ and $f(5)$ have opposite signs then $f(x)$ has a zero between 4 and 5.

47. The graph of $y = x$, or $y = x + 0$, has y-intercept $(0, 0)$, so (d) is the correct answer.

49. The graph of $y - 2x = 6$, or $y = 2x + 6$, has y-intercept $(0, 6)$, so (e) is the correct answer.

51. The graph of $y = 1 - x$, or $y = -x + 1$, has y-intercept $(0, 1)$, so (b) is the correct answer.

53.
$$2x - \frac{1}{2} = 4 - 3x$$

$$5x - \frac{1}{2} = 4 \qquad \text{Adding } 3x$$

$$5x = \frac{9}{2} \qquad \text{Adding } \frac{1}{2}$$

$$x = \frac{1}{5} \cdot \frac{9}{2} \qquad \text{Multiplying by } \frac{1}{5}$$

$$x = \frac{9}{10}$$

The solution is $\frac{9}{10}$.

55.
$$6x^2 - 23x - 55 = 0$$

$$(3x + 5)(2x - 11) = 0$$

$$3x + 5 = 0 \quad or \quad 2x - 11 = 0$$

$$3x = -5 \quad or \qquad 2x = 11$$

$$x = -\frac{5}{3} \quad or \qquad x = \frac{11}{2}$$

The solutions are $-\frac{5}{3}$ and $\frac{11}{2}$.

Exercise Set 4.3

1. a)

$$
\begin{array}{r}
x^3 - 7x^2 + 8x + 16 \\
x + 1 \overline{\smash{\big)}\ x^4 - 6x^3 + x^2 + 24x - 20} \\
\underline{x^4 + x^3} \\
-7x^3 + x^2 \\
\underline{-7x^3 - 7x^2} \\
8x^2 + 24x \\
\underline{8x^2 + 8x} \\
16x - 20 \\
\underline{16x + 16} \\
-4
\end{array}
$$

Since the remainder is not 0, $x + 1$ is not a factor of $f(x)$.

b)

$$
\begin{array}{r}
x^3 - 4x^2 - 7x + 10 \\
x - 2 \overline{\smash{\big)}\ x^4 - 6x^3 + x^2 + 24x - 20} \\
\underline{x^4 - 2x^3} \\
-4x^3 + x^2 \\
\underline{-4x^3 + 8x^2} \\
-7x^2 + 24x \\
\underline{-7x^2 + 14x} \\
10x - 20 \\
\underline{10x - 20} \\
0
\end{array}
$$

Since the remainder is 0, $x - 2$ is a factor of $f(x)$.

c)

$$
\begin{array}{r}
x^3 - 11x^2 + 56x - 256 \\
x + 5 \overline{\smash{\big)}\ x^4 - 6x^3 + x^2 + 24x - 20} \\
\underline{x^4 + 5x^3} \\
-11x^3 + x^2 \\
\underline{-11x^3 - 55x^2} \\
56x^2 + 24x \\
\underline{56x^2 + 280x} \\
-256x - 20 \\
\underline{-256x - 1280} \\
1260
\end{array}
$$

Since the remainder is not 0, $x + 5$ is not a factor of $f(x)$.

3. a)

$$
\begin{array}{r}
x^2 + 2x - 3 \\
x - 4 \overline{\smash{\big)}\ x^3 - 2x^2 - 11x + 12} \\
\underline{x^3 - 4x^2} \\
2x^2 - 11x \\
\underline{2x^2 - 8x} \\
-3x + 12 \\
\underline{-3x + 12} \\
0
\end{array}
$$

Since the remainder is 0, $x - 4$ is a factor of $g(x)$.

b)

$$
\begin{array}{r}
x^2 + x - 8 \\
x - 3 \overline{\smash{\big)}\ x^3 - 2x^2 - 11x + 12} \\
\underline{x^3 - 3x^2} \\
x^2 - 11x \\
\underline{x^2 - 3x} \\
-8x + 12 \\
\underline{-8x + 24} \\
-12
\end{array}
$$

Since the remainder is not 0, $x - 3$ is not a factor of $g(x)$.

c)

$$
\begin{array}{r}
x^2 - x - 12 \\
x - 1 \overline{\smash{\big)}\ x^3 - 2x^2 - 11x + 12} \\
\underline{x^3 - x^2} \\
-x^2 - 11x \\
\underline{-x^2 + x} \\
-12x + 12 \\
\underline{-12x + 12} \\
0
\end{array}
$$

Since the remainder is 0, $x - 1$ is a factor of $g(x)$.

5.
$$
\begin{array}{r}
x^2 - 2x + 4 \\
x+2 \overline{\smash{\big)}\ x^3 + 0x^2 + 0x - 8} \\
\underline{x^3 + 2x^2} \\
-2x^2 + 0x \\
\underline{-2x^2 - 4x} \\
4x - 8 \\
\underline{4x + 8} \\
-16
\end{array}
$$

$x^3 - 8 = (x+2)(x^2 - 2x + 4) - 16$

7.
$$
\begin{array}{r}
x^2 - 3x + 2 \\
x+9 \overline{\smash{\big)}\ x^3 + 6x^2 - 25x + 18} \\
\underline{x^3 + 9x^2} \\
-3x^2 - 25x \\
\underline{-3x^2 - 27x} \\
2x + 18 \\
\underline{2x + 18} \\
0
\end{array}
$$

$x^3 + 6x^2 - 25x + 18 = (x+9)(x^2 - 3x + 2) + 0$

9.
$$
\begin{array}{r}
x^3 - 2x^2 + 2x - 4 \\
x+2 \overline{\smash{\big)}\ x^4 + 0x^3 - 2x^2 + 0x + 3} \\
\underline{x^4 + 2x^3} \\
-2x^3 - 2x^2 \\
\underline{-2x^3 - 4x^2} \\
2x^2 + 0x \\
\underline{2x^2 + 4x} \\
-4x + 3 \\
\underline{-4x - 8} \\
11
\end{array}
$$

$x^4 - 2x^2 + 3 = (x+2)(x^3 - 2x^2 + 2x - 4) + 11$

11. $(2x^4 + 7x^3 + x - 12) \div (x+3)$

$= (2x^4 + 7x^3 + 0x^2 + x - 12) \div [x - (-3)]$

$$
\begin{array}{r|rrrrr}
-3 & 2 & 7 & 0 & 1 & -12 \\
   &   & -6 & -3 & 9 & -30 \\
\hline
   & 2 & 1 & -3 & 10 & -42
\end{array}
$$

The quotient is $2x^3 + x^2 - 3x + 10$. The remainder is -42.

13. $(x^3 - 2x^2 - 8) \div (x+2)$

$= (x^3 - 2x^2 + 0x - 8) \div [x - (-2)]$

$$
\begin{array}{r|rrrr}
-2 & 1 & -2 & 0 & -8 \\
   &   & -2 & 8 & -16 \\
\hline
   & 1 & -4 & 8 & -24
\end{array}
$$

The quotient is $x^2 - 4x + 8$. The remainder is -24.

15. $(3x^3 - x^2 + 4x - 10) \div (x+1)$

$= (3x^3 - x^2 + 4x - 10) \div [x - (-1)]$

$$
\begin{array}{r|rrrr}
-1 & 3 & -1 & 4 & -10 \\
   &   & -3 & 4 & -8 \\
\hline
   & 3 & -4 & 8 & -18
\end{array}
$$

The quotient is $3x^2 - 4x + 8$. The remainder is -18.

17. $(x^5 + x^3 - x) \div (x - 3)$

$= (x^5 + 0x^4 + x^3 + 0x^2 - x + 0) \div (x - 3)$

$$
\begin{array}{r|rrrrrr}
3 & 1 & 0 & 1 & 0 & -1 & 0 \\
  &   & 3 & 9 & 30 & 90 & 267 \\
\hline
  & 1 & 3 & 10 & 30 & 89 & 267
\end{array}
$$

The quotient is $x^4 + 3x^3 + 10x^2 + 30x + 89$.

The remainder is 267.

19. $(x^4 - 1) \div (x - 1)$

$= (x^4 + 0x^3 + 0x^2 + 0x - 1) \div (x - 1)$

$$
\begin{array}{r|rrrrr}
1 & 1 & 0 & 0 & 0 & -1 \\
  &   & 1 & 1 & 1 & 1 \\
\hline
  & 1 & 1 & 1 & 1 & 0
\end{array}
$$

The quotient is $x^3 + x^2 + x + 1$. The remainder is 0.

21. $(2x^4 + 3x^2 - 1) \div \left(x - \dfrac{1}{2}\right)$

$(2x^4 + 0x^3 + 3x^2 + 0x - 1) \div \left(x - \dfrac{1}{2}\right)$

$$
\begin{array}{r|rrrrr}
\frac{1}{2} & 2 & 0 & 3 & 0 & -1 \\
            &   & 1 & \frac{1}{2} & \frac{7}{4} & \frac{7}{8} \\
\hline
            & 2 & 1 & \frac{7}{2} & \frac{7}{4} & -\frac{1}{8}
\end{array}
$$

The quotient is $2x^3 + x^2 + \dfrac{7}{2}x + \dfrac{7}{4}$. The remainder is $-\dfrac{1}{8}$.

23. $f(x) = x^3 - 6x^2 + 11x - 6$

Find $f(1)$.

$$
\begin{array}{r|rrrr}
1 & 1 & -6 & 11 & -6 \\
  &   & 1 & -5 & 6 \\
\hline
  & 1 & -5 & 6 & 0
\end{array}
$$

$f(1) = 0$

Find $f(-2)$.

$$
\begin{array}{r|rrrr}
-2 & 1 & -6 & 11 & -6 \\
   &   & -2 & 16 & -54 \\
\hline
   & 1 & -8 & 27 & -60
\end{array}
$$

$f(-2) = -60$

Find $f(3)$.

$$
\begin{array}{r|rrrr}
3 & 1 & -6 & 11 & -6 \\
  &   & 3 & -9 & 6 \\
\hline
  & 1 & -3 & 2 & 0
\end{array}
$$

$f(3) = 0$

25. $f(x) = x^4 - 3x^3 + 2x + 8$

Find $f(-1)$.

$$
\begin{array}{r|rrrrr}
-1 & 1 & -3 & 0 & 2 & 8 \\
   &   & -1 & 4 & -4 & 2 \\
\hline
   & 1 & -4 & 4 & -2 & 10
\end{array}
$$

$f(-1) = 10$

Find $f(4)$.

$$
\begin{array}{r|rrrrr}
4 & 1 & -3 & 0 & 2 & 8 \\
  &   & 4 & 4 & 16 & 72 \\
\hline
  & 1 & 1 & 4 & 18 & 80
\end{array}
$$

$f(4) = 80$

Find $f(-5)$.

$$
\begin{array}{r|rrrrr}
-5 & 1 & -3 & 0 & 2 & 8 \\
   &   & -5 & 40 & -200 & 990 \\
\hline
   & 1 & -8 & 40 & -198 & 998
\end{array}
$$

$f(-5) = 998$

27. $f(x) = 2x^5 - 3x^4 + 2x^3 - x + 8$

Find $f(20)$.

$$
\begin{array}{r|rrrrrr}
20 & 2 & -3 & 2 & 0 & -1 & 8 \\
  &   & 40 & 740 & 14{,}840 & 296{,}800 & 5{,}935{,}980 \\
\hline
  & 2 & 37 & 742 & 14{,}840 & 296{,}799 & 5{,}935{,}988
\end{array}
$$

$f(20) = 5{,}935{,}988$

Find $f(-3)$.

$$
\begin{array}{r|rrrrrr}
-3 & 2 & -3 & 2 & 0 & -1 & 8 \\
   &   & -6 & 27 & -87 & 261 & -780 \\
\hline
   & 2 & -9 & 29 & -87 & 260 & -772
\end{array}
$$

$f(-3) = -772$

29. $f(x) = x^4 - 16$

Find $f(2)$.

$$
\begin{array}{r|rrrrr}
2 & 1 & 0 & 0 & 0 & -16 \\
  &   & 2 & 4 & 8 & 16 \\
\hline
  & 1 & 2 & 4 & 8 & 0
\end{array}
$$

$f(2) = 0$

Find $f(-2)$.

$$
\begin{array}{r|rrrrr}
-2 & 1 & 0 & 0 & 0 & -16 \\
   &   & -2 & 4 & -8 & 16 \\
\hline
   & 1 & -2 & 4 & -8 & 0
\end{array}
$$

$f(-2) = 0$

Find $f(3)$.

$$
\begin{array}{r|rrrrr}
3 & 1 & 0 & 0 & 0 & -16 \\
  &   & 3 & 9 & 27 & 81 \\
\hline
  & 1 & 3 & 9 & 27 & 65
\end{array}
$$

$f(3) = 65$

Find $f(1-\sqrt{2})$.

$$
\begin{array}{r|rrrrr}
1-\sqrt{2} & 1 & 0 & 0 & 0 & -16 \\
  &   & 1-\sqrt{2} & 3-2\sqrt{2} & 7-5\sqrt{2} & 17-12\sqrt{2} \\
\hline
  & 1 & 1-\sqrt{2} & 3-2\sqrt{2} & 7-5\sqrt{2} & 1-12\sqrt{2}
\end{array}
$$

$f(1-\sqrt{2}) = 1 - 12\sqrt{2}$

31. $f(x) = 3x^3 + 5x^2 - 6x + 18$

If -3 is a zero of $f(x)$, then $f(-3) = 0$. Find $f(-3)$ using synthetic division.

$$
\begin{array}{r|rrrr}
-3 & 3 & 5 & -6 & 18 \\
   &   & -9 & 12 & -18 \\
\hline
   & 3 & -4 & 6 & 0
\end{array}
$$

Since $f(-3) = 0$, -3 is a zero of $f(x)$.

If 2 is a zero of $f(x)$, then $f(2) = 0$. Find $f(2)$ using synthetic division.

$$
\begin{array}{r|rrrr}
2 & 3 & 5 & -6 & 18 \\
  &   & 6 & 22 & 32 \\
\hline
  & 3 & 11 & 16 & 50
\end{array}
$$

Since $f(2) \neq 0$, 2 is not a zero of $f(x)$.

33. $h(x) = x^4 + 4x^3 + 2x^2 - 4x - 3$

If -3 is a zero of $h(x)$, then $h(-3) = 0$. Find $h(-3)$ using synthetic division.

$$
\begin{array}{r|rrrrr}
-3 & 1 & 4 & 2 & -4 & -3 \\
   &   & -3 & -3 & 3 & 3 \\
\hline
   & 1 & 1 & -1 & -1 & 0
\end{array}
$$

Since $h(-3) = 0$, -3 is a zero of $h(x)$.

If 1 is a zero of $h(x)$, then $h(1) = 0$. Find $h(1)$ using synthetic division.

$$
\begin{array}{r|rrrrr}
1 & 1 & 4 & 2 & -4 & -3 \\
  &   & 1 & 5 & 7 & 3 \\
\hline
  & 1 & 5 & 7 & 3 & 0
\end{array}
$$

Since $h(1) = 0$, 1 is a zero of $h(x)$.

35. $g(x) = x^3 - 4x^2 + 4x - 16$

If i is a zero of $g(x)$, then $g(i) = 0$. Find $g(i)$ using synthetic division. Keep in mind that $i^2 = -1$.

$$
\begin{array}{r|rrrr}
i & 1 & -4 & 4 & -16 \\
  &   & i & -4i-1 & 3i+4 \\
\hline
  & 1 & -4+i & 3-4i & -12+3i
\end{array}
$$

Since $g(i) \neq 0$, i is not a zero of $g(x)$.

If $-2i$ is a zero of $g(x)$, then $g(-2i) = 0$. Find $g(-2i)$ using synthetic division. Keep in mind that $i^2 = -1$.

$$
\begin{array}{r|rrrr}
-2i & 1 & -4 & 4 & -16 \\
    &   & -2i & 8i-4 & 16 \\
\hline
    & 1 & -4-2i & 8i & 0
\end{array}
$$

Since $g(-2i) = 0$, $-2i$ is a zero of $g(x)$.

37. $f(x) = x^3 - \dfrac{7}{2}x^2 + x - \dfrac{3}{2}$

If -3 is a zero of $f(x)$, then $f(-3) = 0$. Find $f(-3)$ using synthetic division.

$$
\begin{array}{r|rrrr}
-3 & 1 & -\frac{7}{2} & 1 & -\frac{3}{2} \\
   &   & -3 & \frac{39}{2} & -\frac{123}{2} \\
\hline
   & 1 & -\frac{13}{2} & \frac{41}{2} & -63
\end{array}
$$

Since $f(-3) \neq 0$, -3 is not a zero of $f(x)$.

If $\dfrac{1}{2}$ is a zero of $f(x)$, then $f\left(\dfrac{1}{2}\right) = 0$.

Find $f\left(\dfrac{1}{2}\right)$ using synthetic division.

$$
\begin{array}{r|rrrr}
\frac{1}{2} & 1 & -\frac{7}{2} & 1 & -\frac{3}{2} \\
   &   & \frac{1}{2} & -\frac{3}{2} & -\frac{1}{4} \\
\hline
   & 1 & -3 & -\frac{1}{2} & -\frac{7}{4}
\end{array}
$$

Since $f\left(\dfrac{1}{2}\right) \neq 0$, $\dfrac{1}{2}$ is not a zero of $f(x)$.

39. $f(x) = x^3 + 4x^2 + x - 6$

Try $x - 1$. Use synthetic division to see whether $f(1) = 0$.

$$
\begin{array}{r|rrrr}
1 & 1 & 4 & 1 & -6 \\
  &   & 1 & 5 & 6 \\
\hline
  & 1 & 5 & 6 & 0
\end{array}
$$

Since $f(1) = 0$, $x - 1$ is a factor of $f(x)$. Thus $f(x) = (x-1)(x^2 + 5x + 6)$.

Factoring the trinomial we get

$f(x) = (x-1)(x+2)(x+3)$.

To solve the equation $f(x) = 0$, use the principle of zero products.

$$(x - 1)(x + 2)(x + 3) = 0$$

$x - 1 = 0 \ \ or \ \ x + 2 = 0 \ \ \ or \ \ x + 3 = 0$

$\ \ \ \ x = 1 \ \ or \ \ \ \ \ \ \ x = -2 \ or \ \ \ \ \ \ x = -3$

The solutions are 1, -2, and -3.

41. $f(x) = x^3 - 6x^2 + 3x + 10$

Try $x - 1$. Use synthetic division to see whether $f(1) = 0$.

$$
\begin{array}{r|rrrr}
1 & 1 & -6 & 3 & 10 \\
  &   & 1 & -5 & -2 \\
\hline
  & 1 & -5 & -2 & 8
\end{array}
$$

Since $f(1) \neq 0$, $x - 1$ is not a factor of $P(x)$.

Try $x+1$. Use synthetic division to see whether $f(-1) = 0$.

$$
\begin{array}{r|rrrr}
-1 & 1 & -6 & 3 & 10 \\
  &   & -1 & 7 & -10 \\
\hline
  & 1 & -7 & 10 & 0
\end{array}
$$

Since $f(-1) = 0$, $x + 1$ is a factor of $f(x)$.

Thus $f(x) = (x + 1)(x^2 - 7x + 10)$.

Factoring the trinomial we get

$f(x) = (x + 1)(x - 2)(x - 5)$.

To solve the equation $f(x) = 0$, use the principle of zero products.

$$(x + 1)(x - 2)(x - 5) = 0$$

$x + 1 = 0 \ \ \ or \ \ x - 2 = 0 \ or \ \ x - 5 = 0$

$\ \ \ \ x = -1 \ or \ \ \ \ \ \ x = 2 \ or \ \ \ \ \ \ \ x = 5$

The solutions are -1, 2, and 5.

43. $f(x) = x^3 - x^2 - 14x + 24$

Try $x + 1$, $x - 1$, and $x + 2$. Using synthetic division we find that $f(-1) \neq 0$, $f(1) \neq 0$ and $f(-2) \neq 0$. Thus $x + 1$, $x - 1$, and $x + 2$, are not factors of $f(x)$.

Try $x - 2$. Use synthetic division to see whether $f(2) = 0$.

$$
\begin{array}{r|rrrr}
2 & 1 & -1 & -14 & 24 \\
  &   & 2 & 2 & -24 \\
\hline
  & 1 & 1 & -12 & 0
\end{array}
$$

Since $f(2) = 0$, $x - 2$ is a factor of $f(x)$. Thus $f(x) = (x - 2)(x^2 + x - 12)$.

Factoring the trinomial we get

$f(x) = (x - 2)(x + 4)(x - 3)$

To solve the equation $f(x) = 0$, use the principle of zero products.

$$(x - 2)(x + 4)(x - 3) = 0$$

$x - 2 = 0 \ \ or \ \ x + 4 = 0 \ \ \ or \ \ x - 3 = 0$

$\ \ \ \ x = 2 \ \ or \ \ \ \ \ \ x = -4 \ or \ \ \ \ \ \ x = 3$

The solutions are 2, -4, and 3.

45. $f(x) = x^4 - 7x^3 + 9x^2 + 27x - 54$

Try $x + 1$ and $x - 1$. Using synthetic division we find that $f(-1) \neq 0$ and $f(1) \neq 0$. Thus $x + 1$ and $x - 1$ are not factors of $f(x)$. Try $x + 2$. Use synthetic division to see whether $f(-2) = 0$.

$$
\begin{array}{r|rrrrr}
-2 & 1 & -7 & 9 & 27 & -54 \\
  &   & -2 & 18 & -54 & 54 \\
\hline
  & 1 & -9 & 27 & -27 & 0
\end{array}
$$

Since $f(-2) = 0$, $x + 2$ is a factor of $f(x)$. Thus $f(x) = (x + 2)(x^3 - 9x^2 + 27x - 27)$.

We continue to use synthetic division to factor $g(x) = x^3 - 9x^2 + 27x - 27$. Trying $x + 2$ again and $x - 2$ we find that $g(-2) \neq 0$ and $g(2) \neq 0$. Thus $x + 2$ and $x - 2$ are not factors of $g(x)$. Try $x - 3$.

$$
\begin{array}{r|rrrr}
3 & 1 & -9 & 27 & -27 \\
  &   & 3 & -18 & 27 \\
\hline
  & 1 & -6 & 9 & 0
\end{array}
$$

Since $g(3) = 0$, $x - 3$ is a factor of $x^3 - 9x^2 + 27x - 27$.

Thus $f(x) = (x + 2)(x - 3)(x^2 - 6x + 9)$.

Factoring the trinomial we get

$f(x) = (x + 2)(x - 3)(x - 3)^2$, or $f(x) = (x + 2)(x - 3)^3$.

To solve the equation $f(x) = 0$, use the principle of zero products.

$$(x + 2)(x - 3)(x - 3)(x - 3) = 0$$

$x + 2 = 0 \ \ \ or \ \ x - 3 = 0 \ or \ x - 3 = 0 \ or \ x - 3 = 0$

$\ \ \ \ x = -2 \ or \ \ \ \ \ \ x = 3 \ or \ \ \ \ \ \ x = 3 \ or \ \ \ \ \ \ x = 3$

The solutions are -2 and 3.

47. $f(x) = x^4 - x^3 - 19x^2 + 49x - 30$

Try $x - 1$. Use synthetic division to see whether $f(1) = 0$.

$$
\begin{array}{r|rrrrr}
1 & 1 & -1 & -19 & 49 & -30 \\
  &   & 1 & 0 & -19 & 30 \\
\hline
  & 1 & 0 & -19 & 30 & 0
\end{array}
$$

Since $f(1) = 0$, $x - 1$ is a factor of $f(x)$. Thus $f(x) = (x - 1)(x^3 - 19x + 30)$.

We continue to use synthetic division to factor $g(x) = x^3 - 19x + 30$. Trying $x - 1$, $x + 1$, and $x + 2$ we find that $g(1) \neq 0$, $g(-1) \neq 0$, and $g(-2) \neq 0$. Thus $x - 1$, $x + 1$, and $x + 2$ are not factors of $x^3 - 19x + 30$. Try $x - 2$.

$$
\begin{array}{r|rrrr}
2 & 1 & 0 & -19 & 30 \\
  &   & 2 & 4 & -30 \\
\hline
  & 1 & 2 & -15 & 0
\end{array}
$$

Since $g(2) = 0$, $x - 2$ is a factor of $x^3 - 19x + 30$.

Thus $f(x) = (x - 1)(x - 2)(x^2 + 2x - 15)$.

Factoring the trinomial we get

$f(x) = (x - 1)(x - 2)(x - 3)(x + 5)$.

To solve the equation $f(x) = 0$, use the principle of zero products.

$$(x - 1)(x - 2)(x - 3)(x + 5) = 0$$

$x - 1 = 0 \ or \ x - 2 = 0 \ or \ x - 3 = 0 \ or \ x + 5 = 0$

$\ \ \ \ x = 1 \ or \ \ \ \ \ x = 2 \ or \ \ \ \ \ x = 3 \ or \ \ \ \ \ x = -5$

The solutions are 1, 2, 3, and -5.

49. $f(x) = x^4 - x^3 - 7x^2 + x + 6$

1. The leading term is x^4. The degree, 4, is even and the leading coefficient, 1, is positive so as $x \to \infty$, $f(x) \to \infty$ and as $x \to -\infty$, $f(x) \to \infty$.

2. Find the zeros of the function. We first use synthetic division to determine if $f(1) = 0$.

$$\begin{array}{r|rrrrr} 1 & 1 & -1 & -7 & 1 & 6 \\ & & 1 & 0 & -7 & -6 \\ \hline & 1 & 0 & -7 & -6 & 0 \end{array}$$

1 is a zero of the function and we have $f(x) = (x-1)(x^3 - 7x - 6)$.

Synthetic division shows that -1 is a zero of $g(x) = x^3 - 7x - 6$.

$$\begin{array}{r|rrrr} -1 & 1 & 0 & -7 & -6 \\ & & -1 & 1 & 6 \\ \hline & 1 & -1 & -6 & 0 \end{array}$$

Then we have $f(x) = (x-1)(x+1)(x^2 - x - 6)$.

To find the other zeros we solve the following equation:

$$x^2 - x - 6 = 0$$
$$(x-3)(x+2) = 0$$
$$x - 3 = 0 \;\; or \;\; x + 2 = 0$$
$$x = 3 \;\; or \;\;\;\;\; x = -2$$

The zeros of the function are 1, -1, 3, and -2 so the x-intercepts of the graph are $(1,0)$, $(-1,0)$, $(3,0)$, and $(-2,0)$.

3. The zeros divide the x-axis into five intervals, $(-\infty, -2)$, $(-2, -1)$, $(-1, 1)$, $(1, 3)$, and $(3, \infty)$. We choose a value for x from each interval and find $f(x)$. This tells us the sign of $f(x)$ for all values of x in the interval.

In $(-\infty, -2)$, test -3:

$f(-3) = (-3)^4 - (-3)^3 - 7(-3)^2 + (-3) + 6 = 48 > 0$

In $(-2, -1)$, test -1.5:

$f(-1.5) = (-1.5)^4 - (-1.5)^3 - 7(-1.5)^2 + (-1.5) + 6 = -2.8125 < 0$

In $(-1, 1)$, test 0:

$f(0) = 0^4 - 0^3 - 7 \cdot 0^2 + 0 + 6 = 6 > 0$

In $(1, 3)$, test 2:

$f(2) = 2^4 - 2^3 - 7 \cdot 2^2 + 2 + 6 = -12 < 0$

In $(3, \infty)$, test 4:

$f(4) = 4^4 - 4^3 - 7 \cdot 4^2 + 4 + 6 = 90 > 0$

Thus the graph lies above the x-axis on $(-\infty, -2)$, on $(-1, 1)$, and on $(3, \infty)$. It lies below the x-axis on $(-2, -1)$ and on $(1, 3)$. We also know the points $(-3, 48)$, $(-1.5, -2.8125)$, $(0, 6)$, $(2, -12)$, and $(4, 90)$ are on the graph.

4. From Step 3 we see that $f(0) = 6$ so the y-intercept is $(0, 6)$.

5. We find additional points on the graph and draw the graph.

x	$f(x)$
-2.5	14.3
-0.5	3.9
0.5	4.7
2.5	-11.8

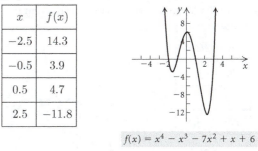

$$f(x) = x^4 - x^3 - 7x^2 + x + 6$$

6. Checking the graph as described on page 240 in the text, we see that it appears to be correct.

51. $f(x) = x^3 - 7x + 6$

1. The leading term is x^3. The degree, 3, is odd and the leading coefficient, 1, is positive so as $x \to \infty$, $f(x) \to \infty$ and as $x \to -\infty$, $f(x) \to -\infty$.

2. Find the zeros of the function. We first use synthetic division to determine if $f(1) = 0$.

$$\begin{array}{r|rrrr} 1 & 1 & 0 & -7 & 6 \\ & & 1 & 1 & -6 \\ \hline & 1 & 1 & -6 & 0 \end{array}$$

1 is a zero of the function and we have $f(x) = (x-1)(x^2 + x - 6)$. To find the other zeros we solve the following equation.

$$x^2 + x - 6 = 0$$
$$(x+3)(x-2) = 0$$
$$x + 3 = 0 \;\; or \;\; x - 2 = 0$$
$$x = -3 \;\; or \;\;\;\;\; x = 2$$

The zeros of the function are 1, -3, and 2 so the x-intercepts of the graph are $(1,0)$, $(-3,0)$, and $(2,0)$.

3. The zeros divide the x-axis into four intervals, $(-\infty, -3)$, $(-3, 1)$, $(1, 2)$, and $(2, \infty)$. We choose a value for x from each interval and find $f(x)$. This tells us the sign of $f(x)$ for all values of x in the interval.

In $(-\infty, -3)$, test -4:

$f(-4) = (-4)^3 - 7(-4) + 6 = -30 < 0$

In $(-3, 1)$, test 0:

$f(0) = 0^3 - 7 \cdot 0 + 6 = 6 > 0$

In $(1, 2)$, test 1.5:

$f(1.5) = (1.5)^3 - 7(1.5) + 6 = -1.125 < 0$

In $(2, \infty)$, test 3:

$f(3) = 3^3 - 7 \cdot 3 + 6 = 12 > 0$

Thus the graph lies below the x-axis on $(-\infty, -3)$ and on $(1, 2)$. It lies above the x-axis on $(-3, 1)$ and on $(2, \infty)$. We also know the points $(-4, -30)$, $(0, 6)$, $(1.5, -1.125)$, and $(3, 12)$ are on the graph.

4. From Step 3 we see that $f(0) = 6$ so the y-intercept is $(0, 6)$.

5. We find additional points on the graph and draw the graph.

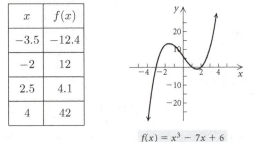

x	$f(x)$
-3.5	-12.4
-2	12
2.5	4.1
4	42

$$f(x) = x^3 - 7x + 6$$

6. Checking the graph as described on page 240 in the text, we see that it appears to be correct.

53. $f(x) = -x^3 + 3x^2 + 6x - 8$

1. The leading term is $-x^3$. The degree, 3, is odd and the leading coefficient, -1, is negative so as $x \to \infty$, $f(x) \to -\infty$ and as $x \to -\infty$, $f(x) \to \infty$.

2. Find the zeros of the function. We first use synthetic division to determine if $f(1) = 0$.

$$\begin{array}{r|rrrr} 1 & -1 & 3 & 6 & -8 \\ & & -1 & 2 & 8 \\ \hline & -1 & 2 & 8 & 0 \end{array}$$

1 is a zero of the function and we have $f(x) = (x - 1)(-x^2 + 2x + 8)$. To find the other zeros we solve the following equation.

$$-x^2 + 2x + 8 = 0$$
$$x^2 - 2x - 8 = 0$$
$$(x - 4)(x + 2) = 0$$
$$x - 4 = 0 \ \text{ or } \ x + 2 = 0$$
$$x = 4 \ \text{ or } \ \quad x = -2$$

The zeros of the function are 1, 4, and -2 so the x-intercepts of the graph are $(1, 0)$, $(4, 0)$, and $(-2, 0)$.

3. The zeros divide the x-axis into four intervals, $(-\infty, -2)$, $(-2, 1)$, $(1, 4)$, and $(4, \infty)$. We choose a value for x from each interval and find $f(x)$. This tells us the sign of $f(x)$ for all values of x in the interval.

In $(-\infty, -2)$, test -3:
$$f(-3) = -(-3)^3 + 3(-3)^2 + 6(-3) - 8 = 28 > 0$$

In $(-2, 1)$, test 0:
$$f(0) = -0^3 + 3 \cdot 0^2 + 6 \cdot 0 - 8 = -8 < 0$$

In $(1, 4)$, test 2:
$$f(2) = -2^3 + 3 \cdot 2^2 + 6 \cdot 2 - 8 = 8 > 0$$

In $(4, \infty)$, test 5:
$$f(5) = -5^3 + 3 \cdot 5^2 + 6 \cdot 5 - 8 = -28 < 0$$

Thus the graph lies above the x-axis on $(-\infty, -2)$ and on $(1, 4)$. It lies below the x-axis on $(-2, 1)$ and on $(4, \infty)$. We also know the points $(-3, 28)$, $(0, -8)$, $(2, 8)$, and $(5, -28)$ are on the graph.

4. From Step 3 we see that $f(0) = -8$ so the y-intercept is $(0, -8)$.

5. We find additional points on the graph and draw the graph.

x	$f(x)$
-2.5	11.4
-1	-10
3	10
4.5	-11.4

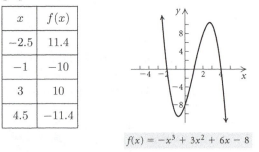

$$f(x) = -x^3 + 3x^2 + 6x - 8$$

6. Checking the graph as described on page 240 in the text, we see that it appears to be correct.

55.
$$2x^2 + 12 = 5x$$
$$2x^2 - 5x + 12 = 0$$
$$a = 2, \ b = -5, \ c = 12$$
$$x = \frac{-b \pm \sqrt{b^2 - 4ac}}{2a}$$
$$x = \frac{-(-5) \pm \sqrt{(-5)^2 - 4 \cdot 2 \cdot 12}}{2 \cdot 2}$$
$$= \frac{5 \pm \sqrt{-71}}{4}$$
$$= \frac{5 \pm i\sqrt{71}}{4} = \frac{5}{4} \pm \frac{\sqrt{71}}{4}i$$

The solutions are $\frac{5}{4} + \frac{\sqrt{71}}{4}i$ and $\frac{5}{4} - \frac{\sqrt{71}}{4}i$, or $\frac{5}{4} \pm \frac{\sqrt{71}}{4}i$.

57. We substitute -14 for $g(x)$ and solve for x.
$$-14 = x^2 + 5x - 14$$
$$0 = x^2 + 5x$$
$$0 = x(x + 5)$$
$$x = 0 \ \text{ or } \ x + 5 = 0$$
$$x = 0 \ \text{ or } \ \quad x = -5$$

When the output is -14, the input is 0 or -5.

59. We substitute -20 for $g(x)$ and solve for x.
$$-20 = x^2 + 5x - 14$$
$$0 = x^2 + 5x + 6$$
$$0 = (x + 3)(x + 2)$$
$$x + 3 = 0 \ \text{ or } \ x + 2 = 0$$
$$x = -3 \ \text{ or } \ \quad x = -2$$

When the output is -20, the input is -3 or -2.

61. Let b and h represent the length of the base and the height of the triangle, respectively.

$b + h = 30$, so $b = 30 - h$.

$$A = \frac{1}{2}bh = \frac{1}{2}(30 - h)h = -\frac{1}{2}h^2 + 15h$$

Find the value of h for which A is a maximum:
$$h = \frac{-15}{2(-1/2)} = 15$$

When $h = 15$, $b = 30 - 15 = 15$.

The area is a maximum when the base and the height are each 15 in.

63. a) -4, -3, 2, and 5 are zeros of the function, so $x + 4$, $x + 3$, $x - 2$, and $x - 5$ are factors.

b) We first write the product of the factors:

$$P(x) = (x + 4)(x + 3)(x - 2)(x - 5)$$

Note that $P(0) = 4 \cdot 3(-2)(-5) > 0$ and the graph shows a positive y-intercept, so this function is a correct one.

c) Yes; two examples are $f(x) = c \cdot P(x)$ for any non-zero constant c and $g(x) = (x - a)P(x)$.

d) No; only the function in part (b) has the given graph.

65. Divide $x^3 - kx^2 + 3x + 7k$ by $x + 2$.

$$
\begin{array}{r|rrrr}
-2 & 1 & -k & 3 & 7k \\
   &   & -2 & 2k+4 & -4k-14 \\
\hline
   & 1 & -k-2 & 2k+7 & 3k-14 \\
\end{array}
$$

Thus $P(-2) = 3k - 14$.

We know that if $x + 2$ is a factor of $f(x)$, then $f(-2) = 0$.

We solve $0 = 3k - 14$ for k.

$$0 = 3k - 14$$
$$\frac{14}{3} = k$$

67. $\dfrac{2x^2}{x^2 - 1} + \dfrac{4}{x + 3} = \dfrac{12x - 4}{x^3 + 3x^2 - x - 3}$,

LCM is $(x + 1)(x - 1)(x + 3)$

$$(x+1)(x-1)(x+3)\left[\frac{2x^2}{(x+1)(x-1)} + \frac{4}{x+3}\right] =$$

$$(x+1)(x-1)(x+3) \cdot \frac{12x-4}{(x+1)(x-1)(x+3)}$$

$$2x^2(x+3) + 4(x+1)(x-1) = 12x - 4$$
$$2x^3 + 6x^2 + 4x^2 - 4 = 12x - 4$$
$$2x^3 + 10x^2 - 12x = 0$$
$$x^3 + 5x^2 - 6x = 0$$
$$x(x^2 + 5x - 6) = 0$$
$$x(x + 6)(x - 1) = 0$$
$$x = 0 \ or \ x + 6 = 0 \quad x - 1 = 0$$
$$x = 0 \ or \qquad x = -6 \quad x \ = 1$$

Only 0 and -6 check. They are the solutions.

69. Answers may vary. One possibility is $P(x) = x^{15} - x^{14}$.

71.
$$
\begin{array}{r|rrrr}
-i & 1 & 3i & -4i & -2 \\
   &   & -i & 2 & -4-2i \\
\hline
   & 1 & 2i & 2-4i & -6-2i \\
\end{array}
$$

$Q(x) = x^2 + 2ix + (2 - 4i)$, $R(x) = -6 - 2i$

73.
$$
\begin{array}{r|rrr}
i & 1 & -3 & 7 \qquad (i^2 = -1) \\
  &   & i & -3i-1 \\
\hline
  & 1 & -3+i & 6-3i \\
\end{array}
$$

The answer is $x - 3 + i$, R $6 - 3i$.

Chapter 4 Mid-Chapter Mixed Review

1. $P(0) = 5 - 2 \cdot 0^3 = 5$, so the y-intercept is $(0, 5)$. The given statement is false.

3. $f(8) = (8 + 7)(8 - 8) = 15 \cdot 0 = 0$

The given statement is true.

5. $f(x) = (x^2 - 10x + 25)^3 = [(x - 5)^2]^3 = (x - 5)^6$

Solving $(x - 5)^6 = 0$, we get $x = 5$.

The factor $x - 5$ occurs 6 times, so the zero has a multiplicity of 6.

7. $g(x) = x^4 - 3x^2 + 2 = (x^2 - 1)(x^2 - 2) = (x+1)(x-1)(x^2-2)$

Solving $(x+1)(x-1)(x^2-2) = 0$, we get $x = -1 \ or \ x = 1 \ or \ x = \pm\sqrt{2}$.

Each factor occurs 1 time, so the multiplicity of each zero is 1.

9. $f(x) = x^4 - x^3 - 6x^2$

The sign of the leading coefficient, 1, is positive and the degree, 4, is even. Thus, graph (d) is the graph of the function.

11. $f(x) = 6x^3 + 8x^2 - 6x - 8$

The sign of the leading coefficient, 6, is positive and the degree, 3, is odd. Thus, graph (b) is the graph of the function.

13. $f(-2) = (-2)^3 - 2(-2)^2 + 3 = -13$

$f(0) = 0^3 - 2 \cdot 0^2 + 3 = 3$

By the intermediate value theorem, since $f(-2)$ and $f(0)$ have opposite signs, $f(x)$ has a zero between -2 and 0.

15.

$$
\begin{array}{r}
x^3 - 5x^2 - 5x - 4 \\
x - 1 \overline{\smash{\big)}\ x^4 - 6x^3 + 0x^2 + x - 2} \\
\underline{x^4 - x^3} \\
-5x^3 + 0x^2 \\
\underline{-5x^3 + 5x^2} \\
-5x^2 + x \\
\underline{-5x^2 + 5x} \\
-4x - 2 \\
\underline{-4x + 4} \\
-6
\end{array}
$$

$$P(x) = (x - 1)(x^3 - 5x^2 - 5x - 4) - \frac{6}{x - 1}$$

17. $(x^5 - 5) \div (x + 1) = (x^5 - 5) \div [x - (-1)]$

$$
\begin{array}{r|rrrrrr}
-1 & 1 & 0 & 0 & 0 & 0 & -5 \\
   &   & -1 & 1 & -1 & 1 & -1 \\
\hline
   & 1 & -1 & 1 & -1 & 1 & -6 \\
\end{array}
$$

$Q(x) = x^4 - x^3 + x^2 - x + 1$, $R(x) = -6$

19.

$$\begin{array}{r} \frac{1}{2}\,\big|\ \ 20 \quad -40 \quad\quad 0 \\ \underline{\quad\quad\quad 10 \quad -15\ } \\ 20 \quad -30 \ \big|\ -15 \end{array}$$

$$f\left(\frac{1}{2}\right) = -15$$

21. $f(x) = x^3 - 4x^2 + 9x - 36$

If $-3i$ is a zero of $f(x)$, then $f(-3i) = 0$. We find $f(-3i)$.

$$\begin{array}{r} -3i\,\big|\ \ 1 \quad\quad -4 \quad\quad\ 9 \quad\quad -36 \\ \underline{\quad\quad\quad -3i \quad -9+12i \quad 36\ } \\ 1 \quad -4-3i \quad\ 12i \ \big|\ \ \ 0 \end{array}$$

Since $f(-3i) = 0$, $-3i$ is a zero of $f(x)$.

If 3 is a zero of $f(x)$, then $f(3) = 0$. We find $f(3)$.

$$\begin{array}{r} 3\,\big|\ \ 1 \quad -4 \quad\ 9 \quad -36 \\ \underline{\quad\quad\ 3 \quad -3 \quad 18\ } \\ 1 \quad -1 \quad\ 6 \ \big|\ {-18} \end{array}$$

Since $f(3) \neq 0$, 3 is not a zero of $f(x)$.

23. $h(x) = x^3 - 2x^2 - 55x + 56$

Try $x - 1$.

$$\begin{array}{r} 1\,\big|\ \ 1 \quad -2 \quad -55 \quad\ 56 \\ \underline{\quad\quad\ 1 \quad -1 \quad -56\ } \\ 1 \quad -1 \quad -56 \ \big|\ \ \ 0 \end{array}$$

Since $h(1) = 0$, $x - 1$ is a factor of $h(x)$. Then $h(x) = (x-1)(x^2 - x - 56)$. Factoring the trinomial, we get $h(x) = (x-1)(x-8)(x+7)$.

Now we solve $h(x) = 0$.

$$(x-1)(x-8)(x+7) = 0$$

$$x - 1 = 0 \ \ or \ \ x - 8 = 0 \ \ or \ \ x + 7 = 0$$

$$x = 1 \ \ or \ \ \quad x = 8 \ \ or \ \ \quad\ x = -7$$

The solutions are 1, 8, and -7.

25. The range of a polynomial function with an odd degree is $(-\infty, \infty)$. The range of a polynomial function with an even degree is $[s, \infty)$ for some real number s if $a_n > 0$ and is $(-\infty, s]$ for some real number s if $a_n < 0$.

27. If function values change from positive to negative or from negative to positive in an interval, there would have to be a zero in the interval. Thus, between a pair of consecutive zeros, all the function values must have the same sign.

Exercise Set 4.4

1. Find a polynomial function of degree 3 with -2, 3, and 5 as zeros.

Such a function has factors $x + 2$, $x - 3$, and $x - 5$, so we have $f(x) = a_n(x+2)(x-3)(x-5)$.

The number a_n can be any nonzero number. The simplest polynomial will be obtained if we let it be 1. Multiplying the factors, we obtain

$$\begin{aligned} f(x) &= (x+2)(x-3)(x-5) \\ &= (x^2 - x - 6)(x-5) \\ &= x^3 - 6x^2 - x + 30. \end{aligned}$$

3. Find a polynomial function of degree 3 with -3, $2i$, and $-2i$ as zeros.

Such a function has factors $x + 3$, $x - 2i$, and $x + 2i$, so we have $f(x) = a_n(x+3)(x-2i)(x+2i)$.

The number a_n can be any nonzero number. The simplest polynomial will be obtained if we let it be 1. Multiplying the factors, we obtain

$$\begin{aligned} f(x) &= (x+3)(x-2i)(x+2i) \\ &= (x+3)(x^2 + 4) \\ &= x^3 + 3x^2 + 4x + 12. \end{aligned}$$

5. Find a polynomial function of degree 3 with $\sqrt{2}$, $-\sqrt{2}$, and 3 as zeros.

Such a function has factors $x - \sqrt{2}$, $x + \sqrt{2}$, and $x - 3$, so we have $f(x) = a_n(x - \sqrt{2})(x + \sqrt{2})(x - 3)$.

The number a_n can be any nonzero number. The simplest polynomial will be obtained if we let it be 1. Multiplying the factors, we obtain

$$\begin{aligned} f(x) &= (x - \sqrt{2})(x + \sqrt{2})(x - 3) \\ &= (x^2 - 2)(x - 3) \\ &= x^3 - 3x^2 - 2x + 6. \end{aligned}$$

7. Find a polynomial function of degree 3 with $1 - \sqrt{3}$, $1 + \sqrt{3}$, and -2 as zeros.

Such a function has factors $x - (1 - \sqrt{3})$, $x - (1 + \sqrt{3})$, and $x + 2$, so we have

$$f(x) = a_n[x - (1 - \sqrt{3})][x - (1 + \sqrt{3})](x + 2).$$

The number a_n can be any nonzero number. The simplest polynomial will be obtained if we let it be 1. Multiplying the factors, we obtain

$$\begin{aligned} f(x) &= [x - (1 - \sqrt{3})][x - (1 + \sqrt{3})](x + 2) \\ &= [(x - 1) + \sqrt{3}][(x - 1) - \sqrt{3}](x + 2) \\ &= [(x - 1)^2 - (\sqrt{3})^2](x + 2) \\ &= (x^2 - 2x + 1 - 3)(x + 2) \\ &= (x^2 - 2x - 2)(x + 2) \\ &= x^3 - 2x^2 - 2x + 2x^2 - 4x - 4 \\ &= x^3 - 6x - 4. \end{aligned}$$

9. Find a polynomial function of degree 3 with $1 + 6i$, $1 - 6i$, and -4 as zeros.

Such a function has factors $x - (1 + 6i)$, $x - (1 - 6i)$, and $x + 4$, so we have

$$f(x) = a_n[x - (1 + 6i)][x - (1 - 6i)](x + 4).$$

The number a_n can be any nonzero number. The simplest polynomial will be obtained if we let it be 1. Multiplying the factors, we obtain

$$\begin{aligned} f(x) &= [x - (1 + 6i)][x - (1 - 6i)](x + 4) \\ &= [(x - 1) - 6i][(x - 1) + 6i](x + 4) \\ &= [(x - 1)^2 - (6i)^2](x + 4) \\ &= (x^2 - 2x + 1 + 36)(x + 4) \\ &= (x^2 - 2x + 37)(x + 4) \\ &= x^3 - 2x^2 + 37x + 4x^2 - 8x + 148 \\ &= x^3 + 2x^2 + 29x + 148. \end{aligned}$$

11. Find a polynomial function of degree 3 with $-\frac{1}{3}$, 0, and 2 as zeros.

Such a function has factors $x + \frac{1}{3}$, $x - 0$ (or x), and $x - 2$ so we have

$$f(x) = a_n \left(x + \frac{1}{3} \right)(x)(x - 2).$$

The number a_n can be any nonzero number. The simplest polynomial will be obtained if we let it be 1. Multiplying the factors, we obtain

$$f(x) = \left(x + \frac{1}{3} \right)(x)(x - 2)$$
$$= \left(x^2 + \frac{1}{3}x \right)(x - 2)$$
$$= x^3 - \frac{5}{3}x^2 - \frac{2}{3}x.$$

13. A polynomial function of degree 5 has at most 5 real zeros. Since 5 zeros are given, these are all of the zeros of the desired function. We proceed as in Exercises 1-11, letting $a_n = 1$.

$$f(x) = (x + 1)^3(x - 0)(x - 1)$$
$$= (x^3 + 3x^2 + 3x + 1)(x^2 - x)$$
$$= x^5 + 2x^4 - 2x^2 - x$$

15. A polynomial function of degree 4 has at most 4 real zeros. Since 4 zeros are given, these are all of the zeros of the desired function. We proceed as in Exercises 1-11, letting $a_n = 1$.

$$f(x) = (x + 1)^3(x - 0)$$
$$= (x^3 + 3x^2 + 3x + 1)(x)$$
$$= x^4 + 3x^3 + 3x^2 + x$$

17. A polynomial function of degree 4 can have at most 4 zeros. Since $f(x)$ has rational coefficients, in addition to the three zeros given, the other zero is the conjugate of $\sqrt{3}$, or $-\sqrt{3}$.

19. A polynomial function of degree 4 can have at most 4 zeros. Since $f(x)$ has rational coefficients, the other zeros are the conjugates of the given zeros. They are i and $2 + \sqrt{5}$.

21. A polynomial function of degree 4 can have at most 4 zeros. Since $f(x)$ has rational coefficients, in addition to the three zeros given, the other zero is the conjugate of $3i$, or $-3i$.

23. A polynomial function of degree 4 can have at most 4 zeros. Since $f(x)$ has rational coefficients, the other zeros are the conjugates of the given zeros. They are $-4 + 3i$ and $2 + \sqrt{3}$.

25. A polynomial function $f(x)$ of degree 5 has at most 5 zeros. Since $f(x)$ has rational coefficients, in addition to the 3 given zeros, the other zeros are the conjugates of $\sqrt{5}$ and $-4i$, or $-\sqrt{5}$ and $4i$.

27. A polynomial function $f(x)$ of degree 5 has at most 5 zeros. Since $f(x)$ has rational coefficients, the other zero is the conjugate of $2 - i$, or $2 + i$.

29. A polynomial function $f(x)$ of degree 5 has at most 5 zeros. Since $f(x)$ has rational coefficients, in addition to the 3 given zeros, the other zeros are the conjugates of $-3 + 4i$ and $4 - \sqrt{5}$, or $-3 - 4i$ and $4 + \sqrt{5}$.

31. A polynomial function $f(x)$ of degree 5 has at most 5 zeros. Since $f(x)$ has rational coefficients, the other zero is the conjugate of $4 - i$, or $4 + i$.

33. Find a polynomial function of lowest degree with rational coefficients that has $1 + i$ and 2 as some of its zeros. $1 - i$ is also a zero.

Thus the polynomial function is
$$f(x) = a_n(x - 2)[x - (1 + i)][x - (1 - i)].$$
If we let $a_n = 1$, we obtain
$$f(x) = (x - 2)[(x - 1) - i][(x - 1) + i]$$
$$= (x - 2)[(x - 1)^2 - i^2]$$
$$= (x - 2)(x^2 - 2x + 1 + 1)$$
$$= (x - 2)(x^2 - 2x + 2)$$
$$= x^3 - 4x^2 + 6x - 4.$$

35. Find a polynomial function of lowest degree with rational coefficients that has $4i$ as one of its zeros. $-4i$ is also a zero.

Thus the polynomial function is
$$f(x) = a_n(x - 4i)(x + 4i).$$
If we let $a_n = 1$, we obtain
$$f(x) = (x - 4i)(x + 4i) = x^2 + 16.$$

37. Find a polynomial function of lowest degree with rational coefficients that has $-4i$ and 5 as some of its zeros.

$4i$ is also a zero.

Thus the polynomial function is
$$f(x) = a_n(x - 5)(x + 4i)(x - 4i).$$
If we let $a_n = 1$, we obtain
$$f(x) = (x - 5)[x^2 - (4i)^2]$$
$$= (x - 5)(x^2 + 16)$$
$$= x^3 - 5x^2 + 16x - 80$$

39. Find a polynomial function of lowest degree with rational coefficients that has $1 - i$ and $-\sqrt{5}$ as some of its zeros. $1 + i$ and $\sqrt{5}$ are also zeros.

Thus the polynomial function is
$$f(x) = a_n[x - (1 - i)][x - (1 + i)](x + \sqrt{5})(x - \sqrt{5}).$$
If we let $a_n = 1$, we obtain
$$f(x) = [x - (1 - i)][x - (1 + i)](x + \sqrt{5})(x - \sqrt{5})$$
$$= [(x - 1) + i][(x - 1) - i](x + \sqrt{5})(x - \sqrt{5})$$
$$= (x^2 - 2x + 1 + 1)(x^2 - 5)$$
$$= (x^2 - 2x + 2)(x^2 - 5)$$
$$= x^4 - 2x^3 + 2x^2 - 5x^2 + 10x - 10$$
$$= x^4 - 2x^3 - 3x^2 + 10x - 10$$

41. Find a polynomial function of lowest degree with rational coefficients that has $\sqrt{5}$ and $-3i$ as some of its zeros.

$-\sqrt{5}$ and $3i$ are also zeros.

Thus the polynomial function is
$$f(x) = a_n(x - \sqrt{5})(x + \sqrt{5})(x + 3i)(x - 3i).$$

If we let $a_n = 1$, we obtain
$$f(x) = (x^2 - 5)(x^2 + 9)$$
$$= x^4 + 4x^2 - 45$$

43. $f(x) = x^3 + 5x^2 - 2x - 10$

Since -5 is a zero of $f(x)$, we have $f(x) = (x + 5) \cdot Q(x)$. We use synthetic division to find $Q(x)$.

$$\begin{array}{r|rrrr} -5 & 1 & 5 & -2 & -10 \\ & & -5 & 0 & 10 \\ \hline & 1 & 0 & -2 & \vert \quad 0 \end{array}$$

Then $f(x) = (x + 5)(x^2 - 2)$. To find the other zeros we solve $x^2 - 2 = 0$.
$$x^2 - 2 = 0$$
$$x^2 = 2$$
$$x = \pm\sqrt{2}$$
The other zeros are $-\sqrt{2}$ and $\sqrt{2}$.

45. If $-i$ is a zero of $f(x) = x^4 - 5x^3 + 7x^2 - 5x + 6$, i is also a zero. Thus $x + i$ and $x - i$ are factors of the polynomial. Since $(x + i)(x - i) = x^2 + 1$, we know that $f(x) = (x^2 + 1) \cdot Q(x)$. Divide $x^4 - 5x^3 + 7x^2 - 5x + 6$ by $x^2 + 1$.

$$\begin{array}{r} x^2 - 5x + 6 \\ x^2 + 1 \overline{\smash{\big)}\ x^4 - 5x^3 + 7x^2 - 5x + 6} \\ \underline{x^4 + x^2} \\ -5x^3 + 6x^2 - 5x \\ \underline{-5x^3 - 5x} \\ 6x^2 + 6 \\ \underline{6x^2 + 6} \\ 0 \end{array}$$

Thus
$$x^4 - 5x^3 + 7x^2 - 5x + 6 = (x + i)(x - i)(x^2 - 5x + 6)$$
$$= (x + i)(x - i)(x - 2)(x - 3)$$

Using the principle of zero products we find the other zeros to be i, 2, and 3.

47. $x^3 - 6x^2 + 13x - 20 = 0$

If 4 is a zero, then $x - 4$ is a factor. Use synthetic division to find another factor.

$$\begin{array}{r|rrrr} 4 & 1 & -6 & 13 & -20 \\ & & 4 & -8 & 20 \\ \hline & 1 & -2 & 5 & \vert \quad 0 \end{array}$$

$$(x - 4)(x^2 - 2x + 5) = 0$$
$x - 4 = 0$ or $x^2 - 2x + 5 = 0$ Principle of zero products

$x = 4$ or $\quad x = \dfrac{2 \pm \sqrt{4 - 20}}{2}$
                     Quadratic formula

$x = 4$ or $\quad x = \dfrac{2 \pm 4i}{2} = 1 \pm 2i$

The other zeros are $1 + 2i$ and $1 - 2i$.

49. $f(x) = x^5 - 3x^2 + 1$

According to the rational zeros theorem, any rational zero of f must be of the form p/q, where p is a factor of the constant term, 1, and q is a factor of the coefficient of x^5, 1.

$$\frac{\text{Possibilities for } p}{\text{Possibilities for } q} : \frac{\pm 1}{\pm 1}$$

Possibilities for p/q: $1, -1$

51. $f(x) = 2x^4 - 3x^3 - x + 8$

According to the rational zeros theorem, any rational zero of f must be of the form p/q, where p is a factor of the constant term, 8, and q is a factor of the coefficient of x^4, 2.

$$\frac{\text{Possibilities for } p}{\text{Possibilities for } q} : \frac{\pm 1, \pm 2, \pm 4, \pm 8}{\pm 1, \pm 2}$$

Possibilities for p/q: $1, -1, 2, -2, 4, -4, 8, -8, \dfrac{1}{2}, -\dfrac{1}{2}$

53. $f(x) = 15x^6 + 47x^2 + 2$

According to the rational zeros theorem, any rational zero of f must be of the form p/q, where p is a factor of 2 and q is a factor of 15.

$$\frac{\text{Possibilities for } p}{\text{Possibilities for } q} : \frac{\pm 1, \pm 2}{\pm 1, \pm 3, \pm 5, \pm 15}$$

Possibilities for p/q: $1, -1, 2, -2, \dfrac{1}{3}, -\dfrac{1}{3}, \dfrac{2}{3}, -\dfrac{2}{3}, \dfrac{1}{5},$

$\quad -\dfrac{1}{5}, \dfrac{2}{5}, -\dfrac{2}{5}, \dfrac{1}{15}, -\dfrac{1}{15}, \dfrac{2}{15}, -\dfrac{2}{15}$

55. $f(x) = x^3 + 3x^2 - 2x - 6$

a) $\dfrac{\text{Possibilities for } p}{\text{Possibilities for } q} : \dfrac{\pm 1, \pm 2, \pm 3, \pm 6}{\pm 1}$

Possibilities for p/q: $1, -1, 2, -2, 3, -3, 6, -6$

We use synthetic division to find a zero. We find that one zero is -3 as shown below.

$$\begin{array}{r|rrrr} -3 & 1 & 3 & -2 & -6 \\ & & -3 & 0 & 6 \\ \hline & 1 & 0 & -2 & \vert \quad 0 \end{array}$$

Then we have $f(x) = (x + 3)(x^2 - 2)$.

We find the other zeros:
$$x^2 - 2 = 0$$
$$x^2 = 2$$
$$x = \pm\sqrt{2}.$$

There is only one rational zero, -3. The other zeros are $\pm\sqrt{2}$. (Note that we could have used factoring by grouping to find this result.)

b) $f(x) = (x + 3)(x - \sqrt{2})(x + \sqrt{2})$

57. $f(x) = 3x^3 - x^2 - 15x + 5$

a) $\dfrac{\text{Possibilities for } p}{\text{Possibilities for } q} : \dfrac{\pm 1, \pm 5}{\pm 1, \pm 3}$

Possibilities for p/q: $1, -1, 5, -5, \dfrac{1}{3}, -\dfrac{1}{3}, \dfrac{5}{3}, -\dfrac{5}{3}$

We use synthetic division to find a zero. We find that one zero is $\dfrac{1}{3}$ as shown below.

$$\underline{\tfrac{1}{3}\,|}\quad 3 \quad -1 \quad -15 \quad 5$$
$$\phantom{\tfrac{1}{3}}\qquad\quad 1 \quad\;\; 0 \quad -5$$
$$\qquad\;\; 3 \quad\;\; 0 \quad -15\,|\;\; 0$$

Then we have $f(x) = \left(x - \dfrac{1}{3}\right)(3x^2 - 15)$, or

$3\left(x - \dfrac{1}{3}\right)(x^2 - 5)$.

Now $x^2 - 5 = 0$ for $x = \pm\sqrt{5}$. Thus, there is only one rational zero, $\dfrac{1}{3}$. The other zeros are $\pm\sqrt{5}$. (Note that we could have used factoring by grouping to find this result.)

b) $f(x) = 3\left(x - \dfrac{1}{3}\right)(x + \sqrt{5})(x - \sqrt{5})$

59. $f(x) = x^3 - 3x + 2$

a) $\dfrac{\text{Possibilities for } p}{\text{Possibilities for } q} : \dfrac{\pm 1, \pm 2}{\pm 1}$

Possibilities for p/q: $1, -1, 2, -2$

We use synthetic division to find a zero. We find that -2 is a zero as shown below.

$$\underline{-2\,|}\quad 1 \quad\;\; 0 \quad -3 \quad\;\; 2$$
$$\qquad\quad -2 \quad\;\; 4 \quad -2$$
$$\qquad\;\; 1 \quad -2 \quad\;\; 1\,|\;\; 0$$

Then we have $f(x) = (x + 2)(x^2 - 2x + 1) = (x + 2)(x - 1)^2$.

Now $(x - 1)^2 = 0$ for $x = 1$. Thus, the rational zeros are -2 and 1. (The zero 1 has a multiplicity of 2.) These are the only zeros.

b) $f(x) = (x + 2)(x - 1)^2$

61. $f(x) = 2x^3 + 3x^2 + 18x + 27$

a) $\dfrac{\text{Possibilities for } p}{\text{Possibilities for } q} : \dfrac{\pm 1, \pm 3, \pm 9, \pm 27}{\pm 1, \pm 2}$

Possibilities for p/q: $1, -1, 3, -3, 9, -9, 27, -27,$ $\dfrac{1}{2}, -\dfrac{1}{2}, \dfrac{3}{2}, -\dfrac{3}{2}, \dfrac{9}{2}, -\dfrac{9}{2}, \dfrac{27}{2}, -\dfrac{27}{2}$

We use synthetic division to find a zero. We find that $-\dfrac{3}{2}$ is a zero as shown below.

$$\underline{-\tfrac{3}{2}\,|}\quad 2 \quad\;\; 3 \quad\;\; 18 \quad\;\; 27$$
$$\phantom{-\tfrac{3}{2}|}\qquad\quad -3 \quad\;\; 0 \quad -27$$
$$\qquad\;\; 2 \quad\;\; 0 \quad\;\; 18\,|\;\; 0$$

Then we have $f(x) = \left(x + \dfrac{3}{2}\right)(2x^2 + 18)$, or

$2\left(x + \dfrac{3}{2}\right)(x^2 + 9)$.

Now $x^2 + 9 = 0$ for $x = \pm 3i$. Thus, the only rational zero is $-\dfrac{3}{2}$. The other zeros are $\pm 3i$. (Note that we could have used factoring by grouping to find this result.)

b) $f(x) = 2\left(x + \dfrac{3}{2}\right)(x + 3i)(x - 3i)$

63. $f(x) = 5x^4 - 4x^3 + 19x^2 - 16x - 4$

a) $\dfrac{\text{Possibilities for } p}{\text{Possibilities for } q} : \dfrac{\pm 1, \pm 2, \pm 4}{\pm 1, \pm 5}$

Possibilities for p/q: $1, -1, 2, -2, 4, -4, \dfrac{1}{5}, -\dfrac{1}{5},$ $\dfrac{2}{5}, -\dfrac{2}{5}, \dfrac{4}{5}, -\dfrac{4}{5}$

We use synthetic division to find a zero. We find that 1 is a zero as shown below.

$$\underline{1\,|}\quad 5 \quad -4 \quad\;\; 19 \quad -16 \quad -4$$
$$\qquad\quad 5 \quad\;\; 1 \quad\;\; 20 \quad\;\; 4$$
$$\qquad\;\; 5 \quad\;\; 1 \quad\;\; 20 \quad\;\; 4\,|\;\; 0$$

Then we have
$$f(x) = (x - 1)(5x^3 + x^2 + 20x + 4)$$
$$= (x - 1)[x^2(5x + 1) + 4(5x + 1)]$$
$$= (x - 1)(5x + 1)(x^2 + 4).$$

We find the other zeros:
$$5x + 1 = 0 \quad or \quad x^2 + 4 = 0$$
$$5x = -1 \quad or \quad x^2 = -4$$
$$x = -\dfrac{1}{5} \quad or \quad x = \pm 2i$$

The rational zeros are $-\dfrac{1}{5}$ and 1. The other zeros are $\pm 2i$.

b) From part (a) we see that
$$f(x) = (5x + 1)(x - 1)(x + 2i)(x - 2i), \text{ or}$$
$$5\left(x + \dfrac{1}{5}\right)(x - 1)(x + 2i)(x - 2i).$$

65. $f(x) = x^4 - 3x^3 - 20x^2 - 24x - 8$

a) $\dfrac{\text{Possibilities for } p}{\text{Possibilities for } q} : \dfrac{\pm 1, \pm 2, \pm 4, \pm 8}{\pm 1}$

Possibilities for p/q: $1, -1, 2, -2, 4, -4, 8, -8$

We use synthetic division to find a zero. We find that -2 is a zero as shown below.

$$\underline{-2\,|}\quad 1 \quad -3 \quad -20 \quad -24 \quad -8$$
$$\qquad\quad -2 \quad\;\; 10 \quad\;\; 20 \quad\;\; 8$$
$$\qquad\;\; 1 \quad -5 \quad -10 \quad -4\,|\;\; 0$$

Now we determine whether -1 is a zero.

$$\underline{-1\,|}\quad 1 \quad -5 \quad -10 \quad -4$$
$$\qquad\quad -1 \quad\;\; 6 \quad\;\; 4$$
$$\qquad\;\; 1 \quad -6 \quad -4\,|\;\; 0$$

Then we have $f(x) = (x + 2)(x + 1)(x^2 - 6x - 4)$.

Use the quadratic formula to find the other zeros.
$$x^2 - 6x - 4 = 0$$
$$x = \dfrac{-(-6) \pm \sqrt{(-6)^2 - 4 \cdot 1 \cdot (-4)}}{2 \cdot 1}$$
$$= \dfrac{6 \pm \sqrt{52}}{2} = \dfrac{6 \pm 2\sqrt{13}}{2}$$
$$= 3 \pm \sqrt{13}$$

The rational zeros are -2 and -1. The other zeros are $3 \pm \sqrt{13}$.

b) $f(x) = (x+2)(x+1)[x-(3+\sqrt{13})][x-(3-\sqrt{13})]$
$$= (x+2)(x+1)(x-3-\sqrt{13})(x-3+\sqrt{13})$$

67. $f(x) = x^3 - 4x^2 + 2x + 4$

a) $\dfrac{\text{Possibilities for } p}{\text{Possibilities for } q} : \dfrac{\pm 1, \pm 2, \pm 4}{\pm 1}$

Possibilities for p/q: $1, -1, 2, -2, 4, -4$

Synthetic division shows that neither -1 nor 1 is a zero. Try 2.

$$\begin{array}{r|rrrr} 2 & 1 & -4 & 2 & 4 \\ & & 2 & -4 & -4 \\ \hline & 1 & -2 & -2 & 0 \end{array}$$

Then we have $f(x) = (x-2)(x^2 - 2x - 2)$. Use the quadratic formula to find the other zeros.

$x^2 - 2x - 2 = 0$

$$x = \frac{-(-2) \pm \sqrt{(-2)^2 - 4 \cdot 1 \cdot (-2)}}{2 \cdot 1}$$

$$= \frac{2 \pm \sqrt{12}}{2} = \frac{2 \pm 2\sqrt{3}}{2}$$

$$= 1 \pm \sqrt{3}$$

The only rational zero is 2. The other zeros are $1 \pm \sqrt{3}$.

b) $f(x) = (x-2)[x - (1 + \sqrt{3})][x - (1 - \sqrt{3})]$
$= (x-2)(x - 1 - \sqrt{3})(x - 1 + \sqrt{3})$

69. $f(x) = x^3 + 8$

a) $\dfrac{\text{Possibilities for } p}{\text{Possibilities for } q} : \dfrac{\pm 1, \pm 2, \pm 4, \pm 8}{\pm 1}$

Possibilities for p/q: $1, -1, 2, -2, 4, -4, 8, -8$

We use synthetic division to find a zero. We find that -2 is a zero as shown below.

$$\begin{array}{r|rrrr} -2 & 1 & 0 & 0 & 8 \\ & & -2 & 4 & -8 \\ \hline & 1 & -2 & 4 & 0 \end{array}$$

We have $f(x) = (x+2)(x^2 - 2x + 4)$. Use the quadratic formula to find the other zeros.

$x^2 - 2x + 4 = 0$

$$x = \frac{-(-2) \pm \sqrt{(-2)^2 - 4 \cdot 1 \cdot 4}}{2 \cdot 1}$$

$$= \frac{2 \pm \sqrt{-12}}{2} = \frac{2 \pm 2\sqrt{3}i}{2}$$

$$= 1 \pm \sqrt{3}i$$

The only rational zero is -2. The other zeros are $1 \pm \sqrt{3}i$.

b) $f(x) = (x+2)[x - (1 + \sqrt{3}i)][x - (1 - \sqrt{3}i)]$
$= (x+2)(x - 1 - \sqrt{3}i)(x - 1 + \sqrt{3}i)$

71. $f(x) = \dfrac{1}{3}x^3 - \dfrac{1}{2}x^2 - \dfrac{1}{6}x + \dfrac{1}{6}$

$= \dfrac{1}{6}(2x^3 - 3x^2 - x + 1)$

a) The second form of the equation is equivalent to the first and has the advantage of having integer coefficients. Thus, we can use the rational zeros theorem for $g(x) = 2x^3 - 3x^2 - x + 1$. The zeros of $g(x)$ are the same as the zeros of $f(x)$. We find the zeros of $g(x)$.

$\dfrac{\text{Possibilities for } p}{\text{Possibilities for } q} : \dfrac{\pm 1}{\pm 1, \pm 2}$

Possibilities for p/q: $1, -1, \dfrac{1}{2}, -\dfrac{1}{2}$

Synthetic division shows that $-\dfrac{1}{2}$ is not a zero.

Try $\dfrac{1}{2}$.

$$\begin{array}{r|rrrr} \frac{1}{2} & 2 & -3 & -1 & 1 \\ & & 1 & -1 & -1 \\ \hline & 2 & -2 & -2 & 0 \end{array}$$

We have $g(x) = \left(x - \dfrac{1}{2}\right)(2x^2 - 2x - 2) =$
$\left(x - \dfrac{1}{2}\right)(2)(x^2 - x - 1)$. Use the quadratic formula to find the other zeros.

$x^2 - x - 1 = 0$

$$x = \frac{-(-1) \pm \sqrt{(-1)^2 - 4 \cdot 1 \cdot (-1)}}{2 \cdot 1}$$

$$= \frac{1 \pm \sqrt{5}}{2}$$

The only rational zero is $\dfrac{1}{2}$. The other zeros are $\dfrac{1 \pm \sqrt{5}}{2}$.

b) $f(x) = \dfrac{1}{6}g(x)$

$= \dfrac{1}{6}\left(x - \dfrac{1}{2}\right)(2)\left[x - \dfrac{1+\sqrt{5}}{2}\right]\left[x - \dfrac{1-\sqrt{5}}{2}\right]$

$= \dfrac{1}{3}\left(x - \dfrac{1}{2}\right)\left(x - \dfrac{1+\sqrt{5}}{2}\right)\left(x - \dfrac{1-\sqrt{5}}{2}\right)$

73. $f(x) = x^4 + 2x^3 - 5x^2 - 4x + 6$

According to the rational zeros theorem, the possible rational zeros are ± 1, ± 2, ± 3, and ± 6. Synthetic division shows that only 1 and -3 are zeros.

75. $f(x) = x^3 - x^2 - 4x + 3$

According to the rational zeros theorem, the possible rational zeros are ± 1 and ± 3. Synthetic division shows that none of these is a zero. Thus, there are no rational zeros.

77. $f(x) = x^4 + 2x^3 + 2x^2 - 4x - 8$

According to the rational zeros theorem, the possible rational zeros are ± 1, ± 2, ± 4, and ± 8. Synthetic division shows that none of the possibilities is a zero. Thus, there are no rational zeros.

79. $f(x) = x^5 - 5x^4 + 5x^3 + 15x^2 - 36x + 20$

According to the rational zeros theorem, the possible rational zeros are ± 1, ± 2, ± 4, ± 5, ± 10, and ± 20. We try -2.

$$\begin{array}{r|rrrrrr} -2 & 1 & -5 & 5 & 15 & -36 & 20 \\ & & -2 & 14 & -38 & 46 & -20 \\ \hline & 1 & -7 & 19 & -23 & 10 & 0 \end{array}$$

Thus, -2 is a zero. Now try 1.

$$\begin{array}{r|rrrrr} 1 & 1 & -7 & 19 & -23 & 10 \\ & & 1 & -6 & 13 & -10 \\ \hline & 1 & -6 & 13 & -10 & 0 \end{array}$$

1 is also a zero. Try 2.

$$\begin{array}{r|rrrr} 2 & 1 & -6 & 13 & -10 \\ & & 2 & -8 & 10 \\ \hline & 1 & -4 & 5 & 0 \end{array}$$

2 is also a zero.

We have $f(x) = (x + 2)(x - 1)(x - 2)(x^2 - 4x + 5)$. The discriminant of $x^2 - 4x + 5$ is $(-4)^2 - 4 \cdot 1 \cdot 5$, or $4 < 0$, so $x^2 - 4x + 5$ has two nonreal zeros. Thus, the rational zeros are -2, 1, and 2.

81. $f(x) = 3x^5 - 2x^2 + x - 1$

The number of variations in sign in $f(x)$ is 3. Then the number of positive real zeros is either 3 or less than 3 by 2, 4, 6, and so on. Thus, the number of positive real zeros is 3 or 1.

$$f(-x) = 3(-x)^5 - 2(-x)^2 + (-x) - 1$$
$$= -3x^5 - 2x^2 - x - 1$$

There are no variations in sign in $f(-x)$, so there are 0 negative real zeros.

83. $h(x) = 6x^7 + 2x^2 + 5x + 4$

There are no variations in sign in $h(x)$, so there are 0 positive real zeros.

$$h(-x) = 6(-x)^7 + 2(-x)^2 + 5(-x) + 4$$
$$= -6x^7 + 2x^2 - 5x + 4$$

The number of variations in sign in $h(-x)$ is 3. Thus, there are 3 or 1 negative real zeros.

85. $F(p) = 3p^{18} + 2p^4 - 5p^2 + p + 3$

There are 2 variations in sign in $F(p)$, so there are 2 or 0 positive real zeros.

$$F(-p) = 3(-p)^{18} + 2(-p)^4 - 5(-p)^2 + (-p) + 3$$
$$= 3p^{18} + 2p^4 - 5p^2 - p + 3$$

There are 2 variations in sign in $F(-p)$, so there are 2 or 0 negative real zeros.

87. $C(x) = 7x^6 + 3x^4 - x - 10$

There is 1 variation in sign in $C(x)$, so there is 1 positive real zero.

$$C(-x) = 7(-x)^6 + 3(-x)^4 - (-x) - 10$$
$$= 7x^6 + 3x^4 + x - 10$$

There is 1 variation in sign in $C(-x)$, so there is 1 negative real zero.

89. $h(t) = -4t^5 - t^3 + 2t^2 + 1$

There is 1 variation in sign in $h(t)$, so there is 1 positive real zero.

$$h(-t) = -4(-t)^5 - (-t)^3 + 2(-t)^2 + 1$$
$$= 4t^5 + t^3 + 2t^2 + 1$$

There are no variations in sign in $h(-t)$, so there are 0 negative real zeros.

91. $f(y) = y^4 + 13y^3 - y + 5$

There are 2 variations in sign in $f(y)$, so there are 2 or 0 positive real zeros.

$$f(-y) = (-y)^4 + 13(-y)^3 - (-y) + 5$$
$$= y^4 - 13y^3 + y + 5$$

There are 2 variations in sign in $f(-y)$, so there are 2 or 0 negative real zeros.

93. $r(x) = x^4 - 6x^2 + 20x - 24$

There are 3 variations in sign in $r(x)$, so there are 3 or 1 positive real zeros.

$$r(-x) = (-x)^4 - 6(-x)^2 + 20(-x) - 24$$
$$= x^4 - 6x^2 - 20x - 24$$

There is 1 variation in sign in $r(-x)$, so there is 1 negative real zero.

95. $R(x) = 3x^5 - 5x^3 - 4x$

There is 1 variation in sign in $R(x)$, so there is 1 positive real zero.

$$R(-x) = 3(-x)^5 - 5(-x)^3 - 4(-x)$$
$$= -3x^5 + 5x^3 + 4x$$

There is 1 variation in sign in $R(-x)$, so there is 1 negative real zero.

97. $f(x) = 4x^3 + x^2 - 8x - 2$

1. The leading term is $4x^3$. The degree, 3, is odd and the leading coefficient, 4, is positive so as $x \to \infty$, $f(x) \to \infty$ and $x \to -\infty$, $f(x) \to -\infty$.

2. We find the rational zeros p/q of $f(x)$.

$$\frac{\text{Possibilities for } p}{\text{Possibilities for } q} : \frac{\pm 1, \pm 2}{\pm 1, \pm 2, \pm 4}$$

Possibilities for p/q: $1, -1, 2, -2, \dfrac{1}{2}, -\dfrac{1}{2}, \dfrac{1}{4}, -\dfrac{1}{4}$

Synthetic division shows that $-\dfrac{1}{4}$ is a zero.

$$\begin{array}{r|rrrr} -\frac{1}{4} & 4 & 1 & -8 & -2 \\ & & -1 & 0 & 2 \\ \hline & 4 & 0 & -8 & 0 \end{array}$$

We have $f(x) = \left(x + \dfrac{1}{4}\right)(4x^2 - 8) =$

$4\left(x + \dfrac{1}{4}\right)(x^2 - 2)$. Solving $x^2 - 2 = 0$ we get

$x = \pm\sqrt{2}$. Thus the zeros of the function are $-\dfrac{1}{4}$,

$-\sqrt{2}$, and $\sqrt{2}$ so the x-intercepts of the graph are

$\left(-\dfrac{1}{4}, 0\right)$, $(-\sqrt{2}, 0)$, and $(\sqrt{2}, 0)$.

3. The zeros divide the x-axis into 4 intervals, $(-\infty, -\sqrt{2})$, $\left(-\sqrt{2}, -\frac{1}{4}\right)$, $\left(-\frac{1}{4}, \sqrt{2}\right)$, and $(\sqrt{2}, \infty)$. We choose a value for x from each interval and find $f(x)$. This tells us the sign of $f(x)$ for all values of x in that interval.

In $(-\infty, -\sqrt{2})$, test -2:
$$f(-2) = 4(-2)^3 + (-2)^2 - 8(-2) - 2 = -14 < 0$$

In $\left(-\sqrt{2}, -\frac{1}{4}\right)$, test -1:
$$f(-1) = 4(-1)^3 + (-1)^2 - 8(-1) - 2 = 3 > 0$$

In $\left(-\frac{1}{4}, \sqrt{2}\right)$, test 0:
$$f(0) = 4 \cdot 0^3 + 0^2 - 8 \cdot 0 - 2 = -2 < 0$$

In $(\sqrt{2}, \infty)$, test 2:
$$f(2) = 4 \cdot 2^3 + 2^2 - 8 \cdot 2 - 2 = 18 > 0$$

Thus the graph lies below the x-axis on $(-\infty, -\sqrt{2})$ and on $\left(-\frac{1}{4}, \sqrt{2}\right)$. It lies above the x-axis on $\left(-\sqrt{2}, -\frac{1}{4}\right)$ and on $(\sqrt{2}, \infty)$. We also know the points $(-2, -14)$, $(-1, 3)$, $(0, -2)$, and $(2, 18)$ are on the graph.

4. From Step 3 we see that $f(0) = -2$ so the y-intercept is $(0, -2)$.

5. We find additional points on the graph and then draw the graph.

x	$f(x)$
-1.5	-1.25
-0.5	1.75
1	-5
1.5	1.75

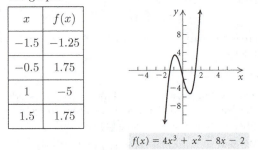

$f(x) = 4x^3 + x^2 - 8x - 2$

6. Checking the graph as described on page 240 in the text, we see that it appears to be correct.

99. $f(x) = 2x^4 - 3x^3 - 2x^2 + 3x$

1. The leading term is $2x^4$. The degree, 4, is even and the leading coefficient, 2, is positive so as $x \to \infty$, $f(x) \to \infty$ and as $x \to -\infty$, $f(x) \to \infty$.

2. We find the rational zeros p/q of $f(x)$. First note that $f(x) = x(2x^3 - 3x^2 - 2x + 3)$, so 0 is a zero. Now consider $g(x) = 2x^3 - 3x^2 - 2x + 3$.

$$\frac{\text{Possibilities for } p}{\text{Possibilities for } q} : \frac{\pm 1, \pm 3}{\pm 1, \pm 2}$$

Possibilities for p/q: $1, -1, 3, -3, \frac{1}{2}, -\frac{1}{2}, \frac{3}{2}, -\frac{3}{2}$

We try 1.

$$\begin{array}{r|rrrr} 1 & 2 & -3 & -2 & 3 \\ & & 2 & -1 & -3 \\ \hline & 2 & -1 & -3 & 0 \end{array}$$

Then $f(x) = x(x-1)(2x^2 - x - 3)$. Using the principle of zero products to solve $2x^2 - x - 3 = 0$, we get $x = \frac{3}{2}$ or $x = -1$.

Thus the zeros of the function are 0, 1, $\frac{3}{2}$, and -1 so the x-intercepts of the graph are $(0, 0)$, $(1, 0)$, $\left(\frac{3}{2}, 0\right)$, and $(-1, 0)$.

3. The zeros divide the x-axis into 5 intervals, $(-\infty, -1)$, $(-1, 0)$, $(0, 1)$, $\left(1, \frac{3}{2}\right)$, and $\left(\frac{3}{2}, \infty\right)$. We choose a value for x from each interval and find $f(x)$. This tells us the sign of $f(x)$ for all values of x in that interval.

In $(-\infty, -1)$, test -2:
$$f(-2) = 2(-2)^4 - 3(-2)^3 - 2(-2)^2 + 3(-2) = 42 > 0$$

In $(-1, 0)$, test -0.5:
$$f(-0.5) = 2(-0.5)^4 - 3(-0.5)^3 - 2(-0.5)^2 + 3(-0.5) = -1.5 < 0$$

In $(0, 1)$, test 0.5:
$$f(0.5) = 2(0.5)^4 - 3(0.5)^3 - 2(0.5)^2 + 3(0.5) = 0.75 > 0$$

In $\left(1, \frac{3}{2}\right)$, test 1.25:
$$f(1.25) = 2(1.25)^4 - 3(1.25)^3 - 2(1.25)^2 + 3(1.25) = -0.3515625 < 0$$

In $\left(\frac{3}{2}, \infty\right)$, test 2:
$$f(2) = 2 \cdot 2^4 - 3 \cdot 2^3 - 2 \cdot 2^2 + 3 \cdot 2 = 6 > 0$$

Thus the graph lies above the x-axis on $(-\infty, -1)$, on $(0, 1)$, and on $\left(\frac{3}{2}, \infty\right)$. It lies below the x-axis on $(-1, 0)$ and on $\left(1, \frac{3}{2}\right)$. We also know the points $(-2, 42)$, $(-0.5, -1.5)$, $(0.5, 0.75)$, $(1.25, -0.3515625)$, and $(2, 6)$ are on the graph.

4. From Step 2 we know that $f(0) = 0$ so the y-intercept is $(0, 0)$.

5. We find additional points on the graph and then draw the graph.

x	$f(x)$
-1.5	11.25
2.5	26.25
3	72

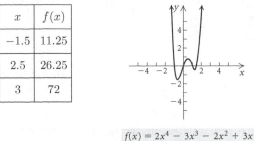

$f(x) = 2x^4 - 3x^3 - 2x^2 + 3x$

6. Checking the graph as described on page 240 in the text, we see that it appears to be correct.

101. $f(x) = x^2 - 8x + 10$

a) $-\dfrac{b}{2a} = -\dfrac{-8}{2 \cdot 1} = -(-4) = 4$

$f(4) = 4^2 - 8 \cdot 4 + 10 = -6$

The vertex is $(4, -6)$.

b) The axis of symmetry is $x = 4$.

c) Since the coefficient of x^2 is positive, there is a minimum function value. It is the second coordinate of the vertex, -6. It occurs when $x = 4$.

103. $\qquad -\dfrac{4}{5}x + 8 = 0$

$\qquad -\dfrac{4}{5}x = -8 \qquad\qquad$ Subtracting 8

$-\dfrac{5}{4}\left(-\dfrac{4}{5}x\right) = -\dfrac{5}{4}(-8) \quad$ Multiplying by $-\dfrac{5}{4}$

$\qquad\qquad x = 10$

The zero is 10.

105. $g(x) = -x^3 - 2x^2$

Leading term: $-x^3$; leading coefficient: -1

The degree is 3, so the function is cubic.

Since the degree is odd and the leading coefficient is negative, as $x \to \infty$, $g(x) \to -\infty$ and as $x \to -\infty$, $g(x) \to \infty$.

107. $f(x) = -\dfrac{4}{9}$

Leading term: $-\dfrac{4}{9}$; leading coefficient: $-\dfrac{4}{9}$;

for all x, $f(x) = -\dfrac{4}{9}$

The degree is 0, so this is a constant function.

109. $g(x) = x^4 - 2x^3 + x^2 - x + 2$

Leading term: x^4; leading coefficient: 1

The degree is 4, so the function is quartic.

Since the degree is even and the leading coefficient is positive, as $x \to \infty$, $g(x) \to \infty$ and as $x \to -\infty$, $g(x) \to \infty$.

111. $f(x) = 2x^3 - 5x^2 - 4x + 3$

a) $2x^3 - 5x^2 - 4x + 3 = 0$

$\dfrac{\text{Possibilities for } p}{\text{Possibilities for } q} : \dfrac{\pm 1, \pm 3}{\pm 1, \pm 2}$

Possibilities for p/q: $1, -1, 3, -3, \dfrac{1}{2}, -\dfrac{1}{2}, \dfrac{3}{2}, -\dfrac{3}{2}$

The first possibility that is a solution of $f(x) = 0$ is -1:

$$\begin{array}{r|rrrr} -1 & 2 & -5 & -4 & 3 \\ & & -2 & 7 & -3 \\ \hline & 2 & -7 & 3 & 0 \end{array}$$

Thus, -1 is a solution.

Then we have:

$(x + 1)(2x^2 - 7x + 3) = 0$

$(x + 1)(2x - 1)(x - 3) = 0$

The other solutions are $\dfrac{1}{2}$ and 3.

b) The graph of $y = f(x - 1)$ is the graph of $y = f(x)$ shifted 1 unit right. Thus, we add 1 to each solution of $f(x) = 0$ to find the solutions of $f(x - 1) = 0$. The solutions are $-1 + 1$, or 0; $\dfrac{1}{2} + 1$, or $\dfrac{3}{2}$; and $3 + 1$, or 4.

c) The graph of $y = f(x + 2)$ is the graph of $y = f(x)$ shifted 2 units left. Thus, we subtract 2 from each solution of $f(x) = 0$ to find the solutions of $f(x + 2) = 0$. The solutions are $-1 - 2$, or -3; $\dfrac{1}{2} - 2$, or $-\dfrac{3}{2}$; and $3 - 2$, or 1.

d) The graph of $y = f(2x)$ is a horizontal shrinking of the graph of $y = f(x)$ by a factor of 2. We divide each solution of $f(x) = 0$ by 2 to find the solutions of $f(2x) = 0$. The solutions are $\dfrac{-1}{2}$ or $-\dfrac{1}{2}$; $\dfrac{1/2}{2}$, or $\dfrac{1}{4}$; and $\dfrac{3}{2}$.

113. $P(x) = 2x^5 - 33x^4 - 84x^3 + 2203x^2 - 3348x - 10,080$

a) $2x^5 - 33x^4 - 84x^3 + 2203x^2 - 3348x - 10,080 = 0$

Trying some of the many possibilities for p/q, we find that 4 is a zero.

$$\begin{array}{r|rrrrrr} 4 & 2 & -33 & -84 & 2203 & -3348 & -10,080 \\ & & 8 & -100 & -736 & 5868 & 10,080 \\ \hline & 2 & -25 & -184 & 1467 & 2520 & 0 \end{array}$$

Then we have:

$(x - 4)(2x^4 - 25x^3 - 184x^2 + 1467x + 2520) = 0$

We now use the fourth degree polynomial above to find another zero. Synthetic division shows that 4 is not a double zero, but 7 is a zero.

$$\begin{array}{r|rrrrr} 7 & 2 & -25 & -184 & 1467 & 2520 \\ & & 14 & -77 & -1827 & -2520 \\ \hline & 2 & -11 & -261 & -360 & 0 \end{array}$$

Now we have:

$(x - 4)(x - 7)(2x^3 - 11x^2 - 261x - 360) = 0$

Use the third degree polynomial above to find a third zero. Synthetic division shows that 7 is not a double zero, but 15 is a zero.

$$\begin{array}{r|rrrr} 15 & 2 & -11 & -261 & -360 \\ & & 30 & 285 & 360 \\ \hline & 2 & 19 & 24 & 0 \end{array}$$

We have:

$P(x) = (x - 4)(x - 7)(x - 15)(2x^2 + 19x + 24)$

$\qquad\quad = (x - 4)(x - 7)(x - 15)(2x + 3)(x + 8)$

The rational zeros are 4, 7, 15, $-\dfrac{3}{2}$, and -8.

Exercise Set 4.5

1. $f(x) = \dfrac{x^2}{2 - x}$

We find the value(s) of x for which the denominator is 0.

$\qquad 2 - x = 0$

$\qquad\quad 2 = x$

The domain is $\{x | x \neq 2\}$, or $(-\infty, 2) \cup (2, \infty)$.

3. $f(x) = \dfrac{x+1}{x^2 - 6x + 5}$

We find the value(s) of x for which the denominator is 0.

$$x^2 - 6x + 5 = 0$$
$$(x-1)(x-5) = 0$$
$$x - 1 = 0 \;\; or \;\; x - 5 = 0$$
$$x = 1 \;\; or \;\;\;\;\;\; x = 5$$

The domain is $\{x | x \neq 1 \; and \; x \neq 5\}$, or $(-\infty, 1) \cup (1, 5) \cup (5, \infty)$.

5. $f(x) = \dfrac{3x - 4}{3x + 15}$

We find the value(s) of x for which the denominator is 0.

$$3x + 15 = 0$$
$$3x = -15$$
$$x = -5$$

The domain is $\{x | x \neq -5\}$, or $(-\infty, -5) \cup (-5, \infty)$.

7. Graph (d) is the graph of $f(x) = \dfrac{8}{x^2 - 4}$.

$x^2 - 4 = 0$ when $x = \pm 2$, so $x = -2$ and $x = 2$ are vertical asymptotes.

The x-axis, $y = 0$, is the horizontal asymptote because the degree of the numerator is less than the degree of the denominator.

There is no oblique asymptote.

9. Graph (e) is the graph of $f(x) = \dfrac{8x}{x^2 - 4}$.

As in Exercise 7, $x = -2$ and $x = 2$ are vertical asymptotes.

The x-axis, $y = 0$, is the horizontal asymptote because the degree of the numerator is less than the degree of the denominator.

There is no oblique asymptote.

11. Graph (c) is the graph of $f(x) = \dfrac{8x^3}{x^2 - 4}$.

As in Exercise 7, $x = -2$ and $x = 2$ are vertical asymptotes.

The degree of the numerator is greater than the degree of the denominator, so there is no horizontal asymptote but there is an oblique asymptote. To find it we first divide to find an equivalent expression.

$$
\begin{array}{r}
8x \\
x^2 - 4 \overline{\smash{\big)}\, 8x^3 } \\
\underline{8x^3 - 32x} \\
32x
\end{array}
$$

$\dfrac{8x^3}{x^2 - 4} = 8x + \dfrac{32x}{x^2 - 4}$

Now we multiply by 1, using $(1/x^2)/(1/x^2)$.

$$\dfrac{32x}{x^2 - 4} \cdot \dfrac{\frac{1}{x^2}}{\frac{1}{x^2}} = \dfrac{\frac{32}{x}}{1 - \frac{4}{x^2}}$$

As $|x|$ becomes very large, each expression with x in the denominator tends toward zero.

Then, as $|x| \to \infty$, we have

$$\dfrac{\frac{32}{x}}{1 - \frac{4}{x^2}} \to \dfrac{0}{1 - 0}, \text{ or } 0.$$

Thus, as $|x|$ becomes very large, the graph of $f(x)$ gets very close to the graph of $y = 8x$, so $y = 8x$ is the oblique asymptote.

13. $g(x) = \dfrac{1}{x^2}$

The numerator and the denominator have no common factors. The zero of the denominator is 0, so the vertical asymptote is $x = 0$.

15. $h(x) = \dfrac{x + 7}{2 - x}$

The numerator and the denominator have no common factors. $2 - x = 0$ when $x = 2$, so the vertical asymptote is $x = 2$.

17. $f(x) = \dfrac{3 - x}{(x - 4)(x + 6)}$

The numerator and the denominator have no common factors. The zeros of the denominator are 4 and -6, so the vertical asymptotes are $x = 4$ and $x = -6$.

19. $g(x) = \dfrac{x^3}{2x^3 - x^2 - 3x} = \dfrac{x^3}{x(2x - 3)(x + 1)}$

The numerator and the denominator have a common factor, x. The zeros of the denominator that are not also zeros of the numerator are $\dfrac{3}{2}$ and -1, so the vertical asymptotes are $x = \dfrac{3}{2}$ and $x = -1$.

21. $f(x) = \dfrac{3x^2 + 5}{4x^2 - 3}$

The numerator and the denominator have the same degree and the ratio of the leading coefficients is $\dfrac{3}{4}$, so $y = \dfrac{3}{4}$ is the horizontal asymptote.

23. $h(x) = \dfrac{x^2 - 4}{2x^4 + 3}$

The degree of the numerator is less than the degree of the denominator, so $y = 0$ is the horizontal asymptote.

25. $g(x) = \dfrac{x^3 - 2x^2 + x - 1}{x^2 - 16}$

The degree of the numerator is greater than the degree of the denominator, so there is no horizontal asymptote.

27. $g(x) = \dfrac{x^2 + 4x - 1}{x + 3}$

$$
\begin{array}{r}
x + 1 \\
x + 3 \overline{\smash{\big)}\, x^2 + 4x - 1} \\
\underline{x^2 + 3x } \\
x - 1 \\
\underline{x + 3} \\
-4
\end{array}
$$

Then $g(x) = x + 1 + \dfrac{-4}{x + 3}$. The oblique asymptote is $y = x + 1$.

29. $h(x) = \dfrac{x^4 - 2}{x^3 + 1}$

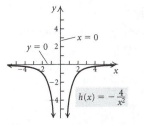

$$
\begin{array}{r}
x \\
x^3 + 1\;\overline{\smash{\big)}\;x^4 + 0x^3 + 0x^2 + 0x} \\
\underline{x^4 \qquad\qquad\quad + x} \\
- x
\end{array}
$$

Then $h(x) = x + \dfrac{-x}{x^3 + 1}$. The oblique asymptote is $y = x$.

31. $f(x) = \dfrac{x^3 - x^2 + x - 4}{x^2 + 2x - 1}$

$$
\begin{array}{r}
x - 3 \\
x^2 + 2x - 1\;\overline{\smash{\big)}\;x^3 - x^2 + x - 4} \\
\underline{x^3 + 2x^2 - x} \\
-3x^2 + 2x - 4 \\
\underline{-3x^2 - 6x + 3} \\
8x - 7
\end{array}
$$

Then $f(x) = x - 3 + \dfrac{8x - 7}{x^2 + 2x - 1}$. The oblique asymptote is $y = x - 3$.

33. $f(x) = \dfrac{1}{x}$

1. The numerator and the denominator have no common factors. 0 is the zero of the denominator, so the domain excludes 0. It is $(-\infty, 0) \cup (0, \infty)$. The line $x = 0$, or the y-axis, is the vertical asymptote.

2. Because the degree of the numerator is less than the degree of the denominator, the x-axis, or $y = 0$, is the horizontal asymptote. There are no oblique asymptotes.

3. The numerator has no zeros, so there is no x-intercept.

4. Since 0 is not in the domain of the function, there is no y-intercept.

5. Find other function values to determine the shape of the graph and then draw it.

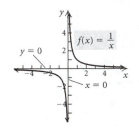

35. $h(x) = -\dfrac{4}{x^2}$

1. The numerator and the denominator have no common factors. 0 is the zero of the denominator, so the domain excludes 0. It is $(-\infty, 0) \cup (0, \infty)$. The line $x = 0$, or the y-axis, is the vertical asymptote.

2. Because the degree of the numerator is less than the degree of the denominator, the x-axis, or $y = 0$, is the horizontal asymptote. There is no oblique asymptote.

3. The numerator has no zeros, so there is no x-intercept.

4. Since 0 is not in the domain of the function, there is no y-intercept.

5. Find other function values to determine the shape of the graph and then draw it.

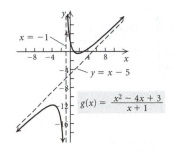

37. $g(x) = \dfrac{x^2 - 4x + 3}{x + 1} = \dfrac{(x - 1)(x - 3)}{x + 1}$

1. The numerator and the denominator have no common factors. The denominator, $x + 1$, is 0 when $x = -1$, so the domain excludes -1. It is $(-\infty, -1) \cup (-1, \infty)$. The line $x = -1$ is the vertical asymptote.

2. The degree of the numerator is 1 greater than the degree of the denominator, so we divide to find the oblique asymptote.

$$
\begin{array}{r}
x - 5 \\
x + 1\;\overline{\smash{\big)}\;x^2 - 4x + 3} \\
\underline{x^2 + x} \\
-5x + 3 \\
\underline{-5x - 5} \\
8
\end{array}
$$

The oblique asymptote is $y = x - 5$. There is no horizontal asymptote.

3. The zeros of the numerator are 1 and 3. Thus the x-intercepts are $(1, 0)$ and $(3, 0)$.

4. $g(0) = \dfrac{0^2 - 4 \cdot 0 + 3}{0 + 1} = 3$, so the y-intercept is $(0, 3)$.

5. Find other function values to determine the shape of the graph and then draw it.

39. $f(x) = \dfrac{-2}{x - 5}$

1. The numerator and the denominator have no common factors. 5 is the zero of the denominator, so the domain excludes 5. It is $(-\infty, 5) \cup (5, \infty)$. The line $x = 5$ is the vertical asymptote.

2. Because the degree of the numerator is less than the degree of the denominator, the x-axis, or $y = 0$, is the horizontal asymptote. There is no oblique asymptote.

3. The numerator has no zeros, so there is no x-intercept.

4. $f(0) = \dfrac{-2}{0-5} = \dfrac{2}{5}$, so $\left(0, \dfrac{2}{5}\right)$ is the y-intercept.

5. Find other function values to determine the shape of the graph and then draw it.

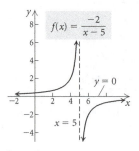

41. $f(x) = \dfrac{2x+1}{x}$

1. The numerator and the denominator have no common factors. 0 is the zero of the denominator, so the domain excludes 0. It is $(-\infty, 0) \cup (0, \infty)$. The line $x = 0$, or the y-axis, is the vertical asymptote.

2. The numerator and denominator have the same degree, so the horizontal asymptote is determined by the ratio of the leading coefficients, $2/1$, or 2. Thus, $y = 2$ is the horizontal asymptote. There is no oblique asymptote.

3. The zero of the numerator is the solution of $2x+1 = 0$, or $-\dfrac{1}{2}$. The x-intercept is $\left(-\dfrac{1}{2}, 0\right)$.

4. Since 0 is not in the domain of the function, there is no y-intercept.

5. Find other function values to determine the shape of the graph and then draw it.

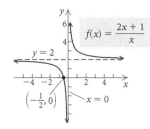

43. $f(x) = \dfrac{x+3}{x^2-9} = \dfrac{x+3}{(x+3)(x-3)}$

1. The domain of the function is $(-\infty, -3) \cup (-3, 3) \cup (3, \infty)$. The numerator and denominator have the common factor $x+3$. The zeros of the denominator are -3 and 3, and the zero of the numerator is -3. Since 3 is the only zero of the denominator that is not a zero of the numerator, the only vertical asymptote is $x = 3$.

2. Because the degree of the numerator is less than the degree of the denominator, the x-axis, or $y = 0$, is the horizontal asymptote. There are no oblique asymptotes.

3. The zero of the numerator, -3, is not in the domain of the function, so there is no x-intercept.

4. $f(0) = \dfrac{0+3}{0^2-9} = -\dfrac{1}{3}$, so the y-intercept is $\left(0, -\dfrac{1}{3}\right)$.

5. Find other function values to determine the shape of the graph and then draw it.

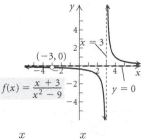

45. $f(x) = \dfrac{x}{x^2+3x} = \dfrac{x}{x(x+3)}$

1. The zeros of the denominator are 0 and -3, so the domain is $(-\infty, -3) \cup (-3, 0) \cup (0, \infty)$. The zero of the numerator is 0. Since -3 is the only zero of the denominator that is not also a zero of the numerator, the only vertical asymptote is $x = -3$.

2. Because the degree of the numerator is less than the degree of the denominator, the x-axis, or $y = 0$, is the horizontal asymptote. There is no oblique asymptote.

3. The zero of the numerator is 0, but 0 is not in the domain of the function, so there is no x-intercept.

4. Since 0 is not in the domain of the function, there is no y-intercept.

5. Find other function values to determine the shape of the graph and then draw it, indicating the "hole" when $x = 0$ with an open circle.

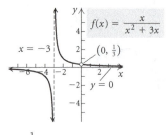

47. $f(x) = \dfrac{1}{(x-2)^2}$

1. The numerator and the denominator have no common factors. 2 is the zero of the denominator, so the domain excludes 2. It is $(-\infty, 2) \cup (2, \infty)$. The line $x = 2$ is the vertical asymptote.

2. Because the degree of the numerator is less than the degree of the denominator, the x-axis, or $y = 0$, is the horizontal asymptote. There is no oblique asymptote.

3. The numerator has no zeros, so there is no x-intercept.

4. $f(0) = \dfrac{1}{(0-2)^2} = \dfrac{1}{4}$, so $\left(0, \dfrac{1}{4}\right)$ is the y-intercept.

5. Find other function values to determine the shape of the graph and then draw it.

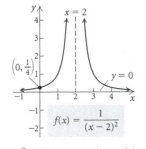

49. $f(x) = \dfrac{x^2 + 2x - 3}{x^2 + 4x + 3} = \dfrac{(x+3)(x-1)}{(x+3)(x+1)}$

1. The zeros of the denominator are -3 and -1, so the domain is $(-\infty, -3) \cup (-3, -1) \cup (-1, \infty)$. The zeros of the numerator are -3 and 1. Since -1 is the only zero of the denominator that is not also a zero of the numerator, the only vertical asymptote is $x = -1$.

2. The numerator and the denominator have the same degree, so the horizontal asymptote is determined by the ratio of the leading coefficients, $1/1$, or 1. Thus, $y = 1$ is the horizontal asymptote. There is no oblique asymptote.

3. The only zero of the numerator that is in the domain of the function is 1, so the only x-intercept is $(1, 0)$.

4. $f(0) = \dfrac{0^2 + 2 \cdot 0 - 3}{0^2 + 4 \cdot 0 + 3} = \dfrac{-3}{3} = -1$, so the y-intercept is $(0, -1)$.

5. Find other function values to determine the shape of the graph and then draw it, indicating the "hole" when $x = -3$ with an open circle.

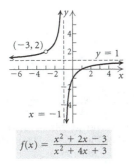

51. $f(x) = \dfrac{1}{x^2 + 3}$

1. The numerator and the denominator have no common factors. The denominator has no real-number zeros, so the domain is $(-\infty, \infty)$ and there is no vertical asymptote.

2. Because the degree of the numerator is less than the degree of the denominator, the x-axis, or $y = 0$, is the horizontal asymptote. There is no oblique asymptote.

3. The numerator has no zeros, so there is no x-intercept.

4. $f(0) = \dfrac{1}{0^2 + 3} = \dfrac{1}{3}$, so $\left(0, \dfrac{1}{3}\right)$ is the y-intercept.

5. Find other function values to determine the shape of the graph and then draw it.

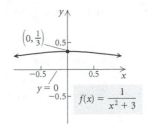

53. $f(x) = \dfrac{x^2 - 4}{x - 2} = \dfrac{(x+2)(x-2)}{x - 2} = x + 2,\ x \neq 2$

The graph is the same as the graph of $f(x) = x + 2$ except at $x = 2$, where there is a hole. Thus the domain is $(-\infty, 2) \cup (2, \infty)$. The zero of $f(x) = x + 2$ is -2, so the x-intercept is $(-2, 0)$; $f(0) = 2$, so the y-intercept is $(0, 2)$.

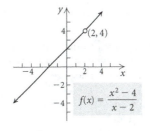

55. $f(x) = \dfrac{x - 1}{x + 2}$

1. The numerator and the denominator have no common factors. -2 is the zero of the denominator, so the domain excludes -2. It is $(-\infty, -2) \cup (-2, \infty)$. The line $x = -2$ is the vertical asymptote.

2. The numerator and denominator have the same degree, so the horizontal asymptote is determined by the ratio of the leading coefficients, $1/1$, or 1. Thus, $y = 1$ is the horizontal asymptote. There is no oblique asymptote.

3. The zero of the numerator is 1, so the x-intercept is $(1, 0)$.

4. $f(0) = \dfrac{0 - 1}{0 + 2} = -\dfrac{1}{2}$, so $\left(0, -\dfrac{1}{2}\right)$ is the y-intercept.

5. Find other function values to determine the shape of the graph and then draw it.

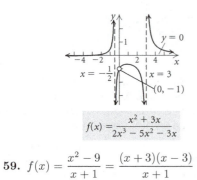

57. $f(x) = \dfrac{x^2 + 3x}{2x^3 - 5x^2 - 3x} = \dfrac{x(x+3)}{x(2x^2 - 5x - 3)} = \dfrac{x(x+3)}{x(2x+1)(x-3)}$

1. The zeros of the denominator are 0, $-\dfrac{1}{2}$, and 3, so the domain is $\left(-\infty, -\dfrac{1}{2}\right) \cup \left(-\dfrac{1}{2}, 0\right) \cup (0, 3) \cup (3, \infty)$. The zeros of the numerator are 0 and -3. Since $-\dfrac{1}{2}$ and 3 are the only zeros of the denominator that are not also zeros of the numerator, the vertical asymptotes are $x = -\dfrac{1}{2}$ and $x = 3$.

2. Because the degree of the numerator is less than the degree of the denominator, the x-axis, or $y = 0$, is the horizontal asymptote. There is no oblique asymptote.

3. The only zero of the numerator that is in the domain of the function is -3 so the only x-intercept is $(-3, 0)$.

4. 0 is not in the domain of the function, so there is no y-intercept.

5. Find other function values to determine the shape of the graph and then draw it, indicating the "hole" when $x = 0$ with an open circle.

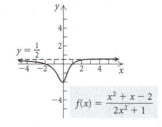

59. $f(x) = \dfrac{x^2 - 9}{x + 1} = \dfrac{(x + 3)(x - 3)}{x + 1}$

1. The numerator and the denominator have no common factors. -1 is the zero of the denominator, so the domain is $(-\infty, -1) \cup (-1, \infty)$. The line $x = -1$ is the vertical asymptote.

2. Because the degree of the numerator is one greater than the degree of the denominator, there is an oblique asymptote. Using division, we find that $\dfrac{x^2 - 9}{x + 1} = x - 1 + \dfrac{-8}{x + 1}$. As $|x|$ becomes very large, the graph of $f(x)$ gets close to the graph of $y = x - 1$. Thus, the line $y = x - 1$ is the oblique asymptote.

3. The zeros of the numerator are -3 and 3. Thus, the x-intercepts are $(-3, 0)$ and $(3, 0)$.

4. $f(0) = \dfrac{0^2 - 9}{0 + 1} = -9$, so $(0, -9)$ is the y-intercept.

5. Find other function values to determine the shape of the graph and then draw it.

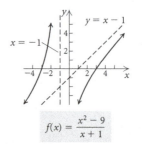

61. $f(x) = \dfrac{x^2 + x - 2}{2x^2 + 1} = \dfrac{(x + 2)(x - 1)}{2x^2 + 1}$

1. The numerator and the denominator have no common factors. The denominator has no real-number zeros, so the domain is $(-\infty, \infty)$ and there is no vertical asymptote.

2. The numerator and the denominator have the same degree, so the horizontal asymptote is determined by the ratio of the leading coefficients, $1/2$. Thus, $y = 1/2$ is the horizontal asymptote. There is no oblique asymptote.

3. The zeros of the numerator are -2 and 1. Thus, the x-intercepts are $(-2, 0)$ and $(1, 0)$.

4. $f(0) = \dfrac{0^2 + 0 - 2}{2 \cdot 0^2 + 1} = -2$, so $(0, -2)$ is the y-intercept.

5. Find other function values to determine the shape of the graph and then draw it.

63. $g(x) = \dfrac{3x^2 - x - 2}{x - 1} = \dfrac{(3x + 2)(x - 1)}{x - 1} = 3x + 2, \; x \neq 1$

The graph is the same as the graph of $g(x) = 3x + 2$ except at $x = 1$, where there is a hole. Thus the domain is $(-\infty, 1) \cup (1, \infty)$.

The zero of $g(x) = 3x + 2$ is $-\dfrac{2}{3}$, so the x-intercept is $\left(-\dfrac{2}{3}, 0\right)$; $g(0) = 2$, so the y-intercept is $(0, 2)$.

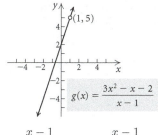

$g(x) = \dfrac{3x^2 - x - 2}{x - 1}$

65. $f(x) = \dfrac{x - 1}{x^2 - 2x - 3} = \dfrac{x - 1}{(x + 1)(x - 3)}$

1. The numerator and the denominator have no common factors. The zeros of the denominator are -1 and 3. Thus, the domain is $(-\infty, -1) \cup (-1, 3) \cup (3, \infty)$ and the lines $x = -1$ and $x = 3$ are the vertical asymptotes.

2. Because the degree of the numerator is less than the degree of the denominator, the x-axis, or $y = 0$, is the horizontal asymptote. There is no oblique asymptote.

3. 1 is the zero of the numerator, so $(1, 0)$ is the x-intercept.

4. $f(0) = \dfrac{0 - 1}{0^2 - 2 \cdot 0 - 3} = \dfrac{1}{3}$, so $\left(0, \dfrac{1}{3}\right)$ is the y-intercept.

5. Find other function values to determine the shape of the graph and then draw it.

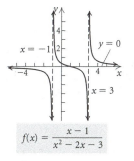

$f(x) = \dfrac{x - 1}{x^2 - 2x - 3}$

67. $f(x) = \dfrac{3x^2 + 11x - 4}{x^2 + 2x - 8} = \dfrac{(3x - 1)(x + 4)}{(x + 4)(x - 2)}$

1. The domain of the function is $(-\infty, -4) \cup (-4, 2) \cup (2, \infty)$. The numerator and the denominator have the common factor $x + 4$. The zeros of the denominator are -4 and 2, and the zeros of the numerator are $\dfrac{1}{3}$ and -4. Since 2 is the only zero of the denominator that is not a zero of the numerator, the only vertical asymptote is $x = 2$.

2. The numerator and the denominator have the same degree, so the horizontal asymptote is determined by the ratio of the leading coefficients, $3/1$, or 3. Thus, $y = 3$ is the horizontal asymptote. There is no oblique asymptote.

3. The only zero of the numerator that is in the domain of the function is $\dfrac{1}{3}$, so the x-intercept is $\left(\dfrac{1}{3}, 0\right)$.

4. $f(0) = \dfrac{3 \cdot 0^2 + 11 \cdot 0 - 4}{0^2 + 2 \cdot 0 - 8} = \dfrac{-4}{-8} = \dfrac{1}{2}$, so the y-intercept is $\left(0, \dfrac{1}{2}\right)$.

5. Find other function values to determine the shape of the graph and then draw it.

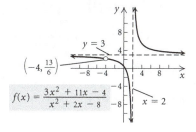

$f(x) = \dfrac{3x^2 + 11x - 4}{x^2 + 2x - 8}$

69. $f(x) = \dfrac{x - 3}{(x + 1)^3}$

1. The numerator and the denominator have no common factors. -1 is the zero of the denominator, so the domain excludes -1. It is $(-\infty, -1) \cup (-1, \infty)$. The line $x = -1$ is the vertical asymptote.

2. Because the degree of the numerator is less than the degree of the denominator, the x-axis, or $y = 0$, is the horizontal asymptote. There is no oblique asymptote.

3. 3 is the zero of the numerator, so $(3, 0)$ is the x-intercept.

4. $f(0) = \dfrac{0 - 3}{(0 + 1)^3} = -3$, so $(0, -3)$ is the y-intercept.

5. Find other function values to determine the shape of the graph and then draw it.

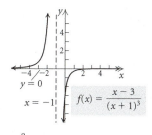

$f(x) = \dfrac{x - 3}{(x + 1)^3}$

71. $f(x) = \dfrac{x^3 + 1}{x}$

1. The numerator and the denominator have no common factors. 0 is the zero of the denominator, so the domain excludes 0. It is $(-\infty, 0) \cup (0, \infty)$. The line $x = 0$, or the y-axis, is the vertical asymptote.

2. Because the degree of the numerator is more than one greater than the degree of the denominator, there is no horizontal or oblique asymptote.

3. The real-number zero of the numerator is -1, so the x-intercept is $(-1, 0)$.

4. Since 0 is not in the domain of the function, there is no y-intercept.

5. Find other function values to determine the shape of the graph and then draw it.

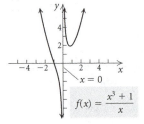

73. $f(x) = \dfrac{x^3 + 2x^2 - 15x}{x^2 - 5x - 14} = \dfrac{x(x+5)(x-3)}{(x+2)(x-7)}$

1. The numerator and the denominator have no common factors. The zeros of the denominator are -2 and 7. Thus, the domain is $(-\infty, -2) \cup (-2, 7) \cup (7, \infty)$ and the lines $x = -2$ and $x = 7$ are the vertical asymptotes.

2. Because the degree of the numerator is one greater than the degree of the denominator, there is an oblique asymptote. Using division, we find that $\dfrac{x^3 + 2x^2 - 15x}{x^2 - 5x - 14} = x + 7 + \dfrac{34x + 98}{x^2 - 5x - 14}$. As $|x|$ becomes very large, the graph of $f(x)$ gets close to the graph of $y = x + 7$. Thus, the line $y = x + 7$ is the oblique asymptote.

3. The zeros of the numerator are 0, -5, and 3. Thus, the x-intercepts are $(-5, 0)$, $(0, 0)$, and $(3, 0)$.

4. From part (3) we see that $(0, 0)$ is the y-intercept.

5. Find other function values to determine the shape of the graph and then draw it.

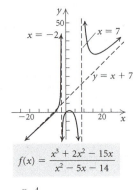

75. $f(x) = \dfrac{5x^4}{x^4 + 1}$

1. The numerator and the denominator have no common factors. The denominator has no real-number zeros, so the domain is $(-\infty, \infty)$ and there is no vertical asymptote.

2. The numerator and denominator have the same degree, so the horizontal asymptote is determined by the ratio of the leading coefficients, $5/1$, or 5. Thus, $y = 5$ is the horizontal asymptote. There is no oblique asymptote.

3. The zero of the numerator is 0, so $(0, 0)$ is the x-intercept.

4. From part (3) we see that $(0, 0)$ is the y-intercept.

5. Find other function values to determine the shape of the graph and then draw it.

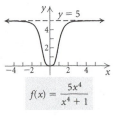

77. $f(x) = \dfrac{x^2}{x^2 - x - 2} = \dfrac{x^2}{(x+1)(x-2)}$

1. The numerator and the denominator have no common factors. The zeros of the denominator are -1 and 2. Thus, the domain is $(-\infty, -1) \cup (-1, 2) \cup (2, \infty)$ and the lines $x = -1$ and $x = 2$ are the vertical asymptotes.

2. The numerator and denominator have the same degree, so the horizontal asymptote is determined by the ratio of the leading coefficients, $1/1$, or 1. Thus, $y = 1$ is the horizontal asymptote. There is no oblique asymptote.

3. The zero of the numerator is 0, so the x-intercept is $(0, 0)$.

4. From part (3) we see that $(0, 0)$ is the y-intercept.

5. Find other function values to determine the shape of the graph and then draw it.

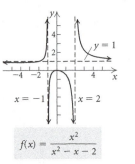

79. Answers may vary. The numbers -4 and 5 must be zeros of the denominator but not of the numerator. A function that satisfies these conditions is
$$f(x) = \dfrac{1}{(x+4)(x-5)}, \text{ or } f(x) = \dfrac{1}{x^2 - x - 20}.$$

81. Answers may vary. The numbers -4 and 5 must be zeros of the denominator but not of the numerator and -2 must be a zero of the numerator but not of the denominator. In addition, the numerator and denominator must have the same degree and the ratio of their leading coefficients must be $3/2$. A function that satisfies these conditions is
$$f(x) = \dfrac{3x(x+2)}{2(x+4)(x-5)}, \text{ or } f(x) = \dfrac{3x^2 + 6x}{2x^2 - 2x - 40}.$$
Another function that satisfies these conditions is
$$g(x) = \dfrac{3(x+2)^2}{2(x+4)(x-5)}, \text{ or } g(x) = \dfrac{3x^2 + 12x + 12}{2x^2 - 2x - 40}.$$

83. a) The horizontal asymptote of $N(t)$ is the ratio of the leading coefficients of the numerator and denominator, $0.8/5$, or 0.16. Thus, $N(t) \to 0.16$ as $t \to \infty$.

   b) The medication never completely disappears from the body; a trace amount remains.

85. a) $P(0) = 0$; $P(1) = 45.455$ thousand, or $45,455$;

   $P(3) = 55.556$ thousand, or $55,556$;

   $P(8) = 29.197$ thousand, or $29,197$

   b) The degree of the numerator is less than the degree of the denominator, so the x-axis is the horizontal asymptote. Thus, $P(t) \to 0$ as $t \to \infty$.

   c) Eventually, no one will live in this community.

87. slope

89. point-slope equation

91. $f(-x) = -f(x)$

93. midpoint formula

95. $f(x) = \dfrac{x^5 + 2x^3 + 4x^2}{x^2 + 2} = x^3 + 4 + \dfrac{-8}{x^2 + 2}$

   As $|x| \to \infty$, $\dfrac{-8}{x^2 + 2} \to 0$ and the value of $f(x) \to x^3 + 4$.
   Thus, the nonlinear asymptote is $y = x^3 + 4$.

97.

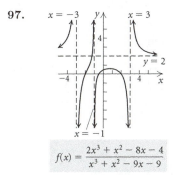

$$f(x) = \frac{2x^3 + x^2 - 8x - 4}{x^3 + x^2 - 9x - 9}$$

Exercise Set 4.6

1. $x^2 + 2x - 15 = 0$

   $(x + 5)(x - 3) = 0$

   $x + 5 = 0 \quad or \quad x - 3 = 0$

   $x = -5 \ or \qquad x = 3$

   The solution set is $\{-5, 3\}$.

3. Solve $x^2 + 2x - 15 \le 0$.

   From Exercise 2 we know the solution set of $x^2 + 2x - 15 < 0$ is $(-5, 3)$. The solution set of $x^2 + 2x - 15 \le 0$ includes the endpoints of this interval. Thus the solution set is $[-5, 3]$.

5. Solve $x^2 + 2x - 15 \ge 0$.

   From Exercise 4 we know the solution set of $x^2 + 2x - 15 > 0$ is $(-\infty, -5) \cup (3, \infty)$. The solution set of $x^2 + 2x - 15 \ge 0$ includes the endpoints -5 and 3. Thus the solution set is $(-\infty, -5] \cup [3, \infty)$.

7. Solve $\dfrac{x - 2}{x + 4} > 0$.

   The denominator tells us that $g(x)$ is not defined when $x = -4$. From Exercise 6 we know that $g(2) = 0$. The critical values of -4 and 2 divide the x-axis into three intervals $(-\infty, -4)$, $(-4, 2)$, and $(2, \infty)$. We test a value in each interval.

   $(-\infty, -4)$: $g(-5) = 7 > 0$

   $(-4, 2)$: $g(0) = -\dfrac{1}{2} < 0$

   $(2, \infty)$: $g(3) = \dfrac{1}{7} > 0$

   Function values are positive on $(-\infty, -4)$ and on $(2, \infty)$. The solution set is $(-\infty, -4) \cup (2, \infty)$.

9. Solve $\dfrac{x - 2}{x + 4} \ge 0$.

   From Exercise 7 we know that the solution set of $\dfrac{x - 2}{x + 4} > 0$ is $(-\infty, -4) \cup (2, \infty)$. We include the zero of the function, 2, since the inequality symbol is $\ge$. The critical value -4 is not included because it is not in the domain of the function. The solution set is $(-\infty, -4) \cup [2, \infty)$.

11. $\dfrac{7x}{(x - 1)(x + 5)} = 0$

   $7x = 0$ Multiplying by $(x - 1)(x + 5)$

   $x = 0$

   The solution set is $\{0\}$.

13. Solve $\dfrac{7x}{(x - 1)(x + 5)} \ge 0$.

   From our work in Exercise 12 we see that function values are positive on $(-5, 0)$ and on $(1, \infty)$. We also include the zero of the function, 0, in the solution set because the inequality symbol is $\ge$. The critical values -5 and 1 are not included because they are not in the domain of the function. The solution set is $(-5, 0] \cup (1, \infty)$.

15. Solve $\dfrac{7x}{(x - 1)(x + 5)} < 0$.

   From our work in Exercise 12 we see that the solution set is $(-\infty, -5) \cup (0, 1)$.

17. Solve $x^5 - 9x^3 < 0$.

   From Exercise 16 we know the solutions of the related equation are -3, 0, and 3. These numbers divide the x-axis into the intervals $(-\infty, -3)$, $(-3, 0)$, $(0, 3)$, and $(3, \infty)$. We test a value in each interval.

   $(-\infty, -3)$: $g(-4) = -448 < 0$

   $(-3, 0)$: $g(-1) = 8 > 0$

   $(0, 3)$: $g(1) = -8 < 0$

   $(3, \infty)$: $g(4) = 448 > 0$

   Function values are negative on $(-\infty, -3)$ and on $(0, 3)$. The solution set is $(-\infty, -3) \cup (0, 3)$.

19. Solve $x^5 - 9x^3 > 0$.

   From our work in Exercise 17 we see that the solution set is $(-3, 0) \cup (3, \infty)$.

21. First we find an equivalent inequality with 0 on one side.

$$x^3 + 6x^2 < x + 30$$

$$x^3 + 6x^2 - x - 30 < 0$$

From the graph we see that the x-intercepts of the related function occur at $x = -5$, $x = -3$, and $x = 2$. They divide the x-axis into the intervals $(-\infty, -5)$, $(-5, -3)$, $(-3, 2)$, and $(2, \infty)$. From the graph we see that the function has negative values only on $(-\infty, -5)$ and $(-3, 2)$. Thus, the solution set is $(-\infty, -5) \cup (-3, 2)$.

23. By observing the graph or the denominator of the function, we see that the function is not defined for $x = -2$ or $x = 2$. We also see that 0 is a zero of the function. These numbers divide the x-axis into the intervals $(-\infty, -2)$, $(-2, 0)$, $(0, 2)$, and $(2, \infty)$. From the graph we see that the function has positive values only on $(-2, 0)$ and $(2, \infty)$. Since the inequality symbol is $\geq$, 0 must be included in the solution set. It is $(-2, 0] \cup (2, \infty)$.

25. $(x - 1)(x + 4) < 0$

The related equation is $(x - 1)(x + 4) = 0$. Using the principle of zero products, we find that the solutions of the related equation are 1 and -4. These numbers divide the x-axis into the intervals $(-\infty, -4)$, $(-4, 1)$, and $(1, \infty)$. We let $f(x) = (x - 1)(x + 4)$ and test a value in each interval.

$(-\infty, -4)$: $f(-5) = 6 > 0$

$(-4, 1)$: $f(0) = -4 < 0$

$(1, \infty)$: $f(2) = 6 > 0$

Function values are negative only in the interval $(-4, 1)$. The solution set is $(-4, 1)$.

27.
$$\begin{aligned} x^2 + x - 2 &> 0 \quad \text{Polynomial inequality} \\ x^2 + x - 2 &= 0 \quad \text{Related equation} \\ (x + 2)(x - 1) &= 0 \quad \text{Factoring} \end{aligned}$$

Using the principle of zero products, we find that the solutions of the related equation are -2 and 1. These numbers divide the x-axis into the intervals $(-\infty, -2)$, $(-2, 1)$, and $(1, \infty)$. We let $f(x) = x^2 + x - 2$ and test a value in each interval.

$(-\infty, -2)$: $f(-3) = 4 > 0$

$(-2, 1)$: $f(0) = -2 < 0$

$(1, \infty)$: $f(2) = 4 > 0$

Function values are positive on $(-\infty, -2)$ and $(1, \infty)$. The solution set is $(-\infty, -2) \cup (1, \infty)$.

29.
$$\begin{aligned} x^2 - x - 5 &\geq x - 2 \\ x^2 - 2x - 3 &\geq 0 \quad \text{Polynomial inequality} \\ x^2 - 2x - 3 &= 0 \quad \text{Related equation} \\ (x + 1)(x - 3) &= 0 \quad \text{Factoring} \end{aligned}$$

Using the principle of zero products, we find that the solutions of the related equation are -1 and 3. The numbers divide the x-axis into the intervals $(-\infty, -1)$, $(-1, 3)$, and $(3, \infty)$. We let $f(x) = x^2 - 2x - 3$ and test a value in each interval.

$(-\infty, -1)$: $f(-2) = 5 > 0$

$(-1, 3)$: $f(0) = -3 < 0$

$(3, \infty)$: $f(4) = 5 > 0$

Function values are positive on $(-\infty, -1)$ and on $(3, \infty)$. Since the inequality symbol is $\geq$, the endpoints of the intervals must be included in the solution set. It is $(-\infty, -1] \cup [3, \infty)$.

31.
$$\begin{aligned} x^2 &> 25 \quad \text{Polynomial inequality} \\ x^2 - 25 &> 0 \quad \begin{array}{l}\text{Equivalent inequality with} \\ \text{0 on one side}\end{array} \\ x^2 - 25 &= 0 \quad \text{Related equation} \\ (x + 5)(x - 5) &= 0 \quad \text{Factoring} \end{aligned}$$

Using the principle of zero products, we find that the solutions of the related equation are -5 and 5. These numbers divide the x-axis into the intervals $(-\infty, -5)$, $(-5, 5)$, and $(5, \infty)$. We let $f(x) = x^2 - 25$ and test a value in each interval.

$(-\infty, -5)$: $f(-6) = 11 > 0$

$(-5, 5)$: $f(0) = -25 < 0$

$(5, \infty)$: $f(6) = 11 > 0$

Function values are positive on $(-\infty, -5)$ and $(5, \infty)$. The solution set is $(-\infty, -5) \cup (5, \infty)$.

33.
$$\begin{aligned} 4 - x^2 &\leq 0 \quad \text{Polynomial inequality} \\ 4 - x^2 &= 0 \quad \text{Related equation} \\ (2 + x)(2 - x) &= 0 \quad \text{Factoring} \end{aligned}$$

Using the principle of zero products, we find that the solutions of the related equation are -2 and 2. These numbers divide the x-axis into the intervals $(-\infty, -2)$, $(-2, 2)$, and $(2, \infty)$. We let $f(x) = 4 - x^2$ and test a value in each interval.

$(-\infty, -2)$: $f(-3) = -5 < 0$

$(-2, 2)$: $f(0) = 4 > 0$

$(2, \infty)$: $f(3) = -5 < 0$

Function values are negative on $(-\infty, -2)$ and $(2, \infty)$. Since the inequality symbol is $\leq$, the endpoints of the intervals must be included in the solution set. It is $(-\infty, -2] \cup [2, \infty)$.

35.
$$\begin{aligned} 6x - 9 - x^2 &< 0 \quad \text{Polynomial inequality} \\ 6x - 9 - x^2 &= 0 \quad \text{Related equation} \\ -(x^2 - 6x + 9) &= 0 \quad \begin{array}{l}\text{Factoring out } -1 \text{ and} \\ \text{rearranging}\end{array} \\ -(x - 3)(x - 3) &= 0 \quad \text{Factoring} \end{aligned}$$

Using the principle of zero products, we find that the solution of the related equation is 3. This number divides the x-axis into the intervals $(-\infty, 3)$ and $(3, \infty)$. We let $f(x) = 6x - 9 - x^2$ and test a value in each interval.

$(-\infty, 3)$: $f(-4) = -49 < 0$

$(3, \infty)$: $f(4) = -1 < 0$

Function values are negative on both intervals. The solution set is $(-\infty, 3) \cup (3, \infty)$.

37. $x^2 + 12 < 4x$ Polynomial inequality

$x^2 - 4x + 12 < 0$ Equivalent inequality with 0 on one side

$x^2 - 4x + 12 = 0$ Related equation

Using the quadratic formula, we find that the related equation has no real-number solutions. The graph lies entirely above the x-axis, so the inequality has no solution. We could determine this algebraically by letting $f(x) = x^2 - 4x + 12$ and testing any real number (since there are no real-number solutions of $f(x) = 0$ to divide the x-axis into intervals). For example, $f(0) = 12 > 0$, so we see algebraically that the inequality has no solution. The solution set is $\emptyset$.

39. $4x^3 - 7x^2 \le 15x$ Polynomial inequality

$4x^3 - 7x^2 - 15x \le 0$ Equivalent inequality with 0 on one side

$4x^3 - 7x^2 - 15x = 0$ Related equation

$x(4x + 5)(x - 3) = 0$ Factoring

Using the principle of zero products, we find that the solutions of the related equation are 0, $-\dfrac{5}{4}$, and 3. These numbers divide the x-axis into the intervals $\left(-\infty, -\dfrac{5}{4}\right)$, $\left(-\dfrac{5}{4}, 0\right)$, $(0, 3)$, and $(3, \infty)$. We let $f(x) = 4x^3 - 7x^2 - 15x$ and test a value in each interval.

$\left(-\infty, -\dfrac{5}{4}\right)$: $f(-2) = -30 < 0$

$\left(-\dfrac{5}{4}, 0\right)$: $f(-1) = 4 > 0$

$(0, 3)$ $f(1) = -18 < 0$

$(3, \infty)$: $f(4) = 84 > 0$

Function values are negative on $\left(-\infty, -\dfrac{5}{4}\right)$ and $(0, 3)$. Since the inequality symbol is $\le$, the endpoints of the intervals must be included in the solution set. It is $\left(-\infty, -\dfrac{5}{4}\right] \cup [0, 3]$.

41. $x^3 + 3x^2 - x - 3 \ge 0$ Polynomial inequality

$x^3 + 3x^2 - x - 3 = 0$ Related equation

$x^2(x + 3) - (x + 3) = 0$ Factoring

$(x^2 - 1)(x + 3) = 0$

$(x + 1)(x - 1)(x + 3) = 0$

Using the principle of zero products, we find that the solutions of the related equation are -1, 1, and -3. These numbers divide the x-axis into the intervals $(-\infty, -3)$, $(-3, -1)$, $(-1, 1)$, and $(1, \infty)$. We let $f(x) = x^3 + 3x^2 - x - 3$ and test a value in each interval.

$(-\infty, -3)$: $f(-4) = -15 < 0$

$(-3, -1)$: $f(-2) = 3 > 0$

$(-1, 1)$: $f(0) = -3 < 0$

$(1, \infty)$: $f(2) = 15 > 0$

Function values are positive on $(-3, -1)$ and $(1, \infty)$. Since the inequality symbol is $\ge$, the endpoints of the intervals must be included in the solution set. It is $[-3, -1] \cup [1, \infty)$.

43. $x^3 - 2x^2 < 5x - 6$ Polynomial inequality

$x^3 - 2x^2 - 5x + 6 < 0$ Equivalent inequality with 0 on one side

$x^3 - 2x^2 - 5x + 6 = 0$ Related equation

Using the techniques of Section 3.3, we find that the solutions of the related equation are -2, 1, and 3. They divide the x-axis into the intervals $(-\infty, -2)$, $(-2, 1)$, $(1, 3)$, and $(3, \infty)$. Let $f(x) = x^3 - 2x^2 - 5x + 6$ and test a value in each interval.

$(-\infty, -2)$: $f(-3) = -24 < 0$

$(-2, 1)$: $f(0) = 6 > 0$

$(1, 3)$: $f(2) = -4 < 0$

$(3, \infty)$: $f(4) = 18 > 0$

Function values are negative on $(-\infty, -2)$ and $(1, 3)$. The solution set is $(-\infty, -2) \cup (1, 3)$.

45. $x^5 + x^2 \ge 2x^3 + 2$ Polynomial inequality

$x^5 - 2x^3 + x^2 - 2 \ge 0$ Related inequality with 0 on one side

$x^5 - 2x^3 + x^2 - 2 = 0$ Related equation

$x^3(x^2 - 2) + x^2 - 2 = 0$ Factoring

$(x^3 + 1)(x^2 - 2) = 0$

Using the principle of zero products, we find that the real-number solutions of the related equation are -1, $-\sqrt{2}$, and $\sqrt{2}$. These numbers divide the x-axis into the intervals $(-\infty, -\sqrt{2})$, $(-\sqrt{2}, -1)$, $(-1, \sqrt{2})$, and $(\sqrt{2}, \infty)$. We let $f(x) = x^5 - 2x^3 + x^2 - 2$ and test a value in each interval.

$(-\infty, -\sqrt{2})$: $f(-2) = -14 < 0$

$(-\sqrt{2}, -1)$: $f(-1.3) \approx 0.37107 > 0$

$(-1, \sqrt{2})$: $f(0) = -2 < 0$

$(\sqrt{2}, \infty)$: $f(2) = 18 > 0$

Function values are positive on $(-\sqrt{2}, -1)$ and $(\sqrt{2}, \infty)$. Since the inequality symbol is $\ge$, the endpoints of the intervals must be included in the solution set. It is $[-\sqrt{2}, -1] \cup [\sqrt{2}, \infty)$.

47. $2x^3 + 6 \le 5x^2 + x$ Polynomial inequality

$2x^3 - 5x^2 - x + 6 \le 0$ Equivalent inequality with 0 on one side

$2x^3 - 5x^2 - x + 6 = 0$ Related equation

Using the techniques of Section 3.3, we find that the solutions of the related equation are -1, $\dfrac{3}{2}$, and 2. We can also use the graph of $y = 2x^3 - 5x^2 - x + 6$ to find these solutions. They divide the x-axis into the intervals $(-\infty, -1)$, $\left(-1, \dfrac{3}{2}\right)$, $\left(\dfrac{3}{2}, 2\right)$, and $(2, \infty)$. Let $f(x) = 2x^3 - 5x^2 - x + 6$ and test a value in each interval.

$(-\infty, -1)$: $f(-2) = -28 < 0$

$\left(-1, \dfrac{3}{2}\right)$: $f(0) = 6 > 0$

$\left(\dfrac{3}{2}, 2\right)$: $f(1.6) = -0.208 < 0$

$(2, \infty)$: $f(3) = 12 > 0$

Function values are negative in $(-\infty, -1)$ and $\left(\dfrac{3}{2}, 2\right)$. Since the inequality symbol is $\leq$, the endpoints of the intervals must be included in the solution set. The solution set is $(-\infty, -1] \cup \left[\dfrac{3}{2}, 2\right]$.

49.
$$x^3 + 5x^2 - 25x \leq 125 \quad \text{Polynomial inequality}$$
$$x^3 + 5x^2 - 25x - 125 \leq 0 \quad \text{Equivalent inequality with 0 on one side}$$
$$x^3 + 5x^2 - 25x - 125 = 0 \quad \text{Related equation}$$
$$x^2(x + 5) - 25(x + 5) = 0 \quad \text{Factoring}$$
$$(x^2 - 25)(x + 5) = 0$$
$$(x + 5)(x - 5)(x + 5) = 0$$

Using the principle of zero products, we find that the solutions of the related equation are -5 and 5. These numbers divide the x-axis into the intervals $(-\infty, -5)$, $(-5, 5)$, and $(5, \infty)$. We let $f(x) = x^3 + 5x^2 - 25x - 125$ and test a value in each interval.

$(-\infty, -5)$: $f(-6) = -11 < 0$

$(-5, 5)$: $f(0) = -125 < 0$

$(5, \infty)$: $f(6) = 121 > 0$

Function values are negative on $(-\infty, -5)$ and $(-5, 5)$. Since the inequality symbol is $\leq$, the endpoints of the intervals must be included in the solution set. It is $(-\infty, -5] \cup [-5, 5]$ or $(-\infty, 5]$.

51.
$$0.1x^3 - 0.6x^2 - 0.1x + 2 < 0 \quad \text{Polynomial inequality}$$
$$0.1x^3 - 0.6x^2 - 0.1x + 2 = 0 \quad \text{Related equation}$$

After trying all the possibilities, we find that the related equation has no rational zeros. Using the graph of $y = 0.1x^3 - 0.6x^2 - 0.1x + 2$, we find that the only real-number solutions of the related equation are approximately -1.680, 2.154, and 5.526. These numbers divide the x-axis into the intervals $(-\infty, -1.680)$, $(-1.680, 2.154)$, $(2.154, 5.526)$, and $(5.526, \infty)$. We let $f(x) = 0.1x^3 - 0.6x^2 - 0.1x + 2$ and test a value in each interval.

$(-\infty, -1.680)$: $f(-2) = -1 < 0$

$(-1.680, 2.154)$: $f(0) = 2 > 0$

$(2.154, 5.526)$: $f(3) = -1 < 0$

$(5.526, \infty)$: $f(6) = 1.4 > 0$

Function values are negative on $(-\infty, -1.680)$ and $(2.154, 5.526)$. The graph can also be used to determine this. The solution set is $(-\infty, -1.680) \cup (2.154, 5.526)$.

53. $\dfrac{1}{x + 4} > 0 \quad$ Rational inequality

$\dfrac{1}{x + 4} = 0 \quad$ Related equation

The denominator of $f(x) = \dfrac{1}{x + 4}$ is 0 when $x = -4$, so the function is not defined for $x = -4$. The related equation has no solution. Thus, the only critical value is -4. It divides the x-axis into the intervals $(-\infty, -4)$ and $(-4, \infty)$. We test a value in each interval.

$(-\infty, -4)$: $f(-5) = -1 < 0$

$(-4, \infty)$: $f(0) = \dfrac{1}{4} > 0$

Function values are positive on $(-4, \infty)$. This can also be determined from the graph of $y = \dfrac{1}{x + 4}$. The solution set is $(-4, \infty)$.

55. $\dfrac{-4}{2x + 5} < 0 \quad$ Rational inequality

$\dfrac{-4}{2x + 5} = 0 \quad$ Related equation

The denominator of $f(x) = \dfrac{-4}{2x + 5}$ is 0 when $x = -\dfrac{5}{2}$, so the function is not defined for $x = -\dfrac{5}{2}$. The related equation has no solution. Thus, the only critical value is $-\dfrac{5}{2}$. It divides the x-axis into the intervals $\left(-\infty, -\dfrac{5}{2}\right)$ and $\left(-\dfrac{5}{2}, \infty\right)$. We test a value in each interval.

$\left(-\infty, -\dfrac{5}{2}\right)$: $f(-3) = 4 > 0$

$\left(-\dfrac{5}{2}, \infty\right)$: $f(0) = -\dfrac{4}{5} < 0$

Function values are negative on $\left(-\dfrac{5}{2}, \infty\right)$. The solution set is $\left(-\dfrac{5}{2}, \infty\right)$.

57. $\dfrac{2x}{x - 4} \geq 0 \quad$ Rational inequality

$\dfrac{2x}{x - 4} = 0 \quad$ Related equation

The denominator of $f(x) = \dfrac{2x}{x - 4}$ is 0 when $x = 4$, so the function is not defined for $x = 4$.

We solve the related equation $f(x) = 0$.

$$\frac{2x}{x - 4} = 0$$
$$2x = 0 \quad \text{Multiplying by } x - 4$$
$$x = 0$$

The critical values are 0 and 4. They divide the x-axis into the intervals $(-\infty, 0)$, $(0, 4)$, and $(4, \infty)$. We test a value in each interval.

$(-\infty, 0)$: $f(-1) = \dfrac{2}{5} > 0$

$(0, 4)$: $f(1) = -\dfrac{2}{3} < 0$

$(4, \infty)$: $f(5) = 10 > 0$

Function values are positive on $(-\infty, 0)$ and $(4, \infty)$. Since the inequality symbol is $\geq$ and $f(0) = 0$, then 0 must be included in the solution set. And since 4 is not in the domain of $f(x)$, 4 is not included in the solution set. It is $(-\infty, 0] \cup (4, \infty)$.

59. $\dfrac{x+1}{x-2} \geq 3$ Rational inequality

$\dfrac{x+1}{x-2} - 3 \geq 0$ Equivalent inequality
                              with 0 on one side

The denominator of $f(x) = \dfrac{x+1}{x-2} - 3$ is 0 when $x = 2$, so the function is not defined for this value of x. We solve the related equation $f(x) = 0$.

$$\dfrac{x+1}{x-2} - 3 = 0$$

$$(x-2)\left(\dfrac{x+1}{x-2} - 3\right) = (x-2) \cdot 0$$

$$x + 1 - 3(x-2) = 0$$

$$x + 1 - 3x + 6 = 0$$

$$-2x + 7 = 0$$

$$-2x = -7$$

$$x = \dfrac{7}{2}$$

The critical values are 2 and $\dfrac{7}{2}$. They divide the x-axis into the intervals $(-\infty, 2)$, $\left(2, \dfrac{7}{2}\right)$, and $\left(\dfrac{7}{2}, \infty\right)$. We test a value in each interval.

$(-\infty, 2)$: $f(0) = -3.5 < 0$

$\left(2, \dfrac{7}{2}\right)$: $f(3) = 1 > 0$

$\left(\dfrac{7}{2}, \infty\right)$: $f(4) = -0.5 < 0$

Function values are positive on $\left(2, \dfrac{7}{2}\right)$. Note that since the inequality symbol is $\geq$ and $f\left(\dfrac{7}{2}\right) = 0$, then $\dfrac{7}{2}$ must be included in the solution set. Note also that since 2 is not in the domain of $f(x)$, it is not included in the solution set. It is $\left(2, \dfrac{7}{2}\right]$.

61. $\dfrac{x-4}{x+3} - \dfrac{x+2}{x-1} \leq 0$

The denominator of $f(x) = \dfrac{x-4}{x+3} - \dfrac{x+2}{x-1}$ is 0 when $x = -3$ or $x = 1$, so the function is not defined for these values of x. We solve the related equation $f(x) = 0$.

$$\dfrac{x-4}{x+3} - \dfrac{x+2}{x-1} = 0$$

$$(x+3)(x-1)\left(\dfrac{x-4}{x+3} - \dfrac{x+2}{x-1}\right) = (x+3)(x-1) \cdot 0$$

$$(x-1)(x-4) - (x+3)(x+2) = 0$$

$$x^2 - 5x + 4 - (x^2 + 5x + 6) = 0$$

$$-10x - 2 = 0$$

$$-10x = 2$$

$$x = -\dfrac{1}{5}$$

The critical values are -3, $-\dfrac{1}{5}$, and 1. They divide the x-axis into the intervals $(-\infty, -3)$, $\left(-3, -\dfrac{1}{5}\right)$, $\left(-\dfrac{1}{5}, 1\right)$, and $(1, \infty)$. We test a value in each interval.

$(-\infty, -3)$: $f(-4) = 7.6 > 0$

$\left(-3, -\dfrac{1}{5}\right)$: $f(-1) = -2 < 0$

$\left(-\dfrac{1}{5}, 1\right)$: $f(0) = \dfrac{2}{3} > 0$

$(1, \infty)$: $f(2) = -4.4 < 0$

Function values are negative on $\left(-3, -\dfrac{1}{5}\right)$ and $(1, \infty)$. Note that since the inequality symbol is $\leq$ and $f\left(-\dfrac{1}{5}\right) = 0$, then $-\dfrac{1}{5}$ must be included in the solution set. Note also that since neither -3 nor 1 is in the domain of $f(x)$, they are not included in the solution set. It is $\left(-3, -\dfrac{1}{5}\right] \cup (1, \infty)$.

63. $\dfrac{x+6}{x-2} > \dfrac{x-8}{x-5}$ Rational inequality

$\dfrac{x+6}{x-2} - \dfrac{x-8}{x-5} > 0$ Equivalent inequality
                                          with 0 on one side

The denominator of $f(x) = \dfrac{x+6}{x-2} - \dfrac{x-8}{x-5}$ is 0 when $x = 2$ or $x = 5$, so the function is not defined for these values of x. We solve the related equation $f(x) = 0$.

$$\dfrac{x+6}{x-2} - \dfrac{x-8}{x-5} = 0$$

$$(x-2)(x-5)\left(\dfrac{x+6}{x-2} - \dfrac{x-8}{x-5}\right) = (x-2)(x-5) \cdot 0$$

$$(x-5)(x+6) - (x-2)(x-8) = 0$$

$$x^2 + x - 30 - (x^2 - 10x + 16) = 0$$

$$x^2 + x - 30 - x^2 + 10x - 16 = 0$$

$$11x - 46 = 0$$

$$11x = 46$$

$$x = \dfrac{46}{11}$$

The critical values are 2, $\dfrac{46}{11}$, and 5. They divide the x-axis into the intervals $(-\infty, 2)$, $\left(2, \dfrac{46}{11}\right)$, $\left(\dfrac{46}{11}, 5\right)$, and $(5, \infty)$. We test a value in each interval.

$(-\infty, 2)$: $f(0) = -4.6 < 0$

$\left(2, \dfrac{46}{11}\right)$: $f(4) = 1 > 0$

$\left(\dfrac{46}{11}, 5\right)$: $f(4.5) = -2.8 < 0$

$(5, \infty)$: $f(6) = 5 > 0$

Function values are positive on $\left(2, \dfrac{46}{11}\right)$ and $(5, \infty)$. The solution set is $\left(2, \dfrac{46}{11}\right) \cup (5, \infty)$.

65. $x - 2 > \dfrac{1}{x}$ Rational inequality

$x - 2 - \dfrac{1}{x} > 0$ Equivalent inequality with 0 on one side

The denominator of $f(x) = x - 2 - \dfrac{1}{x}$ is 0 when $x = 0$, so the function is not defined for this value of x. We solve the related equation $f(x) = 0$.

$$x - 2 - \frac{1}{x} = 0$$

$$x\left(x - 2 - \frac{1}{x}\right) = x \cdot 0$$

$$x^2 - 2x - x \cdot \frac{1}{x} = 0$$

$$x^2 - 2x - 1 = 0$$

Using the quadratic formula we find that $x = 1 \pm \sqrt{2}$. The critical values are $1 - \sqrt{2}$, 0, and $1 + \sqrt{2}$. They divide the x-axis into the intervals $(-\infty, 1 - \sqrt{2})$, $(1 - \sqrt{2}, 0)$, $(0, 1 + \sqrt{2})$, and $(1 + \sqrt{2}, \infty)$. We test a value in each interval.

$(-\infty, 1 - \sqrt{2})$: $f(-1) = -2 < 0$

$(1 - \sqrt{2}, 0)$: $f(-0.1) = 7.9 > 0$

$(0, 1 + \sqrt{2})$: $f(1) = -2 < 0$

$(1 + \sqrt{2}, \infty)$: $f(3) = \dfrac{2}{3} > 0$

Function values are positive on $(1 - \sqrt{2}, 0)$ and $(1 + \sqrt{2}, \infty)$. The solution set is $(1 - \sqrt{2}, 0) \cup (1 + \sqrt{2}, \infty)$.

67.

$$\frac{2}{x^2 - 4x + 3} \leq \frac{5}{x^2 - 9}$$

$$\frac{2}{x^2 - 4x + 3} - \frac{5}{x^2 - 9} \leq 0$$

$$\frac{2}{(x-1)(x-3)} - \frac{5}{(x+3)(x-3)} \leq 0$$

The denominator of $f(x) = \dfrac{2}{(x-1)(x-3)} - \dfrac{5}{(x+3)(x-3)}$ is 0 when $x = 1$, 3, or -3, so the function is not defined for these values of x. We solve the related equation $f(x) = 0$.

$$\frac{2}{(x-1)(x-3)} - \frac{5}{(x+3)(x-3)} = 0$$

$$(x-1)(x-3)(x+3)\left(\frac{2}{(x-1)(x-3)} - \frac{5}{(x+3)(x-3)}\right)$$
$$= (x-1)(x-3)(x+3) \cdot 0$$

$$2(x+3) - 5(x-1) = 0$$

$$2x + 6 - 5x + 5 = 0$$

$$-3x + 11 = 0$$

$$-3x = -11$$

$$x = \frac{11}{3}$$

The critical values are -3, 1, 3, and $\dfrac{11}{3}$. They divide the x-axis into the intervals $(-\infty, -3)$, $(-3, 1)$, $(1, 3)$, $\left(3, \dfrac{11}{3}\right)$, and $\left(\dfrac{11}{3}, \infty\right)$. We test a value in each interval.

$(-\infty, -3)$: $f(-4) \approx -0.6571 < 0$

$(-3, 1)$: $f(0) \approx 1.2222 > 0$

$(1, 3)$: $f(2) = -1 < 0$

$\left(3, \dfrac{11}{3}\right)$: $f(3.5) \approx 0.6154 > 0$

$\left(\dfrac{11}{3}, \infty\right)$: $f(4) \approx -0.0476 < 0$

Function values are negative on $(-\infty, -3)$, $(1, 3)$, and $\left(\dfrac{11}{3}, \infty\right)$. Note that since the inequality symbol is $\leq$ and $f\left(\dfrac{11}{3}\right) = 0$, then $\dfrac{11}{3}$ must be included in the solution set. Note also that since -3, 1, and 3 are not in the domain of $f(x)$, they are not included in the solution set. It is $(-\infty, -3) \cup (1, 3) \cup \left[\dfrac{11}{3}, \infty\right)$.

69.

$$\frac{3}{x^2 + 1} \geq \frac{6}{5x^2 + 2}$$

$$\frac{3}{x^2 + 1} - \frac{6}{5x^2 + 2} \geq 0$$

The denominator of $f(x) = \dfrac{3}{x^2 + 1} - \dfrac{6}{5x^2 + 2}$ has no real-number zeros. We solve the related equation $f(x) = 0$.

$$\frac{3}{x^2 + 1} - \frac{6}{5x^2 + 2} = 0$$

$$(x^2 + 1)(5x^2 + 2)\left(\frac{3}{x^2 + 1} - \frac{6}{5x^2 + 2}\right) =$$
$$(x^2 + 1)(5x^2 + 2) \cdot 0$$

$$3(5x^2 + 2) - 6(x^2 + 1) = 0$$

$$15x^2 + 6 - 6x^2 - 6 = 0$$

$$9x^2 = 0$$

$$x^2 = 0$$

$$x = 0$$

The only critical value is 0. It divides the x-axis into the intervals $(-\infty, 0)$ and $(0, \infty)$. We test a value in each interval.

$(-\infty, 0)$: $f(-1) \approx 0.64286 > 0$

$(0, \infty)$: $f(1) \approx 0.64286 > 0$

Function values are positive on both intervals. Note that since the inequality symbol is $\geq$ and $f(0) = 0$, then 0 must be included in the solution set. It is $(-\infty, 0] \cup [0, \infty)$, or $(-\infty, \infty)$.

71.
$$\frac{5}{x^2 + 3x} < \frac{3}{2x + 1}$$

$$\frac{5}{x^2 + 3x} - \frac{3}{2x + 1} < 0$$

$$\frac{5}{x(x + 3)} - \frac{3}{2x + 1} < 0$$

The denominator of $f(x) = \dfrac{5}{x(x + 3)} - \dfrac{3}{2x + 1}$ is 0 when $x = 0$, -3, or $-\dfrac{1}{2}$, so the function is not defined for these values of x. We solve the related equation $f(x) = 0$.

$$\frac{5}{x(x + 3)} - \frac{3}{2x + 1} = 0$$

$$x(x+3)(2x+1)\left(\frac{5}{x(x+3)} - \frac{3}{2x+1}\right) =$$
$$x(x+3)(2x+1) \cdot 0$$

$$5(2x + 1) - 3x(x + 3) = 0$$

$$10x + 5 - 3x^2 - 9x = 0$$

$$-3x^2 + x + 5 = 0$$

Using the quadratic formula we find that $x = \dfrac{1 \pm \sqrt{61}}{6}$. The critical values are -3, $\dfrac{1 - \sqrt{61}}{6}$, $-\dfrac{1}{2}$, 0, and $\dfrac{1 + \sqrt{61}}{6}$. They divide the x-axis into the intervals $(-\infty, -3)$, $\left(-3, \dfrac{1 - \sqrt{61}}{6}\right)$, $\left(\dfrac{1 - \sqrt{61}}{6}, -\dfrac{1}{2}\right)$, $\left(-\dfrac{1}{2}, 0\right)$, $\left(0, \dfrac{1 + \sqrt{61}}{6}\right)$, and $\left(\dfrac{1 + \sqrt{61}}{6}, \infty\right)$.

We test a value in each interval.

$(-\infty, -3)$: $f(-4) \approx 1.6786 > 0$

$\left(-3, \dfrac{1 - \sqrt{61}}{6}\right)$: $f(-2) = -1.5 < 0$

$\left(\dfrac{1 - \sqrt{61}}{6}, -\dfrac{1}{2}\right)$: $f(-1) = 0.5 > 0$

$\left(-\dfrac{1}{2}, 0\right)$: $f(-0.1) \approx -20.99 < 0$

$\left(0, \dfrac{1 + \sqrt{61}}{6}\right)$: $f(1) = 0.25 > 0$

$\left(\dfrac{1 + \sqrt{61}}{6}, \infty\right)$: $f(2) = -0.1 < 0$

Function values are negative on $\left(-3, \dfrac{1 - \sqrt{61}}{6}\right)$, $\left(-\dfrac{1}{2}, 0\right)$ and $\left(\dfrac{1 + \sqrt{61}}{6}, \infty\right)$. The solution set is

$$\left(-3, \frac{1 - \sqrt{61}}{6}\right) \cup \left(-\frac{1}{2}, 0\right) \cup \left(\frac{1 + \sqrt{61}}{6}, \infty\right).$$

73.
$$\frac{5x}{7x - 2} > \frac{x}{x + 1}$$

$$\frac{5x}{7x - 2} - \frac{x}{x + 1} > 0$$

The denominator of $f(x) = \dfrac{5x}{7x - 2} - \dfrac{x}{x + 1}$ is 0 when $x = \dfrac{2}{7}$ or $x = -1$, so the function is not defined for these values of x. We solve the related equation $f(x) = 0$.

$$\frac{5x}{7x - 2} - \frac{x}{x + 1} = 0$$

$$(7x-2)(x+1)\left(\frac{5x}{7x-2} - \frac{x}{x+1}\right) = (7x-2)(x+1) \cdot 0$$

$$5x(x + 1) - x(7x - 2) = 0$$

$$5x^2 + 5x - 7x^2 + 2x = 0$$

$$-2x^2 + 7x = 0$$

$$-x(2x - 7) = 0$$

$$x = 0 \quad or \quad x = \frac{7}{2}$$

The critical values are -1, 0, $\dfrac{2}{7}$, and $\dfrac{7}{2}$. They divide the x-axis into the intervals $(-\infty, -1)$, $(-1, 0)$, $\left(0, \dfrac{2}{7}\right)$, $\left(\dfrac{2}{7}, \dfrac{7}{2}\right)$, and $\left(\dfrac{7}{2}, \infty\right)$. We test a value in each interval.

$(-\infty, -1)$: $f(-2) = -1.375 < 0$

$(-1, 0)$: $f(-0.5) \approx 1.4545 > 0$

$\left(0, \dfrac{2}{7}\right)$: $f(0.1) \approx -0.4755 < 0$

$\left(\dfrac{2}{7}, \dfrac{7}{2}\right)$: $f(1) = 0.5 > 0$

$\left(\dfrac{7}{2}, \infty\right)$: $f(4) \approx -0.0308 < 0$

Function values are positive on $(-1, 0)$ and $\left(\dfrac{2}{7}, \dfrac{7}{2}\right)$. The solution set is $(-1, 0) \cup \left(\dfrac{2}{7}, \dfrac{7}{2}\right)$.

75.
$$\frac{x}{x^2 + 4x - 5} + \frac{3}{x^2 - 25} \leq$$
$$\frac{2x}{x^2 - 6x + 5}$$

$$\frac{x}{x^2 + 4x - 5} + \frac{3}{x^2 - 25} - \frac{2x}{x^2 - 6x + 5} \leq 0$$

$$\frac{x}{(x+5)(x-1)} + \frac{3}{(x+5)(x-5)} - \frac{2x}{(x-5)(x-1)} \leq 0$$

The denominator of
$$f(x) = \frac{x}{(x+5)(x-1)} + \frac{3}{(x+5)(x-5)} - \frac{2x}{(x-5)(x-1)}$$
is 0 when $x = -5$, 1, or 5, so the function is not defined for these values of x. We solve the related equation $f(x) = 0$.

$$\frac{x}{(x+5)(x-1)} + \frac{3}{(x+5)(x-5)} - \frac{2x}{(x-5)(x-1)} = 0$$

$$x(x-5) + 3(x-1) - 2x(x+5) = 0$$

Multiplying by $(x+5)(x-1)(x-5)$

$$x^2 - 5x + 3x - 3 - 2x^2 - 10x = 0$$

$$-x^2 - 12x - 3 = 0$$

$$x^2 + 12x + 3 = 0$$

Using the quadratic formula, we find that $x = -6 \pm \sqrt{33}$. The critical values are $-6 - \sqrt{33}$, -5, $-6 + \sqrt{33}$, 1, and 5. They divide the x-axis into the intervals $(-\infty, -6 - \sqrt{33})$, $(-6 - \sqrt{33}, -5)$, $(-5, -6 + \sqrt{33})$, $(-6 + \sqrt{33}, 1)$, $(1, 5)$, and $(5, \infty)$. We test a value in each interval.

$(-\infty, -6 - \sqrt{33})$: $f(-12) \approx 0.00194 > 0$

$(-6 - \sqrt{33}, -5)$: $f(-6) \approx -0.4286 < 0$

$(-5, -6 + \sqrt{33})$: $f(-1) \approx 0.16667 > 0$

$(-6 + \sqrt{33}, 1)$: $f(0) = -0.12 < 0$

$(1, 5)$: $f(2) \approx 1.4762 > 0$

$(5, \infty)$: $f(6) \approx -2.018 < 0$

Function values are negative on $(-6 - \sqrt{33}, -5)$, $(-6 + \sqrt{33}, 1)$, and $(5, \infty)$. Note that since the inequality symbol is $\leq$ and $f(-6 \pm \sqrt{33}) = 0$, then $-6 - \sqrt{33}$ and $-6 + \sqrt{33}$ must be included in the solution set. Note also that since -5, 1, and 5 are not in the domain of $f(x)$, they are not included in the solution set. It is $[-6 - \sqrt{33}, -5) \cup [-6 + \sqrt{33}, 1) \cup (5, \infty)$.

77. We write and solve a rational inequality.

$$\frac{4t}{t^2 + 1} + 98.6 > 100$$

$$\frac{4t}{t^2 + 1} - 1.4 > 0$$

The denominator of $f(t) = \dfrac{4t}{t^2 + 1} - 1.4$ has no real-number zeros. We solve the related equation $f(t) = 0$.

$$\frac{4t}{t^2 + 1} - 1.4 = 0$$

$$4t - 1.4(t^2 + 1) = 0 \quad \text{Multiplying by } t^2 + 1$$

$$4t - 1.4t^2 - 1.4 = 0$$

Using the quadratic formula, we find that $t = \dfrac{4 \pm \sqrt{8.16}}{2.8}$; that is, $t \approx 0.408$ or $t \approx 2.449$. These numbers divide the t-axis into the intervals $(-\infty, 0.408)$, $(0.408, 2.449)$, and $(2.449, \infty)$. We test a value in each interval.

$(-\infty, 0.408)$: $f(0) = -1.4 < 0$

$(0.408, 2.449)$: $f(1) = 0.6 > 0$

$(2.449, \infty)$: $f(3) = -0.2 < 0$

Function values are positive on $(0.408, 2.449)$. The solution set is $(0.408, 2.449)$.

79. a) We write and solve a polynomial inequality.

$$-3x^2 + 630x - 6000 > 0 \quad (x \geq 0)$$

We first solve the related equation.

$$-3x^2 + 630x - 6000 = 0$$

$$x^2 - 210x + 2000 = 0 \quad \text{Dividing by } -3$$

$$(x - 10)(x - 200) = 0 \quad \text{Factoring}$$

Using the principle of zero products or by observing the graph of $y = -3x^2 + 630 - 6000$, we see that the solutions of the related equation are 10 and 200. These numbers divide the x-axis into the intervals $(-\infty, 10)$, $(10, 200)$, and $(200, \infty)$. Since we are restricting our discussion to nonnegative values of x, we consider the intervals $[0, 10)$, $(10, 200)$, and $(200, \infty)$.

We let $f(x) = -3x^2 + 630x - 6000$ and test a value in each interval.

$[0, 10)$: $f(0) = -6000 < 0$

$(10, 200)$: $f(11) = 567 > 0$

$(200, \infty)$: $f(201) = -573 < 0$

Function values are positive only on $(10, 200)$. The solution set is $\{x | 10 < x < 200\}$, or $(10, 200)$.

b) From part (a), we see that function values are negative on $[0, 10)$ and $(200, \infty)$. Thus, the solution set is $\{x | 0 < x < 10 \text{ or } x > 200\}$, or $(0, 10) \cup (200, \infty)$.

81. We write an inequality.

$$27 \leq \frac{n(n-3)}{2} \leq 230$$

$$54 \leq n(n-3) \leq 460 \quad \text{Multiplying by 2}$$

$$54 \leq n^2 - 3n \leq 460$$

We write this as two inequalities.

$$54 \leq n^2 - 3n \quad and \quad n^2 - 3n \leq 460$$

Solve each inequality.

$$n^2 - 3n \geq 54$$

$$n^2 - 3n - 54 \geq 0$$

$$n^2 - 3n - 54 = 0 \quad \text{Related equation}$$

$$(n + 6)(n - 9) = 0$$

$$n = -6 \quad or \quad n = 9$$

Since only positive values of n have meaning in this application, we consider the intervals $(0, 9)$ and $(9, \infty)$. Let $f(n) = n^2 - 3n - 54$ and test a value in each interval.

$(0, 9)$: $f(1) = -56 < 0$

$(9, \infty)$: $f(10) = 16 > 0$

Function values are positive on $(9, \infty)$. Since the inequality symbol is $\geq$, 9 must also be included in the solution set for this portion of the inequality. It is $\{n | n \geq 9\}$.

Now solve the second inequality.

$$n^2 - 3n \leq 460$$

$$n^2 - 3n - 460 \leq 0$$

$$n^2 - 3n - 460 = 0 \quad \text{Related equation}$$

$$(n + 20)(n - 23) = 0$$

$n = -20$ *or* $n = 23$

We consider only positive values of n as above. Thus, we consider the intervals $(0, 23)$ and $(23, \infty)$. Let $f(n) = n^2 - 3n - 460$ and test a value in each interval.

$(0, 23)$: $f(1) = -462 < 0$

$(23, \infty)$: $f(24) = 44 > 0$

Function values are negative on $(0, 23)$. Since the inequality symbol is $\leq$, 23 must also be included in the solution set for this portion of the inequality. It is $\{n | 0 < n \leq 23\}$.

The solution set of the original inequality is $\{n | n \geq 9 \ and \ 0 < n \leq 23\}$, or $\{n | 9 \leq n \leq 23\}$.

83.
$$(x - h)^2 + (y - k)^2 = r^2$$
$$[x - (-2)]^2 + (y - 4)^2 = 3^2$$
$$(x + 2)^2 + (y - 4)^2 = 9$$

85. $h(x) = -2x^2 + 3x - 8$

a) $-\dfrac{b}{2a} = -\dfrac{3}{2(-2)} = \dfrac{3}{4}$

$h\left(\dfrac{3}{4}\right) = -2\left(\dfrac{3}{4}\right)^2 + 3 \cdot \dfrac{3}{4} - 8 = -\dfrac{55}{8}$

The vertex is $\left(\dfrac{3}{4}, -\dfrac{55}{8}\right)$.

b) The coefficient of x^2 is negative, so there is a maximum value. It is the second coordinate of the vertex, $-\dfrac{55}{8}$. It occurs at $x = \dfrac{3}{4}$.

c) The range is $\left(-\infty, -\dfrac{55}{8}\right]$.

87. $|x^2 - 5| = |5 - x^2| = 5 - x^2$ when $5 - x^2 \geq 0$. Thus we solve $5 - x^2 \geq 0$.

$$5 - x^2 \geq 0$$
$$5 - x^2 = 0 \quad \text{Related equation}$$
$$5 = x^2$$
$$\pm\sqrt{5} = x$$

Let $f(x) = 5 - x^2$ and test a value in each of the intervals determined by the solutions of the related equation.

$(-\infty, -\sqrt{5})$: $f(-3) = -4 < 0$
$(-\sqrt{5}, \sqrt{5})$: $f(0) = 5 > 0$
$(\sqrt{5}, \infty)$: $f(3) = -4 < 0$

Function values are positive on $(-\sqrt{5}, \sqrt{5})$. Since the inequality symbol is $\geq$, the endpoints of the interval must be included in the solution set. It is $\left[-\sqrt{5}, \sqrt{5}\right]$.

89.
$$2|x|^2 - |x| + 2 \leq 5$$
$$2|x|^2 - |x| - 3 \leq 0$$
$$2|x|^2 - |x| - 3 = 0 \quad \text{Related equation}$$
$$(2|x| - 3)(|x| + 1) = 0 \quad \text{Factoring}$$
$$2|x| - 3 = 0 \quad or \quad |x| + 1 = 0$$
$$|x| = \dfrac{3}{2} \quad or \quad |x| = -1$$

The solution of the first equation is $x = -\dfrac{3}{2}$ or $x = \dfrac{3}{2}$. The second equation has no solution. Let $f(x) = 2|x|^2 - |x| - 3$ and test a value in each interval determined by the solutions of the related equation.

$\left(-\infty, -\dfrac{3}{2}\right)$: $f(-2) = 3 > 0$

$\left(-\dfrac{3}{2}, \dfrac{3}{2}\right)$: $f(0) = -3 < 0$

$\left(\dfrac{3}{2}, \infty\right)$: $f(2) = 3 > 0$

Function values are negative on $\left(-\dfrac{3}{2}, \dfrac{3}{2}\right)$. Since the inequality symbol is $\leq$, the endpoints of the interval must also be included in the solution set. It is $\left[-\dfrac{3}{2}, \dfrac{3}{2}\right]$.

91.
$$\left|1 + \dfrac{1}{x}\right| < 3$$
$$-3 < 1 + \dfrac{1}{x} < 3$$
$$-3 < 1 + \dfrac{1}{x} \quad and \quad 1 + \dfrac{1}{x} < 3$$

First solve $-3 < 1 + \dfrac{1}{x}$.

$$0 < 4 + \dfrac{1}{x}, \quad or \quad \dfrac{1}{x} + 4 > 0$$

The denominator of $f(x) = \dfrac{1}{x} + 4$ is 0 when $x = 0$, so the function is not defined for this value of x. Now solve the related equation.

$$\dfrac{1}{x} + 4 = 0$$
$$1 + 4x = 0 \quad \text{Multiplying by } x$$
$$x = -\dfrac{1}{4}$$

The critical values are $-\dfrac{1}{4}$ and 0. Test a value in each of the intervals determined by them.

$\left(-\infty, -\dfrac{1}{4}\right)$: $f(-1) = 3 > 0$

$\left(-\dfrac{1}{4}, 0\right)$: $f(-0.1) = -6 < 0$

$(0, \infty)$: $f(1) = 5 > 0$

The solution set for this portion of the inequality is $\left(-\infty, -\dfrac{1}{4}\right) \cup (0, \infty)$.

Next solve $1 + \dfrac{1}{x} < 3$, or $\dfrac{1}{x} - 2 < 0$. The denominator of $f(x) = \dfrac{1}{x} - 2$ is 0 when $x = 0$, so the function is not defined for this value of x. Now solve the related equation.

$$\dfrac{1}{x} - 2 = 0$$
$$1 - 2x = 0 \quad \text{Multiplying by } x$$
$$x = \dfrac{1}{2}$$

The critical values are 0 and $\frac{1}{2}$. Test a value in each of the intervals determined by them.

$(-\infty, 0)$: $f(-1) = -3 < 0$

$\left(0, \frac{1}{2}\right)$: $f(0.1) = 8 > 0$

$\left(\frac{1}{2}, \infty\right)$: $f(1) = -1 < 0$

The solution set for this portion of the inequality is $(-\infty, 0) \cup \left(\frac{1}{2}, \infty\right)$.

The solution set of the original inequality is

$$\left(\left(-\infty, -\frac{1}{4}\right) \cup (0, \infty)\right) \ and \ \left((-\infty, 0) \cup \left(\frac{1}{2}, \infty\right)\right),$$

or $\left(-\infty, -\frac{1}{4}\right) \cup \left(\frac{1}{2}, \infty\right)$.

93. First find a quadratic equation with solutions -4 and 3.

$$(x + 4)(x - 3) = 0$$
$$x^2 + x - 12 = 0$$

Test a point in each of the three intervals determined by -4 and 3.

$(-\infty, -4)$: $(-5 + 4)(-5 - 3) = 8 > 0$

$(-4, 3)$: $(0 + 4)(0 - 3) = -12 < 0$

$(3, \infty)$: $(4 + 4)(4 - 3) = 8 > 0$

Then a quadratic inequality for which the solution set is $(-4, 3)$ is $x^2 + x - 12 < 0$. Answers may vary.

95. $f(x) = \sqrt{\dfrac{72}{x^2 - 4x - 21}}$

The radicand must be nonnegative and the denominator must be nonzero. Thus, the values of x for which $x^2 - 4x - 21 > 0$ comprise the domain. By inspecting the graph of $y = x^2 - 4x - 21$ we see that the domain is $\{x | x < -3 \ or \ x > 7\}$, or $(-\infty, -3) \cup (7, \infty)$.

Chapter 4 Review Exercises

1. $f(-b) = (-b + a)(-b + b)(-b - c) = (-b + a) \cdot 0 \cdot (-b - c) = 0$, so the statement is true.

3. In addition to the given possibilities, 9 and -9 are also possible rational zeros. The statement is false.

5. The domain of the function is the set of all real numbers except -2 and 3, or $\{x | x \neq -2 \ and \ x \neq 3\}$. The statement is false.

7. $h(x) = -25$

The leading term is -25 and the leading coefficient is -25. The degree of the polynomial is 0, so the polynomial is constant.

9. $f(x) = \dfrac{1}{3}x^3 - 2x + 3$

The leading term is $\frac{1}{3}x^3$ and the leading coefficient is $\frac{1}{3}$. The degree of the polynomial is 3, so the polynomial is cubic.

11. $f(x) = x^5 + 2x^3 - x^2 + 5x + 4$

The leading term is x^5. The degree, 5, is odd and the leading coefficient, 1, is positive. As $x \to \infty$, $f(x) \to \infty$, and as $x \to -\infty$, $f(x) \to -\infty$.

13. $f(x) = x^4 - 26x^2 + 25$
$$= (x^2 - 1)(x^2 - 25)$$
$$= (x + 1)(x - 1)(x + 5)(x - 5)$$

± 1, ± 5; each has multiplicity 1

15. $A = P(1 + r)^t$

a) $6760 = 6250(1 + r)^2$

$$\frac{6760}{6250} = (1 + r)^2$$

$$\pm 1.04 = 1 + r$$

$$-1 \pm 1.04 = r$$

$$-2.04 = r \ \ or \ \ 0.04 = r$$

Only 0.04 has meaning in this application. The interest rate is 0.04 or 4%.

b) $1,215,506.25 = 1,000,000(1 + r)^4$

$$\frac{1,215,506.25}{1,000,000} = (1 + r)^4$$

$$\pm 1.05 = 1 + r$$

$$-1 \pm 1.05 = r$$

$$-2.05 = r \ \ or \ \ 0.05 = r$$

Only 0.05 has meaning in this application. The interest rate is 0.05 or 5%.

17. $g(x) = (x - 1)^3(x + 2)^2$

1. The leading term is $x \cdot x \cdot x \cdot x \cdot x$, or x^5. The degree, 5, is odd and the leading coefficient, 1, is positive so as $x \to \infty$, $g(x) \to \infty$ and as $x \to -\infty$, $g(x) \to -\infty$.

2. We see that the zeros of the function are 1 and -2, so the x-intercepts of the graph are $(1, 0)$ and $(-2, 0)$.

3. The zeros divide the x-axis into 3 intervals, $(-\infty, -2)$, $(-2, 1)$, and $(1, \infty)$. We choose a value for x from each interval and find $g(x)$. This tells us the sign of $g(x)$ for all values of x in that interval.

   In $(-\infty, -2)$, test -3:

   $g(-3) = (-3 - 1)^3(-3 + 2)^2 = -64 < 0$

   In $(-2, 1)$, test 0:

   $g(0) = (0 - 1)^3(0 + 2)^2 = -4 < 0$

   In $(1, \infty)$, test 2:

   $g(2) = (2 - 1)^3(2 + 2)^2 = 16 > 0$

   Thus the graph lies below the x-axis on $(-\infty, -2)$ and on $(-2, 1)$ and above the x-axis on $(1, \infty)$. We also know that the points $(-3, -64)$, $(0, -4)$, and $(2, 16)$ are on the graph.

4. From Step 3 we know that $g(0) = -4$, so the y-intercept is $(0, -4)$.

5. We find additional points on the graph and then draw the graph.

x	$g(x)$
-2.5	-10.7
-1	-8
-0.5	-7.6
0.5	-0.8

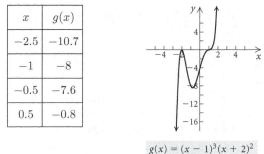

$$g(x) = (x - 1)^3(x + 2)^2$$

6. Checking the graph as described on page 240 in the text, we see that it appears to be correct.

19. $f(x) = x^4 - 5x^3 + 6x^2 + 4x - 8$

1. The leading term is x^4. The degree, 4, is even and the leading coefficient, 1, is positive so as $x \to \infty$, $f(x) \to \infty$ and as $x \to -\infty$, $f(x) \to \infty$.

2. We solve $f(x) = 0$, or $x^4 - 5x^3 + 6x^2 + 4x - 8 = 0$. The possible rational zeros are ± 1, ± 2, ± 4, and ± 8. We try -1.

$$\begin{array}{r|rrrrr}
-1 & 1 & -5 & 6 & 4 & -8 \\
   &   & -1 & 6 & -12 & 8 \\
\hline
   & 1 & -6 & 12 & -8 & 0
\end{array}$$

Now we have $(x + 1)(x^3 - 6x^2 + 12x - 8) = 0$. We use synthetic division to determine if 2 is a zero of $x^3 - 6x^2 + 12x - 8 = 0$.

$$\begin{array}{r|rrrr}
2 & 1 & -6 & 12 & -8 \\
  &   & 2 & -8 & 8 \\
\hline
  & 1 & -4 & 4 & 0
\end{array}$$

We have $(x + 1)(x - 2)(x^2 - 4x + 4) = 0$, or $(x + 1)(x - 2)(x - 2)^2 = 0$. Thus the zeros of $f(x)$ are -1 and 2 and the x-intercepts of the graph are $(-1, 0)$ and $(2, 0)$.

3. The zeros divide the x-axis into 3 intervals, $(-\infty, -1)$, $(-1, 2)$, and $(2, \infty)$. We choose a value for x from each interval and find $f(x)$. This tells us the sign of $f(x)$ for all values of x in that interval.

In $(-\infty, -1)$, test -2:
$f(-2) = (-2)^4 - 5(-2)^3 + 6(-2)^2 + 4(-2) - 8 = 64 > 0$

In $(-1, 2)$, test 0:
$f(0) = 0^4 - 5 \cdot 0^3 + 6 \cdot 0^2 + 4 \cdot 0 - 8 = -8 < 0$

In $(2, \infty)$, test 3:
$f(3) = 3^4 - 5 \cdot 3^3 + 6 \cdot 3^2 + 4 \cdot 3 - 8 = 4 > 0$

Thus the graph lies above the x-axis on $(-\infty, -1)$ and on $(2, \infty)$ and below the x-axis on $(-1, 2)$. We also know that the points $(-2, 64)$, $(0, -8)$, and $(3, 4)$ are on the graph.

4. From Step 3 we know that $f(0) = -8$, so the y-intercept is $(0, -8)$.

5. We find additional points on the graph and then draw the graph.

x	$f(x)$
-1.5	21.4
-0.5	-7.8
1	-2
4	40

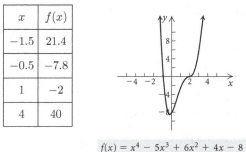

$$f(x) = x^4 - 5x^3 + 6x^2 + 4x - 8$$

6. Checking the graph as described on page 240 in the text, we see that it appears to be correct.

21. $f(1) = 4 \cdot 1^2 - 5 \cdot 1 - 3 = -4$

$f(2) = 4 \cdot 2^2 - 5 \cdot 2 - 3 = 3$

By the intermediate value theorem, since $f(1)$ and $f(2)$ have opposite signs, $f(x)$ has a zero between 1 and 2.

23.

$$\begin{array}{r}
6x^2 + 16x + 52 \\
x - 3 \overline{\smash{\big)}\ 6x^3 - 2x^2 + 4x - 1} \\
\underline{6x^3 - 18x^2} \\
16x^2 + 4x \\
\underline{16x^2 - 48x} \\
52x - 1 \\
\underline{52x - 156} \\
155
\end{array}$$

$Q(x) = 6x^2 + 16x + 52$; $R(x) = 155$;

$P(x) = (x - 3)(6x^2 + 16x + 52) + 155$

25.
$$\begin{array}{r|rrrr}
5 & 1 & 2 & -13 & 10 \\
  &   & 5 & 35 & 110 \\
\hline
  & 1 & 7 & 22 & 120
\end{array}$$

The quotient is $x^2 + 7x + 22$; the remainder is 120.

27.
$$\begin{array}{r|rrrrrr}
-1 & 1 & 0 & 0 & 0 & -2 & 0 \\
   &   & -1 & 1 & -1 & 1 & 1 \\
\hline
   & 1 & -1 & 1 & -1 & -1 & 1
\end{array}$$

The quotient is $x^4 - x^3 + x^2 - x - 1$; the remainder is 1.

29.
$$\begin{array}{r|rrrrr}
-2 & 1 & 0 & 0 & 0 & -16 \\
   &   & -2 & 4 & -8 & 16 \\
\hline
   & 1 & -2 & 4 & -8 & 0
\end{array}$$

$f(-2) = 0$

31.
$$\begin{array}{r|rrrr}
-i & 1 & -5 & 1 & -5 \\
   &   & -i & -1 + 5i & 5 \\
\hline
   & 1 & -5 - i & 5i & 0
\end{array}$$

$f(-i) = 0$, so $-i$ is a zero of $f(x)$.

$$\begin{array}{r|rrrr}
-5 & 1 & -5 & 1 & -5 \\
   &   & -5 & 50 & -255 \\
\hline
   & 1 & -10 & 51 & -260
\end{array}$$

$f(-5) \neq 0$, so -5 is not a zero of $f(x)$.

33.

$$\frac{1}{3} \;\big|\; \begin{array}{cccc} 1 & -\frac{4}{3} & -\frac{5}{3} & \frac{2}{3} \\ & \frac{1}{3} & -\frac{1}{3} & -\frac{2}{3} \\ \hline 1 & -1 & -2 & 0 \end{array}$$

$f\left(\dfrac{1}{3}\right) = 0$, so $\dfrac{1}{3}$ is a zero of $f(x)$.

$$1 \;\big|\; \begin{array}{cccc} 1 & -\frac{4}{3} & -\frac{5}{3} & \frac{2}{3} \\ & 1 & -\frac{1}{3} & -2 \\ \hline 1 & -\frac{1}{3} & -2 & -\frac{4}{3} \end{array}$$

$f(1) \neq 0$, so 1 is not a zero of $f(x)$.

35. $f(x) = x^3 + 2x^2 - 7x + 4$

Try $x + 1$ and $x + 2$. Using synthetic division we find that $f(-1) \neq 0$ and $f(-2) \neq 0$. Thus $x + 1$ and $x + 2$ are not factors of $f(x)$. Try $x + 4$.

$$-4 \;\big|\; \begin{array}{cccc} 1 & 2 & -7 & 4 \\ & -4 & 8 & -4 \\ \hline 1 & -2 & 1 & 0 \end{array}$$

Since $f(-4) = 0$, $x + 4$ is a factor of $f(x)$. Thus $f(x) = (x + 4)(x^2 - 2x + 1) = (x + 4)(x - 1)^2$.

Now we solve $f(x) = 0$.

$$x + 4 = 0 \quad or \quad (x - 1)^2 = 0$$
$$x = -4 \quad or \quad x - 1 = 0$$
$$x = -4 \quad or \quad x = 1$$

The solutions of $f(x) = 0$ are -4 and 1.

37. $f(x) = x^4 - 4x^3 - 21x^2 + 100x - 100$

Using synthetic division we find that $f(2) = 0$:

$$2 \;\big|\; \begin{array}{ccccc} 1 & -4 & -21 & 100 & -100 \\ & 2 & -4 & -50 & 100 \\ \hline 1 & -2 & -25 & 50 & 0 \end{array}$$

Then we have:

$$f(x) = (x - 2)(x^3 - 2x^2 - 25x + 50)$$
$$= (x - 2)[x^2(x - 2) - 25(x - 2)]$$
$$= (x - 2)(x - 2)(x^2 - 25)$$
$$= (x - 2)^2(x + 5)(x - 5)$$

Now solve $f(x) = 0$.

$$(x - 2)^2 = 0 \quad or \quad x + 5 = 0 \quad or \quad x - 5 = 0$$
$$x - 2 = 0 \quad or \quad x = -5 \quad or \quad x = 5$$
$$x = 2 \quad or \quad x = -5 \quad or \quad x = 5$$

The solutions of $f(x) = 0$ are 2, -5, and 5.

39. A polynomial function of degree 3 with -4, -1, and 2 as zeros has factors $x + 4$, $x + 1$, and $x - 2$ so we have $f(x) = a_n(x + 4)(x + 1)(x - 2)$.

The simplest polynomial is obtained if we let $a_n = 1$.

$$f(x) = (x + 4)(x + 1)(x - 2)$$
$$= (x^2 + 5x + 4)(x - 2)$$
$$= x^3 - 2x^2 + 5x^2 - 10x + 4x - 8$$
$$= x^3 + 3x^2 - 6x - 8$$

41. A polynomial function of degree 3 with $\dfrac{1}{2}$, $1 - \sqrt{2}$, and $1 + \sqrt{2}$ as zeros has factors $x - \dfrac{1}{2}$, $x - (1 - \sqrt{2})$, and $x - (1 + \sqrt{2})$ so we have

$$f(x) = a_n\left(x - \frac{1}{2}\right)[x - (1 - \sqrt{2})][x - (1 + \sqrt{2})].$$

Let $a_n = 1$.

$$f(x) = \left(x - \frac{1}{2}\right)[x - (1 - \sqrt{2})][x - (1 + \sqrt{2})]$$
$$= \left(x - \frac{1}{2}\right)[(x - 1) + \sqrt{2}][(x - 1) - \sqrt{2}]$$
$$= \left(x - \frac{1}{2}\right)(x^2 - 2x + 1 - 2)$$
$$= \left(x - \frac{1}{2}\right)(x^2 - 2x - 1)$$
$$= x^3 - 2x^2 - x - \frac{1}{2}x^2 + x + \frac{1}{2}$$
$$= x^3 - \frac{5}{2}x^2 + \frac{1}{2}$$

If we let $a_n = 2$, we obtain $f(x) = 2x^3 - 5x^2 + 1$.

43. A polynomial function of degree 5 has at most 5 real zeros. Since 5 zeros are given, these are all of the zeros of the desired function. We proceed as in Exercise 39 above, letting $a_n = 1$.

$$f(x) = (x + 3)^2(x - 2)(x - 0)^2$$
$$= (x^2 + 6x + 9)(x^3 - 2x^2)$$
$$= x^5 + 6x^4 + 9x^3 - 2x^4 - 12x^3 - 18x^2$$
$$= x^5 + 4x^4 - 3x^3 - 18x^2$$

45. A polynomial function of degree 5 can have at most 5 zeros. Since $f(x)$ has rational coefficients, in addition to the 3 given zeros, the other zeros are the conjugates of $1 + \sqrt{3}$ and $-\sqrt{3}$, or $1 - \sqrt{3}$ and $\sqrt{3}$.

47. $-\sqrt{11}$ is also a zero.

$$f(x) = (x - \sqrt{11})(x + \sqrt{11})$$
$$= x^2 - 11$$

49. $1 - i$ is also a zero.

$$f(x) = (x + 1)(x - 4)[x - (1 + i)][x - (1 - i)]$$
$$= (x^2 - 3x - 4)(x^2 - 2x + 2)$$
$$= x^4 - 2x^3 + 2x^2 - 3x^3 + 6x^2 - 6x - 4x^2 + 8x - 8$$
$$= x^4 - 5x^3 + 4x^2 + 2x - 8$$

51. $f(x) = \left(x - \dfrac{1}{3}\right)(x - 0)(x + 3)$

$$= \left(x^2 - \frac{1}{3}x\right)(x + 3)$$
$$= x^3 + \frac{8}{3}x^2 - x$$

53. $g(x) = 3x^4 - x^3 + 5x^2 - x + 1$

$$\frac{\text{Possibilities for } p}{\text{Possibilities for } q} : \frac{\pm 1}{\pm 1, \pm 3}$$

Possibilities for p/q: $\pm 1, \pm \dfrac{1}{3}$

55. $f(x) = 3x^5 + 2x^4 - 25x^3 - 28x^2 + 12x$

   a) We know that 0 is a zero since

$$f(x) = x(3x^4 + 2x^3 - 25x^2 - 28x + 12).$$

   Now consider $g(x) = 3x^4 + 2x^3 - 25x^2 - 28x + 12$.

   Possibilities for p/q: $\pm 1, \pm 2, \pm 3, \pm 4, \pm 6, \pm 12$,

$$\pm\frac{1}{3}, \pm\frac{2}{3}, \pm\frac{4}{3}$$

   From the graph of $y = 3x^4 + 2x^3 - 25x^2 - 28x + 12$, we see that, of all the possibilities above, only -2, $\frac{1}{3}$, $\frac{2}{3}$, and 3 might be zeros. We use synthetic division to determine if -2 is a zero.

```
-2 |  3    2   -25   -28    12
   |      -6     8    34   -12
   ------------------------------
      3   -4   -17     6  |  0
```

   Now try 3 in the quotient above.

```
 3 |  3   -4   -17    6
   |       9    15   -6
   --------------------------
      3    5    -2 |  0
```

   We have $f(x) = (x+2)(x-3)(3x^2 + 5x - 2)$.

   We find the other zeros.

$$3x^3 + 5x - 2 = 0$$
$$(3x - 1)(x + 2) = 0$$
$$3x - 1 = 0 \quad or \quad x + 2 = 0$$
$$3x = 1 \quad or \quad \quad x = -2$$
$$x = \frac{1}{3} \quad or \quad \quad x = -2$$

   The rational zeros of $g(x) = 3x^4 + 2x^3 - 25x^2 - 28x + 12$ are -2, 3, and $\frac{1}{3}$. Since 0 is also a zero of $f(x)$, the zeros of $f(x)$ are -2, 3, $\frac{1}{3}$, and 0. (The zero -2 has multiplicity 2.) These are the only zeros.

   b) From our work above we see

$$f(x) = x(x+2)(x-3)(3x-1)(x+2), \text{ or}$$
$$x(x+2)^2(x-3)(3x-1).$$

57. $f(x) = x^4 - 6x^3 + 9x^2 + 6x - 10$

   a) Possibilities for p/q: $\pm 1, \pm 2, \pm 5, \pm 10$

   From the graph of $f(x)$, we see that -1 and 1 might be zeros.

```
-1 |  1   -6    9     6   -10
   |      -1    7   -16    10
   --------------------------------
      1   -7   16   -10  |  0
```

```
 1 |  1   -7   16   -10
   |       1   -6    10
   ----------------------------
      1   -6   10  |  0
```

$$f(x) = (x+1)(x-1)(x^2 - 6x + 10)$$

   Using the quadratic formula, we find that the other zeros are $3 \pm i$.

   The rational zeros are -1 and 1. The other zeros are $3 \pm i$.

   b) $\quad f(x) = (x+1)(x-1)[x-(3+i)][x-(3-i)]$

$$= (x+1)(x-1)(x-3-i)(x-3+i)$$

59. $f(x) = 3x^3 - 8x^2 + 7x - 2$

   a) Possibilities for p/q: $\pm 1, \pm 2, \pm\frac{1}{3}, \pm\frac{2}{3}$

   From the graph of $f(x)$, we see that $\frac{2}{3}$ and 1 might be zeros.

```
 1 |  3   -8    7   -2
   |       3   -5    2
   ----------------------
      3   -5    2 |  0
```

   We have $f(x) = (x-1)(3x^2 - 5x + 2)$.

   We find the other zeros.

$$3x^2 - 5x + 2 = 0$$
$$(3x - 2)(x - 1) = 0$$
$$3x - 2 = 0 \quad or \quad x - 1 = 0$$
$$x = \frac{2}{3} \quad or \quad \quad x = 1$$

   The rational zeros are 1 and $\frac{2}{3}$. (The zero 1 has multiplicity 2.) These are the only zeros.

   b) $f(x) = (x-1)^2(3x-2)$

61. $f(x) = x^6 + x^5 - 28x^4 - 16x^3 + 192x^2$

   a) We know that 0 is a zero since

$$f(x) = x^2(x^4 + x^3 - 28x^2 - 16x + 192).$$

   Consider $g(x) = x^4 + x^3 - 28x^2 - 16x + 192$.

   Possibilities for p/q: $\pm 1, \pm 2, \pm 3, \pm 4, \pm 6, \pm 8, \pm 12,$

$$\pm 16, \pm 24, \pm 32, \pm 48, \pm 64, \pm 96,$$
$$\pm 192$$

   From the graph of $y = g(x)$, we see that -4, 3 and 4 might be zeros.

```
-4 |  1    1   -28   -16   192
   |      -4    12    64  -192
   ------------------------------
      1   -3   -16    48  |  0
```

   We have $f(x) = x^2 \cdot g(x) =$
$x^2(x+4)(x^3 - 3x^2 - 16x + 48)$.

   We find the other zeros.

$$x^3 - 3x^2 - 16x + 48 = 0$$
$$x^2(x - 3) - 16(x - 3) = 0$$
$$(x - 3)(x^2 - 16) = 0$$
$$(x - 3)(x + 4)(x - 4) = 0$$
$$x - 3 = 0 \quad or \quad x + 4 = 0 \quad or \quad x - 4 = 0$$
$$x = 3 \quad or \quad \quad x = -4 \quad or \quad \quad x = 4$$

   The rational zeros are 0, -4, 3, and 4. (The zeros 0 and -4 each have multiplicity 2.) These are the only zeros.

   b) $f(x) = x^2(x+4)^2(x-3)(x-4)$

63. $f(x) = 2x^6 - 7x^3 + x^2 - x$

   There are 3 variations in sign in $f(x)$, so there are 3 or 1 positive real zeros.

$$f(-x) = 2(-x)^6 - 7(-x)^3 + (-x)^2 - (-x)$$
$$= 2x^6 + 7x^3 + x^2 + x$$

   There are no variations in sign in $f(-x)$, so there are no negative real zeros.

65. $g(x) = 5x^5 - 4x^2 + x - 1$

There are 3 variations in sign in $g(x)$, so there are 3 or 1 positive real zeros.

$$g(-x) = 5(-x)^5 - 4(-x)^2 + (-x) - 1$$
$$= -5x^5 - 4x^2 - x - 1$$

There is no variation in sign in $g(-x)$, so there are 0 negative real zeros.

67. $f(x) = \dfrac{5}{(x-2)^2}$

1. The numerator and the denominator have no common factors. The denominator is zero when $x = 2$, so the domain excludes 2. It is $(-\infty, 2) \cup (2, \infty)$. The line $x = 2$ is the vertical asymptote.

2. Because the degree of the numerator is less than the degree of the denominator, the x-axis, or $y = 0$, is the horizontal asymptote. There is no oblique asymptote.

3. The numerator has no zeros, so there is no x-intercept.

4. $f(0) = \dfrac{5}{(0-2)^2} = \dfrac{5}{4}$, so the y-intercept is $\left(0, \dfrac{5}{4}\right)$.

5. Find other function values to determine the shape of the graph and then draw it.

$$f(x) = \frac{5}{(x-2)^2}$$

69. $f(x) = \dfrac{x-2}{x^2 - 2x - 15} = \dfrac{x-2}{(x+3)(x-5)}$

1. The numerator and the denominator have no common factors. The denominator is zero when $x = -3$, or $x = 5$, so the domain excludes -3 and 5. It is $(-\infty, -3) \cup (-3, 5) \cup (5, \infty)$. The lines $x = -3$ and $x = 5$ are vertical asymptotes.

2. Because the degree of the numerator is less than the degree of the denominator, the x-axis, or $y = 0$, is the horizontal asymptote. There is no oblique asymptote.

3. The numerator is zero when $x = 2$, so the x-intercept is $(2, 0)$.

4. $f(0) = \dfrac{0-2}{0^2 - 2 \cdot 0 - 15} = \dfrac{2}{15}$, so the y-intercept is $\left(0, \dfrac{2}{15}\right)$.

5. Find other function values to determine the shape of the graph and then draw it.

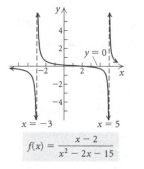

$$f(x) = \frac{x-2}{x^2 - 2x - 15}$$

71. Answers may vary. The numbers -2 and 3 must be zeros of the denominator but not of the numerator, and -3 must be zero of the numerator but not of the denominator. In addition, the numerator and denominator must have the same degree and the ratio of the leading coefficients must be 4.

$$f(x) = \frac{4x(x+3)}{(x+2)(x-3)}, \text{ or } f(x) = \frac{4x^2 + 12x}{x^2 - x - 6}$$

73.
$$x^2 - 9 < 0 \quad \text{Polynomial inequality}$$
$$x^2 - 9 = 0 \quad \text{Related equation}$$
$$(x+3)(x-3) = 0 \quad \text{Factoring}$$

The solutions of the related equation are -3 and 3. These numbers divide the x-axis into the intervals $(-\infty, -3)$, $(-3, 3)$, and $(3, \infty)$.

We let $f(x) = (x+3)(x-3)$ and test a value in each interval.

$$(-\infty, -3): \; f(-4) = 7 > 0$$
$$(-3, 3): \; f(0) = -9 < 0$$
$$(3, \infty): \; f(4) = 7 > 0$$

Function values are negative only on $(-3, 3)$. The solution set is $(-3, 3)$.

75.
$$(1-x)(x+4)(x-2) \leq 0 \quad \text{Polynomial inequality}$$
$$(1-x)(x+4)(x-2) = 0 \quad \text{Related equation}$$

The solutions of the related equation are 1, -4 and 2. These numbers divide the x-axis into the intervals $(-\infty, -4)$, $(-4, 1)$, $(1, 2)$ and $(2, \infty)$.

We let $f(x) = (1-x)(x+4)(x-2)$ and test a value in each interval.

$$(-\infty, -4): \; f(-5) = 42 > 0$$
$$(-4, 1): \; f(0) = -8 < 0$$
$$(1, 2): \; f\left(\frac{3}{2}\right) = \frac{11}{8} > 0$$
$$(2, \infty): \; f(3) = -14 < 0$$

Function values are negative on $(-4, 1)$ and $(2, \infty)$. Since the inequality symbol is $\leq$, the endpoints of the intervals must be included in the solution set. It is $[-4, 1] \cup [2, \infty)$.

77. a) We write and solve a polynomial equation.

$$-16t^2 + 80t + 224 = 0$$
$$-16(t^2 - 5t - 14) = 0$$
$$-16(t + 2)(t - 7) = 0$$

The solutions are $t = -2$ and $t = 7$. Only $t = 7$ has meaning in this application. The rocket reaches the ground at $t = 7$ seconds.

b) We write and solve a polynomial inequality.

$$-16t^2 + 80t + 224 > 320 \quad \text{Polynomial inequality}$$
$$-16t^2 + 80t - 96 > 0 \quad \text{Equivalent inequality}$$
$$-16t^2 + 80t - 96 = 0 \quad \text{Related equation}$$
$$-16(t^2 - 5t + 6) = 0$$
$$-16(t - 2)(t - 3) = 0$$

The solutions of the related equation are 2 and 3. These numbers divide the t-axis into the intervals $(-\infty, 2)$, $(2, 3)$ and $(3, \infty)$. We restrict our discussion to values of t such that $0 \le t \le 7$ since we know from part (a) the rocket is in the air for 7 sec. We consider the intervals $[0, 2)$, $(2, 3)$ and $(3, 7]$. We let $f(t) = -16t^2 + 80t - 96$ and test a value in each interval.

$$[0, 2): \ f(1) = -32 < 0$$
$$(2, 3): \ f\left(\frac{5}{2}\right) = 4 > 0$$
$$(3, 7]: \ f(4) = -32 < 0$$

Function values are positive on $(2, 3)$. The solution set is $(2, 3)$

79. $g(x) = \dfrac{x^2 + 2x - 3}{x^2 - 5x + 6} = \dfrac{x^2 + 2x - 3}{(x - 2)(x - 3)}$

The values of x that make the denominator 0 are 2 and 3, so the domain is $(-\infty, 2) \cup (2, 3) \cup (3, \infty)$. Answer A is correct.

81. $f(x) = -\dfrac{1}{2}x^4 + x^3 + 1$

The degree of the function is even and the leading coefficient is negative, so as $x \to \infty$, $f(x) \to -\infty$ and as $x \to -\infty$, $f(x) \to -\infty$. In addition, $f(0) = 1$, so the y-intercept is $(0, 1)$. Thus B is the correct graph.

83.
$$\left| 1 - \frac{1}{x^2} \right| < 3$$
$$-3 < 1 - \frac{1}{x^2} < 3$$
$$-3 < \frac{x^2 - 1}{x^2} < 3$$
$$-3 < \frac{(x + 1)(x - 1)}{x^2} < 3$$
$$-3 < \frac{(x + 1)(x - 1)}{x^2} \ and \ \frac{(x + 1)(x - 1)}{x^2} < 3$$

First, solve
$$-3 < \frac{(x + 1)(x - 1)}{x^2}$$
$$0 < \frac{(x + 1)(x - 1)}{x^2} + 3$$

The denominator of $f(x) = \dfrac{(x + 1)(x - 1)}{x^2} + 3$ is zero when $x = 0$, so the function is not defined for this value of x. Solve the related equation.

$$\frac{(x + 1)(x - 1)}{x^2} + 3 = 0$$
$$(x + 1)(x - 1) + 3x^2 = 0 \quad \text{Multiplying by } x^2$$
$$x^2 - 1 + 3x^2 = 0$$
$$4x^2 = 1$$
$$x^2 = \frac{1}{4}$$
$$x = \pm\frac{1}{2}$$

The critical values are $-\dfrac{1}{2}$, 0 and $\dfrac{1}{2}$. Test a value in each of the intervals determined by them.

$$\left(-\infty, -\frac{1}{2} \right): \ f(-1) = 3 > 0$$
$$\left(-\frac{1}{2}, 0 \right): \ f\left(-\frac{1}{4} \right) = -12 < 0$$
$$\left(0, \frac{1}{2} \right): \ f\left(\frac{1}{4} \right) = -12 < 0$$
$$\left(\frac{1}{2}, \infty \right): \ f(1) = 3 > 0$$

The solution set for this portion of the inequality is $\left(-\infty, -\dfrac{1}{2} \right) \cup \left(\dfrac{1}{2}, \infty \right)$.

Next, solve
$$\frac{(x + 1)(x - 1)}{x^2} < 3$$
$$\frac{(x + 1)(x - 1)}{x^2} - 3 < 0$$

The denominator of $f(x) = \dfrac{(x + 1)(x - 1)}{x^2} - 3$ is zero when $x = 0$, so the function is not defined for this value of x. Now solve the related equation.

$$\frac{(x + 1)(x - 1)}{x^2} - 3 = 0$$
$$(x + 1)(x - 1) - 3x^2 = 0 \quad \text{Multiplying by } x^2$$
$$x^2 - 1 - 3x^2 = 0$$
$$2x^2 = -1$$
$$x^2 = -\frac{1}{2}$$

There are no real solutions for this portion of the inequality. The solution set of the original inequality is $\left(-\infty, -\dfrac{1}{2} \right) \cup \left(\dfrac{1}{2}, \infty \right)$.

85.
$$(x - 2)^{-3} < 0$$
$$\frac{1}{(x - 2)^3} < 0$$

The denominator of $f(x) = \dfrac{1}{(x - 2)^3}$ is zero when $x = 2$, so the function is not defined for this value of x. The related equation $\dfrac{1}{(x - 2)^3} = 0$ has no solution, so 2 is the only critical point. Test a value in each of the intervals determined by this critical point.

$(-\infty, 2): \; f(1) = -1 < 0$

$(2, \infty): \; f(3) = 1 > 0$

Function values are negative on $(-\infty, 2)$. The solution set is $(-\infty, 2)$.

87. Divide $x^3 + kx^2 + kx - 15$ by $x + 3$.

$$
\begin{array}{r|rrrr}
-3 & 1 & k & k & -15 \\
  &   & -3 & 9-3k & -27+6k \\
\hline
  & 1 & -3+k & 9-2k & -42+6k
\end{array}
$$

Thus $f(-3) = -42 + 6k$.

We know that if $x + 3$ is a factor of $f(x)$, then $f(-3) = 0$. We solve $-42 + 6k = 0$ for k.

$$-42 + 6k = 0$$
$$6k = 42$$
$$k = 7$$

89. $f(x) = \sqrt{x^2 + 3x - 10}$

Since we cannot take the square root of a negative number, then $x^2 + 3x - 10 \geq 0$.

$$x^2 + 3x - 10 \geq 0 \quad \text{Polynomial inequality}$$
$$x^2 + 3x - 10 = 0 \quad \text{Related equation}$$
$$(x + 5)(x - 2) = 0 \quad \text{Factoring}$$

The solutions of the related equation are -5 and 2. These numbers divide the x-axis into the intervals $(-\infty, -5)$, $(-5, 2)$ and $(2, \infty)$.

We let $g(x) = (x + 5)(x - 2)$ and test a value in each interval.

$(-\infty, -5): \; g(-6) = 8 > 0$

$(-5, 2): \; g(0) = -10 < 0$

$(2, \infty): \; g(3) = 8 > 0$

Functions values are positive on $(-\infty, -5)$ and $(2, \infty)$. Since the equality symbol is $\geq$, the endpoints of the intervals must be included in the solution set. It is $(-\infty, -5] \cup [2, \infty)$.

91. $f(x) = \dfrac{1}{\sqrt{5 - |7x + 2|}}$

We cannot take the square root of a negative number; neither can the denominator be zero. Thus we have $5 - |7x + 2| > 0$.

$$5 - |7x + 2| > 0 \quad \text{Polynomial inequality}$$
$$|7x + 2| < 5$$
$$-5 < 7x + 2 < 5$$
$$-7 < 7x < 3$$
$$-1 < x < \frac{3}{7}$$

The solution set is $\left(-1, \dfrac{3}{7}\right)$.

93. No; since imaginary zeros of polynomials with rational coefficients occur in conjugate pairs, a third-degree polynomial with rational coefficients can have at most two imaginary zeros. Thus, there must be at least one real zero.

95. If $P(x)$ is an even function, then $P(-x) = P(x)$ and thus $P(-x)$ has the same number of sign changes as $P(x)$. Hence, $P(x)$ has one negative real zero also.

97. A quadratic inequality $ax^2 + bx + c \leq 0$, $a > 0$, or $ax^2 + bx + c \geq 0$, $a < 0$, has a solution set that is a closed interval.

Chapter 4 Test

1. $f(x) = 2x^3 + 6x^2 - x^4 + 11$
$$= -x^4 + 2x^3 + 6x^2 + 11$$

The leading term is $-x^4$ and the leading coefficient is -1. The degree of the polynomial is 4, so the polynomial is quartic.

2. $h(x) = -4.7x + 29$

The leading term is $-4.7x$ and the leading coefficient is -4.7. The degree of the polynomial is 1, so the polynomial is linear.

3. $f(x) = x(3x - 5)(x - 3)^2(x + 1)^3$

The zeros of the function are 0, $\dfrac{5}{3}$, 3, and -1.

The factors x and $3x - 5$ each occur once, so the zeros 0 and $\dfrac{5}{3}$ have multiplicity 1.

The factor $x - 3$ occurs twice, so the zero 3 has multiplicity 2.

The factor $x + 1$ occurs three times, so the zero -1 has multiplicity 3.

4. For 2008, $x = 2008 - 2004 = 4$.

$f(4) = 897.690(4)^4 - 10,349.487(4)^3 + 19,202.137(4)^2 + 91,597.838(4) + 88,209.580 \approx 329,277$ hybrid automobiles

For 2011, $x = 2011 - 2004 = 7$.

$f(7) = 897.690(7)^4 - 10,349.487(7)^3 + 19,202.137(7)^2 + 91,597.838(7) + 88,209.580 \approx 275,779$ hybrid automobiles

5. $f(x) = x^3 - 5x^2 + 2x + 8$

1. The leading term is x^3. The degree, 3, is odd and the leading coefficient, 1, is positive so as $x \to \infty$, $f(x) \to \infty$ and as $x \to -\infty$, $f(x) \to -\infty$.

2. We solve $f(x) = 0$. By the rational zeros theorem, we know that the possible rational zeros are 1, -1, 2, -2, 4, -4, 8, and -8. Synthetic division shows that 1 is not a zero. We try -1.

$$
\begin{array}{r|rrrr}
-1 & 1 & -5 & 2 & 8 \\
   &   & -1 & 6 & -8 \\
\hline
   & 1 & -6 & 8 & 0
\end{array}
$$

We have $f(x) = (x + 1)(x^2 - 6x + 8) = (x + 1)(x - 2)(x - 4)$.

Now we find the zeros of $f(x)$.

$$x + 1 = 0 \quad or \quad x - 2 = 0 \quad or \quad x - 4 = 0$$
$$x = -1 \quad or \qquad x = 2 \quad or \qquad x = 4$$

The zeros of the function are -1, 2, and 4, so the x-intercepts are $(-1, 0)$, $(2, 0)$, and $(4, 0)$.

3. The zeros divide the x-axis into 4 intervals, $(-\infty, -1)$, $(-1, 2)$, $(2, 4)$, and $(4, \infty)$. We choose a value for x in each interval and find $f(x)$. This tells us the sign of $f(x)$ for all values of x in that interval.

In $(-\infty, -1)$, test -3:

$f(-3) = (-3)^3 - 5(-3)^2 + 2(-3) + 8 = -70 < 0$

In $(-1, 2)$, test 0:

$f(0) = 0^3 - 5(0)^2 + 2(0) + 8 = 8 > 0$

In $(2, 4)$, test 3:

$f(3) = 3^3 - 5(3)^2 + 2(3) + 8 = -4 < 0$

In $(4, \infty)$, test 5:

$f(5) = 5^3 - 5(5)^2 + 2(5) + 8 = 18 > 0$

Thus the graph lies below the x-axis on $(-\infty, -1)$ and on $(2, 4)$ and above the x-axis on $(-1, 2)$ and $(4, \infty)$. We also know the points $(-3, -70)$, $(0, 8)$, $(3, -4)$, and $(5, 18)$ are on the graph.

4. From Step 3 we know that $f(0) = 8$, so the y-intercept is $(0, 8)$.

5. We find additional points on the graph and draw the graph.

x	$f(x)$
-0.5	5.625
0.5	7.875
2.5	-2.625
6	56

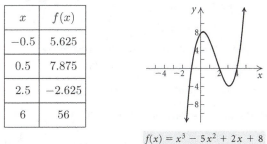

$f(x) = x^3 - 5x^2 + 2x + 8$

6. Checking the graph as described on page 240 in the text, we see that it appears to be correct.

6. $f(x) = -2x^4 + x^3 + 11x^2 - 4x - 12$

1. The leading term is $-2x^4$. The degree, 4, is even and the leading coefficient, -2, is negative so $x \to \infty$, $f(x) \to -\infty$ and as $x \to -\infty$, $f(x) \to -\infty$.

2. We solve $f(x) = 0$.

The possible rational zeros are ± 1, ± 2, ± 3, ± 4, ± 6, ± 12, $\pm \dfrac{1}{2}$, and $\pm \dfrac{3}{2}$. We try -1.

$$\begin{array}{r|rrrrr}
-1 & -2 & 1 & 11 & -4 & -12 \\
   &   & 2 & -3 & -8 & 12 \\
\hline
   & -2 & 3 & 8 & -12 & 0
\end{array}$$

We have $f(x) = (x+1)(-2x^3 + 3x^2 + 8x - 12)$. Find the other zeros.

$-2x^3 + 3x^2 + 8x - 12 = 0$

$2x^3 - 3x^2 - 8x + 12 = 0$ Multiplying by -1

$x^2(2x - 3) - 4(2x - 3) = 0$

$(2x - 3)(x^2 - 4) = 0$

$(2x - 3)(x + 2)(x - 2) = 0$

$2x - 3 = 0 \quad or \quad x + 2 = 0 \quad or \quad x - 2 = 0$

$2x = 3 \quad or \qquad x = -2 \quad or \qquad x = 2$

$x = \dfrac{3}{2} \quad or \qquad x = -2 \quad or \qquad x = 2$

The zeros of the function are -1, $\dfrac{3}{2}$, -2, and 2, so the x-intercepts are $(-2, 0)$, $(-1, 0)$, $\left(\dfrac{3}{2}, 0\right)$, and $(2, 0)$.

3. The zeros divide the x-axis into 5 intervals, $(-\infty, -2)$, $(-2, -1)$, $\left(-1, \dfrac{3}{2}\right)$, $\left(\dfrac{3}{2}, 2\right)$, and $(2, \infty)$. We choose a value for x in each interval and find $f(x)$. This tells us the sign of $f(x)$ for all values of x in that interval.

In $(-\infty, -2)$, test -3: $f(-3) = -90 < 0$

In $(-2, -1)$, test -1.5: $f(-1.5) = 5.25 > 0$

In $\left(-1, \dfrac{3}{2}\right)$, test 0: $f(0) = -12 < 0$

In $\left(\dfrac{3}{2}, 2\right)$, test 1.75: $f(1.75) \approx 1.29 > 0$

In $(2, \infty)$, test 3: $f(3) = -60 < 0$

Thus the graph lies below the x-axis on $(-\infty, -2)$, on $\left(-1, \dfrac{3}{2}\right)$, and on $(2, \infty)$ and above the x-axis on $(-2, -1)$ and on $\left(\dfrac{3}{2}, 2\right)$. We also know the points $(-3, -90)$, $(-1.5, 5.25)$, $(0, -12)$, $(1.75, 1.29)$, and $(3, -60)$ are on the graph.

4. From Step 3 we know that $f(0) = -12$, so the y-intercept is $(0, -12)$.

5. We find additional points on the graph and draw the graph.

x	$f(x)$
-0.5	-7.5
0.5	-11.25
2.5	-15.75

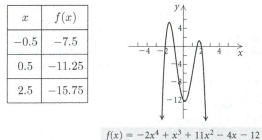

$f(x) = -2x^4 + x^3 + 11x^2 - 4x - 12$

6. Checking the graph as described on page 240 in the text, we see that it appears to be correct.

7. $f(0) = -5 \cdot 0^2 + 3 = 3$

$f(2) = -5 \cdot 2^2 + 3 = -17$

By the intermediate value theorem, since $f(0)$ and $f(2)$ have opposite signs, $f(x)$ has a zero between 0 and 2.

8. $g(-2) = 2(-2)^3 + 6(-2)^2 - 3 = 5$

$g(-1) = 2(-1)^3 + 6(-1)^2 - 3 = 1$

Since both $g(-2)$ and $g(-1)$ are positive, we cannot use the intermediate value theorem to determine if there is a zero between -2 and -1.

9.
$$
\begin{array}{r}
x^3 + 4x^2 + 4x + 6 \\
x - 1 \overline{\smash{\big)}\ x^4 + 3x^3 + 0x^2 + 2x - 5} \\
\underline{x^4 - x^3} \\
4x^3 + 0x^2 \\
\underline{4x^3 - 4x^2} \\
4x^2 + 2x \\
\underline{4x^2 - 4x} \\
6x - 5 \\
\underline{6x - 6} \\
1
\end{array}
$$

The quotient is $x^3 + 4x^2 + 4x + 6$; the remainder is 1.

$P(x) = (x-1)(x^3 + 4x^2 + 4x + 6) + 1$

10.
$$
\begin{array}{r|rrrr}
5 & 3 & 0 & -12 & 7 \\
  &   & 15 & 75 & 315 \\
\hline
  & 3 & 15 & 63 & 322
\end{array}
$$

$Q(x) = 3x^2 + 15x + 63;\ R(x) = 322$

11.
$$
\begin{array}{r|rrrr}
-3 & 2 & -6 & 1 & -4 \\
   &   & -6 & 36 & -111 \\
\hline
   & 2 & -12 & 37 & -115
\end{array}
$$

$P(-3) = -115$

12.
$$
\begin{array}{r|rrrr}
-2 & 1 & 4 & 1 & -6 \\
   &   & -2 & -4 & 6 \\
\hline
   & 1 & 2 & -3 & 0
\end{array}
$$

$f(-2) = 0$, so -2 is a zero of $f(x)$.

13. The function can be written in the form
$$f(x) = a_n(x+3)^2(x)(x-6).$$
The simplest polynomial is obtained if we let $a_n = 1$.
$$
\begin{aligned}
f(x) &= (x+3)^2(x)(x-6) \\
&= (x^2 + 6x + 9)(x^2 - 6x) \\
&= x^4 + 6x^3 + 9x^2 - 6x^3 - 36x^2 - 54x \\
&= x^4 - 27x^2 - 54x
\end{aligned}
$$

14. A polynomial function of degree 5 can have at most 5 zeros. Since $f(x)$ has rational coefficients, in addition to the 3 given zeros, the other zeros are the conjugates of $\sqrt{3}$ and $2 - i$, or $-\sqrt{3}$ and $2 + i$.

15. $-3i$ is also a zero.
$$
\begin{aligned}
f(x) &= (x+10)(x-3i)(x+3i) \\
&= (x+10)(x^2 + 9) \\
&= x^3 + 10x^2 + 9x + 90
\end{aligned}
$$

16. $\sqrt{3}$ and $1 + i$ are also zeros.
$$
\begin{aligned}
f(x) &= (x-0)(x+\sqrt{3})(x-\sqrt{3})[x-(1-i)][x-(1+i)] \\
&= x(x^2 - 3)[(x-1)+i][(x-1)-i] \\
&= (x^3 - 3x)(x^2 - 2x + 1 + 1) \\
&= (x^3 - 3x)(x^2 - 2x + 2) \\
&= x^5 - 2x^4 + 2x^3 - 3x^3 + 6x^2 - 6x \\
&= x^5 - 2x^4 - x^3 + 6x^2 - 6x
\end{aligned}
$$

17. $f(x) = 2x^3 + x^2 - 2x + 12$

$\dfrac{\text{Possibilities for } p}{\text{Possibilities for } q} : \dfrac{\pm 1, \pm 2, \pm 3, \pm 4, \pm 6, \pm 12}{\pm 1, \pm 2}$

Possibilities for p/q: $\pm 1, \pm 2, \pm 3, \pm 4, \pm 6, \pm 12, \pm\dfrac{1}{2}, \pm\dfrac{3}{2}$

18. $h(x) = 10x^4 - x^3 + 2x - 5$

$\dfrac{\text{Possibilities for } p}{\text{Possibilities for } q} : \dfrac{\pm 1, \pm 5}{\pm 1, \pm 2, \pm 5, \pm 10}$

Possibilities for p/q: $\pm 1, \pm 5, \pm\dfrac{1}{2}, \pm\dfrac{5}{2}, \pm\dfrac{1}{5}, \pm\dfrac{1}{10}$

19. $f(x) = x^3 + x^2 - 5x - 5$

a) Possibilities for p/q: $\pm 1, \pm 5$

From the graph of $y = f(x)$, we see that -1 might be a zero.
$$
\begin{array}{r|rrrr}
-1 & 1 & 1 & -5 & -5 \\
   &   & -1 & 0 & 5 \\
\hline
   & 1 & 0 & -5 & 0
\end{array}
$$
We have $f(x) = (x+1)(x^2 - 5)$. We find the other zeros.
$$
\begin{aligned}
x^2 - 5 &= 0 \\
x^2 &= 5 \\
x &= \pm\sqrt{5}
\end{aligned}
$$
The rational zero is -1. The other zeros are $\pm\sqrt{5}$.

b) $f(x) = (x+1)(x-\sqrt{5})(x+\sqrt{5})$

20. $g(x) = 2x^4 - 11x^3 + 16x^2 - x - 6$

a) Possibilities for p/q: $\pm 1, \pm 2, \pm 3, \pm 6, \pm\dfrac{1}{2}, \pm\dfrac{3}{2}$

From the graph of $y = g(x)$, we see that $-\dfrac{1}{2}$, 1, 2, and 3 might be zeros. We try $-\dfrac{1}{2}$.
$$
\begin{array}{r|rrrrr}
-\frac{1}{2} & 2 & -11 & 16 & -1 & -6 \\
             &   & -1 & 6 & -11 & 6 \\
\hline
             & 2 & -12 & 22 & -12 & 0
\end{array}
$$
Now we try 1.
$$
\begin{array}{r|rrrr}
1 & 2 & -12 & 22 & -12 \\
  &   & 2 & -10 & 12 \\
\hline
  & 2 & -10 & 12 & 0
\end{array}
$$
We have $g(x) = \left(x + \dfrac{1}{2}\right)(x-1)(2x^2 - 10x + 12) = 2\left(x + \dfrac{1}{2}\right)(x-1)(x^2 - 5x + 6)$. We find the other zeros.
$$
\begin{aligned}
x^2 - 5x + 6 &= 0 \\
(x-2)(x-3) &= 0 \\
x - 2 = 0 \ &or\ x - 3 = 0 \\
x = 2 \ &or\ x = 3
\end{aligned}
$$
The rational zeros are $-\dfrac{1}{2}$, 1, 2, and 3. These are the only zeros.

b) $g(x) = 2\left(x + \dfrac{1}{2}\right)(x-1)(x-2)(x-3)$
$ = (2x+1)(x-1)(x-2)(x-3)$

21. $h(x) = x^3 + 4x^2 + 4x + 16$

a) Possibilities for p/q: $\pm 1, \pm 2, \pm 4, \pm 8, \pm 16$

From the graph of $h(x)$, we see that -4 might be a zero.

$$\begin{array}{r|rrrr} -4 & 1 & 4 & 4 & 16 \\ & & -4 & 0 & -16 \\ \hline & 1 & 0 & 4 & 0 \end{array}$$

We have $h(x) = (x + 4)(x^2 + 4)$. We find the other zeros.

$$x^2 + 4 = 0$$
$$x^2 = -4$$
$$x = \pm 2i$$

The rational zero is -4. The other zeros are $\pm 2i$.

b) $h(x) = (x + 4)(x + 2i)(x - 2i)$

22. $f(x) = 3x^4 - 11x^3 + 15x^2 - 9x + 2$

a) Possibilities for p/q: $\pm 1, \pm 2, \pm\dfrac{1}{3}, \pm\dfrac{2}{3}$

From the graph of $f(x)$, we see that $\dfrac{2}{3}$ and 1 might be zeros. We try $\dfrac{2}{3}$.

$$\begin{array}{r|rrrrr} \frac{2}{3} & 3 & -11 & 15 & -9 & 2 \\ & & 2 & -6 & 6 & -2 \\ \hline & 3 & -9 & 9 & -3 & 0 \end{array}$$

Now we try 1.

$$\begin{array}{r|rrrr} 1 & 3 & -9 & 9 & -3 \\ & & 3 & -6 & 3 \\ \hline & 3 & -6 & 3 & 0 \end{array}$$

We have $f(x) = \left(x - \dfrac{2}{3}\right)(x-1)(3x^2 - 6x + 3) =$

$3\left(x - \dfrac{2}{3}\right)(x - 1)(x^2 - 2x + 1)$. We find the other zeros.

$$x^2 - 2x + 1 = 0$$
$$(x - 1)(x - 1) = 0$$
$$x - 1 = 0 \ \ or \ \ x - 1 = 0$$
$$x = 1 \ \ or \ \ \ \ \ \ x = 1$$

The rational zeros are $\dfrac{2}{3}$ and 1. (The zero 1 has multiplicity 3.) These are the only zeros.

b) $f(x) = 3\left(x - \dfrac{2}{3}\right)(x - 1)(x - 1)(x - 1)$

$= (3x - 2)(x - 1)^3$

23. $g(x) = -x^8 + 2x^6 - 4x^3 - 1$

There are 2 variations in sign in $g(x)$, so there are 2 or 0 positive real zeros.

$$g(-x) = -(-x)^8 + 2(-x)^6 - 4(-x)^3 - 1$$
$$= -x^8 + 2x^6 + 4x^3 - 1$$

There are 2 variations in sign in $g(-x)$, so there are 2 or 0 negative real zeros.

24. $f(x) = \dfrac{2}{(x - 3)^2}$

1. The numerator and the denominator have no common factors. The denominator is zero when $x = 3$, so the domain excludes 3. it is $(-\infty, 3) \cup (3, \infty)$. The line $x = 3$ is the vertical asymptote.

2. Because the degree of the numerator is less than the degree of the denominator, the x-axis, or $y = 0$, is the horizontal asymptote.

3. The numerator has no zeros, so there is no x-intercept.

4. $f(0) = \dfrac{2}{(0 - 3)^2} = \dfrac{2}{9}$, so the y-intercept is $\left(0, \dfrac{2}{9}\right)$.

5. Find other function values to determine the shape of the graph and then draw it.

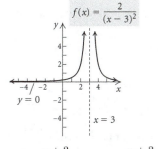

25. $f(x) = \dfrac{x + 3}{x^2 - 3x - 4} = \dfrac{x + 3}{(x + 1)(x - 4)}$

1. The numerator and the denominator have no common factors. The denominator is zero when $x = -1$ or $x = 4$, so the domain excludes -1 and 4. It is $(-\infty, -1) \cup (-1, 4) \cup (4, \infty)$. The lines $x = -1$ and $x = 4$ are vertical asymptotes.

2. Because the degree of the numerator is less than the degree of the denominator, the x-axis, or $y = 0$, is the horizontal asymptote.

3. The numerator is zero at $x = -3$, so the x-intercept is $(-3, 0)$.

4. $f(0) = \dfrac{0 + 3}{(0 + 1)(0 - 4)} = -\dfrac{3}{4}$, so the y-intercept is $\left(0, -\dfrac{3}{4}\right)$.

5. Find other function values to determine the shape of the graph and then draw it.

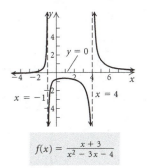

26. Answers may vary. The numbers -1 and 2 must be zeros of the denominator but not of the numerator, and -4 must be a zero of the numerator but not of the denominator.

$$f(x) = \frac{x+4}{(x+1)(x-2)}, \text{ or } f(x) = \frac{x+4}{x^2-x-2}$$

27.
$$2x^2 > 5x + 3 \quad \text{Polynomial inequality}$$
$$2x^2 - 5x - 3 > 0 \quad \text{Equivalent inequality}$$
$$2x^2 - 5x - 3 = 0 \quad \text{Related equation}$$
$$(2x+1)(x-3) = 0 \quad \text{Factoring}$$

The solutions of the related equation are $-\dfrac{1}{2}$ and 3. These numbers divide the x-axis into the intervals $\left(-\infty, -\dfrac{1}{2}\right)$, $\left(-\dfrac{1}{2}, 3\right)$, and $(3, \infty)$.

We let $f(x) = (2x+1)(x-3)$ and test a value in each interval.

$$\left(-\infty, -\frac{1}{2}\right): \ f(-1) = 4 > 0$$
$$\left(-\frac{1}{2}, 3\right): \ f(0) = -3 < 0$$
$$(3, \infty): \ f(4) = 9 > 0$$

Function values are positive on $\left(-\infty, -\dfrac{1}{2}\right)$ and $(3, \infty)$. The solution set is $\left(-\infty, -\dfrac{1}{2}\right) \cup (3, \infty)$.

28.
$$\frac{x+1}{x-4} \le 3 \quad \text{Rational inequality}$$
$$\frac{x+1}{x-4} - 3 \le 0 \quad \text{Equivalent inequality}$$

The denominator of $f(x) = \dfrac{x+1}{x-4} - 3$ is zero when $x = 4$, so the function is not defined for this value of x. We solve the related equation $f(x) = 0$.

$$\frac{x+1}{x-4} - 3 = 0$$
$$(x-4)\left(\frac{x+1}{x-4} - 3\right) = (x-4) \cdot 0$$
$$x+1 - 3(x-4) = 0$$
$$x+1 - 3x + 12 = 0$$
$$-2x + 13 = 0$$
$$2x = 13$$
$$x = \frac{13}{2}$$

The critical values are 4 and $\dfrac{13}{2}$. They divide the x-axis into the intervals $(-\infty, 4)$, $\left(4, \dfrac{13}{2}\right)$ and $\left(\dfrac{13}{2}, \infty\right)$. We test a value in each interval.

$$(-\infty, 4): \ f(3) = -7 < 0$$
$$\left(4, \frac{13}{2}\right): \ f(5) = 3 > 0$$
$$\left(\frac{13}{2}, \infty\right): \ f(9) = -1 < 0$$

Function values are negative for $(-\infty, 4)$ and $\left(\dfrac{13}{2}, \infty\right)$. Since the inequality symbol is $\le$, the endpoint of the interval $\left(\dfrac{13}{2}, \infty\right)$ must be included in the solution set. It is $(-\infty, 4) \cup \left[\dfrac{13}{2}, \infty\right)$.

29. a) We write and solve a polynomial equation.
$$-16t^2 + 64t + 192 = 0$$
$$-16(t^2 - 4t - 12) = 0$$
$$-16(t+2)(t-6) = 0$$

The solutions are $t = -2$ and $t = 6$. Only $t = 6$ has meaning in this application. The rocket reaches the ground at $t = 6$ seconds.

b) We write and solve a polynomial inequality.
$$-16t^2 + 64t + 192 > 240$$
$$-16t^2 + 64t - 48 > 0$$
$$-16t^2 + 64t - 48 = 0 \quad \text{Related equation}$$
$$-16(t^2 - 4t + 3) = 0$$
$$-16(t-1)(t-3) = 0$$

The solutions of the related equation are 1 and 3. These numbers divide the t-axis into the intervals $(-\infty, 1)$, $(1, 3)$ and $(3, \infty)$. Because the rocket returns to the ground at $t = 6$, we restrict our discussion to values of t such that $0 \le t \le 6$. We consider the intervals $[0, 1)$, $(1, 3)$ and $(3, 6]$. We let $f(t) = -16t^2 + 64t - 48$ and test a value in each interval.

$$[0, 1): \ f\left(\frac{1}{2}\right) = -20 < 0$$
$$(1, 3): \ f(2) = 16 > 0$$
$$(3, 6]: \ f(4) = -48 < 0$$

Function values are positive on $(1, 3)$. The solution set is $(1, 3)$.

30. $f(x) = x^3 - x^2 - 2$

The degree of the function is odd and the leading coefficient is positive, so as $x \to \infty$, $f(x) \to \infty$ and as $x \to -\infty$, $f(x) \to -\infty$. In addition, $f(0) = -2$, so the y-intercept is $(0, -2)$. Thus D is the correct graph.

31. $f(x) = \sqrt{x^2 + x - 12}$

Since we cannot take the square root of a negative number, then $x^2 + x - 12 \ge 0$.

$$x^2 + x - 12 \ge 0 \ \text{Polynomial inequality}$$
$$x^2 + x - 12 = 0 \ \text{Related equation}$$
$$(x+4)(x-3) = 0 \ \text{Factoring}$$

The solutions of the related equation are -4 and 3. These numbers divide the x-axis into the intervals $(-\infty, -4)$, $(-4, 3)$ and $(3, \infty)$.

We let $f(x) = (x+4)(x-3)$ and test a value in each interval.

$$(-\infty, -4): \ g(-5) = 8 > 0$$
$$(-4, 3): \ g(0) = -12 < 0$$
$$(3, \infty): \ g(4) = 8 > 0$$

Function values are positive on $(-\infty, -4)$ and $(3, \infty)$. Since the inequality symbol is $\geq$, the endpoints of the intervals must be included in the solution set. It is $(-\infty, -4] \cup [3, \infty)$.

Chapter 5

Exponential and Logarithmic Functions

Exercise Set 5.1

---

1. We interchange the first and second coordinates of each ordered pair to find the inverse of the relation. It is

$$\{(8,7),\ (8,-2),\ (-4,3),\ (-8,8)\}.$$

3. We interchange the first and second coordinates of each ordered pair to find the inverse of the relation. It is

$$\{(-1,-1),\ (4,-3)\}.$$

5. Interchange x and y.

$$y = 4x-5$$
$$\downarrow\quad\downarrow$$
$$x = 4y-5$$

7. Interchange x and y.

$$x^3 y = -5$$
$$\downarrow\downarrow$$
$$y^3 x = -5$$

9. Interchange x and y.

$$x = y^2 - 2y$$
$$\downarrow\quad\downarrow\quad\downarrow$$
$$y = x^2 - 2x$$

11. Graph $x = y^2 - 3$. Some points on the graph are $(-3,0)$, $(-2,-1)$, $(-2,1)$, $(1,-2)$, and $(1,2)$. Plot these points and draw the curve. Then reflect the graph across the line $y = x$.

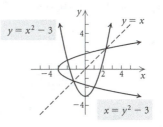

13. Graph $y = 3x - 2$. The intercepts are $(0,-2)$ and $\left(\dfrac{2}{3},0\right)$.

Plot these points and draw the line. Then reflect the graph across the line $y = x$.

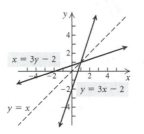

15. Graph $y = |x|$. Some points on the graph are $(0,0)$, $(-2,2)$, $(2,2)$, $(-5,5)$, and $(5,5)$. Plot these points and draw the graph. Then reflect the graph across the line $y = x$.

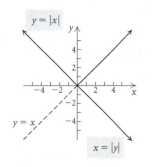

17. We show that if $f(a) = f(b)$, then $a = b$.

$$\frac{1}{3}a - 6 = \frac{1}{3}b - 6$$
$$\frac{1}{3}a = \frac{1}{3}b \qquad \text{Adding 6}$$
$$a = b \qquad \text{Multiplying by 3}$$

Thus f is one-to-one.

19. We show that if $f(a) = f(b)$, then $a = b$.

$$a^3 + \frac{1}{2} = b^3 + \frac{1}{2}$$
$$a^3 = b^3 \qquad \text{Subtracting } \frac{1}{2}$$
$$a = b \qquad \text{Taking cube roots}$$

Thus f is one-to-one.

21. $g(-1) = 1 - (-1)^2 = 1 - 1 = 0$ and $g(1) = 1 - 1^2 = 1 - 1 = 0$, so $g(-1) = g(1)$ but $-1 \neq 1$. Thus the function is not one-to-one.

23. $f(-2) = (-2)^4 - (-2)^2 = 16 - 4 = 12$ and $f(2) = 2^4 - 2^2 = 16 - 4 = 12$, so $f(-2) = f(2)$ but $-2 \neq 2$. Thus the function is not one-to-one.

25. The function is one-to-one, because no horizontal line crosses the graph more than once.

27. The function is not one-to-one, because there are many horizontal lines that cross the graph more than once.

29. The function is not one-to-one, because there are many horizontal lines that cross the graph more than once.

31. The function is one-to-one, because no horizontal line crosses the graph more than once.

33. The graph of $f(x) = 5x - 8$ is shown below.

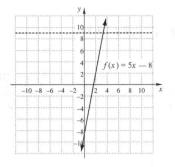

Since there is no horizontal line that crosses the graph more than once, the function is one-to-one.

35. The graph of $f(x) = 1 - x^2$ is shown below.

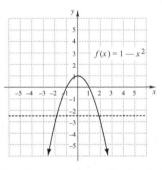

Since there are many horizontal lines that cross the graph more than once, the function is not one-to-one.

37. The graph of $f(x) = |x + 2|$ is shown below.

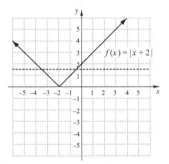

Since there are many horizontal lines that cross the graph more than once, the function is not one-to-one.

39. The graph of $f(x) = -\dfrac{4}{x}$ is shown below.

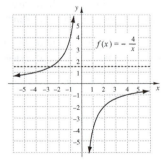

Since there is no horizontal line that crosses the graph more than once, the function is one-to-one.

41. The graph of $f(x) = \dfrac{2}{3}$ is shown below.

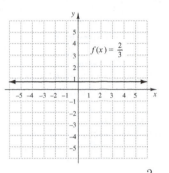

Since the horizontal line $y = \dfrac{2}{3}$ crosses the graph more than once, the function is not one-to-one.

43. The graph of $f(x) = \sqrt{25 - x^2}$ is shown below.

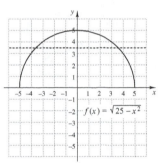

Since there are many horizontal lines that cross the graph more than once, the function is not one-to-one.

45. a) The graph of $f(x) = x + 4$ is shown below. It passes the horizontal-line test, so it is one-to-one.

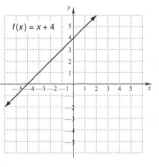

b) Replace $f(x)$ with y: $y = x + 4$

Interchange x and y: $x = y + 4$

Solve for y: $x - 4 = y$

Replace y with $f^{-1}(x)$: $f^{-1}(x) = x - 4$

47. a) The graph of $f(x) = 2x - 1$ is shown below. It passes the horizontal-line test, so it is one-to-one.

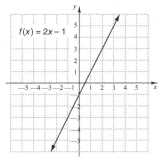

$f(x) = 2x - 1$

b) Replace $f(x)$ with y: $y = 2x - 1$

Interchange x and y: $x = 2y - 1$

Solve for y: $\dfrac{x+1}{2} = y$

Replace y with $f^{-1}(x)$: $f^{-1}(x) = \dfrac{x+1}{2}$

49. a) The graph of $f(x) = \dfrac{4}{x+7}$ is shown below. It passes the horizontal-line test, so the function is one-to-one.

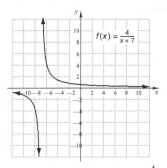

$f(x) = \dfrac{4}{x+7}$

b) Replace $f(x)$ with y: $y = \dfrac{4}{x+7}$

Interchange x and y: $x = \dfrac{4}{y+7}$

Solve for y: $x(y+7) = 4$

$$y + 7 = \frac{4}{x}$$

$$y = \frac{4}{x} - 7$$

Replace y with $f^{-1}(x)$: $f^{-1}(x) = \dfrac{4}{x} - 7$

51. a) The graph of $f(x) = \dfrac{x+4}{x-3}$ is shown below. It passes the horizontal-line test, so the function is one-to-one.

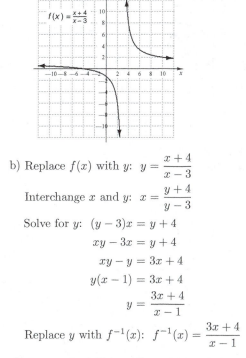

$f(x) = \dfrac{x+4}{x-3}$

b) Replace $f(x)$ with y: $y = \dfrac{x+4}{x-3}$

Interchange x and y: $x = \dfrac{y+4}{y-3}$

Solve for y: $(y-3)x = y + 4$

$$xy - 3x = y + 4$$

$$xy - y = 3x + 4$$

$$y(x - 1) = 3x + 4$$

$$y = \frac{3x + 4}{x - 1}$$

Replace y with $f^{-1}(x)$: $f^{-1}(x) = \dfrac{3x+4}{x-1}$

53. a) The graph of $f(x) = x^3 - 1$ is shown below. It passes the horizontal-line test, so the function is one-to-one.

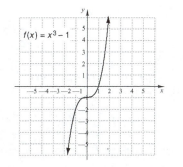

$f(x) = x^3 - 1$

b) Replace $f(x)$ with y: $y = x^3 - 1$

Interchange x and y: $x = y^3 - 1$

Solve for y: $x + 1 = y^3$

$$\sqrt[3]{x + 1} = y$$

Replace y with $f^{-1}(x)$: $f^{-1}(x) = \sqrt[3]{x + 1}$

55. a) The graph of $f(x) = x\sqrt{4 - x^2}$ is shown below. Since there are many horizontal lines that cross the graph more than once, the function is not one-to-one and thus does not have an inverse that is a function.

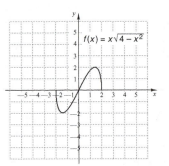

57. a) The graph of $f(x) = 5x^2 - 2$, $x \geq 0$ is shown below. It passes the horizontal-line test, so it is one-to-one.

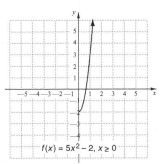

b) Replace $f(x)$ with y: $y = 5x^2 - 2$

Interchange x and y: $x = 5y^2 - 2$

Solve for y: $x + 2 = 5y^2$

$$\frac{x + 2}{5} = y^2$$

$$\sqrt{\frac{x + 2}{5}} = y$$

(We take the principal square root, because $x \geq 0$ in the original equation.)

Replace y with $f^{-1}(x)$: $f^{-1}(x) = \sqrt{\dfrac{x + 2}{5}}$ for

all x in the range of $f(x)$, or $f^{-1}(x) = \sqrt{\dfrac{x + 2}{5}}$,

$x \geq -2$

59. a) The graph of $f(x) = \sqrt{x + 1}$ is shown below. It passes the horizontal-line test, so the function is one-to-one.

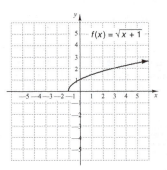

b) Replace $f(x)$ with y: $y = \sqrt{x + 1}$

Interchange x and y: $x = \sqrt{y + 1}$

Solve for y: $x^2 = y + 1$

$x^2 - 1 = y$

Replace y with $f^{-1}(x)$: $f^{-1}(x) = x^2 - 1$ for all x in the range of $f(x)$, or $f^{-1}(x) = x^2 - 1$, $x \geq 0$.

61. $f(x) = 3x$

The function f multiplies an input by 3. Then to reverse this procedure, f^{-1} would divide each of its inputs by 3. Thus, $f^{-1}(x) = \dfrac{x}{3}$, or $f^{-1}(x) = \dfrac{1}{3}x$.

63. $f(x) = -x$

The outputs of f are the opposites, or additive inverses, of the inputs. Then the outputs of f^{-1} are the opposites of its inputs. Thus, $f^{-1}(x) = -x$.

65. $f(x) = \sqrt[3]{x - 5}$

The function f subtracts 5 from each input and then takes the cube root of the result. To reverse this procedure, f^{-1} would raise each input to the third power and then add 5 to the result. Thus, $f^{-1}(x) = x^3 + 5$.

67. We reflect the graph of f across the line $y = x$. The reflections of the labeled points are $(-5, -5)$, $(-3, 0)$, $(1, 2)$, and $(3, 5)$.

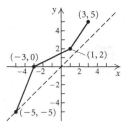

69. We reflect the graph of f across the line $y = x$. The reflections of the labeled points are $(-6, -2)$, $(1, -1)$, $(2, 0)$, and $(5.375, 1.5)$.

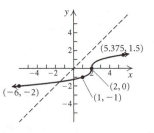

71. We reflect the graph of f across the line $y = x$.

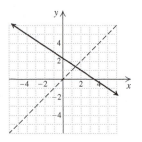

73. We find $(f^{-1} \circ f)(x)$ and $(f \circ f^{-1})(x)$ and check to see that each is x.

$$(f^{-1} \circ f)(x) = f^{-1}(f(x)) = f^{-1}\left(\frac{7}{8}x\right) =$$

$$\frac{8}{7}\left(\frac{7}{8}x\right) = x$$

$$(f \circ f^{-1})(x) = f(f^{-1}(x)) = f\left(\frac{8}{7}x\right) = \frac{7}{8}\left(\frac{8}{7}x\right) = x$$

75. We find $(f^{-1} \circ f)(x)$ and $(f \circ f^{-1})(x)$ and check to see that each is x.

$$(f^{-1} \circ f)(x) = f^{-1}(f(x)) = f^{-1}\left(\frac{1-x}{x}\right) =$$

$$\frac{1}{\dfrac{1-x}{x}+1} = \frac{1}{\dfrac{1-x+x}{x}} = \frac{1}{\dfrac{1}{x}} = x$$

$$(f \circ f^{-1})(x) = f(f^{-1}(x)) = f\left(\frac{1}{x+1}\right) =$$

$$\frac{1 - \dfrac{1}{x+1}}{\dfrac{1}{x+1}} = \frac{\dfrac{x+1-1}{x+1}}{\dfrac{1}{x+1}} = \frac{\dfrac{x}{x+1}}{\dfrac{1}{x+1}} = x$$

77. $(f^{-1} \circ f)(x) = f^{-1}(f(x)) = \dfrac{5\left(\dfrac{2}{5}x+1\right) - 5}{2} =$

$$\frac{2x+5-5}{2} = \frac{2x}{2} = x$$

$$(f \circ f^{-1})(x) = f(f^{-1}(x)) = \frac{2}{5}\left(\frac{5x-5}{2}\right) + 1 =$$

$$x - 1 + 1 = x$$

79. Replace $f(x)$ with y: $y = 5x - 3$

Interchange x and y: $x = 5y - 3$

Solve for y: $x + 3 = 5y$

$$\frac{x+3}{5} = y$$

Replace y with $f^{-1}(x)$: $f^{-1}(x) = \dfrac{x+3}{5}$, or $\dfrac{1}{5}x + \dfrac{3}{5}$

The domain and range of f are $(-\infty, \infty)$, so the domain and range of f^{-1} are also $(-\infty, \infty)$.

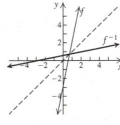

81. Replace $f(x)$ with y: $y = \dfrac{2}{x}$

Interchange x and y: $x = \dfrac{2}{y}$

Solve for y: $xy = 2$

$$y = \frac{2}{x}$$

Replace y with $f^{-1}(x)$: $f^{-1}(x) = \dfrac{2}{x}$

The domain and range of f are $(-\infty, 0) \cup (0, \infty)$, so the domain and range of f^{-1} are also $(-\infty, 0) \cup (0, \infty)$.

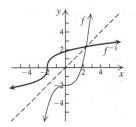

83. Replace $f(x)$ with y: $y = \dfrac{1}{3}x^3 - 2$

Interchange x and y: $x = \dfrac{1}{3}y^3 - 2$

Solve for y: $x + 2 = \dfrac{1}{3}y^3$

$$3x + 6 = y^3$$

$$\sqrt[3]{3x+6} = y$$

Replace y with $f^{-1}(x)$: $f^{-1}(x) = \sqrt[3]{3x+6}$

The domain and range of f are $(-\infty, \infty)$, so the domain and range of f^{-1} are also $(-\infty, \infty)$.

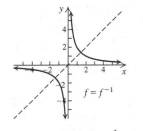

85. Replace $f(x)$ with y: $y = \dfrac{x+1}{x-3}$

Interchange x and y: $x = \dfrac{y+1}{y-3}$

Solve for y: $xy - 3x = y + 1$

$$xy - y = 3x + 1$$

$$y(x - 1) = 3x + 1$$

$$y = \frac{3x+1}{x-1}$$

Replace y with $f^{-1}(x)$: $f^{-1}(x) = \dfrac{3x+1}{x-1}$

The domain of f is $(-\infty, 3) \cup (3, \infty)$ and the range of f is $(-\infty, 1) \cup (1, \infty)$. Thus the domain of f^{-1} is $(-\infty, 1) \cup (1, \infty)$ and the range of f^{-1} is $(-\infty, 3) \cup (3, \infty)$.

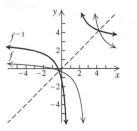

87. Since $f(f^{-1}(x)) = f^{-1}(f(x)) = x$, then $f(f^{-1}(5)) = 5$ and $f^{-1}(f(a)) = a$.

89. a) $C(x) = \dfrac{72 + 2x}{x}$

$$C(2) = \frac{72 + 2 \cdot 2}{2} = \frac{72 + 4}{2} = \frac{76}{2} = \$38$$

$$C(5) = \frac{72 + 2 \cdot 5}{5} = \frac{72 + 10}{5} = \frac{82}{5} = \$16.40$$

$$C(8) = \frac{72 + 2 \cdot 8}{8} = \frac{72 + 16}{8} = \frac{88}{8} = \$11$$

b) The graph of $C(x)$ passes the horizontal-line test and thus has an inverse that is a function.

Replace $C(x)$ with y: $y = \dfrac{72 + 2x}{x}$

Interchange x and y: $x = \dfrac{72 + 2y}{y}$

Solve for y: $xy = 72 + 2y$

$$xy - 2y = 72$$
$$y(x - 2) = 72$$
$$y = \frac{72}{x - 2}$$

Replace y with $C^{-1}(x)$: $C^{-1}(x) = \dfrac{72}{x - 2}$

$C^{-1}(x)$ represents the number of players in the group lesson when x is the cost per player, in dollars.

c) $C^{-1}(74) = \dfrac{72}{74 - 2} = \dfrac{72}{72} = 1$ player

$C^{-1}(20) = \dfrac{72}{20 - 2} = \dfrac{72}{18} = 4$ players

$C^{-1}(11) = \dfrac{72}{11 - 2} = \dfrac{72}{9} = 8$ players

91. a) In 2010, $x = 2010 - 2008 = 2$.

$H(2) = 6.58(2) + 27.7 = \$40.86$ billion

In 2013, $x = 2013 - 2008 = 5$.

$H(5) = 6.58(5) + 27.7 = \$60.6$ billion

b) The graph of $H(x)$ passes the horizontal-line test and thus has an inverse that is a function.

Replace $H(x)$ with y: $y = 6.58x + 27.7$

Interchange x and y: $x = 6.58y + 27.7$

Solve for y: $x - 27.7 = 6.58y$

$$\frac{x - 27.7}{6.58} = y$$

Replace y with $H^{-1}(x)$: $H^{-1}(x) = \dfrac{x - 27.7}{6.58}$

$H^{-1}(x)$ represents the number of years after 2008, where x is the amount of e-commerce holiday season sales, in billions of dollars.

93. The functions for which the coefficient of x^2 is negative have a maximum value. These are (b), (d), (f), and (h).

95. Since $|2| > 1$ the graph of $f(x) = 2x^2$ can be obtained by stretching the graph of $f(x) = x^2$ vertically. Since $0 < \left|\dfrac{1}{4}\right| < 1$, the graph of $f(x) = \dfrac{1}{4}x^2$ can be obtained by shrinking the graph of $y = x^2$ vertically. Thus the graph of $f(x) = 2x^2$, or (a) is narrower.

97. We can write (f) as $f(x) = -2[x - (-3)]^2 + 1$. Thus the graph of (f) has vertex $(-3, 1)$.

99. The graph of $f(x) = x^2 - 3$ is a parabola with vertex $(0, -3)$. If we consider x-values such that $x \geq 0$, then the graph is the right-hand side of the parabola and it passes the horizontal line test. We find the inverse of $f(x) = x^2 - 3$, $x \geq 0$.

Replace $f(x)$ with y: $y = x^2 - 3$

Interchange x and y: $x = y^2 - 3$

Solve for y: $x + 3 = y^2$

$$\sqrt{x + 3} = y$$

(We take the principal square root, because $x \geq 0$ in the original equation.)

Replace y with $f^{-1}(x)$: $f^{-1}(x) = \sqrt{x + 3}$ for all x in the range of $f(x)$, or $f^{-1}(x) = \sqrt{x + 3}$, $x \geq -3$.

Answers may vary. There are other restrictions that also make $f(x)$ one-to-one.

101. Answers may vary. $f(x) = \dfrac{3}{x}$, $f(x) = 1 - x$, $f(x) = x$.

Exercise Set 5.2

1. $e^4 \approx 54.5982$

3. $e^{-2.458} \approx 0.0856$

5. $f(x) = -2^x - 1$

$f(0) = -2^0 - 1 = -1 - 1 = -2$

The only graph with y-intercept $(0, -2)$ is (f).

7. $f(x) = e^x + 3$

This is the graph of $f(x) = e^x$ shifted up 3 units. Then (e) is the correct choice.

9. $f(x) = 3^{-x} - 2$

$f(0) = 3^{-0} - 2 = 1 - 2 = -1$

Since the y-intercept is $(0, -1)$, the correct graph is (a) or (c). Check another point on the graph. $f(-1) = 3^{-(-1)} - 2 = 3 - 2 = 1$, so $(-1, 1)$ is on the graph. Thus (a) is the correct choice.

11. Graph $f(x) = 3^x$.

Compute some function values, plot the corresponding points, and connect them with a smooth curve.

x	$y = f(x)$	(x, y)
-3	$\dfrac{1}{27}$	$\left(-3, \dfrac{1}{27}\right)$
-2	$\dfrac{1}{9}$	$\left(-2, \dfrac{1}{9}\right)$
-1	$\dfrac{1}{3}$	$\left(-1, \dfrac{1}{3}\right)$
0	1	$(0, 1)$
1	3	$(1, 3)$
2	9	$(2, 9)$
3	27	$(3, 27)$

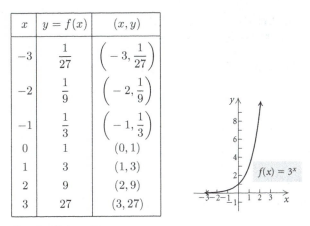

13. Graph $f(x) = 6^x$.

Compute some function values, plot the corresponding points, and connect them with a smooth curve.

x	$y = f(x)$	(x, y)
-3	$\dfrac{1}{216}$	$\left(-3, \dfrac{1}{216}\right)$
-2	$\dfrac{1}{36}$	$\left(-2, \dfrac{1}{36}\right)$
-1	$\dfrac{1}{6}$	$\left(-1, \dfrac{1}{6}\right)$
0	1	$(0, 1)$
1	6	$(1, 6)$
2	36	$(2, 36)$
3	216	$(3, 216)$

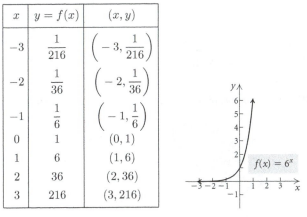

15. Graph $f(x) = \left(\dfrac{1}{4}\right)^x$.

Compute some function values, plot the corresponding points, and connect them with a smooth curve.

x	$y = f(x)$	(x, y)
-3	64	$(-3, 64)$
-2	16	$(-2, 16)$
-1	4	$(-1, 4)$
0	1	$(0, 1)$
1	$\dfrac{1}{4}$	$\left(1, \dfrac{1}{4}\right)$
2	$\dfrac{1}{16}$	$\left(2, \dfrac{1}{16}\right)$
3	$\dfrac{1}{64}$	$\left(3, \dfrac{1}{64}\right)$

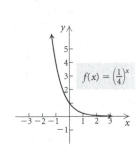

17. Graph $y = -2^x$.

x	y	(x, y)
-3	$-\dfrac{1}{8}$	$\left(-3, -\dfrac{1}{8}\right)$
-2	$-\dfrac{1}{4}$	$\left(-2, -\dfrac{1}{4}\right)$
-1	$-\dfrac{1}{2}$	$\left(-1, -\dfrac{1}{2}\right)$
0	-1	$(0, -1)$
1	-2	$(1, -2)$
2	-4	$(2, -4)$
3	-8	$(3, -8)$

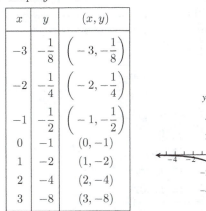

19. Graph $f(x) = -0.25^x + 4$.

x	$y = f(x)$	(x, y)
-3	-60	$(-3, -60)$
-2	-12	$(-2, -12)$
-1	0	$(-1, 0)$
0	3	$(0, 3)$
1	3.75	$(1, 3.75)$
2	3.94	$(2, 3.94)$
3	3.98	$(3, 3.98)$

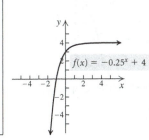

21. Graph $f(x) = 1 + e^{-x}$.

x	$y = f(x)$	(x, y)
-3	21.1	$(-3, 21.1)$
-2	8.4	$(-2, 8.4)$
-1	3.7	$(-1, 3.7)$
0	2	$(0, 2)$
1	1.4	$(1, 1.4)$
2	1.1	$(2, 1.1)$
3	1.0	$(3, 1.0)$

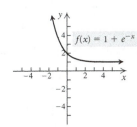

23. Graph $y = \dfrac{1}{4}e^x$.

Choose values for x and compute the corresponding y-values. Plot the points (x, y) and connect them with a smooth curve.

x	y	(x, y)
-3	0.0124	$(-3, 0.0124)$
-2	0.0338	$(-2, 0.0338)$
-1	0.0920	$(-1, 0.0920)$
0	0.25	$(0, 0.25)$
1	0.6796	$(1, 0.6796)$
2	1.8473	$(2, 1.8473)$
3	5.0214	$(3, 5.0214)$

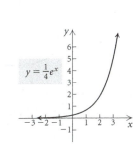

25. Graph $f(x) = 1 - e^{-x}$.

Compute some function values, plot the corresponding points, and connect them with a smooth curve.

x	y	(x, y)
-3	-19.0855	$(-3, -19.0855)$
-2	-6.3891	$(-2, -6.3891)$
-1	-1.7183	$(-1, -1.7183)$
0	0	$(0, 0)$
1	0.6321	$(1, 0.6321)$
2	0.8647	$(2, 0.8647)$
3	0.9502	$(3, 0.9502)$

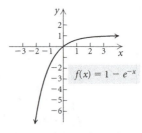

27. Shift the graph of $y = 2^x$ left 1 unit.

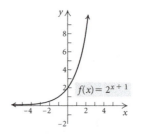

29. Shift the graph of $y = 2^x$ down 3 units.

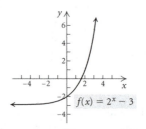

31. Shift the graph of $y = 2^x$ left 1 unit, reflect it across the y-axis, and shift it up 2 units.

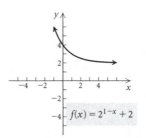

33. Reflect the graph of $y = 3^x$ across the y-axis, then across the x-axis, and then shift it up 4 units.

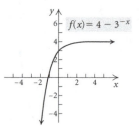

35. Shift the graph of $y = \left(\dfrac{3}{2}\right)^x$ right 1 unit.

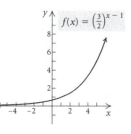

37. Shift the graph of $y = 2^x$ left 3 units and down 5 units.

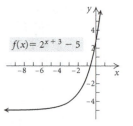

39. Shift the graph of $y = 2^x$ right 1 unit, stretch it vertically, and shift it up 1 unit.

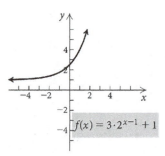

41. Shrink the graph of $y = e^x$ horizontally.

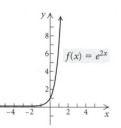

43. Reflect the graph of $y = e^x$ across the x-axis, shift it up 1 unit, and shrink it vertically.

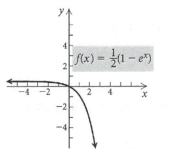

45. Shift the graph of $y = e^x$ left 1 unit and reflect it across the y-axis.

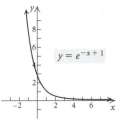

47. Reflect the graph of $y = e^x$ across the y-axis and then across the x-axis; shift it up 1 unit and then stretch it vertically.

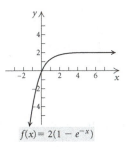

49. We graph $f(x) = e^{-x} - 4$ for $x < -2$, $f(x) = x + 3$ for $-2 \leq x < 1$, and $f(x) = x^2$ for $x \geq 1$.

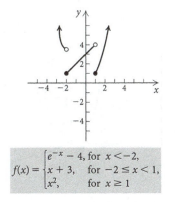

$$f(x) = \begin{cases} e^{-x} - 4, & \text{for } x < -2, \\ x + 3, & \text{for } -2 \leq x < 1, \\ x^2, & \text{for } x \geq 1 \end{cases}$$

51. a) We use the formula $A = P\left(1 + \dfrac{r}{n}\right)^{nt}$ and substitute 82,000 for P, 0.045 for r, and 4 for n.

$$A(t) = 82,000\left(1 + \frac{0.045}{4}\right)^{4t} = 82,000(1.01125)^{4t}$$

b) $A(0) = 82,000(1.01125)^{4 \cdot 0} = \$82,000$

$A(2) = 82,000(1.01125)^{4 \cdot 2} \approx \$89,677.22$

$A(5) = 82,000(1.01125)^{4 \cdot 5} \approx \$102,561.54$

$A(10) = 82,000(1.01125)^{4 \cdot 10} \approx \$128,278.90$

53. We use the formula $A = P\left(1 + \dfrac{r}{n}\right)^{nt}$ and substitute 3000 for P, 0.05 for r, and 4 for n.

$$A(t) = 3000\left(1 + \frac{0.05}{4}\right)^{4t} = 3000(1.0125)^{4t}$$

On Elizabeth's sixteenth birthday, $t = 16 - 6 = 10$.

$A(10) = 3000(1.0125)^{4 \cdot 10} = 4930.86$

When the CD matures \$4930.86 will be available.

55. We use the formula $A = P\left(1 + \dfrac{r}{n}\right)^{nt}$ and substitute 3000 for P, 0.04 for r, 2 for n, and 2 for t.

$$A = 3000\left(1 + \frac{0.04}{2}\right)^{2 \cdot 2} \approx \$3247.30$$

57. We use the formula $A = P\left(1 + \dfrac{r}{n}\right)^{nt}$ and substitute 120,000 for P, 0.025 for r, 1 for n, and 10 for t.

$$A = 120,000\left(1 + \frac{0.025}{1}\right)^{1 \cdot 10} \approx \$153,610.15$$

59. We use the formula $A = P\left(1 + \dfrac{r}{n}\right)^{nt}$ and substitute 53,500 for P, 0.055 for r, 4 for n, and 6.5 for t.

$$A = 53,500\left(1 + \frac{0.055}{4}\right)^{4(6.5)} \approx \$76,305.59$$

61. We use the formula $A = P\left(1 + \dfrac{r}{n}\right)^{nt}$ and substitute 17,400 for P, 0.081 for r, 365 for n, and 5 for t.

$$A = 17,400\left(1 + \frac{0.081}{365}\right)^{365 \cdot 5} \approx \$26,086.69$$

63. In 1998, $x = 1998 - 1995 = 3$.

$A(3) = 246,855(1.0931)^3 \approx 322,420$ vehicles

In 2010, $x = 2010 - 1995 = 15$.

$A(15) = 246,855(1.0931)^{15} \approx 938,297$ vehicles

In 2018, $x = 2018 - 1995 = 23$.

$A(23) = 246,855(1.0931)^{23} \approx 1,912,580$ vehicles

65. In 2011, $x = 2011 - 2008 = 3$.

$S(3) = 20.913(2.236)^3 \approx \$234$ million

In 2015, $x = 2015 - 2008 = 7$.

$S(7) = 20.913(2.236)^7 \approx \$5844$ million, or \$5.844 billion

67. In 2005, $t = 2005 - 2004 = 1$.

$P(1) = 2.307(1.483)^1 \approx 3$ million users

In 2009, $t = 2009 - 2004 = 5$.

$P(5) = 2.307(1.483)^5 \approx 17$ million users

In 2012, $t = 2012 - 2004 = 8$.

$P(8) = 2.307(1.483)^8 \approx 54$ million users

69. In 2020, $t = 2020 - 2015 = 5$.

$H(5) = 80,040.68(1.0481)^5 \approx 101,234$ centenarians

In 2050, $t = 2050 - 2015 = 35$.

$H(35) = 80,040.68(1.0481)^{35} \approx 414,387$ centenarians

71. In 1982, $x = 1982 - 1969 = 13$.

$G(13) = 20.7(1.066)^{13} \approx \$48$ billion

In 1995, $x = 1995 - 1969 = 26$.

$G(26) = 20.7(1.066)^{26} \approx \$109$ billion

In 2010, $x = 2010 - 1969 = 41$.

$G(41) = 20.7(1.066)^{41} \approx \$284$ billion

73. $V(t) = 6982(0.85)^t$

$V(0) = 6982(0.85)^0 = \$6982$

$V(1) = 6982(0.85)^1 \approx \$5935$

$V(2) = 6982(0.85)^2 \approx \$5044$

$V(5) = 6982(0.85)^5 \approx \$3098$

$V(8) = 6982(0.85)^8 \approx \$1903$

75. $f(25) = 100(1 - e^{-0.04(25)}) \approx 63\%$.

77. $(1 - 4i)(7 + 6i) = 7 + 6i - 28i - 24i^2$

$\qquad\qquad\qquad = 7 + 6i - 28i + 24$

$\qquad\qquad\qquad = 31 - 22i$

79. $2x^2 - 13x - 7 = 0$ Setting $f(x) = 0$

$(2x + 1)(x - 7) = 0$

$2x + 1 = 0$ or $x - 7 = 0$

$2x = -1$ or $x = 7$

$x = -\dfrac{1}{2}$ or $x = 7$

The zeros of the function are $-\dfrac{1}{2}$ and 7, and the x-intercepts are $\left(-\dfrac{1}{2}, 0\right)$ and $(7, 0)$.

81. $\qquad x^4 - x^2 = 0$ Setting $h(x) = 0$

$\qquad x^2(x^2 - 1) = 0$

$x^2(x + 1)(x - 1) = 0$

$x^2 = 0$ or $x + 1 = 0$ or $x - 1 = 0$

$x = 0$ or $x = -1$ or $x = 1$

The zeros of the function are 0, -1, and 1, and the x-intercepts are $(0, 0)$, $(-1, 0)$, and $(1, 0)$.

83. $\qquad x^3 + 6x^2 - 16x = 0$

$\qquad x(x^2 + 6x - 16) = 0$

$\qquad x(x + 8)(x - 2) = 0$

$x = 0$ or $x + 8 = 0$ or $x - 2 = 0$

$x = 0$ or $\quad x = -8$ or $\quad x = 2$

The solutions are 0, -8, and 2.

85. $7^\pi \approx 451.8078726$ and $\pi^7 \approx 3020.293228$, so π^7 is larger.

$70^{80} \approx 4.054 \times 10^{147}$ and $80^{70} \approx 1.646 \times 10^{133}$, so 70^{80} is larger.

Exercise Set 5.3

1. Graph $x = 3^y$.

Choose values for y and compute the corresponding x-values. Plot the points (x, y) and connect them with a smooth curve.

x	y	(x, y)
$\dfrac{1}{27}$	-3	$\left(\dfrac{1}{27}, -3\right)$
$\dfrac{1}{9}$	-2	$\left(\dfrac{1}{9}, -2\right)$
$\dfrac{1}{3}$	-1	$\left(\dfrac{1}{3}, -1\right)$
1	0	$(1, 0)$
3	1	$(3, 1)$
9	2	$(9, 2)$
27	3	$(27, 3)$

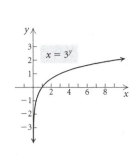

3. Graph $x = \left(\dfrac{1}{2}\right)^y$.

Choose values for y and compute the corresponding x-values. Plot the points (x, y) and connect them with a smooth curve.

x	y	(x, y)
8	-3	$(8, -3)$
4	-2	$(4, -2)$
2	-1	$(2, -1)$
1	0	$(1, 0)$
$\dfrac{1}{2}$	1	$\left(\dfrac{1}{2}, 1\right)$
$\dfrac{1}{4}$	2	$\left(\dfrac{1}{4}, 2\right)$
$\dfrac{1}{8}$	3	$\left(\dfrac{1}{8}, 3\right)$

5. Graph $y = \log_3 x$.

The equation $y = \log_3 x$ is equivalent to $x = 3^y$. We can find ordered pairs that are solutions by choosing values for y and computing the corresponding x-values.

For $y = -2$, $x = 3^{-2} = \frac{1}{9}$.

For $y = -1$, $x = 3^{-1} = \frac{1}{3}$.

For $y = 0$, $x = 3^0 = 1$.

For $y = 1$, $x = 3^1 = 3$.

For $y = 2$, $x = 3^2 = 9$.

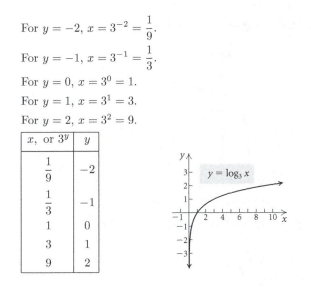

x, or 3^y	y
$\frac{1}{9}$	-2
$\frac{1}{3}$	-1
1	0
3	1
9	2

7. Graph $f(x) = \log x$.

Think of $f(x)$ as y. The equation $y = \log x$ is equivalent to $x = 10^y$. We can find ordered pairs that are solutions by choosing values for y and computing the corresponding x-values.

For $y = -2$, $x = 10^{-2} = 0.01$.

For $y = -1$, $x = 10^{-1} = 0.1$.

For $y = 0$, $x = 10^0 = 1$.

For $y = 1$, $x = 10^1 = 10$.

For $y = 2$, $x = 10^2 = 100$.

x, or 10^y	y
0.01	-2
0.1	-1
1	0
10	1
100	2

9. $\log_2 16 = 4$ because the exponent to which we raise 2 to get 16 is 4.

11. $\log_5 125 = 3$, because the exponent to which we raise 5 to get 125 is 3.

13. $\log 0.001 = -3$, because the exponent to which we raise 10 to get 0.001 is -3.

15. $\log_2 \frac{1}{4} = -2$, because the exponent to which we raise 2 to get $\frac{1}{4}$ is -2.

17. $\ln 1 = 0$, because the exponent to which we raise e to get 1 is 0.

19. $\log 10 = 1$, because the exponent to which we raise 10 to get 10 is 1.

21. $\log_5 5^4 = 4$, because the exponent to which we raise 5 to get 5^4 is 4.

23. $\log_3 \sqrt[4]{3} = \log_3 3^{1/4} = \frac{1}{4}$, because the exponent to which we raise 3 to get $3^{1/4}$ is $\frac{1}{4}$.

25. $\log 10^{-7} = -7$, because the exponent to which we raise 10 to get 10^{-7} is -7.

27. $\log_{49} 7 = \frac{1}{2}$, because the exponent to which we raise 49 to get 7 is $\frac{1}{2}$. ($49^{1/2} = \sqrt{49} = 7$)

29. $\ln e^{3/4} = \frac{3}{4}$, because the exponent to which we raise e to get $e^{3/4}$ is $\frac{3}{4}$.

31. $\log_4 1 = 0$, because the exponent to which we raise 4 to get 1 is 0.

33. $\ln \sqrt{e} = \ln e^{1/2} = \frac{1}{2}$, because the exponent to which we raise e to get $e^{1/2}$ is $\frac{1}{2}$.

35. The exponent is the logarithm.

$$10^3 = 1000 \Rightarrow 3 = \log_{10} 1000$$

The base remains the same.

We could also say $3 = \log 1000$.

37. The exponent is the logarithm.

$$8^{1/3} = 2 \Rightarrow \log_8 2 = \frac{1}{3}$$

The base remains the same.

39. $e^3 = t \Rightarrow \log_e t = 3$, or $\ln t = 3$

41. $e^2 = 7.3891 \Rightarrow \log_e 7.3891 = 2$, or $\ln 7.3891 = 2$

43. $p^k = 3 \Rightarrow \log_p 3 = k$

45. The logarithm is the exponent.

$$\log_5 5 = 1 \Rightarrow 5^1 = 5$$

The base remains the same.

47. $\log 0.01 = -2$ is equivalent to $\log_{10} 0.01 = -2$.

The logarithm is the exponent.

$$\log_{10} 0.01 = -2 \Rightarrow 10^{-2} = 0.01$$

The base remains the same.

49. $\ln 30 = 3.4012 \Rightarrow e^{3.4012} = 30$

51. $\log_a M = -x \Rightarrow a^{-x} = M$

53. $\log_a T^3 = x \Rightarrow a^x = T^3$

55. $\log 3 \approx 0.4771$

57. $\log 532 \approx 2.7259$

59. $\log 0.57 \approx -0.2441$

61. $\log(-2)$ does not exist. (The calculator gives an error message.)

63. $\ln 2 \approx 0.6931$

65. $\ln 809.3 \approx 6.6962$

67. $\ln(-1.32)$ does not exist. (The calculator gives an error message.)

69. Let $a = 10$, $b = 4$, and $M = 100$ and substitute in the change-of-base formula.
$$\log_4 100 = \frac{\log_{10} 100}{\log_{10} 4} \approx 3.3219$$

71. Let $a = 10$, $b = 100$, and $M = 0.3$ and substitute in the change-of-base formula.
$$\log_{100} 0.3 = \frac{\log_{10} 0.3}{\log_{10} 100} \approx -0.2614$$

73. Let $a = 10$, $b = 200$, and $M = 50$ and substitute in the change-of-base formula.
$$\log_{200} 50 = \frac{\log_{10} 50}{\log_{10} 200} \approx 0.7384$$

75. Let $a = e$, $b = 3$, and $M = 12$ and substitute in the change-of-base formula.
$$\log_3 12 = \frac{\ln 12}{\ln 3} \approx 2.2619$$

77. Let $a = e$, $b = 100$, and $M = 15$ and substitute in the change-of-base formula.
$$\log_{100} 15 = \frac{\ln 15}{\ln 100} \approx 0.5880$$

79. Graph $y = 3^x$ and then reflect this graph across the line $y = x$ to get the graph of $y = \log_3 x$.

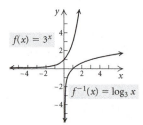

81. Graph $y = \log x$ and then reflect this graph across the line $y = x$ to get the graph of $y = 10^x$.

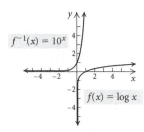

83. Shift the graph of $y = \log_2 x$ left 3 units.

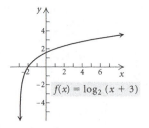

Domain: $(-3, \infty)$

Vertical asymptote: $x = -3$

85. Shift the graph of $y = \log_3 x$ down 1 unit.

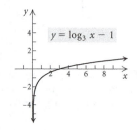

Domain: $(0, \infty)$

Vertical asymptote: $x = 0$

87. Stretch the graph of $y = \ln x$ vertically.

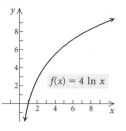

Domain: $(0, \infty)$

Vertical asymptote: $x = 0$

89. Reflect the graph of $y = \ln x$ across the x-axis and then shift it up 2 units.

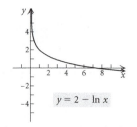

Domain: $(0, \infty)$

Vertical asymptote: $x = 0$

91. Shift the graph of $y = \log x$ right 1 unit, shrink it vertically, and shift it down 2 units.

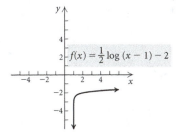

93. Graph $g(x) = 5$ for $x \le 0$ and $g(x) = \log x + 1$ for $x > 0$.

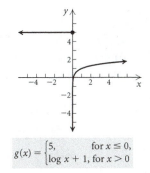

$$g(x) = \begin{cases} 5, & \text{for } x \le 0, \\ \log x + 1, & \text{for } x > 0 \end{cases}$$

95. a) We substitute 672.538 for P, since P is in thousands.

$$w(672.538) = 0.37 \ln 672.538 + 0.05$$
$$\approx 2.5 \text{ ft/sec}$$

b) We substitute 1488.750 for P, since P is in thousands.

$$w(1488.750) = 0.37 \ln 1488.750 + 0.05$$
$$\approx 2.8 \text{ ft/sec}$$

c) We substitute 212.038 for P, since P is in thousands.

$$w(212.038) = 0.37 \ln 212.038 + 0.05$$
$$\approx 2.0 \text{ ft/sec}$$

d) We substitute 598.916 for P, since P is in thousands.

$$w(598.916) = 0.37 \ln 598.916 + 0.05$$
$$\approx 2.4 \text{ ft/sec}$$

e) We substitute 345.610 for P, since P is in thousands.

$$w(345.610) = 0.37 \ln 345.610 + 0.05$$
$$\approx 2.2 \text{ ft/sec}$$

f) We substitute 775.202 for P, since P is in thousands.

$$w(775.202) = 0.37 \ln 775.202 + 0.05$$
$$\approx 2.5 \text{ ft/sec}$$

g) We substitute 421.570 for P, since P is in thousands.

$$w(421.570) = 0.37 \ln 421.570 + 0.05$$
$$\approx 2.3 \text{ ft/sec}$$

h) We substitute 3908.643 for P, since P is in thousands.

$$w(3908.643) = 0.37 \ln 3908.643 + 0.05$$
$$\approx 3.1 \text{ ft/sec}$$

97. a) $R = \log \dfrac{10^{7.7} \cdot I_0}{I_0} = \log 10^{7.7} = 7.7$

b) $R = \log \dfrac{10^{9.5} \cdot I_0}{I_0} = \log 10^{9.5} = 9.5$

c) $R = \log \dfrac{10^{6.6} \cdot I_0}{I_0} = \log 10^{6.6} = 6.6$

d) $R = \log \dfrac{10^{7.6} \cdot I_0}{I_0} = \log 10^{7.6} = 7.6$

e) $R = \log \dfrac{10^{8.0} \cdot I_0}{I_0} = \log 10^{8.0} = 8.0$

f) $R = \log \dfrac{10^{7.9} \cdot I_0}{I_0} = \log 10^{7.9} = 7.9$

g) $R = \log \dfrac{10^{5.1} \cdot I_0}{I_0} = \log 10^{5.1} = 5.1$

h) $R = \log \dfrac{10^{9.3} \cdot I_0}{I_0} = \log 10^{9.3} = 9.3$

99. a)
$$7 = -\log[\text{H}^+]$$
$$-7 = \log[\text{H}^+]$$
$$\text{H}^+ = 10^{-7} \qquad \text{Using the definition of logarithm}$$

b)
$$5.4 = -\log[\text{H}^+]$$
$$-5.4 = \log[\text{H}^+]$$
$$\text{H}^+ = 10^{-5.4} \qquad \text{Using the definition of logarithm}$$
$$\text{H}^+ \approx 4.0 \times 10^{-6}$$

c)
$$3.2 = -\log[\text{H}^+]$$
$$-3.2 = \log[\text{H}^+]$$
$$\text{H}^+ = 10^{-3.2} \qquad \text{Using the definition of logarithm}$$
$$\text{H}^+ \approx 6.3 \times 10^{-4}$$

d)
$$4.8 = -\log[\text{H}^+]$$
$$-4.8 = \log[\text{H}^+]$$
$$\text{H}^+ = 10^{-4.8} \qquad \text{Using the definition of logarithm}$$
$$\text{H}^+ \approx 1.6 \times 10^{-5}$$

101. a) $L = 10 \log \dfrac{10^{14} \cdot I_0}{I_0}$
$$= 10 \log 10^{14} = 10 \cdot 14$$
$$\approx 140 \text{ decibels}$$

b) $L = 10 \log \dfrac{10^{11.5} \cdot I_0}{I_0}$
$$= 10 \log 10^{11.5} = 10 \cdot 11.5$$
$$\approx 115 \text{ decibels}$$

c) $L = 10 \log \dfrac{10^{4} \cdot I_0}{I_0}$
$$= 10 \log 10^{4} = 10 \cdot 4$$
$$= 40 \text{ decibels}$$

d) $L = 10 \log \dfrac{10^{6.5} \cdot I_0}{I_0}$

$= 10 \log 10^{6.5} = 10 \cdot 6.5$

$= 65$ decibels

e) $L = 10 \log \dfrac{10^{12} \cdot I_0}{I_0}$

$= 10 \log 10^{12} = 10 \cdot 12$

$= 120$ decibels

f) $L = 10 \log \dfrac{10^{19.4} \cdot I_0}{I_0}$

$= 10 \log 10^{19.4} = 10 \cdot 19.4$

$= 194$ decibels

103. $y = 6 = 0 \cdot x + 6$

Slope: 0; y-intercept $(0, 6)$

105.

$$
\begin{array}{r|rrrr}
-5 & 1 & -6 & 3 & 10 \\
   &   & -5 & 55 & -290 \\
\hline
   & 1 & -11 & 58 & -280
\end{array}
$$

The remainder is -280, so $f(-5) = -280$.

107. $f(x) = (x - \sqrt{7})(x + \sqrt{7})(x - 0)$

$= (x^2 - 7)(x)$

$= x^3 - 7x$

109. Using the change-of-base formula, we get

$\dfrac{\log_5 8}{\log_2 8} = \log_2 8 = 3.$

111. $f(x) = \log_5 x^3$

x^3 must be positive. Since $x^3 > 0$ for $x > 0$, the domain is $(0, \infty)$.

113. $f(x) = \ln |x|$

$|x|$ must be positive. Since $|x| > 0$ for $x \neq 0$, the domain is $(-\infty, 0) \cup (0, \infty)$.

115. Graph $y = \log_2(2x + 5) = \dfrac{\log(2x + 5)}{\log 2}$. Observe

that outputs are negative for inputs between $-\dfrac{5}{2}$ and -2.

Thus, the solution set is $\left(-\dfrac{5}{2}, -2 \right)$.

117. Graph (d) is the graph of $f(x) = \ln |x|$.

119. Graph (b) is the graph of $f(x) = \ln x^2$.

Chapter 5 Mid-Chapter Mixed Review

1. The statement is false. The domain of $y = \log x$, for instance, is $(0, \infty)$.

3. $f(0) = e^{-0} = 1$, so the y-intercept is $(0, 1)$. The given statement is false.

5. The graph of $f(x) = 3 + x^2$ is shown below. Since there are many horizontal lines that cross the graph more than once, the function is not one-to-one and thus does not have an inverse that is a function.

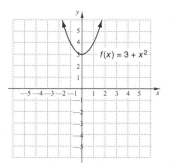

7. $(f^{-1} \circ f)(x) = f^{-1}(f(x)) = (\sqrt{x - 5})^2 + 5 = x - 5 + 5 = x$

$(f \circ f^{-1})(x) = f(f^{-1}(x)) = \sqrt{x^2 + 5 - 5} = \sqrt{x^2} = x$

(This assumes that the domain of $f^{-1}(x)$ is restricted to $\{x | x \geq 0\}$.)

Since $(f^{-1} \circ f)(x) = x = (f \circ f^{-1})(x)$, we know that $f^{-1}(x) = x^2 + 5$.

9. The graph of $y = \log_2 x$ is (d).

11. The graph of $f(x) = e^{x-1}$ is (c).

13. The graph of $f(x) = \ln(x - 2)$ is (b).

15. The graph of $f(x) = |\log x|$ is (e).

17. $A = P\left(1 + \dfrac{r}{n}\right)^{nt}$

$A = 3200\left(1 + \dfrac{0.045}{4}\right)^{4 \cdot 6} \approx \$4185.57$

19. $\ln e^{-4/5}$ is $-\dfrac{4}{5}$ because the exponent to which we raise e to get $e^{-4/5}$ is $-\dfrac{4}{5}$.

21. $\ln e^2 = 2$ because the exponent to which we raise e to get e^2 is 2.

23. $\log_2 \dfrac{1}{16} = -4$ because the exponent to which we raise 2 to get $\dfrac{1}{16}$, or 2^{-4}, is -4.

25. $\log_3 27 = 3$ because the exponent to which we raise 3 to get 27 is 3.

27. $\ln e = 1$ because the exponent to which we raise e to get e is 1.

29. $\log T = r$ is equivalent to $10^r = T$.

31. $\log_\pi 10 = \dfrac{\log 10}{\log \pi} = \dfrac{1}{\log \pi} \approx 2.0115$

33. The most interest will be earned the eighth year, because the principle is greatest during that year.

35. If $\log b < 0$, then $b < 1$.

Exercise Set 5.4

1. Use the product rule.

$\log_3(81 \cdot 27) = \log_3 81 + \log_3 27 = 4 + 3 = 7$

3. Use the product rule.

$\log_5(5 \cdot 125) = \log_5 5 + \log_5 125 = 1 + 3 = 4$

5. Use the product rule.

$\log_t 8Y = \log_t 8 + \log_t Y$

7. Use the product rule.

$\ln xy = \ln x + \ln y$

9. Use the power rule.

$\log_b t^3 = 3 \log_b t$

11. Use the power rule.

$\log y^8 = 8 \log y$

13. Use the power rule.

$\log_c K^{-6} = -6 \log_c K$

15. Use the power rule.

$\ln \sqrt[3]{4} = \ln 4^{1/3} = \dfrac{1}{3} \ln 4$

17. Use the quotient rule.

$\log_t \dfrac{M}{8} = \log_t M - \log_t 8$

19. Use the quotient rule.

$\log \dfrac{x}{y} = \log x - \log y$

21. Use the quotient rule.

$\ln \dfrac{r}{s} = \ln r - \ln s$

23.
$$\log_a 6xy^5z^4$$
$$= \log_a 6 + \log_a x + \log_a y^5 + \log_a z^4$$
$$\text{Product rule}$$
$$= \log_a 6 + \log_a x + 5 \log_a y + 4 \log_a z$$
$$\text{Power rule}$$

25.
$$\log_b \dfrac{p^2 q^5}{m^4 b^9}$$
$$= \log_b p^2 q^5 - \log_b m^4 b^9 \quad \text{Quotient rule}$$
$$= \log_b p^2 + \log_b q^5 - (\log_b m^4 + \log_b b^9)$$
$$\text{Product rule}$$
$$= \log_b p^2 + \log_b q^5 - \log_b m^4 - \log_b b^9$$
$$= \log_b p^2 + \log_b q^5 - \log_b m^4 - 9 \quad (\log_b b^9 = 9)$$
$$= 2 \log_b p + 5 \log_b q - 4 \log_b m - 9 \quad \text{Power rule}$$

27.
$$\ln \dfrac{2}{3x^3 y}$$
$$= \ln 2 - \ln 3x^3 y \qquad \text{Quotient rule}$$
$$= \ln 2 - (\ln 3 + \ln x^3 + \ln y) \quad \text{Product rule}$$
$$= \ln 2 - \ln 3 - \ln x^3 - \ln y$$
$$= \ln 2 - \ln 3 - 3 \ln x - \ln y \qquad \text{Power rule}$$

29.
$$\log \sqrt{r^3 t}$$
$$= \log(r^3 t)^{1/2}$$
$$= \dfrac{1}{2} \log r^3 t \qquad \text{Power rule}$$
$$= \dfrac{1}{2}(\log r^3 + \log t) \quad \text{Product rule}$$
$$= \dfrac{1}{2}(3 \log r + \log t) \quad \text{Power rule}$$
$$= \dfrac{3}{2} \log r + \dfrac{1}{2} \log t$$

31.
$$\log_a \sqrt{\dfrac{x^6}{p^5 q^8}}$$
$$= \dfrac{1}{2} \log_a \dfrac{x^6}{p^5 q^8}$$
$$= \dfrac{1}{2}[\log_a x^6 - \log_a(p^5 q^8)] \qquad \text{Quotient rule}$$
$$= \dfrac{1}{2}[\log_a x^6 - (\log_a p^5 + \log_a q^8)] \quad \text{Product rule}$$
$$= \dfrac{1}{2}(\log_a x^6 - \log_a p^5 - \log_a q^8)$$
$$= \dfrac{1}{2}(6 \log_a x - 5 \log_a p - 8 \log_a q) \quad \text{Power rule}$$
$$= 3 \log_a x - \dfrac{5}{2} \log_a p - 4 \log_a q$$

33.
$$\log_a \sqrt[4]{\dfrac{m^8 n^{12}}{a^3 b^5}}$$
$$= \dfrac{1}{4} \log_a \dfrac{m^8 n^{12}}{a^3 b^5} \qquad \text{Power rule}$$
$$= \dfrac{1}{4}(\log_a m^8 n^{12} - \log_a a^3 b^5) \quad \text{Quotient rule}$$
$$= \dfrac{1}{4}[\log_a m^8 + \log_a n^{12} - (\log_a a^3 + \log_a b^5)]$$
$$\text{Product rule}$$
$$= \dfrac{1}{4}(\log_a m^8 + \log_a n^{12} - \log_a a^3 - \log_a b^5)$$
$$= \dfrac{1}{4}(\log_a m^8 + \log_a n^{12} - 3 - \log_a b^5)$$
$$(\log_a a^3 = 3)$$
$$= \dfrac{1}{4}(8 \log_a m + 12 \log_a n - 3 - 5 \log_a b)$$
$$\text{Power rule}$$
$$= 2 \log_a m + 3 \log_a n - \dfrac{3}{4} - \dfrac{5}{4} \log_a b$$

35.
$$\log_a 75 + \log_a 2$$
$$= \log_a(75 \cdot 2) \qquad \text{Product rule}$$
$$= \log_a 150$$

37. $\log 10,000 - \log 100$

$= \log \dfrac{10,000}{100}$ Quotient rule

$= \log 100$

$= 2$

39. $\dfrac{1}{2} \log n + 3 \log m$

$= \log n^{1/2} + \log m^3$ Power rule

$= \log n^{1/2} m^3,$ or   Product rule

    $\log m^3 \sqrt{n}$ $n^{1/2} = \sqrt{n}$

41. $\dfrac{1}{2} \log_a x + 4 \log_a y - 3 \log_a x$

$= \log_a x^{1/2} + \log_a y^4 - \log_a x^3$ Power rule

$= \log_a x^{1/2} y^4 - \log_a x^3$ Product rule

$= \log_a \dfrac{x^{1/2} y^4}{x^3}$ Quotient rule

$= \log_a x^{-5/2} y^4,$ or $\log_a \dfrac{y^4}{x^{5/2}}$ Simplifying

43. $\ln x^2 - 2 \ln \sqrt{x}$

$= \ln x^2 - \ln \left(\sqrt{x}\right)^2$ Power rule

$= \ln x^2 - \ln x$ $[(\sqrt{x})^2 = x]$

$= \ln \dfrac{x^2}{x}$ Quotient rule

$= \ln x$

45. $\ln(x^2 - 4) - \ln(x + 2)$

$= \ln \dfrac{x^2 - 4}{x + 2}$ Quotient rule

$= \ln \dfrac{(x + 2)(x - 2)}{x + 2}$ Factoring

$= \ln(x - 2)$ Removing a factor of 1

47. $\log(x^2 - 5x - 14) - \log(x^2 - 4)$

$= \log \dfrac{x^2 - 5x - 14}{x^2 - 4}$ Quotient rule

$= \log \dfrac{(x + 2)(x - 7)}{(x + 2)(x - 2)}$ Factoring

$= \log \dfrac{x - 7}{x - 2}$ Removing a factor of 1

49. $\ln x - 3[\ln(x - 5) + \ln(x + 5)]$

$= \ln x - 3 \ln[(x - 5)(x + 5)]$ Product rule

$= \ln x - 3 \ln(x^2 - 25)$

$= \ln x - \ln(x^2 - 25)^3$ Power rule

$= \ln \dfrac{x}{(x^2 - 25)^3}$ Quotient rule

51. $\dfrac{3}{2} \ln 4x^6 - \dfrac{4}{5} \ln 2y^{10}$

$= \dfrac{3}{2} \ln 2^2 x^6 - \dfrac{4}{5} \ln 2y^{10}$ Writing 4 as 2^2

$= \ln(2^2 x^6)^{3/2} - \ln(2y^{10})^{4/5}$ Power rule

$= \ln(2^3 x^9) - \ln(2^{4/5} y^8)$

$= \ln \dfrac{2^3 x^9}{2^{4/5} y^8}$ Quotient rule

$= \ln \dfrac{2^{11/5} x^9}{y^8}$

53. $\log_a \dfrac{2}{11} = \log_a 2 - \log_a 11$ Quotient rule

$\approx 0.301 - 1.041$

≈ -0.74

55. $\log_a 98 = \log_a(7^2 \cdot 2)$

$= \log_a 7^2 + \log_a 2$ Product rule

$= 2 \log_a 7 + \log_a 2$ Power rule

$\approx 2(0.845) + 0.301$

≈ 1.991

57. $\dfrac{\log_a 2}{\log_a 7} \approx \dfrac{0.301}{0.845} \approx 0.356$

59. $\log_b 125 = \log_b 5^3$

$= 3 \log_b 5$ Power rule

$\approx 3(1.609)$

≈ 4.827

61. $\log_b \dfrac{1}{6} = \log_b 1 - \log_b 6$ Quotient rule

$= \log_b 1 - \log_b(2 \cdot 3)$

$= \log_b 1 - (\log_b 2 + \log_b 3)$ Product rule

$= \log_b 1 - \log_b 2 - \log_b 3$

$\approx 0 - 0.693 - 1.099$

≈ -1.792

63. $\log_b \dfrac{3}{b} = \log_b 3 - \log_b b$ Quotient rule

$\approx 1.099 - 1$

≈ 0.099

65. $\log_p p^3 = 3$ $(\log_a a^x = x)$

67. $\log_e e^{|x-4|} = |x - 4|$ $(\log_a a^x = x)$

69. $3^{\log_3 4x} = 4x$ $(a^{\log_a x} = x)$

71. $10^{\log w} = w$ $(a^{\log_a x} = x)$

73. $\ln e^{8t} = 8t$ $(\log_a a^x = x)$

75. $\log_b \sqrt{b} = \log_b b^{1/2}$

$\qquad = \dfrac{1}{2} \log_b b \quad$ Power rule

$\qquad = \dfrac{1}{2} \cdot 1 \qquad (\log_b b = 1)$

$\qquad = \dfrac{1}{2}$

77. The degree of $f(x) = 5 - x^2 + x^4$ is 4, so the function is quartic.

79. $f(x) = -\dfrac{3}{4}$ is of the form $f(x) = mx + b$ $\left(\text{with } m = 0 \text{ and } b = -\dfrac{3}{4}\right)$, so it is a linear function. In fact, it is a constant function.

81. $f(x) = -\dfrac{3}{x}$ is of the form $f(x) = \dfrac{p(x)}{q(x)}$ where $p(x)$ and $q(x)$ are polynomials and $q(x)$ is not the zero polynomial, so $f(x)$ is a rational function.

83. The degree of $f(x) = -\dfrac{1}{3}x^3 - 4x^2 + 6x + 42$ is 3, so the function is cubic.

85. $f(x) = \dfrac{1}{2}x + 3$ is of the form $f(x) = mx + b$, so it is a linear function.

87. $5^{\log_5 8} = 2x$

$\qquad 8 = 2x \quad (a^{\log_a x} = x)$

$\qquad 4 = x$

The solution is 4.

89. $\log_a(x^2 + xy + y^2) + \log_a(x - y)$

$= \log_a[(x^2 + xy + y^2)(x - y)] \quad$ Product rule

$= \log_a(x^3 - y^3) \qquad\qquad$ Multiplying

91. $\log_a \dfrac{x - y}{\sqrt{x^2 - y^2}}$

$= \log_a \dfrac{x - y}{(x^2 - y^2)^{1/2}}$

$= \log_a(x - y) - \log_a(x^2 - y^2)^{1/2} \quad$ Quotient rule

$= \log_a(x - y) - \dfrac{1}{2}\log_a(x^2 - y^2) \quad$ Power rule

$= \log_a(x - y) - \dfrac{1}{2}\log_a[(x + y)(x - y)]$

$= \log_a(x - y) - \dfrac{1}{2}[\log_a(x + y) + \log_a(x - y)]$

$\qquad\qquad\qquad\qquad\qquad\qquad$ Product rule

$= \log_a(x - y) - \dfrac{1}{2}\log_a(x + y) - \dfrac{1}{2}\log_a(x - y)$

$= \dfrac{1}{2}\log_a(x - y) - \dfrac{1}{2}\log_a(x + y)$

93. $\log_a \dfrac{\sqrt[4]{y^2 z^5}}{\sqrt[4]{x^3 z^{-2}}}$

$= \log_a \sqrt[4]{\dfrac{y^2 z^5}{x^3 z^{-2}}}$

$= \log_a \sqrt[4]{\dfrac{y^2 z^7}{x^3}}$

$= \log_a \left(\dfrac{y^2 z^7}{x^3}\right)^{1/4}$

$= \dfrac{1}{4}\log_a \left(\dfrac{y^2 z^7}{x^3}\right) \qquad\qquad$ Power rule

$= \dfrac{1}{4}(\log_a y^2 z^7 - \log_a x^3) \qquad$ Quotient rule

$= \dfrac{1}{4}(\log_a y^2 + \log_a z^7 - \log_a x^3) \quad$ Product rule

$= \dfrac{1}{4}(2\log_a y + 7\log_a z - 3\log_a x) \quad$ Power rule

$= \dfrac{1}{4}(2 \cdot 3 + 7 \cdot 4 - 3 \cdot 2)$

$= \dfrac{1}{4} \cdot 28$

$= 7$

95. $\log_a M - \log_a N = \log_a \dfrac{M}{N}$

This is the quotient rule, so it is true.

97. $\dfrac{\log_a M}{x} = \dfrac{1}{x}\log_a M = \log_a M^{1/x}$. The statement is true by the power rule.

99. $\log_a 8x = \log_a 8 + \log_a x = \log_a x + \log_a 8$. The statement is true by the product rule and the commutative property of addition.

101. $\log_a \left(\dfrac{1}{x}\right) = \log_a x^{-1} = -1 \cdot \log_a x = -1 \cdot 2 = -2$

103. We use the change-of-base formula.

$\log_{10} 11 \cdot \log_{11} 12 \cdot \log_{12} 13 \cdots$

$\qquad\qquad \log_{998} 999 \cdot \log_{999} 1000$

$= \log_{10} 11 \cdot \dfrac{\log_{10} 12}{\log_{10} 11} \cdot \dfrac{\log_{10} 13}{\log_{10} 12} \cdots$

$\qquad\qquad \dfrac{\log_{10} 999}{\log_{10} 998} \cdot \dfrac{\log_{10} 1000}{\log_{10} 999}$

$= \dfrac{\log_{10} 11}{\log_{10} 11} \cdot \dfrac{\log_{10} 12}{\log_{10} 12} \cdots \dfrac{\log_{10} 999}{\log_{10} 999} \cdot \log_{10} 1000$

$= \log_{10} 1000$

$= 3$

105. $\ln a - \ln b + xy = 0$

$\qquad \ln a - \ln b = -xy$

$\qquad\qquad \ln \dfrac{a}{b} = -xy$

Then, using the definition of a logarithm, we have $e^{-xy} = \dfrac{a}{b}$.

107.

$$\log_a\left(\frac{x+\sqrt{x^2-5}}{5}\right)$$

$$=\log_a\left(\frac{x+\sqrt{x^2-5}}{5}\cdot\frac{x-\sqrt{x^2-5}}{x-\sqrt{x^2-5}}\right)$$

$$=\log_a\left(\frac{5}{5(x-\sqrt{x^2-5})}\right)=\log_a\left(\frac{1}{x-\sqrt{x^2-5}}\right)$$

$$=\log_a 1-\log_a(x-\sqrt{x^2-5})$$

$$=-\log_a(x-\sqrt{x^2-5})$$

Exercise Set 5.5

1. $3^x=81$

$3^x=3^4$

$x=4$ The exponents are the same.

The solution is 4.

3. $2^{2x}=8$

$2^{2x}=2^3$

$2x=3$ The exponents are the same.

$x=\dfrac{3}{2}$

The solution is $\dfrac{3}{2}$.

5. $2^x=33$

$\log 2^x=\log 33$ Taking the common logarithm on both sides

$x\log 2=\log 33$ Power rule

$x=\dfrac{\log 33}{\log 2}$

$x\approx\dfrac{1.5185}{0.3010}$

$x\approx 5.044$

The solution is 5.044.

7. $5^{4x-7}=125$

$5^{4x-7}=5^3$

$4x-7=3$

$4x=10$

$x=\dfrac{10}{4}=\dfrac{5}{2}$

The solution is $\dfrac{5}{2}$.

9. $27=3^{5x}\cdot 9^{x^2}$

$3^3=3^{5x}\cdot(3^2)^{x^2}$

$3^3=3^{5x}\cdot 3^{2x^2}$

$3^3=3^{5x+2x^2}$

$3=5x+2x^2$

$0=2x^2+5x-3$

$0=(2x-1)(x+3)$

$x=\dfrac{1}{2}$ or $x=-3$

The solutions are -3 and $\dfrac{1}{2}$.

11. $84^x=70$

$\log 84^x=\log 70$

$x\log 84=\log 70$

$x=\dfrac{\log 70}{\log 84}$

$x\approx\dfrac{1.8451}{1.9243}$

$x\approx 0.959$

The solution is 0.959.

13. $10^{-x}=5^{2x}$

$\log 10^{-x}=\log 5^{2x}$

$-x=2x\log 5$

$0=x+2x\log 5$

$0=x(1+2\log 5)$

$0=x$ Dividing by $1+2\log 5$

The solution is 0.

15. $e^{-c}=5^{2c}$

$\ln e^{-c}=\ln 5^{2c}$

$-c=2c\ln 5$

$0=c+2c\ln 5$

$0=c(1+2\ln 5)$

$0=c$ Dividing by $1+2\ln 5$

The solution is 0.

17. $e^t=1000$

$\ln e^t=\ln 1000$

$t=\ln 1000$ Using $\log_a a^x=x$

$t\approx 6.908$

The solution is 6.908.

19. $e^{-0.03t}=0.08$

$\ln e^{-0.03t}=\ln 0.08$

$-0.03t=\ln 0.08$

$t=\dfrac{\ln 0.08}{-0.03}$

$t\approx\dfrac{-2.5257}{-0.03}$

$t\approx 84.191$

The solution is 84.191.

21.
$$3^x = 2^{x-1}$$
$$\ln 3^x = \ln 2^{x-1}$$
$$x \ln 3 = (x-1)\ln 2$$
$$x \ln 3 = x \ln 2 - \ln 2$$
$$\ln 2 = x \ln 2 - x \ln 3$$
$$\ln 2 = x(\ln 2 - \ln 3)$$
$$\frac{\ln 2}{\ln 2 - \ln 3} = x$$
$$\frac{0.6931}{0.6931 - 1.0986} \approx x$$
$$-1.710 \approx x$$

The solution is -1.710.

23.
$$(3.9)^x = 48$$
$$\log(3.9)^x = \log 48$$
$$x \log 3.9 = \log 48$$
$$x = \frac{\log 48}{\log 3.9}$$
$$x \approx \frac{1.6812}{0.5911}$$
$$x \approx 2.844$$

The solution is 2.844.

25.
$$e^x + e^{-x} = 5$$
$$e^{2x} + 1 = 5e^x \qquad \text{Multiplying by } e^x$$
$$e^{2x} - 5e^x + 1 = 0 \qquad \text{This equation is quadratic in } e^x.$$
$$e^x = \frac{5 \pm \sqrt{21}}{2}$$
$$x = \ln\left(\frac{5 \pm \sqrt{21}}{2}\right) \approx \pm 1.567$$

The solutions are -1.567 and 1.567.

27.
$$3^{2x-1} = 5^x$$
$$\log 3^{2x-1} = \log 5^x$$
$$(2x-1)\log 3 = x \log 5$$
$$2x \log 3 - \log 3 = x \log 5$$
$$-\log 3 = x \log 5 - 2x \log 3$$
$$-\log 3 = x(\log 5 - 2\log 3)$$
$$\frac{-\log 3}{\log 5 - 2\log 3} = x$$
$$1.869 \approx x$$

The solution is 1.869.

29.
$$2e^x = 5 - e^{-x}$$
$$2e^x - 5 + e^{-x} = 0$$
$$e^x(2e^x - 5 + e^{-x}) = e^x \cdot 0 \qquad \text{Multiplying by } e^x$$
$$2e^{2x} - 5e^x + 1 = 0$$

Let $u = e^x$.
$$2u^2 - 5u + 1 = 0 \qquad \text{Substituting}$$
$$a = 2, \; b = -5, \; c = 1$$

$$u = \frac{-b \pm \sqrt{b^2 - 4ac}}{2a}$$
$$u = \frac{-(-5) \pm \sqrt{(-5)^2 - 4 \cdot 2 \cdot 1}}{2 \cdot 2}$$
$$u = \frac{5 \pm \sqrt{17}}{4}$$

Replace u with e^x.

$$e^x = \frac{5 - \sqrt{17}}{4} \qquad or \qquad e^x = \frac{5 + \sqrt{17}}{4}$$
$$\ln e^x = \ln\left(\frac{5 - \sqrt{17}}{4}\right) \quad or \quad \ln e^x = \ln\left(\frac{5 + \sqrt{17}}{4}\right)$$
$$x \approx -1.518 \qquad or \qquad x \approx 0.825$$

The solutions are -1.518 and 0.825.

31.
$$\log_5 x = 4$$
$$x = 5^4 \qquad \text{Writing an equivalent exponential equation}$$
$$x = 625$$

The solution is 625.

33.
$$\log x = -4 \qquad \text{The base is 10.}$$
$$x = 10^{-4}, \text{ or } 0.0001$$

The solution is 0.0001.

35.
$$\ln x = 1 \qquad \text{The base is } e.$$
$$x = e^1 = e$$

The solution is e.

37.
$$\log_{64} \frac{1}{4} = x$$
$$\frac{1}{4} = 64^x$$
$$\frac{1}{4} = (4^3)^x$$
$$4^{-1} = 4^{3x}$$
$$-1 = 3x$$
$$-\frac{1}{3} = x$$

The solution is $-\dfrac{1}{3}$.

39.
$$\log_2(10 + 3x) = 5$$
$$2^5 = 10 + 3x$$
$$32 = 10 + 3x$$
$$22 = 3x$$
$$\frac{22}{3} = x$$

The answer checks. The solution is $\dfrac{22}{3}$.

41.
$$\log x + \log(x - 9) = 1 \qquad \text{The base is 10.}$$
$$\log_{10}[x(x-9)] = 1$$
$$x(x-9) = 10^1$$
$$x^2 - 9x = 10$$
$$x^2 - 9x - 10 = 0$$
$$(x-10)(x+1) = 0$$

$x = 10$ or $x = -1$

Check: For 10:

$$\log x + \log(x - 9) = 1$$

$$\underline{\quad\qquad\qquad\qquad\qquad}$$

$$\log 10 + \log(10 - 9) \ ? \ 1$$
$$\log 10 + \log 1 \ \Big|$$
$$1 + 0 \ \Big|$$
$$1 \ \Big| \ 1 \qquad \text{TRUE}$$

For -1:

$$\log x + \log(x - 9) = 1$$

$$\underline{\quad\qquad\qquad\qquad\qquad}$$

$$\log(-1) + \log(-1 - 9) \ ? \ 1$$

The number -1 does not check, because negative numbers do not have logarithms. The solution is 10.

43. $\log_2(x + 20) - \log_2(x + 2) = \log_2 x$

$$\log_2 \frac{x + 20}{x + 2} = \log_2 x$$

$$\frac{x + 20}{x + 2} = x \quad \begin{array}{l}\text{Using the property of} \\ \text{logarithmic equality}\end{array}$$

$$x + 20 = x^2 + 2x \quad \begin{array}{l}\text{Multiplying by} \\ x + 2\end{array}$$

$$0 = x^2 + x - 20$$

$$0 = (x + 5)(x - 4)$$

$x + 5 = 0$ or $x - 4 = 0$

$x = -5$ or $x = 4$

Check: For -5:

$$\log_2(x + 20) - \log_2(x + 2) = \log_2 x$$

$$\underline{\quad\qquad\qquad\qquad\qquad\qquad}$$

$$\log_2(-5 + 20) - \log_2(-5 + 2) \ ? \ \log_2(-5)$$

The number -5 does not check, because negative numbers do not have logarithms.

For 4:

$$\log_2(x + 20) - \log_2(x + 2) = \log_2 x$$

$$\underline{\quad\qquad\qquad\qquad\qquad\qquad}$$

$$\log_2(4 + 20) - \log_2(4 + 2) \ ? \ \log_2 4$$
$$\log_2 24 - \log_2 6 \ \Big|$$
$$\log_2 \frac{24}{6} \ \Big|$$
$$\log_2 4 \ \Big| \ \log_2 4 \qquad \text{TRUE}$$

The solution is 4.

45. $\log_8(x + 1) - \log_8 x = 2$

$$\log_8 \left(\frac{x + 1}{x} \right) = 2 \quad \text{Quotient rule}$$

$$\frac{x + 1}{x} = 8^2$$

$$\frac{x + 1}{x} = 64$$

$$x + 1 = 64x$$

$$1 = 63x$$

$$\frac{1}{63} = x$$

The answer checks. The solution is $\frac{1}{63}$.

47. $\log x + \log(x + 4) = \log 12$

$$\log x(x + 4) = \log 12$$

$$x(x + 4) = 12 \quad \begin{array}{l}\text{Using the property of} \\ \text{logarithmic equality}\end{array}$$

$$x^2 + 4x = 12$$

$$x^2 + 4x - 12 = 0$$

$$(x + 6)(x - 2) = 0$$

$x + 6 = 0$ or $x - 2 = 0$

$x = -6$ or $x = 2$

Check: For -6:

$$\log x + \log(x + 4) = \log 12$$

$$\underline{\quad\qquad\qquad\qquad\qquad\qquad}$$

$$\log(-6) + \log(-6 + 4) \ ? \ \log 12$$

The number -6 does not check, because negative numbers do not have logarithms.

For 2:

$$\log x + \log(x + 4) = \log 12$$

$$\underline{\quad\qquad\qquad\qquad\qquad\qquad}$$

$$\log 2 + \log(2 + 4) \ ? \ \log 12$$
$$\log 2 + \log 6 \ \Big|$$
$$\log(2 \cdot 6) \ \Big|$$
$$\log 12 \ \Big| \ \log 12 \qquad \text{TRUE}$$

The solution is 2.

49. $\log(x + 8) - \log(x + 1) = \log 6$

$$\log \frac{x + 8}{x + 1} = \log 6 \quad \text{Quotient rule}$$

$$\frac{x + 8}{x + 1} = 6 \quad \begin{array}{l}\text{Using the property of} \\ \text{logarithmic equality}\end{array}$$

$$x + 8 = 6x + 6 \quad \text{Multiplying by } x + 1$$

$$2 = 5x$$

$$\frac{2}{5} = x$$

The answer checks. The solution is $\frac{2}{5}$.

51. $\log_4(x + 3) + \log_4(x - 3) = 2$

$$\log_4[(x + 3)(x - 3)] = 2 \quad \text{Product rule}$$

$$(x + 3)(x - 3) = 4^2$$

$$x^2 - 9 = 16$$

$$x^2 = 25$$

$$x = \pm 5$$

The number 5 checks, but -5 does not. The solution is 5.

53. $\log(2x + 1) - \log(x - 2) = 1$

$$\log \left(\frac{2x + 1}{x - 2} \right) = 1 \quad \text{Quotient rule}$$

$$\frac{2x + 1}{x - 2} = 10^1 = 10$$

$$2x + 1 = 10x - 20$$

$$\text{Multiplying by } x - 2$$

$$21 = 8x$$

$$\frac{21}{8} = x$$

The answer checks. The solution is $\frac{21}{8}$.

55. $\ln(x+8) + \ln(x-1) = 2\ln x$

$\qquad \ln(x+8)(x-1) = \ln x^2$

$\qquad\quad (x+8)(x-1) = x^2 \qquad$ Using the property of logarithmic equality

$\qquad\qquad x^2 + 7x - 8 = x^2$

$\qquad\qquad\qquad 7x - 8 = 0$

$\qquad\qquad\qquad\quad 7x = 8$

$\qquad\qquad\qquad\quad x = \frac{8}{7}$

The answer checks. The solution is $\frac{8}{7}$.

57. $\qquad\qquad \log_6 x = 1 - \log_6(x-5)$

$\log_6 x + \log_6(x-5) = 1$

$\qquad \log_6 x(x-5) = 1$

$\qquad\qquad 6^1 = x(x-5)$

$\qquad\qquad\quad 6 = x^2 - 5x$

$\qquad\qquad\quad 0 = x^2 - 5x - 6$

$\qquad\qquad\quad 0 = (x+1)(x-6)$

$x + 1 = 0 \quad or \quad x - 6 = 0$

$\quad x = -1 \quad or \qquad x = 6$

The number -1 does not check, but 6 does. The answer is 6.

59. $\qquad\qquad 9^{x-1} = 100(3^x)$

$\qquad\quad (3^2)^{x-1} = 100(3^x)$

$\qquad\quad 3^{2x-2} = 100(3^x)$

$\qquad\quad \dfrac{3^{2x-2}}{3^x} = 100$

$\qquad\quad 3^{x-2} = 100$

$\qquad \log\, 3^{x-2} = \log\, 100$

$\quad (x-2)\log 3 = 2$

$\qquad\quad x - 2 = \dfrac{2}{\log 3}$

$\qquad\qquad x = 2 + \dfrac{2}{\log 3}$

$\qquad\qquad x \approx 6.192$

The solution is 6.192.

61. $\qquad e^x - 2 = -e^{-x}$

$\qquad e^x - 2 = -\dfrac{1}{e^x}$

$\quad e^{2x} - 2e^x = -1 \qquad$ Multiplying by e^x

$e^{2x} - 2e^x + 1 = 0$

Let $u = e^x$.

$\quad u^2 - 2u + 1 = 0$

$\quad (u-1)(u-1) = 0$

$u - 1 = 0 \quad or \quad u - 1 = 0$

$\quad u = 1 \quad or \qquad u = 1$

$\quad e^x = 1 \quad or \qquad e^x = 1 \qquad$ Replacing u with e^x

$\quad x = 0 \quad or \qquad x = 0$

The solution is 0.

63. $g(x) = x^2 - 6$

a) $-\dfrac{b}{2a} = -\dfrac{0}{2\cdot 1} = 0$

$\quad g(0) = 0^2 - 6 = -6$

The vertex is $(0, -6)$.

b) The axis of symmetry is $x = 0$.

c) Since the coefficient of the x^2-term is positive, the function has a minimum value. It is the second coordinate of the vertex, -6, and it occurs when $x = 0$.

65. $G(x) = -2x^2 - 4x - 7$

a) $-\dfrac{b}{2a} = -\dfrac{-4}{2(-2)} = -1$

$\quad G(-1) = -2(-1)^2 - 4(-1) - 7 = -5$

The vertex is $(-1, -5)$.

b) The axis of symmetry is $x = -1$.

c) Since the coefficient of the x^2-term is negative, the function has a maximum value. It is the second coordinate of the vertex, -5, and it occurs when $x = -1$.

67. $\dfrac{e^x + e^{-x}}{e^x - e^{-x}} = 3$

$\quad e^x + e^{-x} = 3e^x - 3e^{-x} \qquad$ Multiplying by $e^x - e^{-x}$

$\qquad\qquad 4e^{-x} = 2e^x \qquad$ Subtracting e^x and adding $3e^{-x}$

$\qquad\qquad 2e^{-x} = e^x$

$\qquad\qquad\quad 2 = e^{2x} \qquad$ Multiplying by e^x

$\qquad\quad \ln 2 = \ln e^{2x}$

$\qquad\quad \ln 2 = 2x$

$\qquad\quad \dfrac{\ln 2}{2} = x$

$\qquad 0.347 \approx x$

The solution is 0.347.

69. $\sqrt{\ln x} = \ln \sqrt{x}$

$\quad \sqrt{\ln x} = \dfrac{1}{2}\ln x \qquad\qquad$ Power rule

$\qquad \ln x = \dfrac{1}{4}(\ln x)^2 \qquad$ Squaring both sides

$\qquad\quad 0 = \dfrac{1}{4}(\ln x)^2 - \ln x$

Let $u = \ln x$ and substitute.

$\qquad \dfrac{1}{4}u^2 - u = 0$

$\qquad u\left(\dfrac{1}{4}u - 1\right) = 0$

$u = 0 \qquad or \quad \dfrac{1}{4}u - 1 = 0$

$u = 0 \qquad or \qquad \dfrac{1}{4}u = 1$

$u = 0 \qquad or \qquad u = 4$

$\ln x = 0 \qquad or \qquad \ln x = 4$

$x = e^0 = 1 \quad or \qquad x = e^4 \approx 54.598$

Both answers check. The solutions are 1 and e^4, or 1 and 54.598.

71. $\qquad (\log_3 x)^2 - \log_3 x^2 = 3$

$(\log_3 x)^2 - 2\log_3 x - 3 = 0$

Let $u = \log_3 x$ and substitute:

$u^2 - 2u - 3 = 0$

$(u - 3)(u + 1) = 0$

$u = 3 \quad or \qquad u = -1$

$\log_3 x = 3 \quad or \ \log_3 x = -1$

$x = 3^3 \quad or \qquad x = 3^{-1}$

$x = 27 \quad or \qquad x = \dfrac{1}{3}$

Both answers check. The solutions are $\dfrac{1}{3}$ and 27.

73. $\ln x^2 = (\ln x)^2$

$2\ln x = (\ln x)^2$

$0 = (\ln x)^2 - 2\ln x$

Let $u = \ln x$ and substitute.

$0 = u^2 - 2u$

$0 = u(u - 2)$

$u = 0 \quad or \quad u = 2$

$\ln x = 0 \quad or \ \ln x = 2$

$x = 1 \quad or \quad x = e^2 \approx 7.389$

Both answers check. The solutions are 1 and e^2, or 1 and 7.389.

75. $5^{2x} - 3 \cdot 5^x + 2 = 0$

$(5^x - 1)(5^x - 2) = 0 \quad$ This equation is quadratic in 5^x.

$5^x = 1 \qquad or \qquad 5^x = 2$

$\log 5^x = \log 1 \ \ or \ \ \log 5^x = \log 2$

$x \log 5 = 0 \qquad or \ \ x \log 5 = \log 2$

$x = 0 \qquad or \qquad x = \dfrac{\log 2}{\log 5} \approx 0.431$

The solutions are 0 and 0.431.

77. $\ln x^{\ln x} = 4$

$\ln x \cdot \ln x = 4$

$(\ln x)^2 = 4$

$\ln x = \pm 2$

$\ln x = -2 \quad or \ \ \ln x = 2$

$x = e^{-2} \quad or \qquad x = e^2$

$x \approx 0.135 \quad or \qquad x \approx 7.389$

Both answers check. The solutions are e^{-2} and e^2, or 0.135 and 7.389.

79. $\dfrac{\sqrt{(e^{2x} \cdot e^{-5x})^{-4}}}{e^x \div e^{-x}} = e^7$

$\dfrac{\sqrt{e^{12x}}}{e^{x-(-x)}} = e^7$

$\dfrac{e^{6x}}{e^{2x}} = e^7$

$e^{4x} = e^7$

$4x = 7$

$x = \dfrac{7}{4}$

The solution is $\dfrac{7}{4}$.

81. $a = \log_8 225$, so $8^a = 225 = 15^2$.

$b = \log_2 15$, so $2^b = 15$.

Then $\qquad 8^a = (2^b)^2$

$(2^3)^a = 2^{2b}$

$2^{3a} = 2^{2b}$

$3a = 2b$

$a = \dfrac{2}{3}b.$

Exercise Set 5.6

---

1. a) Substitute 6.18 for P_0 and 0.0214 for k in $P(t) = P_0 e^{kt}$. We have:

$P(t) = 6.18e^{0.0214t}$, where $P(t)$ is in millions and t is the number of years after 2012.

b) In 2018, $t = 2018 - 2012 = 6$.

$P(6) = 6.18e^{0.0214(6)} \approx 7.0$ million

c) Substitute 8 for $P(t)$ and solve for t.

$8 = 6.18e^{0.0214t}$

$\dfrac{8}{6.18} = e^{0.0214t}$

$\ln \dfrac{8}{6.18} = \ln e^{0.0214t}$

$\ln \dfrac{8}{6.18} = 0.0214t$

$\dfrac{\ln \dfrac{8}{6.18}}{0.0214} = t$

$12.1 \approx t$

The population of the area will be 8 million about 12.1 years after 2012.

d) $T = \dfrac{\ln 2}{0.0214} \approx 32.4$ years

3. a) $k = \dfrac{\ln 2}{77.0} \approx 0.90\%$

b) $k = \dfrac{\ln 2}{42.5} \approx 1.63\%$

c) $T = \dfrac{\ln 2}{0.0332} \approx 20.9$ years

d) $T = \dfrac{\ln 2}{0.0111} \approx 62.4$ years

e) $k = \dfrac{\ln 2}{385} \approx 0.18\%$

f) $T = \dfrac{\ln 2}{0.0232} \approx 29.9$ years

g) $T = \dfrac{\ln 2}{0.0128} \approx 54.2$ years

h) $k = \dfrac{\ln 2}{150.7} \approx 0.46\%$

i) $k = \dfrac{\ln 2}{26.3} \approx 2.64\%$

j) $T = \dfrac{\ln 2}{0.0039} \approx 177.7$ years

5.
$$P(t) = P_0 e^{kt}$$
$$32{,}961{,}561{,}600 = 9{,}893{,}934 e^{0.0099t}$$
$$\frac{32{,}961{,}561{,}600}{9{,}893{,}934} = e^{0.0099t}$$
$$\ln\left(\frac{32{,}961{,}561{,}600}{9{,}893{,}934}\right) = \ln e^{0.0099t}$$
$$\ln\left(\frac{32{,}961{,}561{,}600}{9{,}893{,}934}\right) = 0.0099t$$
$$\frac{\ln\left(\dfrac{32{,}961{,}561{,}600}{9{,}893{,}934}\right)}{0.0099} = t$$
$$819 \approx t$$

There will be one person for every square yard of land about 819 yr after 2013.

7. a) Substitute 10,000 for P_0 and 5.4%, or 0.054 for k.
$$P(t) = 10{,}000 e^{0.054t}$$

b)
$$P(1) = 10{,}000 e^{0.054(1)} \approx \$10{,}554.85$$
$$P(2) = 10{,}000 e^{0.054(2)} \approx \$11{,}140.48$$
$$P(5) = 10{,}000 e^{0.054(5)} \approx \$13{,}099.64$$
$$P(10) = 10{,}000 e^{0.054(10)} \approx \$17{,}160.07$$

c) $T = \dfrac{\ln 2}{0.054} \approx 12.8$ yr

9. We use the function found in Example 5. If the bones have lost 77.2% of their carbon-14 from an initial amount P_0, then $22.8\% P_0$, or $0.228 P_0$ remains. We substitute in the function.
$$0.228 P_0 = P_0 e^{-0.00012t}$$
$$0.228 = e^{-0.00012t}$$
$$\ln 0.228 = \ln e^{-0.00012t}$$
$$\ln 0.228 = -0.00012t$$
$$\frac{\ln 0.228}{-0.00012} = t$$
$$12{,}320 \approx t$$

The bones are about 12,320 years old.

11. a) $K = \dfrac{\ln 2}{3.1} \approx 0.224$, or 22.4% per min

b) $k = \dfrac{\ln 2}{22.3} \approx 0.031$, or 3.1% per yr

c) $T = \dfrac{\ln 2}{0.0115} \approx 60.3$ days

d) $T = \dfrac{\ln 2}{0.065} \approx 10.7$ yr

e) $k = \dfrac{\ln 2}{29.1} \approx 0.024$, or 2.4% per yr

f) $k = \dfrac{\ln 2}{70.0} \approx 0.010$, or 1.0% per yr

g) $k = \dfrac{\ln 2}{24{,}100} \approx 0.000029$, or 0.0029% per yr

13. a) $t = 2012 - 1960 = 52$
$$M(t) = M_0 e^{-kt}$$
$$50.5 = 72.2 e^{-k \cdot 52}$$
$$\frac{50.5}{72.2} = e^{-52k}$$
$$\ln \frac{50.5}{72.2} = \ln e^{-52k}$$
$$\ln \frac{50.5}{72.2} = -52k$$
$$\frac{\ln \dfrac{50.5}{72.2}}{-52} = k$$
$$0.0069 \approx k$$

Then we have $M(t) = 72.2 e^{-0.0069t}$, where $M(t)$ is a percent and t is the number of years after 1960.

b) In 2015, $t = 2015 - 1960 = 55$.
$$M(55) = 72.2 e^{-0.0069(55)} \approx 49.4\%$$
In 2018, $t = 2018 - 1960 = 58$.
$$M(58) = 72.2 e^{-0.0069(58)} \approx 48.4\%$$

c)
$$40 = 72.2 e^{-0.0069t}$$
$$\frac{40}{72.2} = e^{-0.0069t}$$
$$\ln \frac{40}{72.2} = \ln e^{-0.0069t}$$
$$\ln \frac{40}{72.2} = -0.0069t$$
$$\frac{\ln \dfrac{40}{72.2}}{-0.0069} = t$$
$$86 \approx t$$

40% of adults will be married about 86 years after 1960, or in 2046.

15. a) $t = 2012 - 1980 = 32$

$$C(t) = C_0 e^{kt}$$
$$10.28 = 1.85 e^{k \cdot 32}$$
$$\frac{10.28}{1.85} = e^{32k}$$
$$\ln \frac{10.28}{1.85} = \ln e^{32k}$$
$$\ln \frac{10.28}{1.85} = 32k$$
$$\frac{\ln \dfrac{10.28}{1.85}}{32} = k$$
$$0.0536 \approx k$$

Then we have $C(t) = 1.85 e^{0.0536t}$, where $C(t)$ is in millions and t is the number of years after 1980.

b) In 2005, $t = 2005 - 1980 = 25$.

$C(25) = 1.85 e^{0.0536(25)} \approx 7.07$ million barrels of oil per day

c) $T = \dfrac{\ln 2}{0.0536} \approx 12.9$ years

d)
$$13 = 1.85 e^{0.0536t}$$
$$\frac{13}{1.85} = e^{0.0536t}$$
$$\ln \frac{13}{1.85} = \ln e^{0.0536t}$$
$$\ln \frac{13}{1.85} = 0.0536t$$
$$\frac{\ln \dfrac{13}{1.85}}{0.0536} = t$$
$$36.4 \approx t$$

The consumption of oil in China will be 13 million barrels of oil per day about 36.4 years after 1980.

17. a) $N(0) = \dfrac{3500}{1 + 19.9 e^{-0.6(0)}} \approx 167$

b) $N(2) = \dfrac{3500}{1 + 19.9 e^{-0.6(2)}} \approx 500$

$N(5) = \dfrac{3500}{1 + 19.9 e^{-0.6(5)}} \approx 1758$

$N(8) = \dfrac{3500}{1 + 19.9 e^{-0.6(8)}} \approx 3007$

$N(12) = \dfrac{3500}{1 + 19.9 e^{-0.6(12)}} \approx 3449$

$N(16) = \dfrac{3500}{1 + 19.9 e^{-0.6(16)}} \approx 3495$

c) As $t \to \infty$, $N(t) \to 3500$; the number of people infected approaches 3500 but never actually reaches it.

19. To find k we substitute 105 for T_1, 0 for T_0, 5 for t, and 70 for $T(t)$ and solve for k.

$$70 = 0 + (105 - 0)e^{-5k}$$
$$70 = 105 e^{-5k}$$
$$\frac{70}{105} = e^{-5k}$$
$$\ln \frac{70}{105} = \ln e^{-5k}$$
$$\ln \frac{70}{105} = -5k$$
$$\frac{\ln \dfrac{70}{105}}{-5} = k$$
$$0.081 \approx k$$

The function is $T(t) = 105 e^{-0.081t}$.

Now we find $T(10)$.

$T(10) = 105 e^{-0.081(10)} \approx 46.7 \; °\text{F}$

21. To find k we substitute 43 for T_1, 68 for T_0, 12 for t, and 55 for $T(t)$ and solve for k.

$$55 = 68 + (43 - 68)e^{-12k}$$
$$-13 = -25 e^{-12k}$$
$$0.52 = e^{-12k}$$
$$\ln 0.52 = \ln e^{-12k}$$
$$\ln 0.52 = -12k$$
$$0.0545 \approx k$$

The function is $T(t) = 68 - 25 e^{-0.0545t}$.

Now we find $T(20)$.

$T(20) = 68 - 25 e^{-0.0545(20)} \approx 59.6°\text{F}$

23. Multiplication principle for inequalities

25. Principle of zero products

27. Power rule

29. $480 e^{-0.003p} = 150 e^{0.004p}$

$$\frac{480}{150} = \frac{e^{0.004p}}{e^{-0.003p}}$$
$$3.2 = e^{0.007p}$$
$$\ln 3.2 = \ln e^{0.007p}$$
$$\ln 3.2 = 0.007p$$
$$\frac{\ln 3.2}{0.007} = p$$
$$\$166.16 \approx p$$

31.
$$P(t) = P_0 e^{kt}$$
$$50,000 = P_0 e^{0.052(18)}$$
$$\frac{50,000}{e^{0.052(18)}} = P_0$$
$$\$19,609.67 \approx P_0$$

33.

$$i = \frac{V}{R}\left[1 - e^{-(R/L)t}\right]$$

$$\frac{iR}{V} = 1 - e^{-(R/L)t}$$

$$e^{-(R/L)t} = 1 - \frac{iR}{V}$$

$$\ln e^{-(R/L)t} = \ln\left(1 - \frac{iR}{V}\right)$$

$$-\frac{R}{L}t = \ln\left(1 - \frac{iR}{V}\right)$$

$$t = -\frac{L}{R}\left[\ln\left(1 - \frac{iR}{V}\right)\right]$$

35.

$$y = ae^x$$
$$\ln y = \ln(ae^x)$$
$$\ln y = \ln a + \ln e^x$$
$$\ln y = \ln a + x$$
$$Y = x + \ln a$$

This function is of the form $y = mx + b$, so it is linear.

Chapter 5 Review Exercises

1. The statement is true. See page 314 in the text.

3. The graph of f^{-1} is a reflection of the graph of f across $y = x$, so the statement is false.

5. The statement is false. The range of $y = a^x$, for instance, is $(0, \infty)$.

7. We interchange the first and second coordinates of each pair to find the inverse of the relation. It is
$\{(-2.7, 1.3), (-3, 8), (3, -5), (-3, 6), (-5, 7)\}$.

9. The graph of $f(x) = -|x| + 3$ is shown below. The function is not one-to-one, because there are many horizontal lines that cross the graph more than once.

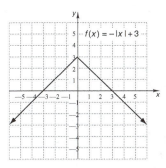

11. The graph of $f(x) = 2x - \dfrac{3}{4}$ is shown below. The function is one-to-one, because no horizontal line crosses the graph more than once.

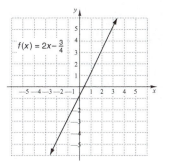

13. a) The graph of $f(x) = 2 - 3x$ is shown below. It passes the horizontal-line test, so the function is one-to-one.

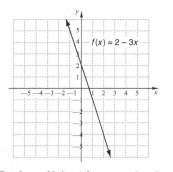

b) Replace $f(x)$ with y: $y = 2 - 3x$

Interchange x and y: $x = 2 - 3y$

Solve for y: $y = \dfrac{-x + 2}{3}$

Replace y with $f^{-1}(x)$: $f^{-1}(x) = \dfrac{-x + 2}{3}$

15. a) The graph of $f(x) = \sqrt{x - 6}$ is shown below. It passes the horizontal-line test, so the function is one-to-one.

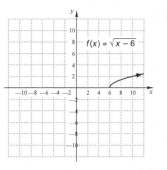

b) Replace $f(x)$ with y: $y = \sqrt{x - 6}$

Interchange x and y: $x = \sqrt{y - 6}$

Solve for y: $x^2 = y - 6$
$$x^2 + 6 = y$$

Replace y with $f^{-1}(x)$: $f^{-1}(x) = x^2 + 6$, for all x in the range of $f(x)$, or $f^{-1}(x) = x^2 + 6$, $x \geq 0$.

17. a) The graph of $f(x) = 3x^2 + 2x - 1$ is shown below. It is not one-to-one since there are many horizontal lines that cross the graph more than once. The function does not have an inverse that is a function.

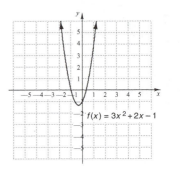

$f(x) = 3x^2 + 2x - 1$

19. We find $(f^{-1} \circ f)(x)$ and $(f \circ f^{-1})(x)$ and check to see that each is x.

$$(f^{-1} \circ f)(x) = f^{-1}(f(x)) = f^{-1}(6x - 5) =$$

$$\frac{(6x - 5) + 5}{6} = \frac{6x}{6} = x$$

$$(f \circ f^{-1})(x) = f(f^{-1}(x)) = f\left(\frac{x + 5}{6}\right) =$$

$$6\left(\frac{x + 5}{6}\right) - 5 = x + 5 - 5 = x$$

21. Replace $f(x)$ with y: $y = 2 - 5x$

Interchange x and y: $x = 2 - 5y$

Solve for y: $y = \dfrac{2 - x}{5}$

Replace y with $f^{-1}(x)$: $f^{-1}(x) = \dfrac{2 - x}{5}$

The domain and range of f are $(-\infty, \infty)$, so the domain and range of f^{-1} are also $(-\infty, \infty)$.

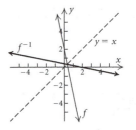

23. Since $f(f^{-1}(x)) = x$, then $f(f^{-1}(657)) = 657$.

25.

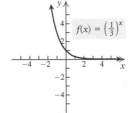

$f(x) = \left(\frac{1}{3}\right)^x$

27.

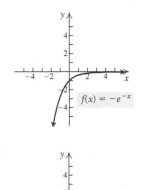

$f(x) = -e^{-x}$

29.

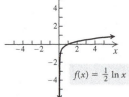

$f(x) = \frac{1}{2} \ln x$

31. $f(x) = e^{x-3}$

This is the graph of $f(x) = e^x$ shifted right 3 units. The correct choice is (c).

33. $f(x) = -\log_3(x + 1)$

This is the graph of $\log_3 x$ shifted left 1 unit and reflected across the y-axis. The correct choice is (b).

35. $f(x) = 3(1 - e^{-x})$, $x \geq 0$

This is the graph of $f(x) = e^x$ reflected across the y-axis, reflected across the x-axis, shifted up 1 unit, and stretched by a factor of 3. The correct choice is (e).

37. $\log_5 125 = 3$ because the exponent to which we raise 5 to get 125 is 3.

39. $\ln e = 1$ because the exponent to which we raise e to get e is 1.

41. $\log 10^{1/4} = \dfrac{1}{4}$ because the exponent to which we raise 10 to get $10^{1/4}$ is $\dfrac{1}{4}$.

43. $\log 1 = 0$ because the exponent to which we raise 10 to get 1 is 0.

45. $\log_2 \sqrt[3]{2} = \log_2 2^{1/3} = \dfrac{1}{3}$ because the exponent to which we raise 2 to get $2^{1/3}$ is $\dfrac{1}{3}$.

47. $\log_4 x = 2 \Rightarrow 4^2 = x$

49. $4^{-3} = \dfrac{1}{64} \Rightarrow \log_4 \dfrac{1}{64} = -3$

51. $\log 11 \approx 1.0414$

53. $\ln 3 \approx 1.0986$

55. $\log(-3)$ does not exist. (The calculator gives an error message.)

57. $\log_5 24 = \dfrac{\log 24}{\log 5} \approx 1.9746$

59. $3 \log_b x - 4 \log_b y + \frac{1}{2} \log_b z$

$= \log_b x^3 - \log_b y^4 + \log_b z^{1/2}$

$= \log_b \dfrac{x^3 z^{1/2}}{y^4}, \text{ or } \log_b \dfrac{x^3 \sqrt{z}}{y^4}$

61. $\ln \sqrt[4]{wr^2} = \ln(wr^2)^{1/4}$

$= \dfrac{1}{4} \ln wr^2$

$= \dfrac{1}{4}(\ln w + \ln r^2)$

$= \dfrac{1}{4}(\ln w + 2 \ln r)$

$= \dfrac{1}{4} \ln w + \dfrac{1}{2} \ln r$

63. $\log_a 3 = \log_a \left(\dfrac{6}{2}\right)$

$= \log_a 6 - \log_a 2$

$\approx 0.778 - 0.301$

≈ 0.477

65. $\log_a \dfrac{1}{5} = \log_a 5^{-1}$

$= -\log_a 5$

≈ -0.699

67. $\ln e^{-5k} = -5k \quad (\log_a a^x = x)$

69. $\log_4 x = 2$

$x = 4^2 = 16$

The solution is 16.

71. $e^x = 80$

$\ln e^x = \ln 80$

$x = \ln 80$

$x \approx 4.382$

The solution is 4.382.

73. $\log_{16} 4 = x$

$16^x = 4$

$(4^2)^x = 4^1$

$4^{2x} = 4^1$

$2x = 1$

$x = \dfrac{1}{2}$

The solution is $\dfrac{1}{2}$.

75. $\log_2 x + \log_2(x-2) = 3$

$\log_2 x(x-2) = 3$

$x(x-2) = 2^3$

$x^2 - 2x = 8$

$x^2 - 2x - 8 = 0$

$(x+2)(x-4) = 0$

$x + 2 = 0 \quad or \quad x - 4 = 0$

$x = -2 \quad or \qquad x = 4$

The number 4 checks, but -2 does not. The solution is 4.

77. $\log x^2 = \log x$

$x^2 = x$

$x^2 - x = 0$

$x(x-1) = 0$

$x = 0 \quad or \quad x - 1 = 0$

$x = 0 \quad or \qquad x = 1$

The number 1 checks, but 0 does not. The solution is 1.

79. a) $A(t) = P\left(1 + \dfrac{r}{n}\right)^{nt}$

$A(t) = 30,000\left(1 + \dfrac{0.042}{4}\right)^{4t} \quad \text{Substituting}$

$A(t) = 30,000(1.0105)^{4t}$

b) $A(0) = 30,000(1.0105)^{4 \cdot 0} = \$30,000$

$A(6) = 30,000(1.0105)^{4 \cdot 6} \approx \$38,547.20$

$A(12) = 30,000(1.0105)^{4 \cdot 12} \approx \$49,529.56$

$A(18) = 30,000(1.0105)^{4 \cdot 18} \approx \$63,640.87$

81. $T = \dfrac{\ln 2}{0.045} \approx 15.4 \text{ years}$

83. $P(t) = P_0 e^{kt}$

$0.73 P_0 = P_0 e^{-0.00012t}$

$0.73 = e^{-0.00012t}$

$\ln 0.73 = \ln e^{-0.00012t}$

$\ln 0.73 = -0.00012t$

$\dfrac{\ln 0.73}{-0.00012} = t$

$2623 \approx t$

The skeleton is about 2623 years old.

85. $R = \log \dfrac{10^{6.3} \cdot I_0}{I_0} = \log 10^{6.3} = 6.3$

87. a) We substitute 353.823 for P, since P is in thousands.

$W(353.823) = 0.37 \ln 353.823 + 0.05$

$\approx 2.2 \text{ ft/sec}$

b) We substitute 3.4 for W and solve for P.

$3.4 = 3.7 \ln P + 0.05$

$3.35 = 0.37 \ln P$

$\dfrac{3.35}{0.37} = \ln P$

$e^{3.35/0.37} = P$

$P \approx 8553.143$

The population is about 8553.143 thousand, or 8,553,143. (Answers may vary due to rounding differences.)

89. a) $P(t) = 15.2e^{0.0167t}$, where $P(t)$ is in millions and t is the number of years after 2013.

b) In 2017, $t = 2017 - 2013 = 4$.

$P(4) = 15.2e^{0.0167(4)} \approx 16.3$ million

In 2020, $t = 2020 - 2013 = 7$.

$P(7) = 15.2e^{0.0167(7)} \approx 17.1$ million

c) $18 = 15.2e^{0.0167t}$

$\dfrac{18}{15.2} = e^{0.0167t}$

$\ln \dfrac{18}{15.2} = \ln e^{0.0167t}$

$\ln \dfrac{18}{15.2} = 0.0167t$

$\dfrac{\ln \dfrac{18}{15.2}}{0.0167} = t$

$10 \approx t$

The population will be 18 million about 10 years after 2013.

d) $T = \dfrac{\ln 2}{0.0167} \approx 41.5$ years

91. We must have $2x - 3 > 0$, or $x > \dfrac{3}{2}$, so answer A is correct.

93. The graph of $f(x) = \log_2 x$ is the graph of $g(x) = 2^x$ reflected across the line $y = x$. Thus B is the correct graph.

95. $\log x = \ln x$

Graph $y_1 = \log x$ and $y_2 = \ln x$ and find the first coordinates of the points of intersection of the graph. We see that the only solution is 1.

97. Measure the atmospheric pressure P at the top of the building. Substitute that value in the equation $P = 14.7e^{-0.00005a}$, and solve for the height, or altitude, a, of the top of the building. Also measure the atmospheric pressure at the base of the building and solve for the altitude of the base. Then subtract to find the height of the building.

99. The inverse of a function $f(x)$ is written $f^{-1}(x)$, whereas $[f(x)]^{-1}$ means $\dfrac{1}{f(x)}$.

Chapter 5 Test

1. We interchange the first and second coordinates of each pair to find the inverse of the relation. It is
$\{(5, -2), (3, 4)(-1, 0), (-3, -6)\}$.

2. The function is not one-to-one, because there are many horizontal lines that cross the graph more than once.

3. The function is one-to-one, because no horizontal line crosses the graph more than once.

4. a) The graph of $f(x) = x^3 + 1$ is shown below. It passes the horizontal-line test, so the function is one-to-one.

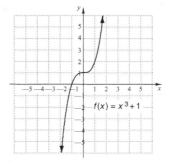

b) Replace $f(x)$ with y: $y = x^3 + 1$

Interchange x and y: $x = y^3 + 1$

Solve for y: $y^3 = x - 1$

$$y = \sqrt[3]{x - 1}$$

Replace y with $f^{-1}(x)$: $f^{-1}(x) = \sqrt[3]{x - 1}$

5. a) The graph of $f(x) = 1 - x$ is shown below. It passes the horizontal-line test, so the function is one-to-one.

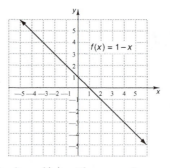

b) Replace $f(x)$ with y: $y = 1 - x$

Interchange x and y: $x = 1 - y$

Solve for y: $y = 1 - x$

Replace y with $f^{-1}(x)$: $f^{-1}(x) = 1 - x$

6. a) The graph of $f(x) = \dfrac{x}{2 - x}$ is shown below. It passes the horizontal-line test, so the function is one-to-one.

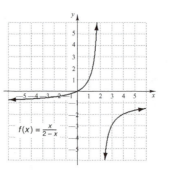

b) Replace $f(x)$ with y: $y = \dfrac{x}{2-x}$

Interchange x and y: $x = \dfrac{y}{2-y}$

Solve for y: $(2-y)x = y$

$$2x - xy = y$$
$$xy + y = 2x$$
$$y(x+1) = 2x$$
$$y = \frac{2x}{x+1}$$

Replace y with $f^{-1}(x)$: $f^{-1}(x) = \dfrac{2x}{x+1}$

7. a) The graph of $f(x) = x^2 + x - 3$ is shown below. It is not one-to-one since there are many horizontal lines that cross the graph more than once. The function does not have an inverse that is a function.

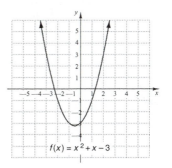

$f(x) = x^2 + x - 3$

8. We find $(f^{-1} \circ f)(x)$ and $(f \circ f^{-1})(x)$ and check to see that each is x.

$(f^{-1} \circ f)(x) = f^{-1}(f(x)) = f^{-1}(-4x+3) =$

$\dfrac{3 - (-4x+3)}{4} = \dfrac{3 + 4x - 3}{4} = \dfrac{4x}{4} = x$

$(f \circ f^{-1})(x) = f(f^{-1}(x)) = f\left(\dfrac{3-x}{4}\right) =$

$-4\left(\dfrac{3-x}{4}\right) + 3 = -3 + x + 3 = x$

9. Replace $f(x)$ with y: $y = \dfrac{1}{x-4}$

Interchange x and y: $x = \dfrac{1}{y-4}$

Solve for y: $x(y-4) = 1$

$$xy - 4x = 1$$
$$xy = 4x + 1$$
$$y = \frac{4x+1}{x}$$

Replace y with $f^{-1}(x)$: $f^{-1}(x) = \dfrac{4x+1}{x}$

The domain of $f(x)$ is $(-\infty, 4) \cup (4, \infty)$ and the range of $f(x)$ is $(-\infty, 0) \cup (0, \infty)$. Thus, the domain of f^{-1} is $(-\infty, 0) \cup (0, \infty)$ and the range of f^{-1} is $(-\infty, 4) \cup (4, \infty)$.

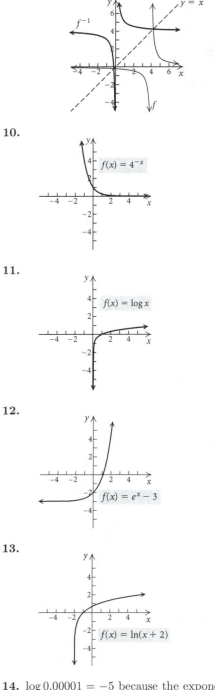

10.

$f(x) = 4^{-x}$

11.

$f(x) = \log x$

12.

$f(x) = e^x - 3$

13.

$f(x) = \ln(x+2)$

14. $\log 0.00001 = -5$ because the exponent to which we raise 10 to get 0.00001 is -5.

15. $\ln e = 1$ because the exponent to which we raise e to get e is 1.

16. $\ln 1 = 0$ because the exponent to which we raise e to get 1 is 0.

17. $\log_4 \sqrt[5]{4} = \log_4 4^{1/5} = \dfrac{1}{5}$ because the exponent to which we raise 4 to get $4^{1/5}$ is $\dfrac{1}{5}$.

18. $\ln x = 4 \Rightarrow x = e^4$

19. $3^x = 5.4 \Rightarrow x = \log_3 5.4$

20. $\ln 16 \approx 2.7726$

21. $\log 0.293 \approx -0.5331$

22. $\log_6 10 = \dfrac{\log 10}{\log 6} \approx 1.2851$

23.
$$2\log_a x - \log_a y + \frac{1}{2}\log_a z$$
$$= \log_a x^2 - \log_a y + \log_a z^{1/2}$$
$$= \log_a \frac{x^2 z^{1/2}}{y}, \text{ or } \log_a \frac{x^2 \sqrt{z}}{y}$$

24.
$$\ln \sqrt[5]{x^2 y} = \ln(x^2 y)^{1/5}$$
$$= \frac{1}{5}\ln x^2 y$$
$$= \frac{1}{5}(\ln x^2 + \ln y)$$
$$= \frac{1}{5}(2\ln x + \ln y)$$
$$= \frac{2}{5}\ln x + \frac{1}{5}\ln y$$

25.
$$\log_a 4 = \log_a \left(\frac{8}{2}\right)$$
$$= \log_a 8 - \log_a 2$$
$$\approx 0.984 - 0.328$$
$$\approx 0.656$$

26. $\ln e^{-4t} = -4t \quad (\log_a a^x = x)$

27.
$$\log_{25} 5 = x$$
$$25^x = 5$$
$$(5^2)^x = 5^1$$
$$5^{2x} = 5^1$$
$$2x = 1$$
$$x = \frac{1}{2}$$
The solution is $\dfrac{1}{2}$.

28.
$$\log_3 x + \log_3(x+8) = 2$$
$$\log_3 x(x+8) = 2$$
$$x(x+8) = 3^2$$
$$x^2 + 8x = 9$$
$$x^2 + 8x - 9 = 0$$
$$(x+9)(x-1) = 0$$
$$x = -9 \ \text{ or } \ x = 1$$
The number 1 checks, but -9 does not. The solution is 1.

29.
$$3^{4-x} = 27^x$$
$$3^{4-x} = (3^3)^x$$
$$3^{4-x} = 3^{3x}$$
$$4 - x = 3x$$
$$4 = 4x$$
$$x = 1$$
The solution is 1.

30.
$$e^x = 65$$
$$\ln e^x = \ln 65$$
$$x = \ln 65$$
$$x \approx 4.174$$
The solution is 4.174.

31. $R = \log \dfrac{10^{6.6} \cdot I_0}{I_0} = \log 10^{6.6} = 6.6$

32. $k = \dfrac{\ln 2}{45} \approx 0.0154 \approx 1.54\%$

33. a)
$$1144.54 = 1000e^{3k}$$
$$1.14454 = e^{3k}$$
$$\ln 1.14454 = \ln e^{3k}$$
$$\ln 1.14454 = 3k$$
$$\frac{\ln 1.14454}{3} = k$$
$$0.045 \approx k$$
The interest rate is about 4.5%.

b) $P(t) = 1000e^{0.045t}$

c) $P(8) = 1000e^{0.045 \cdot 8} \approx \$1433.33$

d) $T = \dfrac{\ln 2}{0.045} \approx 15.4$ yr

34. The graph of $f(x) = 2^{x-1} + 1$ is the graph of $g(x) = 2^x$ shifted right 1 unit and up 1 unit. Thus C is the correct graph.

35.
$$4^{\sqrt[3]{x}} = 8$$
$$(2^2)^{\sqrt[3]{x}} = 2^3$$
$$2^{2\sqrt[3]{x}} = 2^3$$
$$2\sqrt[3]{x} = 3$$
$$\sqrt[3]{x} = \frac{3}{2}$$
$$x = \left(\frac{3}{2}\right)^3$$
$$x = \frac{27}{8}$$
The solution is $\dfrac{27}{8}$.

Chapter 6

Systems of Equations and Matrices

Exercise Set 6.1

1. Graph (c) is the graph of this system.

3. Graph (f) is the graph of this system.

5. Graph (b) is the graph of this system.

7. Graph $x + y = 2$ and $3x + y = 0$ and find the coordinates of the point of intersection.

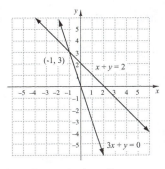

The solution is $(-1, 3)$.

9. Graph $x + 2y = 1$ and $x + 4y = 3$ and find the coordinates of the point of intersection.

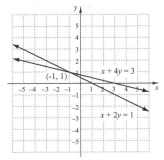

The solution is $(-1, 1)$.

11. Graph $y + 1 = 2x$ and $y - 1 = 2x$ and find the coordinates of the point of intersection.

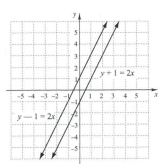

The graphs do not intersect, so there is no solution.

13. Graph $x - y = -6$ and $y = -2x$ and find the coordinates of the point of intersection.

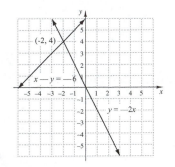

The solution is $(-2, 4)$.

15. Graph $2y = x - 1$ and $3x = 6y + 3$ and find the coordinates of the point of intersection.

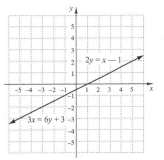

The graphs coincide so there are infinitely many solutions. Solving either equation for y, we get $y = \dfrac{x - 1}{2}$, so the solutions can be expressed as $\left(x, \dfrac{x - 1}{2} \right)$. Similarly, solving either equation for x, we get $x = 2y + 1$, so the solutions can also be expressed as $(2y + 1, y)$.

17.
$$x + y = 9, \quad (1)$$
$$2x - 3y = -2 \quad (2)$$

Solve equation (1) for either x or y. We choose to solve for y.

$$y = 9 - x$$

Then substitute $9 - x$ for y in equation (2) and solve the resulting equation.

$$2x - 3(9 - x) = -2$$
$$2x - 27 + 3x = -2$$
$$5x - 27 = -2$$
$$5x = 25$$
$$x = 5$$

Now substitute 5 for x in either equation (1) or (2) and solve for y.

$5 + y = 9$ Using equation (1).

$y = 4$

The solution is $(5, 4)$.

19. $x - 2y = 7$,   (1)

$x = y + 4$ (2)

Use equation (2) and substitute $y + 4$ for x in equation (1). Then solve for y.

$y + 4 - 2y = 7$

$-y + 4 = 7$

$-y = 3$

$y = -3$

Substitute -3 for y in equation (2) to find x.

$x = -3 + 4 = 1$

The solution is $(1, -3)$.

21. $y = 2x - 6$,   (1)

$5x - 3y = 16$ (2)

Use equation (1) and substitute $2x - 6$ for y in equation (2). Then solve for x.

$5x - 3(2x - 6) = 16$

$5x - 6x + 18 = 16$

$-x + 18 = 16$

$-x = -2$

$x = 2$

Substitute 2 for x in equation (1) to find y.

$y = 2 \cdot 2 - 6 = 4 - 6 = -2$

The solution is $(2, -2)$.

23. $x + y = 3$,   (1)

$y = 4 - x$ (2)

Use equation (2) and substitute $4 - x$ for y in equation (1).

$x + 4 - x = 3$

$4 = 3$

There are no values of x and y for which $4 = 3$, so the system of equations has no solution.

25. $x - 5y = 4$,   (1)

$y = 7 - 2x$ (2)

Use equation (2) and substitute $7 - 2x$ for y in equation (1). Then solve for x.

$x - 5(7 - 2x) = 4$

$x - 35 + 10x = 4$

$11x - 35 = 4$

$11x = 39$

$x = \dfrac{39}{11}$

Substitute $\dfrac{39}{11}$ for x in equation (2) to find y.

$y = 7 - 2 \cdot \dfrac{39}{11} = 7 - \dfrac{78}{11} = -\dfrac{1}{11}$

The solution is $\left(\dfrac{39}{11}, -\dfrac{1}{11} \right)$.

27. $x + 2y = 2$,   (1)

$4x + 4y = 5$ (2)

Solve one equation for either x or y. We choose to solve equation (1) for x since x has a coefficient of 1 in that equation.

$x + 2y = 2$

$x = -2y + 2$

Substitute $-2y + 2$ for x in equation (2) and solve for y.

$4(-2y + 2) + 4y = 5$

$-8y + 8 + 4y = 5$

$-4y + 8 = 5$

$-4y = -3$

$y = \dfrac{3}{4}$

Substitute $\dfrac{3}{4}$ for y in either equation (1) or equation (2) and solve for x.

$x + 2y = 2$ Using equation (1)

$x + 2 \cdot \dfrac{3}{4} = 2$

$x + \dfrac{3}{2} = 2$

$x = \dfrac{1}{2}$

The solution is $\left(\dfrac{1}{2}, \dfrac{3}{4} \right)$.

29. $3x - y = 5$,   (1)

$3y = 9x - 15$ (2)

Solve one equation for x or y. We will solve equation (1) for y.

$3x - y = 5$

$3x = y + 5$

$3x - 5 = y$

Substitute $3x - 5$ for y in equation (2) and solve for x.

$3(3x - 5) = 9x - 15$

$9x - 15 = 9x - 15$

$-15 = -15$

The equation $-15 = -15$ is true for all values of x and y, so the system of equations has infinitely many solutions. We know that $y = 3x - 5$, so we can write the solutions in the form $(x, 3x - 5)$.

If we solve either equation for x, we get $x = \dfrac{1}{3}y + \dfrac{5}{3}$, so we can also write the solutions in the form $\left(\dfrac{1}{3}y + \dfrac{5}{3}, y \right)$.

31. $x + 2y = 7$,   (1)

$x - 2y = -5$ (2)

We add the equations to eliminate y.

$x + 2y = 7$

$\underline{x - 2y = -5}$

$2x \quad\quad = 2$ Adding

$x = 1$

Back-substitute in either equation and solve for y.

$1 + 2y = 7$ Using equation (1)

$\quad 2y = 6$

$\quad\ y = 3$

The solution is $(1, 3)$. Since the system of equations has exactly one solution it is consistent and the equations are independent.

33. $x - 3y = 2,$ (1)

$\quad 6x + 5y = -34$ (2)

Multiply equation (1) by -6 and add it to equation (2) to eliminate x.

$$\begin{array}{r} -6x + 18y = -12 \\ 6x +\ 5y = -34 \\ \hline 23y = -46 \\ y = -2 \end{array}$$

Back-substitute to find x.

$x - 3(-2) = 2$ Using equation (1)

$\quad x + 6 = 2$

$\quad\quad x = -4$

The solution is $(-4, -2)$. Since the system of equations has exactly one solution it is consistent and the equations are independent.

35. $3x - 12y = 6,$ (1)

$\quad 2x -\ 8y = 4$ (2)

Multiply equation (1) by 2 and equation (2) by -3 and add.

$$\begin{array}{r} 6x - 24y = 12 \\ -6x + 24y = -12 \\ \hline 0 = 0 \end{array}$$

The equation $0 = 0$ is true for all values of x and y. Thus, the system of equations has infinitely many solutions. Solving either equation for y, we can write $y = \dfrac{1}{4}x - \dfrac{1}{2}$ so the solutions are ordered pairs of the form $\left(x, \dfrac{1}{4}x - \dfrac{1}{2}\right)$. Equivalently, if we solve either equation for x we get $x = 4y + 2$ so the solutions can also be expressed as $(4y + 2, y)$. Since there are infinitely many solutions, the system of equations is consistent and the equations are dependent.

37. $4x - 2y = 3,$ (1)

$\quad 2x -\ y = 4$ (2)

Multiply equation (2) by -2 and add.

$$\begin{array}{r} 4x - 2y = 3 \\ -4x + 2y = -8 \\ \hline 0 = -5 \end{array}$$

We get a false equation so there is no solution. Since there is no solution the system of equations is inconsistent and the equations are independent.

39. $2x = 5 - 3y,$ (1)

$\quad 4x = 11 - 7y$ (2)

We rewrite the equations.

$2x + 3y = 5,$ (1a)

$4x + 7y = 11$ (2a)

Multiply equation (2a) by -2 and add to eliminate x.

$$\begin{array}{r} -4x - 6y = -10 \\ 4x + 7y = 11 \\ \hline y = 1 \end{array}$$

Back-substitute to find x.

$2x = 5 - 3 \cdot 1$ Using equation (1)

$2x = 2$

$\ x = 1$

The solution is $(1, 1)$. Since the system of equations has exactly one solution it is consistent and the equations are independent.

41. $0.3x - 0.2y = -0.9,$

$\quad 0.2x - 0.3y = -0.6$

First, multiply each equation by 10 to clear the decimals.

$3x - 2y = -9$ (1)

$2x - 3y = -6$ (2)

Now multiply equation (1) by 3 and equation (2) by -2 and add to eliminate y.

$$\begin{array}{r} 9x - 6y = -27 \\ -4x + 6y = 12 \\ \hline 5x\quad\quad = -15 \\ x = -3 \end{array}$$

Back-substitute to find y.

$3(-3) - 2y = -9$ Using equation (1)

$\quad -9 - 2y = -9$

$\quad\quad -2y = 0$

$\quad\quad\quad y = 0$

The solution is $(-3, 0)$. Since the system of equations has exactly one solution it is consistent and the equations are independent.

43. $\dfrac{1}{5}x + \dfrac{1}{2}y = 6,$ (1)

$\quad \dfrac{3}{5}x - \dfrac{1}{2}y = 2$ (2)

We could multiply both equations by 10 to clear fractions, but since the y-coefficients differ only by sign we will just add to eliminate y.

$$\begin{array}{r} \dfrac{1}{5}x + \dfrac{1}{2}y = 6 \\[2mm] \dfrac{3}{5}x - \dfrac{1}{2}y = 2 \\ \hline \dfrac{4}{5}x\quad\quad = 8 \\[2mm] x = 10 \end{array}$$

Back-substitute to find y.

$$\frac{1}{5} \cdot 10 + \frac{1}{2}y = 6 \quad \text{Using equation (1)}$$

$$2 + \frac{1}{2}y = 6$$

$$\frac{1}{2}y = 4$$

$$y = 8$$

The solution is $(10, 8)$. Since the system of equations has exactly one solution it is consistent and the equations are independent.

45. The statement is true. See page 395 in the text.

47. False; a consistent system of equations can have exactly one solution or infinitely many solutions. See page 395 in the text.

49. True; a system of equations that has infinitely many solutions is consistent and dependent. See page 395 in the text.

51. *Familiarize.* Let $x =$ the number of liposuction surgeries and $y =$ the number of breast augmentation surgeries performed in 2013.

Translate. The total number of these surgeries was 677,239, so we have one equation.

$$x + y = 677,239$$

There were 50,585 fewer augmentation surgeries than liposuction surgeries, so we have a second equation.

$$y = x - 50,585$$

Carry out. We solve the system of equations.

$$x + y = 677,239, \quad (1)$$

$$y = x - 50,585 \quad (2)$$

Substitute $x - 50,585$ for y in equation (1) and solve for x.

$$x + (x - 50,585) = 677,239$$

$$2x - 50,585 = 677,239$$

$$2x = 727,824$$

$$x = 363,912$$

Back-substitute in equation (2) to find y.

$$y = 363,912 - 50,585 = 313,327$$

Check. $363,912 + 313,327 = 677,239$ surgeries and 313,327 is 50,585 less than 363,912. The answer checks.

State. There were 363,912 liposuction surgeries and 313,327 breast augmentation surgeries.

53. *Familiarize.* Let $b =$ the monthly rent in Boston and $s =$ the monthly rent in San Francisco.

Translate. The total of the two rents is \$4904, so we have one equation.

$$b + s = 4904$$

The rent in Boston is \$1142 less than the rent in San Francisco, so we have a second equation.

$$b = s - 1142$$

Carry out. We solve the system of equations.

$$b + s = 4904, \quad (1)$$

$$b = s - 1192 \quad (2)$$

Substitute $s - 1142$ for b in equation (1) and solve for s.

$$(s - 1142) + s = 4904$$

$$2s - 1142 = 4904$$

$$2s = 6046$$

$$s = 3023$$

Back-substitute in equation (2) to find b.

$$b = 3023 - 1142 = 1881$$

Check. $\$1881 + \$3023 = \$4904$, and \$1881 is \$1142 less than \$3023. The answer checks.

State. The monthly rent in Boston is \$1881, and the monthly rent in San Francisco is \$3023.

55. *Familiarize.* Let $x =$ the number of Amish who live in Ohio and $y =$ the number of Amish who live in Wisconsin.

Translate. The total number of Amish who live in these two states is 74,060, so we have one equation.

$$x + y = 74,060$$

The number of Amish in Ohio is 14,232 more than three times the number in Wisconsin, so we have a second equation.

$$x = 3y + 14,232$$

Carry out. We solve the system of equations.

$$x + y = 74,060 \quad (1)$$

$$x = 3y + 14,232 \quad (2)$$

Substitute $3y + 14,232$ for x in equation (1) and solve for y.

$$(3y + 14,232) + y = 74,060$$

$$4y + 14,232 = 74,060$$

$$4y = 59,828$$

$$y = 14,957$$

Back-substitute in equation (2) to find x.

$$x = 3(14,957) + 14,232$$

$$= 44,871 + 14,232$$

$$= 59,103$$

Check. $59,103 + 14,957 = 74,060$, and 59,103 is 14,232 more than three times 14,957. The answer checks.

State. 59,103 Amish live in Ohio, and 14,957 Amish live in Wisconsin.

57. *Familiarize.* Let $x =$ the number of standard-delivery packages and $y =$ the number of express-delivery packages.

Translate. A total of 120 packages were shipped, so we have one equation.

$$x + y = 120$$

Total shipping charges were \$934, so we have a second equation.

$$6.50x + 10.00y = 934$$

Carry out. We solve the system of equations:

$$x + \quad y = 120,$$
$$6.50x + 10.00y = 934.$$

First we multiply both sides of the second equation by 10 to clear the decimals. This gives us the system of equations

$$x + \quad y = 120, \quad (1)$$
$$65x + 100y = 9340. \quad (2)$$

Now multiply equation (1) by -65 and add.

$$-65x - \quad 65y = -7800$$
$$\underline{65x + 100y = 9340}$$
$$35y = 1540$$
$$y = 44$$

Back-substitute to find x.

$$x + 44 = 120 \quad \text{Using equation (1)}$$
$$x = 76$$

Check. $76 + 44 = 120$ packages. Total shipping charges are $\$6.50(76) + \$10.00(44) = \$494 + \$440 = \$934$. The answer checks.

State. The business shipped 76 standard-delivery packages and 44 express-delivery packages.

59. **Familiarize**. Let $x =$ the amount invested at 4% and $y =$ the amount invested at 5%. Then the interest from the investments is 4%x and 5%y, or $0.04x$ and $0.05y$.

Translate.

The total investment is $15,000.

$$x + y = 15,000$$

The total interest is $690.

$$0.04x + 0.05y = 690$$

We have a system of equations:

$$x + \quad y = 15,000,$$
$$0.04x + 0.05y = 690$$

Multiplying the second equation by 100 to clear the decimals, we have:

$$x + \quad y = 15,000, \quad (1)$$
$$4x + 5y = 69,000. \quad (2)$$

Carry out. We begin by multiplying equation (1) by -4 and adding.

$$-4x - 4y = -60,000$$
$$\underline{4x + 5y = 69,000}$$
$$y = 9000$$

Back-substitute to find x.

$$x + 9000 = 15,000 \quad \text{Using equation (1)}$$
$$x = 6000$$

Check. The total investment is $\$6000 + \$9000$, or $\$15,000$. The total interest is $0.04(\$6000) + 0.05(\$9000)$, or $\$240 + \$450$, or $\$690$. The solution checks.

State. $6000 was invested at 4% and $9000 was invested at 5%.

61. **Familiarize**. Let $x =$ the number of floral scarves sold and $y =$ the number of chevron scarves sold.

Translate. 39 scarves were sold, so we have one equation.

$$x + y = 39$$

The floral scarves brought in $24x$ dollars, the chevron scarves brought in $18y$ dollars, and the sales totaled $798, so we have a second equation.

$$24x + 18y = 798$$

Carry out. We solve the system of equations.

$$x + \quad y = 39, \quad (1)$$
$$24x + 18y = 798 \quad (2)$$

Multiply equation (1) by -24 and then add.

$$-24x - \quad 24y = -936$$
$$\underline{24x + \quad 18y = 798}$$
$$-6y = -138$$
$$y = 23$$

Back-substitute to find x.

$$x + 23 = 39$$
$$x = 16$$

Check. The number of scarves sold is $16 + 23$, or 39. Sales totaled $\$24 \cdot 16 + \$18 \cdot 23 = \$384 + \$414 = \$798$. The answer checks.

State. 16 floral scarves and 24 chevron scarves were sold.

63. **Familiarize**. Let $x =$ the number of servings of spaghetti and meatballs required and $y =$ the number of servings of iceberg lettuce required. Then x servings of spaghetti contain $260x$ Cal and $32x$ g of carbohydrates; y servings of lettuce contain $5y$ Cal and $1 \cdot y$ or y, g of carbohydrates.

Translate. One equation comes from the fact that 400 Cal are desired:

$$260x + 5y = 400.$$

A second equation comes from the fact that 50g of carbohydrates are required:

$$32x + y = 50.$$

Carry out. We solve the system

$$260x + 5y = 400, \quad (1)$$
$$32x + \quad y = 50. \quad (2)$$

Multiply equation (2) by -5 and add.

$$260x + 5y = 400$$
$$\underline{-160x - 5y = -250}$$
$$100x \quad = 150$$
$$x = 1.5$$

Back-substitute to find y.

$$32(1.5) + y = 50 \quad \text{Using equation (2)}$$
$$48 + y = 50$$
$$y = 2$$

Check. 1.5 servings of spaghetti contain $260(1.5)$, or 390 Cal and $32(1.5)$, or 48 g of carbohydrates; 2 servings of lettuce contain $5 \cdot 2$, or 10 Cal and $1 \cdot 2$, or 2 g of carbohydrates. Together they contain $390 + 10$, or 400 Cal and $48 + 2$, or 50 g of carbohydrates. The solution checks.

State. 1.5 servings of spaghetti and meatballs and 2 servings of iceberg lettuce are required.

65. *Familiarize.* It helps to make a drawing. Then organize the information in a table. Let x = the speed of the boat and y = the speed of the stream. The speed upstream is $x - y$. The speed downstream is $x + y$.

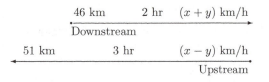

	Distance	Speed	Time
Downstream	46	$x + y$	2
Upstream	51	$x - y$	3

Translate. Using $d = rt$ in each row of the table, we get a system of equations.

$$46 = (x + y)2 \qquad x + y = 23, \quad (1)$$
$$\text{or}$$
$$51 = (x - y)3 \qquad x - y = 17 \quad (2)$$

Carry out. We begin by adding equations (1) and (2).

$$\begin{array}{r} x + y = 23 \\ x - y = 17 \\ \hline 2x \quad\;\; = 40 \\ x = 20 \end{array}$$

Back-substitute to find y.

$$20 + y = 23 \quad \text{Using equation (1)}$$
$$y = 3$$

Check. The speed downstream is $20 + 3$, or 23 km/h. The distance traveled downstream in 2 hr is $23 \cdot 2$, or 46 km. The speed upstream is $20 - 3$, or 17 km/h. The distance traveled upstream in 3 hr is $17 \cdot 3$, or 51 km. The solution checks.

State. The speed of the boat is 20 km/h. The speed of the stream is 3 km/h.

67. *Familiarize.* Let d = the distance the slower plane travels, in km. Then $780 - d$ = the distance the faster plane travels. Let t = the number of hours each plane travels. We organize the information in a table.

	Distance	Speed	Time
Slower plane	d	190	t
Faster plane	$780 - d$	200	t

Translate. Using $d = rt$ in each row of the table, we get a system of equations.

$$d = 190t, \quad (1)$$
$$780 - d = 200t \quad (2)$$

Carry out. We begin by adding the equations.

$$\begin{array}{r} d = 190t \\ 780 - \quad d = 200t \\ \hline 780 = 390t \\ 2 = t \end{array}$$

Check. In 2 hr, the slower plane travels $190 \cdot 2$, or 380 km, and the faster plane travels $200 \cdot 2$, or 400 km. The total distance traveled is 380 km + 400 km, or 780 km, so the answer checks.

State. The planes will meet in 2 hr.

69. *Familiarize and Translate.* We use the system of equations given in the problem.

$$y = 140 + 4x, \quad (1)$$
$$y = 275 - 5x, \quad (2)$$

Carry out. Substitute $275 - 5x$ for y in equation (1) and solve for x.

$$275 - 5x = 140 + 4x$$
$$135 = 9x \quad \text{Adding } 5x \text{ and subtracting } 140$$
$$15 = x$$

Back-substitute in either equation to find y. We choose equation (1).

$$y = 140 + 4 \cdot 15 = 200$$

Check. Substituting 15 for x and 200 for y in both of the original equations yields true equations, so the solution checks.

State. The equilibrium point is $(15, \$200)$.

71. *Familiarize and Translate.* We find the value of x for which $C = R$, where

$$C = 14x + 350,$$
$$R = 16.5x.$$

Carry out. When $C = R$ we have:

$$14x + 350 = 16.5x$$
$$350 = 2.5x$$
$$140 = x$$

Check. When $x = 140$, $C = 14 \cdot 140 + 350$, or 2310 and $R = 16.5(140)$, or 2310. Since $C = R$, the solution checks.

State. 140 units must be produced and sold in order to break even.

73. *Familiarize and Translate.* We find the value of x for which $C = R$, where

$$C = 15x + 12,000,$$
$$R = 18x - 6000.$$

Carry out. When $C = R$ we have:

$$15x + 12,000 = 18x - 6000$$
$$18,000 = 3x \quad \text{Subtracting } 15x \text{ and adding } 6000$$
$$6000 = x$$

Check. When $x = 6000$, $C = 15 \cdot 6000 + 12,000$, or 102,000 and $R = 18 \cdot 6000 - 6000$, or 102,000. Since $C = R$, the solution checks.

State. 6000 units must be produced and sold in order to break even.

75. *Familiarize*. Let $n =$ the number of registered snowmobiles in New York in 2013.

Translate.

$$\underbrace{\begin{array}{c}\text{Number of}\\\text{registered}\\\text{snowmo-}\\\text{biles in MN}\end{array}}_{251,986} \ \underset{=}{\text{was}}\ \underset{21,952}{21,952}\ \underset{+}{\underbrace{\begin{array}{c}\text{more}\\\text{than}\end{array}}}\ \underset{2}{2\text{ times}}\ \underset{\cdot}{}\ \underbrace{\begin{array}{c}\text{number of}\\\text{registered}\\\text{snowmo-}\\\text{biles in NY}\end{array}}_{n}$$

Carry out. We solve the equation.

$$251,986 = 21,952 + 2n$$
$$230,034 = 2n$$
$$115,017 = n$$

Check. We round and estimate.

$$21,952 + 2 \cdot 115,017 \approx 22,000 + 2 \cdot 115,00 \approx$$
$$22,000 + 230,000 \approx 252,000 \approx 251,986$$

The answer is reasonable.

State. There were 115,017 registered snowmobiles in New York in 2013.

77. Substituting 15 for $f(x)$, we solve the following equation.

$$15 = x^2 - 4x + 3$$
$$0 = x^2 - 4x - 12$$
$$0 = (x - 6)(x + 2)$$
$$x - 6 = 0 \ \ or \ \ x + 2 = 0$$
$$x = 6 \ \ or \ \ \ \ \ \ x = -2$$

If the output is 15, the input is 6 or -2.

79. $f(-2) = (-2)^2 - 4(-2) + 3 = 15$

If the input is -2, the output is 15.

81. *Familiarize*. Let x and y represent the number of gallons of gasoline used in city driving and in highway driving, respectively. Then $22x$ and $31y$ represent the number of miles driven in the city and on the highway, respectively.

Translate. The fact that 11 gal of gasoline were used gives us one equation:

$$x + y = 11.$$

A second equation comes from the fact that the car is driven 314 mile:

$$22x + 31y = 314.$$

Carry out. We solve the system of equations

$$x + \ \ y = 11, \ \ \ (1)$$
$$22x + 31y = 314. \ \ \ (2)$$

Multiply equation (1) by -22 and add.

$$\begin{array}{r} -22x - 22y = -242 \\ 22x + 31y = 314 \\ \hline 9y = 72 \\ y = 8 \end{array}$$

Back-substitute to find x.

$$x + 8 = 11$$
$$x = 3$$

Then in the city the car is driven $22(3)$, or 66 mi; on the highway it is driven $31(8)$, or 248 mi.

Check. The number of gallons of gasoline used is $3 + 8$, or 11. The number of miles driven is $66 + 248 = 314$. The answer checks.

State. The car was driven 66 mi in the city and 248 mi on the highway.

83. *Familiarize and Translate*. We let x and y represent the speeds of the trains. Organize the information in a table. Using $d = rt$, we let $3x$, $2y$, $1.5x$, and $3y$ represent the distances the trains travel.

First situation:

3 hours	x km/h	y km/h	2 hours
Union			Central

|— 216 km —|

Second situation:

1.5 hours	x km/h	y km/h	3 hours
Union			Central

|— 216 km —|

	Distance traveled in first situation	Distance traveled in second situation
Train_1 (from Union to Central)	$3x$	$1.5x$
Train_2 (from Central to Union)	$2y$	$3y$
Total	216	216

The total distance in each situation is 216 km. Thus, we have a system of equations.

$$3x + 2y = 216, \ \ \ (1)$$
$$1.5x + 3y = 216 \ \ \ (2)$$

Carry out. Multiply equation (2) by -2 and add.

$$\begin{array}{r} 3x + \ \ 2y = 216 \\ -3x - \ \ 6y = -432 \\ \hline -4y = -216 \\ y = 54 \end{array}$$

Back-substitute to find x.

$$3x + 2 \cdot 54 = 216 \quad \text{Using equation (1)}$$
$$3x + 108 = 216$$
$$3x = 108$$
$$x = 36$$

Check. If $x = 36$ and $y = 54$, the total distance the trains travel in the first situation is $3 \cdot 36 + 2 \cdot 54$, or 216 km. The total distance they travel in the second situation is $1.5 \cdot 36 + 3 \cdot 54$, or 216 km. The solution checks.

State. The speed of the first train is 36 km/h. The speed of the second train is 54 km/h.

85. Substitute the given solutions in the equation $Ax + By = 1$ to get a system of equations.

$$3A - B = 1, \quad (1)$$
$$-4A - 2B = 1 \quad (2)$$

Multiply equation (1) by -2 and add.

$$\begin{array}{r} -6A + 2B = -2 \\ -4A - 2B = 1 \\ \hline -10A = -1 \end{array}$$

$$A = \frac{1}{10}$$

Back-substitute to find B.

$$3\left(\frac{1}{10}\right) - B = 1 \qquad \text{Using equation (1)}$$

$$\frac{3}{10} - B = 1$$

$$-B = \frac{7}{10}$$

$$B = -\frac{7}{10}$$

We have $A = \dfrac{1}{10}$ and $B = -\dfrac{7}{10}$.

Exercise Set 6.2

1.
$$x + y + z = 2, \quad (1)$$
$$6x - 4y + 5z = 31, \quad (2)$$
$$5x + 2y + 2z = 13 \quad (3)$$

Multiply equation (1) by -6 and add it to equation (2). We also multiply equation (1) by -5 and add it to equation (3).

$$x + y + z = 2 \quad (1)$$
$$-10y - z = 19 \quad (4)$$
$$-3y - 3z = 3 \quad (5)$$

Multiply the last equation by 10 to make the y-coefficient a multiple of the y-coefficient in equation (4).

$$x + y + z = 2 \quad (1)$$
$$-10y - z = 19 \quad (4)$$
$$-30y - 30z = 30 \quad (6)$$

Multiply equation (4) by -3 and add it to equation (6).

$$x + y + z = 2 \quad (1)$$
$$-10y - z = 19 \quad (4)$$
$$-27z = -27 \quad (7)$$

Solve equation (7) for z.

$$-27z = -27$$
$$z = 1$$

Back-substitute 1 for z in equation (4) and solve for y.

$$-10y - 1 = 19$$
$$-10y = 20$$
$$y = -2$$

Back-substitute 1 for z for -2 and y in equation (1) and solve for x.

$$x + (-2) + 1 = 2$$
$$x - 1 = 2$$
$$x = 3$$

The solution is $(3, -2, 1)$.

3.
$$x - y + 2z = -3 \quad (1)$$
$$x + 2y + 3z = 4 \quad (2)$$
$$2x + y + z = -3 \quad (3)$$

Multiply equation (1) by -1 and add it to equation (2). We also multiply equation (1) by -2 and add it to equation (3).

$$x - y + 2z = -3 \quad (1)$$
$$3y + z = 7 \quad (4)$$
$$3y - 3z = 3 \quad (5)$$

Multiply equation (4) by -1 and add it to equation (5).

$$x - y + 2z = -3 \quad (1)$$
$$3y + z = 7 \quad (4)$$
$$-4z = -4 \quad (6)$$

Solve equation (6) for z.

$$-4z = -4$$
$$z = 1$$

Back-substitute 1 for z in equation (4) and solve for y.

$$3y + 1 = 7$$
$$3y = 6$$
$$y = 2$$

Back-substitute 1 for z and 2 for y in equation (1) and solve for x.

$$x - 2 + 2 \cdot 1 = -3$$
$$x = -3$$

The solution is $(-3, 2, 1)$.

5.
$$x + 2y - z = 5, \quad (1)$$
$$2x - 4y + z = 0, \quad (2)$$
$$3x + 2y + 2z = 3 \quad (3)$$

Multiply equation (1) by -2 and add it to equation (2). Also, multiply equation (1) by -3 and add it to equation (3).

$$x + 2y - z = 5, \quad (1)$$
$$-8y + 3z = -10, \quad (4)$$
$$-4y + 5z = -12 \quad (5)$$

Multiply equation (5) by 2 to make the y-coefficient a multiple of the y-coefficient of equation (4).

$$x + 2y - z = 5, \quad (1)$$
$$-8y + 3z = -10, \quad (4)$$
$$-8y + 10z = -24 \quad (6)$$

Multiply equation (4) by -1 and add it to equation (6).

$$x + 2y - z = 5, \quad (1)$$
$$-8y + 3z = -10, \quad (4)$$
$$7z = -14 \quad (7)$$

Solve equation (7) for z.

$$7z = -14$$
$$z = -2$$

Back-substitute -2 for z in equation (4) and solve for y.

$$-8y + 3(-2) = -10$$
$$-8y - 6 = -10$$
$$-8y = -4$$
$$y = \frac{1}{2}$$

Back-substitute $\frac{1}{2}$ for y and -2 for z in equation (1) and solve for x.

$$x + 2 \cdot \frac{1}{2} - (-2) = 5$$
$$x + 1 + 2 = 5$$
$$x = 2$$

The solution is $\left(2, \frac{1}{2}, -2\right)$.

7.
$$x + 2y - z = -8, \quad (1)$$
$$2x - y + z = 4, \quad (2)$$
$$8x + y + z = 2 \quad (3)$$

Multiply equation (1) by -2 and add it to equation (2). Also, multiply equation (1) by -8 and add it to equation (3).

$$x + 2y - z = -8, \quad (1)$$
$$-5y + 3z = 20, \quad (4)$$
$$-15y + 9z = 66 \quad (5)$$

Multiply equation (4) by -3 and add it to equation (5).

$$x + 2y - z = -8, \quad (1)$$
$$-5y + 3z = 20, \quad (4)$$
$$0 = 6 \quad (6)$$

Equation (6) is false, so the system of equations has no solution.

9.
$$2x + y - 3z = 1, \quad (1)$$
$$x - 4y + z = 6, \quad (2)$$
$$4x - 7y - z = 13 \quad (3)$$

Interchange equations (1) and (2).

$$x - 4y + z = 6, \quad (2)$$
$$2x + y - 3z = 1, \quad (1)$$
$$4x - 7y - z = 13 \quad (3)$$

Multiply equation (2) by -2 and add it to equation (1). Also, multiply equation (2) by -4 and add it to equation (3).

$$x - 4y + z = 6, \quad (2)$$
$$9y - 5z = -11, \quad (4)$$
$$9y - 5z = -11 \quad (5)$$

Multiply equation (4) by -1 and add it to equation (5).

$$x - 4y + z = 6, \quad (1)$$
$$9y - 5z = -11, \quad (4)$$
$$0 = 0 \quad (6)$$

The equation $0 = 0$ tells us that equation (3) of the original system is dependent on the first two equations. The system of equations has infinitely many solutions and is equivalent to

$$2x + y - 3z = 1, \quad (1)$$
$$x - 4y + z = 6. \quad (2)$$

To find an expression for the solutions, we first solve equation (4) for either y or z. We choose to solve for z.

$$9y - 5z = -11$$
$$-5z = -9y - 11$$
$$z = \frac{9y + 11}{5}$$

Back-substitute in equation (2) to find an expression for x in terms of y.

$$x - 4y + \frac{9y + 11}{5} = 6$$
$$x - 4y + \frac{9}{5}y + \frac{11}{5} = 6$$
$$x = \frac{11}{5}y + \frac{19}{5} = \frac{11y + 19}{5}$$

The solutions are given by $\left(\dfrac{11y + 19}{5}, y, \dfrac{9y + 11}{5}\right)$, where y is any real number.

11.
$$4a + 9b \qquad = 8, \quad (1)$$
$$8a \qquad + 6c = -1, \quad (2)$$
$$6b + 6c = -1 \quad (3)$$

Multiply equation (1) by -2 and add it to equation (2).

$$4a + 9b \qquad = 8, \quad (1)$$
$$-18b + 6c = -17, \quad (4)$$
$$6b + 6c = -1 \quad (3)$$

Multiply equation (3) by 3 to make the b-coefficient a multiple of the b-coefficient in equation (4).

$$4a + 9b \qquad = 8, \quad (1)$$
$$-18b + 6c = -17, \quad (4)$$
$$18b + 18c = -3 \quad (5)$$

Add equation (4) to equation (5).

$$4a + 9b \qquad = 8, \quad (1)$$
$$-18b + 6c = -17, \quad (4)$$
$$24c = -20 \quad (6)$$

Solve equation (6) for c.

$$24c = -20$$
$$c = -\frac{20}{24} = -\frac{5}{6}$$

Back-substitute $-\dfrac{5}{6}$ for c in equation (4) and solve for b.

$$-18b + 6c = -17$$
$$-18b + 6\left(-\frac{5}{6}\right) = -17$$
$$-18b - 5 = -17$$
$$-18b = -12$$
$$b = \frac{12}{18} = \frac{2}{3}$$

Back-substitute $\dfrac{2}{3}$ for b in equation (1) and solve for a.

$$4a + 9b = 8$$
$$4a + 9 \cdot \frac{2}{3} = 8$$
$$4a + 6 = 8$$
$$4a = 2$$
$$a = \frac{1}{2}$$

The solution is $\left(\dfrac{1}{2}, \dfrac{2}{3}, -\dfrac{5}{6}\right)$.

13.
$$2x \qquad + \ z = 1, \qquad (1)$$
$$3y - 2z = 6, \qquad (2)$$
$$x - 2y \qquad = -9 \qquad (3)$$

Interchange equations (1) and (3).

$$x - 2y \qquad = -9, \qquad (3)$$
$$3y - 2z = 6, \qquad (2)$$
$$2x \qquad + \ z = 1 \qquad (1)$$

Multiply equation (3) by -2 and add it to equation (1).

$$x - 2y \qquad = -9, \qquad (3)$$
$$3y - 2z = 6, \qquad (2)$$
$$4y + \ z = 19 \qquad (4)$$

Multiply equation (4) by 3 to make the y-coefficient a multiple of the y-coefficient in equation (2).

$$x - \ 2y \qquad = -9, \qquad (3)$$
$$3y - 2z = 6, \qquad (2)$$
$$12y + 3z = 57 \qquad (5)$$

Multiply equation (2) by -4 and add it to equation (5).

$$x - 2y \qquad = -9, \qquad (3)$$
$$3y - \ 2z = 6, \qquad (2)$$
$$11z = 33 \qquad (6)$$

Solve equation (6) for z.

$$11z = 33$$
$$z = 3$$

Back-substitute 3 for z in equation (2) and solve for y.

$$3y - 2z = 6$$
$$3y - 2 \cdot 3 = 6$$
$$3y - 6 = 6$$
$$3y = 12$$
$$y = 4$$

Back-substitute 4 for y in equation (3) and solve for x.

$$x - 2y = -9$$
$$x - 2 \cdot 4 = -9$$
$$x - 8 = -9$$
$$x = -1$$

The solution is $(-1, 4, 3)$.

15.
$$w + \ x + \ y + \ z = 2 \qquad (1)$$
$$w + 2x + 2y + 4z = 1 \qquad (2)$$
$$-w + \ x - \ y - \ z = -6 \qquad (3)$$
$$-w + 3x + \ y - \ z = -2 \qquad (4)$$

Multiply equation (1) by -1 and add to equation (2). Add equation (1) to equation (3) and to equation (4).

$$w + \ x + \ y + \ z = 2 \qquad (1)$$
$$x + \ y + 3z = -1 \qquad (5)$$
$$2x \qquad \qquad = -4 \qquad (6)$$
$$4x + 2y \qquad = 0 \qquad (7)$$

Solve equation (6) for x.

$$2x = -4$$
$$x = -2$$

Back-substitute -2 for x in equation (7) and solve for y.

$$4(-2) + 2y = 0$$
$$-8 + 2y = 0$$
$$2y = 8$$
$$y = 4$$

Back-substitute -2 for x and 4 for y in equation (5) and solve for z.

$$-2 + 4 + 3z = -1$$
$$3z = -3$$
$$z = -1$$

Back-substitute -2 for x, 4 for y, and -1 for z in equation (1) and solve for w.

$$w - 2 + 4 - 1 = 2$$
$$w = 1$$

The solution is $(1, -2, 4, -1)$.

17. Familiarize. Let x, y, and z represent the number of medals won by the Russian Federation, Ukraine, and the United States, respectively.

Translate. The three countries won a total of 123 medals.

$$x + y + z = 123$$

Ukraine won 7 more medals than the United States.

$$y = z + 7$$

The Russian Federation won 37 more medals than the total number won by Ukraine and the United States.

$$x = y + z + 37$$

We have

$$x + y + z = 123,$$
$$y = z + 7,$$
$$x = y + z + 37,$$

or

$$x + y + z = 123,$$
$$y - z = 7,$$
$$x - y - z = 37.$$

Carry out. Solving the system of equations, we get $(80, 25, 18)$.

Check. The total number of medals is $80+25+18$, or 123. Ukraine won 25 medals which is 7 more than 18 medals, the number won by the United States. Also, Ukraine and the United States won $25+18$, or 43, medals, and $43+37 = 80$, the number of medals won by the Russian Federation. The answer checks.

State. The Russian Federation won 80 medals, Ukraine won 25 medals, and the United States won 18 medals.

19. Familiarize. Let x, y, and z represent the number of restaurant-purchased meals that will be eaten in a restaurant, in a car, and at home, respectively.

Translate. The total number of meals is 170.

$$x + y + z = 170$$

The total number of restaurant-purchased meals eaten in a car or at home is 14 more than the number eaten in a restaurant.

$$y + z = x + 14$$

Twenty more restaurant-purchased meals will be eaten in a restaurant than at home.

$$x = z + 20$$

We have

$$x + y + z = 170,$$
$$y + z = x + 14,$$
$$x = z + 20$$

or

$$x + y + z = 170,$$
$$-x + y + z = 14,$$
$$x \quad - z = 20.$$

Carry out. Solving the system of equations, we get (78, 34, 58).

Check. The total number of meals is $78 + 34 + 58$, or 170. The total number of meals eaten in a car or at home is $34 + 58$, or 92. This is 14 more than 78, the number of meals eaten in a restaurant. Finally, 20 more than the number of restaurant-purchased meals eaten at home, 58, is 78, the number of meals eaten in a restaurant. The answer checks.

State. The number of restaurant-purchased meals eaten in a restaurant, in a car, and at home is 78, 34, and 58, respectively.

21. Familiarize. Let x y, and z represent the prices of the Bacon art, the Warhol art, and the Koons art, respectively, in millions of dollars.

Translate. The prices totaled $306.2 million.

$$x + y + z = 306.2$$

The Warhol art sold for $47 million more than the Koons art.

$$y = z + 47$$

Together, the Warhol art and the Koons art sold for $21.4 million more than the Bacon art.

$$y + z = x + 21.4$$

We have

$$x + y + z = 306.2, \qquad x + y + z = 306.2,$$
$$y = z + 47, \qquad \text{or} \qquad y - z = 47,$$
$$y + z = x + 21.4, \qquad -x + y + z = 21.4.$$

Carry out. Solving the system of equations, we get (142.4, 105.4, 58.4).

Check. The prices total $142.4 + 105.4 + 58.4$, or $306.2 million. The price of the Koons art plus $47 million is $58.4 + 47$, or $105.4 million, the price of the Warhol art. Also, the prices of the Warhol and Bacon art together, $105.4 + 58.4$, or $163.8 million, is $21.4 million more than $142.4 million, the price of the Bacon art. The answer checks.

State. The Bacon art sold for $142.4 million, the Warhol art sold for $105.4 million, and the Koons art sold for $58.4 million.

23. Familiarize. Let x, y, and z represent the number of dogs, cats, and birds Americans own, respectively, in millions.

Translate. The total number of dogs, cats, and birds owned is 152.3 million.

$$x + y + z = 152.3$$

The number of dogs owned is 12.5 million less than the total number of cats and birds owned.

$$x = y + z - 12.5$$

The number of cats owned is 4.2 million more than the number of dogs owned.

$$y = x + 4.2$$

We have

$$x + y + z = 152.3,$$
$$x = y + z - 12.5,$$
$$y = x + 4.2,$$

or

$$x + y + z = 152.3,$$
$$x - y - z = -12.5,$$
$$-x + y \quad = 4.2.$$

Carry out. Solving the system of equations, we get (69.9, 74.1, 8.3).

Check. The total number of dogs, cats, and birds owned is $69.9 + 74.1 + 8.3$, or 152.3 million. The total number of cats and birds owned less 12.5 million is $74.1 + 8.3 - 12.5$, or 69.9 million, the number of dogs owned. Also, 4.2 million more than the number of dogs owned is $69.9 + 4.2$, or 74.1 million, the number of cats owned. The answer checks.

State. Americans own 69.9 million dogs, 74.1 million cats, and 8.3 million birds.

25. Familiarize. Let x, y, and z represent the amount of foreign aid donated by the United States, Great Britain, and Germany, respectively, in billions of dollars.

Translate. The total amount of aid donated by the three countries was $63.5 billion.

$$x + y + z = 63.5$$

Together, Great Britain and Germany donated $0.5 billion more than the United States did.

$$y + z = x + 0.5$$

The United States donated \$17.4 billion more than Germany did.

$$x = z + 17.4$$

We have

$$x + y + z = 63.5,$$
$$y + z = x + 0.5,$$
$$x = z + 17.4$$

or

$$x + y + z = 63.5,$$
$$-x + y + z = 0.5,$$
$$x \quad\ - z = 17.4.$$

Carry out. Solving the system of equations, we get $(31.5, 17.9, 14.1)$.

Check. The aid totals $31.5 + 17.9 + 14.1$, or \$63.5 million. Together, Great Britain and Germany donated $17.9 + 14.1$, or \$32 billion. This is \$0.5 billion more than \$31.5 billion, the amount donated by the United States. The amount Germany donated plus \$17.4 billion is $14.1 + 17.4$, or \$31.5 billion, the amount the United States donated. The answer checks.

State. In 2013, the United States donated \$31.5 billion in foreign aid, Great Britain donated \$17.9 billion, and Germany donated \$14.1 billion.

27. Familiarize. Let x, y, and z represent the number of servings of ground beef, baked potato, and strawberries required, respectively. One serving of ground beef contains $245x$ Cal, $0x$ or 0 g of carbohydrates, and $9x$ mg of calcium. One baked potato contains $145y$ Cal, $34y$ g of carbohydrates, and $8y$ mg of calcium. One serving of strawberries contains $45z$ Cal, $10z$ g of carbohydrates, and $21z$ mg of calcium.

Translate.

The total number of calories is 485.

$$245x + 145y + 45z = 485$$

A total of 41.5 g of carbohydrates is required.

$$34y + 10z = 41.5$$

A total of 35 mg of calcium is required.

$$9x + 8y + 21z = 35$$

We have a system of equations.

$$245x + 145y + 45z = 485,$$
$$34y + 10z = 41.5,$$
$$9x + 8y + 21z = 35$$

Carry out. Solving the system of equations, we get $(1.25, 1, 0.75)$.

Check. 1.25 servings of ground beef contains 306.25 Cal, no carbohydrates, and 11.25 mg of calcium; 1 baked potato contains 145 Cal, 34 g of carbohydrates, and 8 mg of calcium; 0.75 servings of strawberries contains 33.75 Cal, 7.5 g of carbohydrates, and 15.75 mg of calcium. Thus, there are a total of $306.25 + 145 + 33.75$, or 485 Cal, $34 + 7.5$, or 41.5 g of carbohydrates, and $11.25 + 8 + 15.75$, or 35 mg of calcium. The solution checks.

State. 1.25 servings of ground beef, 1 baked potato, and 0.75 serving of strawberries are required.

29. Familiarize. Let x, y, and z represent the amounts invested at 2%, 3%, and 4%, respectively. Then the annual interest from the investments is $2\%x$, $3\%y$, and $4\%z$, or $0.02x$, $0.03y$, and $0.04z$.

Translate. The total interest is \$126.

$$0.02x + 0.03y + 0.04z = 126$$

There is \$500 more invested at 3% than at 2%.

$$y = x + 500$$

The amount invested at 4% is three times the amount invested at 3%.

$$z = 3y$$

We have

$$0.02x + 0.03y + 0.04z = 126,$$
$$y = x + 500,$$
$$z = 3y,$$

or

$$0.02x + 0.03y + 0.04z = 126,$$
$$-x + y = 500,$$
$$-3y + z = 0.$$

Carry out. Solving the system of equations, we get $(300, 800, 2400)$.

Check. The total interest is $0.02(300) + 0.03(800) + 0.04(2400)$, or $6 + 24 + 96$, or \$126. The amount invested at 3%, \$800, is \$500 more than \$300, the amount invested at 2%. Also, the amount invested at 4%, \$2400, is three times \$800, the amount invested at 3%. The answer checks.

State. \$300 is invested at 2%, \$800 is invested at 3%, and \$2400 is invested at 4%.

31. Familiarize. Let x, y, and z represent the prices of orange juice, a raisin bagel, and a cup of coffee, respectively. The new price for orange juice is $x + 25\%x$, or $x + 0.25x$, or $1.25x$; the new price of a bagel is $y + 20\%y$, or $y + 0.2y$, or $1.2y$.

Translate.

Orange juice, a raisin bagel, and a cup of coffee cost \$6.30.

$$x + y + z = 6.30$$

After the price increase, orange juice, a raisin bagel, and a cup of coffee will cost \$7.30.

$$1.25x + 1.2y + z = 7.30$$

After the price increases, orange juice will cost 70¢ (or \$0.70) more than coffee.

$$1.25x = z + 0.70$$

We have a system of equations.

$$x + y + z = 6.30, \qquad\qquad 10x + 10y + 10z = 63,$$
$$1.25x + 1.2y + z = 7.30, \text{ or } 125x + 120y + 100z = 730,$$
$$1.25x = z + 0.70 \qquad\qquad 125x - 100z = 70$$

Carry out. Solving the system of equations, we get $(2, 2.5, 1.8)$.

Check. If orange juice costs $2.00, a bagel costs $2.50, and a cup of coffee costs $1.80, then together they cost $2.00 + $2.50 + $1.80, or $6.30. After the price increases orange juice will cost 1.25($2.00), or $2.50, and a bagel will cost 1.2($2.50) or $3.00. Then orange juice, a bagel, and coffee will cost $2.50 + $3.00 + $1.80, or $7.30. After the price increase the price of orange juice, $2.50, will be 70¢ more than the price of coffee, $1.80. The solution checks.

State. Before the increase orange juice cost $2.00, a raisin bagel cost $2.50, and a cup of coffee cost $1.80.

33. a) Substitute the data points $(0, 43)$, $(10, 33)$, and $(20, 38)$ in the function $f(x) = ax^2 + bx + c$.

$$43 = a \cdot 0^2 + b \cdot 0 + c$$
$$33 = a \cdot 10^2 + b \cdot 10 + c$$
$$38 = a \cdot 20^2 + b \cdot 20 + c$$

We have a system of equations.

$$c = 43,$$
$$100a + 10b + c = 33,$$
$$400a + 20b + c = 38$$

Solving the system of equations, we get

$\left(\dfrac{3}{40}, -\dfrac{7}{4}, 43 \right)$ so $f(x) = \dfrac{3}{40}x^2 - \dfrac{7}{4}x + 43$,

where x is the number of years after 1992 and $f(x)$ is a percent.

b) In 2007, $x = 2007 - 1992 = 15$.

$$f(15) = \frac{3}{40} \cdot 15^2 - \frac{7}{4} \cdot 15 + 43 = 33.625\%$$

In 2014, $x = 2014 - 1992 = 22$.

$$f(22) = \frac{3}{40} \cdot 22^2 - \frac{7}{4} \cdot 22 + 43 = 40.8\%$$

35. a) Substitute the data points $(0, 291)$, $(3, 393)$, and $(6, 369)$ in the function $f(x) = ax^2 + bx + c$.

$$291 = a \cdot 0^2 + b \cdot 0 + c$$
$$393 = a \cdot 3^2 + b \cdot 3 + c$$
$$369 = a \cdot 6^2 + b \cdot 6 + c$$

We have a system of equations.

$$c = 291,$$
$$9a + 3b + c = 393,$$
$$36a + 6b + c = 369$$

Solving the system of equations, we get $(-7, 55, 291)$, so $f(x) = -7x^2 + 55x + 291$, where x is the number of years after 2007 and $f(x)$ is in thousands.

b) In 2014, $x = 2014 - 2007 = 7$.

$f(7) = -7 \cdot 7^2 + 55 \cdot 7 + 291 = 333$, so there were 333,000 deportations in 2014.

37. perpendicular

39. vertical line

41. rational function

43. vertical asymptote

45. $\dfrac{2}{x} - \dfrac{1}{y} - \dfrac{3}{z} = -1,$

$\dfrac{2}{x} - \dfrac{1}{y} + \dfrac{1}{z} = -9,$

$\dfrac{1}{x} + \dfrac{2}{y} - \dfrac{4}{z} = 17$

First substitute u for $\dfrac{1}{x}$, v for $\dfrac{1}{y}$, and w for $\dfrac{1}{z}$ and solve for u, v, and w.

$$2u - v - 3w = -1,$$
$$2u - v + w = -9,$$
$$u + 2v - 4w = 17$$

Solving this system we get $(-1, 5, -2)$.

If $u = -1$, and $u = \dfrac{1}{x}$, then $-1 = \dfrac{1}{x}$, or $x = -1$.

If $v = 5$ and $v = \dfrac{1}{y}$, then $5 = \dfrac{1}{y}$, or $y = \dfrac{1}{5}$.

If $w = -2$ and $w = \dfrac{1}{z}$, then $-2 = \dfrac{1}{z}$, or $z = -\dfrac{1}{2}$.

The solution of the original system is $\left(-1, \dfrac{1}{5}, -\dfrac{1}{2} \right)$.

47. Label the angle measures at the tips of the stars a, b, c, d, and e. Also label the angles of the pentagon p, q, r, s, and t.

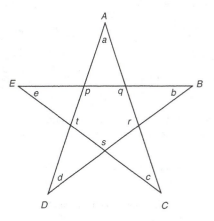

Using the geometric fact that the sum of the angle measures of a triangle is 180°, we get 5 equations.

$$p + b + d = 180$$
$$q + c + e = 180$$
$$r + a + d = 180$$
$$s + b + e = 180$$
$$t + a + c = 180$$

Adding these equations, we get

$$(p + q + r + s + t) + 2a + 2b + 2c + 2d + 2e = 5(180).$$

The sum of the angle measures of any convex polygon with n sides is given by the formula $S = (n - 2)180$. Thus $p + q + r + s + t = (5 - 2)180$, or 540. We substitute and solve for $a + b + c + d + e$.

$$540 + 2(a + b + c + d + e) = 900$$
$$2(a + b + c + d + e) = 360$$
$$a + b + c + d + e = 180$$

The sum of the angle measures at the tips of the star is $180°$.

49. Substituting, we get

$$A + \frac{3}{4}B + 3C = 12,$$
$$\frac{4}{3}A + B + 2C = 12,$$
$$2A + B + C = 12, \text{ or}$$

$$4A + 3B + 12C = 48,$$
$$4A + 3B + 6C = 36, \quad \text{Clearing fractions}$$
$$2A + B + C = 12.$$

Solving the system of equations, we get $(3, 4, 2)$. The equation is $3x + 4y + 2z = 12$.

51. Substituting, we get

$$59 = a(-2)^3 + b(-2)^2 + c(-2) + d,$$
$$13 = a(-1)^3 + b(-1)^2 + c(-1) + d,$$
$$-1 = a \cdot 1^3 + b \cdot 1^2 + c \cdot 1 + d,$$
$$-17 = a \cdot 2^3 + b \cdot 2^2 + c \cdot 2 + d, \text{ or}$$

$$-8a + 4b - 2c + d = 59,$$
$$-a + b - c + d = 13,$$
$$a + b + c + d = -1,$$
$$8a + 4b + 2c + d = -17.$$

Solving the system of equations, we get $(-4, 5, -3, 1)$, so $y = -4x^3 + 5x^2 - 3x + 1$.

53. ***Familiarize and Translate***. Let a, s, and c represent the number of adults, students, and children in attendance, respectively.

The total attendance was 100.

$$a + s + c = 100$$

The total amount of money taken in was $100.

(Express 50 cents as $\frac{1}{2}$ dollar.)

$$10a + 3s + \frac{1}{2}c = 100$$

The resulting system is

$$a + s + c = 100,$$
$$10a + 3s + \frac{1}{2}c = 100.$$

Carry out. Multiply the first equation by -3 and add it to the second equation to obtain

$7a - \frac{5}{2}c = -200$ or $a = \frac{5}{14}(c - 80)$ where $(c - 80)$ is a positive multiple of 14 (because a must be a positive integer). That is $(c - 80) = k \cdot 14$ or $c = 80 + k \cdot 14$, where k is a positive integer. If $k > 1$, then $c > 100$. This is impossible since the total attendance is 100. Thus $k = 1$, so $c = 80 + 1 \cdot 14 = 94$. Then $a = \frac{5}{14}(94 - 80) = \frac{5}{14} \cdot 14 = 5$, and $5 + s + 94 = 100$, or $s = 1$.

Check. The total attendance is $5 + 1 + 94$, or 100.

The total amount of money taken in was

$\$10 \cdot 5 + \$3 \cdot 1 + \$\frac{1}{2} \cdot 94 = \$100$. The result checks.

State. There were 5 adults, 1 student, and 94 children in attendance.

Exercise Set 6.3

1. The matrix has 3 rows and 2 columns, so its order is 3×2.

3. The matrix has 1 row and 4 columns, so its order is 1×4.

5. The matrix has 3 rows and 3 columns, so its order is 3×3.

7. We omit the variables and replace the equals signs with a vertical line.

$$\left[\begin{array}{rr|r} 2 & -1 & 7 \\ 1 & 4 & -5 \end{array}\right]$$

9. We omit the variables, writing zeros for the missing terms, and replace the equals signs with a vertical line.

$$\left[\begin{array}{rrr|r} 1 & -2 & 3 & 12 \\ 2 & 0 & -4 & 8 \\ 0 & 3 & 1 & 7 \end{array}\right]$$

11. Insert variables and replace the vertical line with equals signs.

$$3x - 5y = 1,$$
$$x + 4y = -2$$

13. Insert variables and replace the vertical line with equals signs.

$$2x + y - 4z = 12,$$
$$3x + 5z = -1,$$
$$x - y + z = 2$$

15.
$$4x + 2y = 11,$$
$$3x - y = 2$$

Write the augmented matrix. We will use Gaussian elimination.

$$\left[\begin{array}{rr|r} 4 & 2 & 11 \\ 3 & -1 & 2 \end{array}\right]$$

Multiply row 2 by 4 to make the first number in row 2 a multiple of 4.

$$\left[\begin{array}{rr|r} 4 & 2 & 11 \\ 12 & -4 & 8 \end{array}\right]$$

Multiply row 1 by -3 and add it to row 2.

$$\left[\begin{array}{rr|r} 4 & 2 & 11 \\ 0 & -10 & -25 \end{array}\right]$$

Multiply row 1 by $\frac{1}{4}$ and row 2 by $-\frac{1}{10}$.

$$\begin{bmatrix} 1 & \frac{1}{2} & \Big| & \frac{11}{4} \\ 0 & 1 & \Big| & \frac{5}{2} \end{bmatrix}$$

Write the system of equations that corresponds to the last matrix.

$$x + \frac{1}{2}y = \frac{11}{4}, \quad (1)$$
$$y = \frac{5}{2} \quad (2)$$

Back-substitute in equation (1) and solve for x.

$$x + \frac{1}{2} \cdot \frac{5}{2} = \frac{11}{4}$$
$$x + \frac{5}{4} = \frac{11}{4}$$
$$x = \frac{6}{4} = \frac{3}{2}$$

The solution is $\left(\frac{3}{2}, \frac{5}{2} \right)$.

17. $\quad 5x - 2y = -3,$
$\quad\quad 2x + 5y = -24$

Write the augmented matrix. We will use Gaussian elimination.

$$\begin{bmatrix} 5 & -2 & \Big| & -3 \\ 2 & 5 & \Big| & -24 \end{bmatrix}$$

Multiply row 2 by 5 to make the first number in row 2 a multiple of 5.

$$\begin{bmatrix} 5 & -2 & \Big| & -3 \\ 10 & 25 & \Big| & -120 \end{bmatrix}$$

Multiply row 1 by -2 and add it to row 2.

$$\begin{bmatrix} 5 & -2 & \Big| & -3 \\ 0 & 29 & \Big| & -114 \end{bmatrix}$$

Multiply row 1 by $\frac{1}{5}$ and row 2 by $\frac{1}{29}$.

$$\begin{bmatrix} 1 & -\frac{2}{5} & \Big| & -\frac{3}{5} \\ 0 & 1 & \Big| & -\frac{114}{29} \end{bmatrix}$$

Write the system of equations that corresponds to the last matrix.

$$x - \frac{2}{5}y = -\frac{3}{5}, \quad (1)$$
$$y = -\frac{114}{29} \quad (2)$$

Back-substitute in equation (1) and solve for x.

$$x - \frac{2}{5}\left(-\frac{114}{29} \right) = -\frac{3}{5}$$
$$x + \frac{228}{145} = -\frac{3}{5}$$
$$x = -\frac{315}{145} = -\frac{63}{29}$$

The solution is $\left(-\frac{63}{29}, -\frac{114}{29} \right)$.

19. $\quad 3x + 4y = 7,$
$\quad\quad -5x + 2y = 10$

Write the augmented matrix. We will use Gaussian elimination.

$$\begin{bmatrix} 3 & 4 & \Big| & 7 \\ -5 & 2 & \Big| & 10 \end{bmatrix}$$

Multiply row 2 by 3 to make the first number in row 2 a multiple of 3.

$$\begin{bmatrix} 3 & 4 & \Big| & 7 \\ -15 & 6 & \Big| & 30 \end{bmatrix}$$

Multiply row 1 by 5 and add it to row 2.

$$\begin{bmatrix} 3 & 4 & \Big| & 7 \\ 0 & 26 & \Big| & 65 \end{bmatrix}$$

Multiply row 1 by $\frac{1}{3}$ and row 2 by $\frac{1}{26}$.

$$\begin{bmatrix} 1 & \frac{4}{3} & \Big| & \frac{7}{3} \\ 0 & 1 & \Big| & \frac{5}{2} \end{bmatrix}$$

Write the system of equations that corresponds to the last matrix.

$$x + \frac{4}{3}y = \frac{7}{3}, \quad (1)$$
$$y = \frac{5}{2} \quad (2)$$

Back-substitute in equation (1) and solve for x.

$$x + \frac{4}{3} \cdot \frac{5}{2} = \frac{7}{3}$$
$$x + \frac{10}{3} = \frac{7}{3}$$
$$x = -\frac{3}{3} = -1$$

The solution is $\left(-1, \frac{5}{2} \right)$.

21. $\quad 3x + 2y = 6,$
$\quad\quad 2x - 3y = -9$

Write the augmented matrix. We will use Gauss-Jordan elimination.

$$\begin{bmatrix} 3 & 2 & | & 6 \\ 2 & -3 & | & -9 \end{bmatrix}$$

Multiply row 2 by 3 to make the first number in row 2 a multiple of 3.

$$\begin{bmatrix} 3 & 2 & | & 6 \\ 6 & -9 & | & -27 \end{bmatrix}$$

Multiply row 1 by -2 and add it to row 2.

$$\begin{bmatrix} 3 & 2 & | & 6 \\ 0 & -13 & | & -39 \end{bmatrix}$$

Multiply row 2 by $-\dfrac{1}{13}$.

$$\begin{bmatrix} 3 & 2 & | & 6 \\ 0 & 1 & | & 3 \end{bmatrix}$$

Multiply row 2 by -2 and add it to row 1.

$$\begin{bmatrix} 3 & 0 & | & 0 \\ 0 & 1 & | & 3 \end{bmatrix}$$

Multiply row 1 by $\dfrac{1}{3}$.

$$\begin{bmatrix} 1 & 0 & | & 0 \\ 0 & 1 & | & 3 \end{bmatrix}$$

We have $x = 0$, $y = 3$. The solution is $(0, 3)$.

23. $x - 3y = 8,$
    $2x - 6y = 3$

Write the augmented matrix.

$$\begin{bmatrix} 1 & -3 & | & 8 \\ 2 & -6 & | & 3 \end{bmatrix}$$

Multiply row 1 by -2 and add it to row 2.

$$\begin{bmatrix} 1 & -3 & | & 8 \\ 0 & 0 & | & -13 \end{bmatrix}$$

The last row corresponds to the false equation $0 = -13$, so there is no solution.

25. $-2x + 6y = 4,$
    $3x - 9y = -6$

Write the augmented matrix.

$$\begin{bmatrix} -2 & 6 & | & 4 \\ 3 & -9 & | & -6 \end{bmatrix}$$

Multiply row 1 by $-\dfrac{1}{2}$.

$$\begin{bmatrix} 1 & -3 & | & -2 \\ 3 & -9 & | & -6 \end{bmatrix}$$

Multiply row 1 by -3 and add it to row 2.

$$\begin{bmatrix} 1 & -3 & | & -2 \\ 0 & 0 & | & 0 \end{bmatrix}$$

The last row corresponds to the equation $0 = 0$ which is true for all values of x and y. Thus, the system of equations is dependent and is equivalent to the first equation $-2x + 6y = 4$, or $x - 3y = -2$. Solving for x, we get $x = 3y - 2$. Then the solutions are of the form $(3y - 2, y)$, where y is any real number.

27. $x + 2y - 3z = 9,$
    $2x - y + 2z = -8,$
    $3x - y - 4z = 3$

Write the augmented matrix. We will use Gauss-Jordan elimination.

$$\begin{bmatrix} 1 & 2 & -3 & | & 9 \\ 2 & -1 & 2 & | & -8 \\ 3 & -1 & -4 & | & 3 \end{bmatrix}$$

Multiply row 1 by -2 and add it to row 2. Also, multiply row 1 by -3 and add it to row 3.

$$\begin{bmatrix} 1 & 2 & -3 & | & 9 \\ 0 & -5 & 8 & | & -26 \\ 0 & -7 & 5 & | & -24 \end{bmatrix}$$

Multiply row 2 by $-\dfrac{1}{5}$ to get a 1 in the second row, second column.

$$\begin{bmatrix} 1 & 2 & -3 & | & 9 \\ 0 & 1 & -\dfrac{8}{5} & | & \dfrac{26}{5} \\ 0 & -7 & 5 & | & -24 \end{bmatrix}$$

Multiply row 2 by -2 and add it to row 1. Also, multiply row 2 by 7 and add it to row 3.

$$\begin{bmatrix} 1 & 0 & \dfrac{1}{5} & | & -\dfrac{7}{5} \\ 0 & 1 & -\dfrac{8}{5} & | & \dfrac{26}{5} \\ 0 & 0 & -\dfrac{31}{5} & | & \dfrac{62}{5} \end{bmatrix}$$

Multiply row 3 by $-\dfrac{5}{31}$ to get a 1 in the third row, third column.

$$\begin{bmatrix} 1 & 0 & \dfrac{1}{5} & | & -\dfrac{7}{5} \\ 0 & 1 & -\dfrac{8}{5} & | & \dfrac{26}{5} \\ 0 & 0 & 1 & | & -2 \end{bmatrix}$$

Multiply row 3 by $-\frac{1}{5}$ and add it to row 1. Also, multiply row 3 by $\frac{8}{5}$ and add it to row 2.

$$\left[\begin{array}{ccc|c} 1 & 0 & 0 & -1 \\ 0 & 1 & 0 & 2 \\ 0 & 0 & 1 & -2 \end{array}\right]$$

We have $x = -1$, $y = 2$, $z = -2$. The solution is $(-1, 2, -2)$.

29. $4x - y - 3z = 1,$

$\qquad 8x + y - z = 5,$

$\qquad 2x + y + 2z = 5$

Write the augmented matrix. We will use Gauss-Jordan elimination.

$$\left[\begin{array}{ccc|c} 4 & -1 & -3 & 1 \\ 8 & 1 & -1 & 5 \\ 2 & 1 & 2 & 5 \end{array}\right]$$

First interchange rows 1 and 3 so that each number below the first number in the first row is a multiple of that number.

$$\left[\begin{array}{ccc|c} 2 & 1 & 2 & 5 \\ 8 & 1 & -1 & 5 \\ 4 & -1 & -3 & 1 \end{array}\right]$$

Multiply row 1 by -4 and add it to row 2. Also, multiply row 1 by -2 and add it to row 3.

$$\left[\begin{array}{ccc|c} 2 & 1 & 2 & 5 \\ 0 & -3 & -9 & -15 \\ 0 & -3 & -7 & -9 \end{array}\right]$$

Multiply row 2 by -1 and add it to row 3.

$$\left[\begin{array}{ccc|c} 2 & 1 & 2 & 5 \\ 0 & -3 & -9 & -15 \\ 0 & 0 & 2 & 6 \end{array}\right]$$

Multiply row 2 by $-\frac{1}{3}$ to get a 1 in the second row, second column.

$$\left[\begin{array}{ccc|c} 2 & 1 & 2 & 5 \\ 0 & 1 & 3 & 5 \\ 0 & 0 & 2 & 6 \end{array}\right]$$

Multiply row 2 by -1 and add it to row 1.

$$\left[\begin{array}{ccc|c} 2 & 0 & -1 & 0 \\ 0 & 1 & 3 & 5 \\ 0 & 0 & 2 & 6 \end{array}\right]$$

Multiply row 3 by $\frac{1}{2}$ to get a 1 in the third row, third column.

$$\left[\begin{array}{ccc|c} 2 & 0 & -1 & 0 \\ 0 & 1 & 3 & 5 \\ 0 & 0 & 1 & 3 \end{array}\right]$$

Add row 3 to row 1. Also multiply row 3 by -3 and add it to row 2.

$$\left[\begin{array}{ccc|c} 2 & 0 & 0 & 3 \\ 0 & 1 & 0 & -4 \\ 0 & 0 & 1 & 3 \end{array}\right]$$

Finally, multiply row 1 by $\frac{1}{2}$.

$$\left[\begin{array}{ccc|c} 1 & 0 & 0 & \dfrac{3}{2} \\ 0 & 1 & 0 & -4 \\ 0 & 0 & 1 & 3 \end{array}\right]$$

We have $x = \dfrac{3}{2}$, $y = -4$, $z = 3$. The solution is $\left(\dfrac{3}{2}, -4, 3\right)$.

31. $x - 2y + 3z = 4,$

$\qquad 3x + y - z = 0,$

$\qquad 2x + 3y - 5z = 1$

Write the augmented matrix. We will use Gaussian elimination.

$$\left[\begin{array}{ccc|c} 1 & -2 & 3 & -4 \\ 3 & 1 & -1 & 0 \\ 2 & 3 & -5 & 1 \end{array}\right]$$

Multiply row 1 by -3 and add it to row 2. Also, multiply row 1 by -2 and add it to row 3.

$$\left[\begin{array}{ccc|c} 1 & -2 & 3 & -4 \\ 0 & 7 & -10 & 12 \\ 0 & 7 & -11 & 9 \end{array}\right]$$

Multiply row 2 by -1 and add it to row 3.

$$\left[\begin{array}{ccc|c} 1 & -2 & 3 & -4 \\ 0 & 7 & -10 & 12 \\ 0 & 0 & -1 & -3 \end{array}\right]$$

Multiply row 2 by $\frac{1}{7}$ and multiply row 3 by -1.

$$\left[\begin{array}{ccc|c} 1 & -2 & 3 & -4 \\ 0 & 1 & -\dfrac{10}{7} & \dfrac{12}{7} \\ 0 & 0 & 1 & 3 \end{array}\right]$$

Now write the system of equations that corresponds to the last matrix.

$$\begin{aligned} x - 2y + 3z &= -4, \quad (1) \\ y - \tfrac{10}{7}z &= \tfrac{12}{7}, \quad (2) \\ z &= 3 \quad (3) \end{aligned}$$

Back-substitute 3 for z in equation (2) and solve for y.

$$y - \frac{10}{7} \cdot 3 = \frac{12}{7}$$

$$y - \frac{30}{7} = \frac{12}{7}$$

$$y = \frac{42}{7} = 6$$

Back-substitute 6 for y and 3 for z in equation (1) and solve for x.

$$x - 2 \cdot 6 + 3 \cdot 3 = -4$$

$$x - 3 = -4$$

$$x = -1$$

The solution is $(-1, 6, 3)$.

33. $2x - 4y - 3z = 3,$

$x + 3y + z = -1,$

$5x + y - 2z = 2$

Write the augmented matrix.

$$\begin{bmatrix} 2 & -4 & -3 & 3 \\ 1 & 3 & 1 & -1 \\ 5 & 1 & -2 & 2 \end{bmatrix}$$

Interchange the first two rows to get a 1 in the first row, first column.

$$\begin{bmatrix} 1 & 3 & 1 & -1 \\ 2 & -4 & -3 & 3 \\ 5 & 1 & -2 & 2 \end{bmatrix}$$

Multiply row 1 by -2 and add it to row 2. Also, multiply row 1 by -5 and add it to row 3.

$$\begin{bmatrix} 1 & 3 & 1 & -1 \\ 0 & -10 & -5 & 5 \\ 0 & -14 & -7 & 7 \end{bmatrix}$$

Multiply row 2 by $-\frac{1}{10}$ to get a 1 in the second row, second column.

$$\begin{bmatrix} 1 & 3 & 1 & -1 \\ 0 & 1 & \frac{1}{2} & -\frac{1}{2} \\ 0 & -14 & -7 & 7 \end{bmatrix}$$

Multiply row 2 by 14 and add it to row 3.

$$\begin{bmatrix} 1 & 3 & 1 & -1 \\ 0 & 1 & \frac{1}{2} & -\frac{1}{2} \\ 0 & 0 & 0 & 0 \end{bmatrix}$$

The last row corresponds to the equation $0 = 0$. This indicates that the system of equations is dependent. It is equivalent to

$$x + 3y + z = -1,$$

$$y + \frac{1}{2}z = -\frac{1}{2}.$$

We solve the second equation for y.

$$y = -\frac{1}{2}z - \frac{1}{2}$$

Substitute for y in the first equation and solve for x.

$$x + 3\left(-\frac{1}{2}z - \frac{1}{2}\right) + z = -1$$

$$x - \frac{3}{2}z - \frac{3}{2} + z = -1$$

$$x = \frac{1}{2}z + \frac{1}{2}$$

The solution is $\left(\frac{1}{2}z + \frac{1}{2}, -\frac{1}{2}z - \frac{1}{2}, z\right)$, where z is any real number.

35. $p + q + r = 1,$

$p + 2q + 3r = 4,$

$4p + 5q + 6r = 7$

Write the augmented matrix.

$$\begin{bmatrix} 1 & 1 & 1 & 1 \\ 1 & 2 & 3 & 4 \\ 4 & 5 & 6 & 7 \end{bmatrix}$$

Multiply row 1 by -1 and add it to row 2. Also, multiply row 1 by -4 and add it to row 3.

$$\begin{bmatrix} 1 & 1 & 1 & 1 \\ 0 & 1 & 2 & 3 \\ 0 & 1 & 2 & 3 \end{bmatrix}$$

Multiply row 2 by -1 and add it to row 3.

$$\begin{bmatrix} 1 & 1 & 1 & 1 \\ 0 & 1 & 2 & 3 \\ 0 & 0 & 0 & 0 \end{bmatrix}$$

The last row corresponds to the equation $0 = 0$. This indicates that the system of equations is dependent. It is equivalent to

$$p + q + r = 1,$$

$$q + 2r = 3.$$

We solve the second equation for q.

$$q = -2r + 3$$

Substitute for y in the first equation and solve for p.

$$p - 2r + 3 + r = 1$$

$$p - r + 3 = 1$$

$$p = r - 2$$

The solution is $(r - 2, -2r + 3, r)$, where r is any real number.

37. $a + b - c = 7,$

$a - b + c = 5,$

$3a + b - c = -1$

Write the augmented matrix.

$$\begin{bmatrix} 1 & 1 & -1 & 7 \\ 1 & -1 & 1 & 5 \\ 3 & 1 & -1 & -1 \end{bmatrix}$$

Multiply row 1 by -1 and add it to row 2. Also, multiply row 1 by -3 and add it to row 3.

$$\begin{bmatrix} 1 & 1 & -1 & 7 \\ 0 & -2 & 2 & -2 \\ 0 & -2 & 2 & -22 \end{bmatrix}$$

Multiply row 2 by -1 and add it to row 3.

$$\begin{bmatrix} 1 & 1 & -1 & 7 \\ 0 & -2 & 2 & -2 \\ 0 & 0 & 0 & -20 \end{bmatrix}$$

The last row corresponds to the false equation $0 = -20$. Thus, the system of equations has no solution.

39. $-2w + 2x + 2y - 2z = -10,$
$\quad\ \ w + \ \ x + \ \ y + \ \ z = -5,$
$\quad\ 3w + \ \ x - \ \ y + 4z = -2,$
$\quad\ \ w + 3x - 2y + 2z = -6$

Write the augmented matrix. We will use Gaussian elimination.

$$\begin{bmatrix} -2 & 2 & 2 & -2 & -10 \\ 1 & 1 & 1 & 1 & -5 \\ 3 & 1 & -1 & 4 & -2 \\ 1 & 3 & -2 & 2 & -6 \end{bmatrix}$$

Interchange rows 1 and 2.

$$\begin{bmatrix} 1 & 1 & 1 & 1 & -5 \\ -2 & 2 & 2 & -2 & -10 \\ 3 & 1 & -1 & 4 & -2 \\ 1 & 3 & -2 & 2 & -6 \end{bmatrix}$$

Multiply row 1 by 2 and add it to row 2. Multiply row 1 by -3 and add it to row 3. Multiply row 1 by -1 and add it to row 4.

$$\begin{bmatrix} 1 & 1 & 1 & 1 & -5 \\ 0 & 4 & 4 & 0 & -20 \\ 0 & -2 & -4 & 1 & 13 \\ 0 & 2 & -3 & 1 & -1 \end{bmatrix}$$

Interchange rows 2 and 3.

$$\begin{bmatrix} 1 & 1 & 1 & 1 & -5 \\ 0 & -2 & -4 & 1 & 13 \\ 0 & 4 & 4 & 0 & -20 \\ 0 & 2 & -3 & 1 & -1 \end{bmatrix}$$

Multiply row 2 by 2 and add it to row 3. Add row 2 to row 4.

$$\begin{bmatrix} 1 & 1 & 1 & 1 & -5 \\ 0 & -2 & -4 & 1 & 13 \\ 0 & 0 & -4 & 2 & 6 \\ 0 & 0 & -7 & 2 & 12 \end{bmatrix}$$

Multiply row 4 by 4.

$$\begin{bmatrix} 1 & 1 & 1 & 1 & -5 \\ 0 & -2 & -4 & 1 & 13 \\ 0 & 0 & -4 & 2 & 6 \\ 0 & 0 & -28 & 8 & 48 \end{bmatrix}$$

Multiply row 3 by -7 and add it to row 4.

$$\begin{bmatrix} 1 & 1 & 1 & 1 & -5 \\ 0 & -2 & -4 & 1 & 13 \\ 0 & 0 & -4 & 2 & 6 \\ 0 & 0 & 0 & -6 & 6 \end{bmatrix}$$

Multiply row 2 by $-\dfrac{1}{2}$, row 3 by $-\dfrac{1}{4}$, and row 6 by $-\dfrac{1}{6}$.

$$\begin{bmatrix} 1 & 1 & 1 & 1 & -5 \\ 0 & 1 & 2 & -\dfrac{1}{2} & -\dfrac{13}{2} \\ 0 & 0 & 1 & -\dfrac{1}{2} & -\dfrac{3}{2} \\ 0 & 0 & 0 & 1 & -1 \end{bmatrix}$$

Write the system of equations that corresponds to the last matrix.

$$w + x + \ y + \ z = -5, \quad (1)$$
$$x + 2y - \frac{1}{2}z = -\frac{13}{2}, \quad (2)$$
$$y - \frac{1}{2}z = -\frac{3}{2}, \quad (3)$$
$$z = -1 \quad (4)$$

Back-substitute in equation (3) and solve for y.

$$y - \frac{1}{2}(-1) = -\frac{3}{2}$$
$$y + \frac{1}{2} = -\frac{3}{2}$$
$$y = -2$$

Back-substitute in equation (2) and solve for x.

$$x + 2(-2) - \frac{1}{2}(-1) = -\frac{13}{2}$$
$$x - 4 + \frac{1}{2} = -\frac{13}{2}$$
$$x = -3$$

Back-substitute in equation (1) and solve for w.

$$w - 3 - 2 - 1 = -5$$
$$w = 1$$

The solution is $(1, -3, -2, -1)$.

41. *Familiarize.* Let x, y, and z represent the amounts borrowed at 8%, 10%, and 12%, respectively. Then the annual interest is 8%x, 10%y, and 12%z, or $0.08x$, $0.1y$, and $0.12z$.

Translate.

The total amount borrowed was $30,000.

$$x + y + z = 30,000$$

The total annual interest was $3040.

$$0.08x + 0.1y + 0.12z = 3040$$

The total amount borrowed at 8% and 10% was twice the amount borrowed at 12%.

$$x + y = 2z$$

We have a system of equations.

$$x + y + z = 30,000,$$
$$0.08x + 0.1y + 0.12z = 3040,$$
$$x + y = 2z, \text{ or}$$

$$x + \quad y + \quad z = 30,000,$$
$$0.08x + 0.1y + 0.12z = 3040,$$
$$x + \quad y - \quad 2z = 0$$

Carry out. Using Gaussian elimination or Gauss-Jordan elimination, we find that the solution is $(8000, 12,000, 10,000)$.

Check. The total amount borrowed was $8000 + $12,000 + $10,000$, or $30,000. The total annual interest was $0.08($8000) + 0.1($12,000) + 0.12($10,000)$, or $640 + $1200 + 1200, or $3040. The total amount borrowed at 8% and 10%, $8000 + $12,000 or $20,000, was twice the amount borrowed at 12%, $10,000. The solution checks.

State. The amounts borrowed at 8%, 10%, and 12% were $8000, $12,000 and $10,000, respectively.

43. *Familiarize.* Let x and y represent the number of 49¢ and 21¢ stamps purchased, respectively. Then Olivia paid $0.49x$ for 49¢ stamps and $0.21y$ for the 21¢ stamps.

Translate. The purchase cost $86.80.

$$0.49x + 0.21y = 86.80$$

Oliva bought a total of 200 stamps.

$$x + y = 200$$

Carry out. Use Gaussian elimination or Gauss-Jordan Elimination to solve the system of equations.

$$0.49x + 0.21y = 86.80,$$
$$x + \quad y = 200$$

The solution is $(160, 40)$.

Check. The purchase cost $0.49(160) + 0.21(40) = 78.40 + 8.40 = 86.80. The number of stamps purchased is $160 + 40$, or 200. The answer checks.

State. Olivia bought 160 49¢ stamps and 40 21¢ stamps.

45. The function has a variable in the exponent, so it is an exponential function.

47. The function is the quotient of two polynomials, so it is a rational function.

49. The function is of the form $f(x) = \log_a x$, so it is logarithmic.

51. The function is of the form $f(x) = mx + b$, so it is linear.

53. Substitute to find three equations.

$$12 = a(-3)^2 + b(-3) + c$$
$$-7 = a(-1)^2 + b(-1) + c$$
$$-2 = a \cdot 1^2 + b \cdot 1 + c$$

We have a system of equations.

$$9a - 3b + c = 12,$$
$$a - \quad b + c = -7,$$
$$a + \quad b + c = -2$$

Write the augmented matrix. We will use Gaussian elimination.

$$\begin{bmatrix} 9 & -3 & 1 & | & 12 \\ 1 & -1 & 1 & | & -7 \\ 1 & 1 & 1 & | & -2 \end{bmatrix}$$

Interchange the first two rows.

$$\begin{bmatrix} 1 & -1 & 1 & | & -7 \\ 9 & -3 & 1 & | & 12 \\ 1 & 1 & 1 & | & -2 \end{bmatrix}$$

Multiply row 1 by -9 and add it to row 2. Also, multiply row 1 by -1 and add it to row 3.

$$\begin{bmatrix} 1 & -1 & 1 & | & -7 \\ 0 & 6 & -8 & | & 75 \\ 0 & 2 & 0 & | & 5 \end{bmatrix}$$

Interchange row 2 and row 3.

$$\begin{bmatrix} 1 & -1 & 1 & | & -7 \\ 0 & 2 & 0 & | & 5 \\ 0 & 6 & -8 & | & 75 \end{bmatrix}$$

Multiply row 2 by -3 and add it to row 3.

$$\begin{bmatrix} 1 & -1 & 1 & | & -7 \\ 0 & 2 & 0 & | & 5 \\ 0 & 0 & -8 & | & 60 \end{bmatrix}$$

Multiply row 2 by $\frac{1}{2}$ and row 3 by $-\frac{1}{8}$.

$$\begin{bmatrix} 1 & -1 & 1 & | & -7 \\ 0 & 1 & 0 & | & \frac{5}{2} \\ 0 & 0 & 1 & | & -\frac{15}{2} \end{bmatrix}$$

Write the system of equations that corresponds to the last matrix.

$$x - y + z = -7,$$
$$y = \frac{5}{2},$$
$$z = -\frac{15}{2}$$

Back-substitute $\frac{5}{2}$ for y and $-\frac{15}{2}$ for z in the first equation and solve for x.

$$x - \frac{5}{2} - \frac{15}{2} = -7$$
$$x - 10 = -7$$
$$x = 3$$

The solution is $\left(3, \frac{5}{2}, -\frac{15}{2}\right)$, so the equation is

$$y = 3x^2 + \frac{5}{2}x - \frac{15}{2}.$$

55. $\begin{bmatrix} 1 & 5 \\ 3 & 2 \end{bmatrix}$

Multiply row 1 by -3 and add it to row 2.

$\begin{bmatrix} 1 & 5 \\ 0 & -13 \end{bmatrix}$

Multiply row 2 by $-\frac{1}{13}$.

$\begin{bmatrix} 1 & 5 \\ 0 & 1 \end{bmatrix}$ Row-echelon form

Multiply row 2 by -5 and add it to row 1.

$\begin{bmatrix} 1 & 0 \\ 0 & 1 \end{bmatrix}$ Reduced row-echelon form

57. $y = x + z,$
$3y + 5z = 4,$
$x + 4 = y + 3z,$ or
$x - y + z = 0,$
$3y + 5z = 4,$
$x - y - 3z = -4$

Write the augmented matrix. We will use Gauss-Jordan elimination.

$\left[\begin{array}{rrr|r} 1 & -1 & 1 & 0 \\ 0 & 3 & 5 & 4 \\ 1 & -1 & -3 & -4 \end{array}\right]$

Multiply row 1 by -1 and add it to row 3.

$\left[\begin{array}{rrr|r} 1 & -1 & 1 & 0 \\ 0 & 3 & 5 & 4 \\ 0 & 0 & -4 & -4 \end{array}\right]$

Multiply row 3 by $-\frac{1}{4}$.

$\left[\begin{array}{rrr|r} 1 & -1 & 1 & 0 \\ 0 & 3 & 5 & 4 \\ 0 & 0 & 1 & 1 \end{array}\right]$

Multiply row 3 by -1 and add it to row 1. Also, multiply row 3 by -5 and add it to row 2.

$\left[\begin{array}{rrr|r} 1 & -1 & 0 & -1 \\ 0 & 3 & 0 & -1 \\ 0 & 0 & 1 & 1 \end{array}\right]$

Multiply row 2 by $\frac{1}{3}$.

$\left[\begin{array}{rrr|r} 1 & -1 & 0 & -1 \\ 0 & 1 & 0 & -\frac{1}{3} \\ 0 & 0 & 1 & 1 \end{array}\right]$

Add row 2 to row 1.

$\left[\begin{array}{rrr|r} 1 & 0 & 0 & -\frac{4}{3} \\ 0 & 1 & 0 & -\frac{1}{3} \\ 0 & 0 & 1 & 1 \end{array}\right]$

Read the solution from the last matrix. It is
$\left(-\frac{4}{3}, -\frac{1}{3}, 1\right).$

59. $x - 4y + 2z = 7,$
$3x + y + 3z = -5$

Write the augmented matrix.

$\left[\begin{array}{rrr|r} 1 & -4 & 2 & 7 \\ 3 & 1 & 3 & -5 \end{array}\right]$

Multiply row 1 by -3 and add it to row 2.

$\left[\begin{array}{rrr|r} 1 & -4 & 2 & 7 \\ 0 & 13 & -3 & -26 \end{array}\right]$

Multiply row 2 by $\frac{1}{13}$.

$\left[\begin{array}{rrr|r} 1 & -4 & 2 & 7 \\ 0 & 1 & -\frac{3}{13} & -2 \end{array}\right]$

Write the system of equations that corresponds to the last matrix.

$$x - 4y + 2z = 7,$$
$$y - \frac{3}{13}z = -2$$

Solve the second equation for y.

$$y = \frac{3}{13}z - 2$$

Substitute in the first equation and solve for x.

$$x - 4\left(\frac{3}{13}z - 2\right) + 2z = 7$$

$$x - \frac{12}{13}z + 8 + 2z = 7$$

$$x = -\frac{14}{13}z - 1$$

The solution is $\left(-\frac{14}{13}z - 1, \frac{3}{13}z - 2, z\right)$, where z is any real number.

61.
$$4x + 5y = 3,$$
$$-2x + y = 9,$$
$$3x - 2y = -15$$

Write the augmented matrix.

$$\begin{bmatrix} 4 & 5 & | & 3 \\ -2 & 1 & | & 9 \\ 3 & -2 & | & -15 \end{bmatrix}$$

Multiply row 2 by 2 and row 3 by 4.

$$\begin{bmatrix} 4 & 5 & | & 3 \\ -4 & 2 & | & 18 \\ 12 & -8 & | & -60 \end{bmatrix}$$

Add row 1 to row 2. Also, multiply row 1 by -3 and add it to row 3.

$$\begin{bmatrix} 4 & 5 & | & 3 \\ 0 & 7 & | & 21 \\ 0 & -23 & | & -69 \end{bmatrix}$$

Multiply row 2 by $\frac{1}{7}$ and row 3 by $-\frac{1}{23}$.

$$\begin{bmatrix} 4 & 5 & | & 3 \\ 0 & 1 & | & 3 \\ 0 & 1 & | & 3 \end{bmatrix}$$

Multiply row 2 by -1 and add it to row 3.

$$\begin{bmatrix} 4 & 5 & | & 3 \\ 0 & 1 & | & 3 \\ 0 & 0 & | & 0 \end{bmatrix}$$

The last row corresponds to the equation $0 = 0$. Thus we have a dependent system that is equivalent to

$$4x + 5y = 3, \quad (1)$$
$$y = 3. \quad (2)$$

Back-substitute in equation (1) to find x.

$$4x + 5 \cdot 3 = 3$$
$$4x + 15 = 3$$
$$4x = -12$$
$$x = -3$$

The solution is $(-3, 3)$.

Exercise Set 6.4

1. $\begin{bmatrix} 5 & x \end{bmatrix} = \begin{bmatrix} y & -3 \end{bmatrix}$

Corresponding entries of the two matrices must be equal. Thus we have $5 = y$ and $x = -3$.

3. $\begin{bmatrix} 3 & 2x \\ y & -8 \end{bmatrix} = \begin{bmatrix} 3 & -2 \\ 1 & -8 \end{bmatrix}$

Corresponding entries of the two matrices must be equal. Thus, we have:

$$2x = -2 \quad \text{and} \quad y = 1$$
$$x = -1 \quad \text{and} \quad y = 1$$

5. $\mathbf{A} + \mathbf{B} = \begin{bmatrix} 1 & 2 \\ 4 & 3 \end{bmatrix} + \begin{bmatrix} -3 & 5 \\ 2 & -1 \end{bmatrix}$

$$= \begin{bmatrix} 1 + (-3) & 2 + 5 \\ 4 + 2 & 3 + (-1) \end{bmatrix}$$

$$= \begin{bmatrix} -2 & 7 \\ 6 & 2 \end{bmatrix}$$

7. $\mathbf{E} + \mathbf{0} = \begin{bmatrix} 1 & 3 \\ 2 & 6 \end{bmatrix} + \begin{bmatrix} 0 & 0 \\ 0 & 0 \end{bmatrix}$

$$= \begin{bmatrix} 1 + 0 & 3 + 0 \\ 2 + 0 & 6 + 0 \end{bmatrix}$$

$$= \begin{bmatrix} 1 & 3 \\ 2 & 6 \end{bmatrix}$$

9. $3\mathbf{F} = 3\begin{bmatrix} 3 & 3 \\ -1 & -1 \end{bmatrix}$

$$= \begin{bmatrix} 3 \cdot 3 & 3 \cdot 3 \\ 3 \cdot (-1) & 3 \cdot (-1) \end{bmatrix}$$

$$= \begin{bmatrix} 9 & 9 \\ -3 & -3 \end{bmatrix}$$

11. $3\mathbf{F} = 3\begin{bmatrix} 3 & 3 \\ -1 & -1 \end{bmatrix} = \begin{bmatrix} 9 & 9 \\ -3 & -3 \end{bmatrix},$

$$2\mathbf{A} = 2\begin{bmatrix} 1 & 2 \\ 4 & 3 \end{bmatrix} = \begin{bmatrix} 2 & 4 \\ 8 & 6 \end{bmatrix}$$

$$3\mathbf{F} + 2\mathbf{A} = \begin{bmatrix} 9 & 9 \\ -3 & -3 \end{bmatrix} + \begin{bmatrix} 2 & 4 \\ 8 & 6 \end{bmatrix}$$

$$= \begin{bmatrix} 9 + 2 & 9 + 4 \\ -3 + 8 & -3 + 6 \end{bmatrix}$$

$$= \begin{bmatrix} 11 & 13 \\ 5 & 3 \end{bmatrix}$$

13. $\mathbf{B} - \mathbf{A} = \begin{bmatrix} -3 & 5 \\ 2 & -1 \end{bmatrix} - \begin{bmatrix} 1 & 2 \\ 4 & 3 \end{bmatrix}$

$$= \begin{bmatrix} -3 & 5 \\ 2 & -1 \end{bmatrix} + \begin{bmatrix} -1 & -2 \\ -4 & -3 \end{bmatrix}$$

$$[\mathbf{B} - \mathbf{A} = \mathbf{B} + (-\mathbf{A})]$$

$$= \begin{bmatrix} -3 + (-1) & 5 + (-2) \\ 2 + (-4) & -1 + (-3) \end{bmatrix}$$

$$= \begin{bmatrix} -4 & 3 \\ -2 & -4 \end{bmatrix}$$

15. $\mathbf{BA} = \begin{bmatrix} -3 & 5 \\ 2 & -1 \end{bmatrix} \begin{bmatrix} 1 & 2 \\ 4 & 3 \end{bmatrix}$

$= \begin{bmatrix} -3 \cdot 1 + 5 \cdot 4 & -3 \cdot 2 + 5 \cdot 3 \\ 2 \cdot 1 + (-1)4 & 2 \cdot 2 + (-1)3 \end{bmatrix}$

$= \begin{bmatrix} 17 & 9 \\ -2 & 1 \end{bmatrix}$

17. $\mathbf{CD} = \begin{bmatrix} 1 & -1 \\ -1 & 1 \end{bmatrix} \begin{bmatrix} 1 & 1 \\ 1 & 1 \end{bmatrix}$

$= \begin{bmatrix} 1 \cdot 1 + (-1) \cdot 1 & 1 \cdot 1 + (-1) \cdot 1 \\ -1 \cdot 1 + 1 \cdot 1 & -1 \cdot 1 + 1 \cdot 1 \end{bmatrix}$

$= \begin{bmatrix} 0 & 0 \\ 0 & 0 \end{bmatrix}$

19. $\mathbf{AI} = \begin{bmatrix} 1 & 2 \\ 4 & 3 \end{bmatrix} \begin{bmatrix} 1 & 0 \\ 0 & 1 \end{bmatrix}$

$= \begin{bmatrix} 1 \cdot 1 + 2 \cdot 0 & 1 \cdot 0 + 2 \cdot 1 \\ 4 \cdot 1 + 3 \cdot 0 & 4 \cdot 0 + 3 \cdot 1 \end{bmatrix}$

$= \begin{bmatrix} 1 & 2 \\ 4 & 3 \end{bmatrix}$

21. $\begin{bmatrix} -1 & 0 & 7 \\ 3 & -5 & 2 \end{bmatrix} \begin{bmatrix} 6 \\ -4 \\ 1 \end{bmatrix}$

$= \begin{bmatrix} -1 \cdot 6 + 0(-4) + 7 \cdot 1 \\ 3 \cdot 6 + (-5)(-4) + 2 \cdot 1 \end{bmatrix}$

$= \begin{bmatrix} 1 \\ 40 \end{bmatrix}$

23. $\begin{bmatrix} -2 & 4 \\ 5 & 1 \\ -1 & -3 \end{bmatrix} \begin{bmatrix} 3 & -6 \\ -1 & 4 \end{bmatrix}$

$= \begin{bmatrix} -2 \cdot 3 + 4(-1) & -2(-6) + 4 \cdot 4 \\ 5 \cdot 3 + 1(-1) & 5(-6) + 1 \cdot 4 \\ -1 \cdot 3 + (-3)(-1) & -1(-6) + (-3) \cdot 4 \end{bmatrix}$

$= \begin{bmatrix} -10 & 28 \\ 14 & -26 \\ 0 & -6 \end{bmatrix}$

25. $\begin{bmatrix} 1 \\ -5 \\ 3 \end{bmatrix} \begin{bmatrix} -6 & 5 & 8 \\ 0 & 4 & -1 \end{bmatrix}$

This product is not defined because the number of columns of the first matrix, 1, is not equal to the number of rows of the second matrix, 2.

27. $\begin{bmatrix} 1 & -4 & 3 \\ 0 & 8 & 0 \\ -2 & -1 & 5 \end{bmatrix} \begin{bmatrix} 3 & 0 & 0 \\ 0 & -4 & 0 \\ 0 & 0 & 1 \end{bmatrix}$

$= \begin{bmatrix} 3 + 0 + 0 & 0 + 16 + 0 & 0 + 0 + 3 \\ 0 + 0 + 0 & 0 - 32 + 0 & 0 + 0 + 0 \\ -6 + 0 + 0 & 0 + 4 + 0 & 0 + 0 + 5 \end{bmatrix}$

$= \begin{bmatrix} 3 & 16 & 3 \\ 0 & -32 & 0 \\ -6 & 4 & 5 \end{bmatrix}$

29. a) $\mathbf{A} = \begin{bmatrix} 40 & 20 & 30 \end{bmatrix}$

b) $40 + 10\% \cdot 40 = 1.1(40) = 44$

$20 + 10\% \cdot 20 = 1.1(20) = 22$

$30 + 10\% \cdot 30 = 1.1(30) = 33$

We write the matrix that corresponds to these amounts.

$\mathbf{B} = \begin{bmatrix} 44 & 22 & 33 \end{bmatrix}$

c) $\mathbf{A} + \mathbf{B} = \begin{bmatrix} 40 & 20 & 30 \end{bmatrix} + \begin{bmatrix} 44 & 22 & 33 \end{bmatrix}$

$= \begin{bmatrix} 84 & 42 & 63 \end{bmatrix}$

The entries represent the total amount of each type of produce ordered for both weeks.

31. a) $\mathbf{C} = \begin{bmatrix} 140 & 27 & 3 & 13 & 64 \end{bmatrix}$

$\mathbf{P} = \begin{bmatrix} 180 & 4 & 11 & 24 & 662 \end{bmatrix}$

$\mathbf{B} = \begin{bmatrix} 50 & 5 & 1 & 82 & 20 \end{bmatrix}$

b) $\mathbf{C} + 2\mathbf{P} + 3\mathbf{B}$

$= \begin{bmatrix} 140 & 27 & 3 & 13 & 64 \end{bmatrix} +$
$\begin{bmatrix} 360 & 8 & 22 & 48 & 1324 \end{bmatrix} +$
$\begin{bmatrix} 150 & 15 & 3 & 246 & 60 \end{bmatrix}$

$= \begin{bmatrix} 650 & 50 & 28 & 307 & 1448 \end{bmatrix}$

The entries represent the total nutritional value of one serving of chicken, 1 cup of potato salad, and 3 broccoli spears.

33. a) $\mathbf{M} = \begin{bmatrix} 1.50 & 0.30 & 0.36 & 0.45 & 0.64 \\ 1.55 & 0.28 & 0.48 & 0.57 & 0.75 \\ 1.62 & 0.52 & 0.65 & 0.38 & 0.53 \\ 1.70 & 0.43 & 0.40 & 0.42 & 0.68 \end{bmatrix}$

b) $\mathbf{N} = \begin{bmatrix} 65 & 48 & 93 & 57 \end{bmatrix}$

c) $\mathbf{NM} = \begin{bmatrix} 419.46 & 105.81 & 129.04 & 115.84 & 165.65 \end{bmatrix}$

d) The entries of $\mathbf{NM}$ represent the total cost, in dollars, of each item for the day's meals.

35. a) $\mathbf{S} = \begin{bmatrix} 8 & 15 \\ 6 & 10 \\ 4 & 3 \end{bmatrix}$

b) $\mathbf{C} = \begin{bmatrix} 4 & 2.50 & 3 \end{bmatrix}$

c) $\mathbf{CS} = \begin{bmatrix} 59 & 94 \end{bmatrix}$

d) The entries of $\mathbf{CS}$ represent the total cost, in dollars, of ingredients for each coffee shop.

37. a) $\mathbf{P} = \begin{bmatrix} 7.50 & 4.80 & 6.25 \end{bmatrix}$

b) $\mathbf{PS} = \begin{bmatrix} 7.50 & 4.80 & 6.25 \end{bmatrix} \begin{bmatrix} 8 & 15 \\ 6 & 10 \\ 4 & 3 \end{bmatrix}$

$= \begin{bmatrix} 113.8 & 179.25 \end{bmatrix}$

The profit from Mugsey's Coffee Shop is \$113.80, and the profit from The Coffee Club is \$179.25.

39. $2x - 3y = 7,$

$x + 5y = -6$

Write the coefficients on the left in a matrix. Then write the product of that matrix and the column matrix containing the variables, and set the result equal to the column matrix containing the constants on the right.

$$\begin{bmatrix} 2 & -3 \\ 1 & 5 \end{bmatrix} \begin{bmatrix} x \\ y \end{bmatrix} = \begin{bmatrix} 7 \\ -6 \end{bmatrix}$$

41. $x + y - 2z = 6,$

$3x - y + z = 7,$

$2x + 5y - 3z = 8$

Write the coefficients on the left in a matrix. Then write the product of that matrix and the column matrix containing the variables, and set the result equal to the column matrix containing the constants on the right.

$$\begin{bmatrix} 1 & 1 & -2 \\ 3 & -1 & 1 \\ 2 & 5 & -3 \end{bmatrix} \begin{bmatrix} x \\ y \\ z \end{bmatrix} = \begin{bmatrix} 6 \\ 7 \\ 8 \end{bmatrix}$$

43. $3x - 2y + 4z = 17,$

$2x + y - 5z = 13$

Write the coefficients on the left in a matrix. Then write the product of that matrix and the column matrix containing the variables, and set the result equal to the column matrix containing the constants on the right.

$$\begin{bmatrix} 3 & -2 & 4 \\ 2 & 1 & -5 \end{bmatrix} \begin{bmatrix} x \\ y \\ z \end{bmatrix} = \begin{bmatrix} 17 \\ 13 \end{bmatrix}$$

45. $-4w + x - y + 2z = 12,$

$w + 2x - y - z = 0,$

$-w + x + 4y - 3z = 1,$

$2w + 3x + 5y - 7z = 9$

Write the coefficients on the left in a matrix. Then write the product of that matrix and the column matrix containing the variables, and set the result equal to the column matrix containing the constants on the right.

$$\begin{bmatrix} -4 & 1 & -1 & 2 \\ 1 & 2 & -1 & -1 \\ -1 & 1 & 4 & -3 \\ 2 & 3 & 5 & -7 \end{bmatrix} \begin{bmatrix} w \\ x \\ y \\ z \end{bmatrix} = \begin{bmatrix} 12 \\ 0 \\ 1 \\ 9 \end{bmatrix}$$

47. $f(x) = x^2 - x - 6$

a) $-\dfrac{b}{2a} = -\dfrac{-1}{2 \cdot 1} = \dfrac{1}{2}$

$f\left(\dfrac{1}{2}\right) = \left(\dfrac{1}{2}\right)^2 - \dfrac{1}{2} - 6 = -\dfrac{25}{4}$

The vertex is $\left(\dfrac{1}{2}, -\dfrac{25}{4}\right).$

b) The axis of symmetry is $x = \dfrac{1}{2}.$

c) Since the coefficient of x^2 is positive, the function has a minimum value. It is the second coordinate of the vertex, $-\dfrac{25}{4}.$

d) Plot some points and draw the graph of the function.

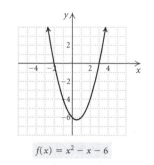

$f(x) = x^2 - x - 6$

49. $f(x) = -x^2 - 3x + 2$

a) $-\dfrac{b}{2a} = -\dfrac{-3}{2(-1)} = -\dfrac{3}{2}$

$f\left(-\dfrac{3}{2}\right) = -\left(-\dfrac{3}{2}\right)^2 - 3\left(-\dfrac{3}{2}\right) + 2 = \dfrac{17}{4}$

The vertex is $\left(-\dfrac{3}{2}, \dfrac{17}{4}\right).$

b) The axis of symmetry is $x = -\dfrac{3}{2}.$

c) Since the coefficient of x^2 is negative, the function has a maximum value. It is the second coordinate of the vertex, $\dfrac{17}{4}.$

d) Plot some points and draw the graph of the function.

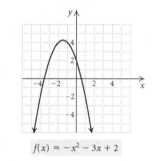

$f(x) = -x^2 - 3x + 2$

51. $\mathbf{A} = \begin{bmatrix} -1 & 0 \\ 2 & 1 \end{bmatrix}, \mathbf{B} = \begin{bmatrix} 1 & -1 \\ 0 & 2 \end{bmatrix}$

$(\mathbf{A} + \mathbf{B})(\mathbf{A} - \mathbf{B}) = \begin{bmatrix} 0 & -1 \\ 2 & 3 \end{bmatrix} \begin{bmatrix} -2 & 1 \\ 2 & -1 \end{bmatrix}$

$= \begin{bmatrix} -2 & 1 \\ 2 & -1 \end{bmatrix}$

$\mathbf{A}^2 - \mathbf{B}^2$

$= \begin{bmatrix} -1 & 0 \\ 2 & 1 \end{bmatrix} \begin{bmatrix} -1 & 0 \\ 2 & 1 \end{bmatrix} - \begin{bmatrix} 1 & -1 \\ 0 & 2 \end{bmatrix} \begin{bmatrix} 1 & -1 \\ 0 & 2 \end{bmatrix}$

$= \begin{bmatrix} 1 & 0 \\ 0 & 1 \end{bmatrix} - \begin{bmatrix} 1 & -3 \\ 0 & 4 \end{bmatrix}$

$= \begin{bmatrix} 0 & 3 \\ 0 & -3 \end{bmatrix}$

Thus $(\mathbf{A} + \mathbf{B})(\mathbf{A} - \mathbf{B}) \neq \mathbf{A}^2 - \mathbf{B}^2.$

53. In Exercise 51 we found that $(\mathbf{A} + \mathbf{B})(\mathbf{A} - \mathbf{B}) =$
$\begin{bmatrix} -2 & 1 \\ 2 & -1 \end{bmatrix}$
and we also found $\mathbf{A}^2$ and $\mathbf{B}^2$.

$\mathbf{BA} = \begin{bmatrix} 1 & -1 \\ 0 & 2 \end{bmatrix} \begin{bmatrix} -1 & 0 \\ 2 & 1 \end{bmatrix} = \begin{bmatrix} -3 & -1 \\ 4 & 2 \end{bmatrix}$

$\mathbf{AB} = \begin{bmatrix} -1 & 0 \\ 2 & 1 \end{bmatrix} \begin{bmatrix} 1 & -1 \\ 0 & 2 \end{bmatrix} = \begin{bmatrix} -1 & 1 \\ 2 & 0 \end{bmatrix}$

$\mathbf{A}^2 + \mathbf{BA} - \mathbf{AB} - \mathbf{B}^2$

$= \begin{bmatrix} 1 & 0 \\ 0 & 1 \end{bmatrix} + \begin{bmatrix} -3 & -1 \\ 4 & 2 \end{bmatrix} - \begin{bmatrix} -1 & 1 \\ 2 & 0 \end{bmatrix} -$

$\qquad\qquad\qquad \begin{bmatrix} 1 & -3 \\ 0 & 4 \end{bmatrix}$

$= \begin{bmatrix} -2 & 1 \\ 2 & -1 \end{bmatrix}$

Thus $(\mathbf{A} + \mathbf{B})(\mathbf{A} - \mathbf{B}) = \mathbf{A}^2 + \mathbf{BA} - \mathbf{AB} - \mathbf{B}^2$.

Chapter 6 Mid-Chapter Mixed Review

1. False; see page 395 in the text.

3. True; see page 429 in the text.

5. $2x + y = -4, \quad (1)$
$x = y - 5 \qquad (2)$
Substitute $y - 5$ for x in equation (1) and solve for y.
$2(y - 5) + y = -4$
$2y - 10 + y = -4$
$3y - 10 = -4$
$3y = 6$
$y = 2$
Back-substitute in equation (2) to find x.
$x = 2 - 5 = -3$
The solution is $(-3, 2)$.

7. $2x - 3y = 8, \quad (1)$
$3x + 2y = -1 \quad (2)$
Multiply equation (1) by 2 and equation (2) by 3 and add.
$\quad 4x - 6y = 16$
$\underline{\quad 9x + 6y = -3 \quad}$
$\quad 13x \qquad = 13$
$\qquad x = 1$
Back-substitute and solve for y.
$3 \cdot 1 + 2y = -1 \quad$ Using equation (2)
$3 + 2y = -1$
$2y = -4$
$y = -2$
The solution is $(1, -2)$.

9. $x + 2y + 3z = 4, \quad (1)$
$x - 2y + z = 2, \quad (2)$
$2x - 6y + 4z = 7 \quad (3)$
Multiply equation (1) by -1 and add it to equation (2).
Multiply equation (1) by -2 and add it to equation (3).
$x + 2y + 3z = 4 \quad (1)$
$ - 4y - 2z = -2 \quad (4)$
$ - 10y - 2z = -1 \quad (5)$
Multiply equation (5) by 2.
$x + 2y + 3z = 4 \quad (1)$
$ - 4y - 2z = -2 \quad (4)$
$ - 20y - 4z = -2 \quad (6)$
Multiply equation (4) by -5 and add it to equation (6).
$x + 2y + 3z = 4 \quad (1)$
$ - 4y - 2z = -2 \quad (4)$
$ 6z = 8 \quad (7)$
Solve equation (7) for z.
$6z = 8$
$z = \dfrac{4}{3}$
Back-substitute $\dfrac{4}{3}$ for z in equation (4) and solve for y.
$-4y - 2\left(\dfrac{4}{3}\right) = -2$
$-4y - \dfrac{8}{3} = -2$
$-4y = \dfrac{2}{3}$
$y = -\dfrac{1}{6}$
Back-substitute $-\dfrac{1}{6}$ for y and $\dfrac{4}{3}$ for z in equation (1) and solve for x.
$x + 2\left(-\dfrac{1}{6}\right) + 3 \cdot \dfrac{4}{3} = 4$
$x - \dfrac{1}{3} + 4 = 4$
$x + \dfrac{11}{3} = 4$
$x = \dfrac{1}{3}$
The solution is $\left(\dfrac{1}{3}, -\dfrac{1}{6}, \dfrac{4}{3}\right)$.

11. $2x + y = 5,$
$3x + 2y = 6$
Write the augmented matrix. We will use Gauss-Jordan elimination.
$\begin{bmatrix} 2 & 1 & | & 5 \\ 3 & 2 & | & 6 \end{bmatrix}$

Multiply row 2 by 2.

$$\begin{bmatrix} 2 & 1 & | & 5 \\ 6 & 4 & | & 12 \end{bmatrix}$$

Multiply row 1 by -3 and add it to row 2.

$$\begin{bmatrix} 2 & 1 & | & 5 \\ 0 & 1 & | & -3 \end{bmatrix}$$

Multiply row 2 by -1 and add it to row 1.

$$\begin{bmatrix} 2 & 0 & | & 8 \\ 0 & 1 & | & -3 \end{bmatrix}$$

Multiply row 1 by $\frac{1}{2}$.

$$\begin{bmatrix} 1 & 0 & | & 4 \\ 0 & 1 & | & -3 \end{bmatrix}$$

The solution is $(4, -3)$.

13. $\mathbf{A} + \mathbf{B} = \begin{bmatrix} 3 & -1 \\ 5 & 4 \end{bmatrix} + \begin{bmatrix} -2 & 6 \\ 1 & -3 \end{bmatrix}$

$= \begin{bmatrix} 3 + (-2) & -1 + 6 \\ 5 + 1 & 4 + (-3) \end{bmatrix}$

$= \begin{bmatrix} 1 & 5 \\ 6 & 1 \end{bmatrix}$

15. $4\mathbf{D} = 4 \begin{bmatrix} -2 & 3 & 0 \\ 1 & -1 & 2 \\ -3 & 4 & 1 \end{bmatrix} = \begin{bmatrix} -8 & 12 & 0 \\ 4 & -4 & 8 \\ -12 & 16 & 4 \end{bmatrix}$

17. $\mathbf{AB} = \begin{bmatrix} 3 & -1 \\ 5 & 4 \end{bmatrix} \begin{bmatrix} -2 & 6 \\ 1 & -3 \end{bmatrix}$

$= \begin{bmatrix} 3(-2) + 9 + (-1) \cdot 1 & 3 \cdot 6 + (-1)(-3) \\ 5(-2) + 4 \cdot 1 & 5 \cdot 6 + 4(-3) \end{bmatrix}$

$= \begin{bmatrix} -7 & 21 \\ -6 & 18 \end{bmatrix}$

19. $\mathbf{BC}$

$= \begin{bmatrix} -2 & 6 \\ 1 & -3 \end{bmatrix} \begin{bmatrix} -4 & 1 & -1 \\ 2 & 3 & -2 \end{bmatrix}$

$= \begin{bmatrix} -2(-4) + 6 \cdot 2 & -2 \cdot 1 + 6 \cdot 3 & -2(-1) + 6(-2) \\ 1(-4) + (-3) \cdot 2 & 1 \cdot 1 + (-3) \cdot 3 & 1(-1) + (-3)(-2) \end{bmatrix}$

$= \begin{bmatrix} 20 & 16 & -10 \\ -10 & -8 & 5 \end{bmatrix}$

21. A matrix equation equivalent to the given system is
$$\begin{bmatrix} 2 & -1 & 3 \\ 1 & 2 & -1 \\ 3 & -4 & 2 \end{bmatrix} \begin{bmatrix} x \\ y \\ z \end{bmatrix} = \begin{bmatrix} 7 \\ 3 \\ 5 \end{bmatrix}.$$

23. Add a non-zero multiple of one equation to a non-zero multiple of the other equation, where the multiples are not opposites.

25. No; for example, let $\mathbf{A} = \begin{bmatrix} 1 & -1 \\ -1 & 1 \end{bmatrix}$ and $\mathbf{B} = \begin{bmatrix} 1 & 1 \\ 1 & 1 \end{bmatrix}$.

Then $\mathbf{AB} = \begin{bmatrix} 0 & 0 \\ 0 & 0 \end{bmatrix}$ and neither $\mathbf{A}$ nor $\mathbf{B}$ is $\begin{bmatrix} 0 & 0 \\ 0 & 0 \end{bmatrix}$.

Exercise Set 6.5

1. $\mathbf{BA} = \begin{bmatrix} 7 & 3 \\ 2 & 1 \end{bmatrix} \begin{bmatrix} 1 & -3 \\ -2 & 7 \end{bmatrix} = \begin{bmatrix} 1 & 0 \\ 0 & 1 \end{bmatrix}$

$\mathbf{AB} = \begin{bmatrix} 1 & -3 \\ -2 & 7 \end{bmatrix} \begin{bmatrix} 7 & 3 \\ 2 & 1 \end{bmatrix} = \begin{bmatrix} 1 & 0 \\ 0 & 1 \end{bmatrix}$

Since $\mathbf{BA} = \mathbf{I} = \mathbf{AB}$, $\mathbf{B}$ is the inverse of $\mathbf{A}$.

3. $\mathbf{BA} = \begin{bmatrix} 2 & 3 & 2 \\ 3 & 3 & 4 \\ 1 & 1 & 1 \end{bmatrix} \begin{bmatrix} -1 & -1 & 6 \\ 1 & 0 & -2 \\ 1 & 0 & -3 \end{bmatrix} =$

$\begin{bmatrix} 3 & -2 & 0 \\ 4 & -3 & 0 \\ 1 & -1 & 1 \end{bmatrix}$

Since $\mathbf{BA} \neq \mathbf{I}$, $\mathbf{B}$ is not the inverse of $\mathbf{A}$.

5. $\mathbf{A} = \begin{bmatrix} 3 & 2 \\ 5 & 3 \end{bmatrix}$

Write the augmented matrix.

$$\begin{bmatrix} 3 & 2 & | & 1 & 0 \\ 5 & 3 & | & 0 & 1 \end{bmatrix}$$

Multiply row 2 by 3.

$$\begin{bmatrix} 3 & 2 & | & 1 & 0 \\ 15 & 9 & | & 0 & 3 \end{bmatrix}$$

Multiply row 1 by -5 and add it to row 2.

$$\begin{bmatrix} 3 & 2 & | & 1 & 0 \\ 0 & -1 & | & -5 & 3 \end{bmatrix}$$

Multiply row 2 by 2 and add it to row 1.

$$\begin{bmatrix} 3 & 0 & | & -9 & 6 \\ 0 & -1 & | & -5 & 3 \end{bmatrix}$$

Multiply row 1 by $\frac{1}{3}$ and row 2 by -1.

$$\begin{bmatrix} 1 & 0 & | & -3 & 2 \\ 0 & 1 & | & 5 & -3 \end{bmatrix}$$

Then $\mathbf{A}^{-1} = \begin{bmatrix} -3 & 2 \\ 5 & -3 \end{bmatrix}$.

7. $\mathbf{A} = \begin{bmatrix} 6 & 9 \\ 4 & 6 \end{bmatrix}$

Write the augmented matrix.

$$\begin{bmatrix} 6 & 9 & | & 1 & 0 \\ 4 & 6 & | & 0 & 1 \end{bmatrix}$$

Multiply row 2 by 3.

$$\begin{bmatrix} 6 & 9 & | & 1 & 0 \\ 12 & 18 & | & 0 & 3 \end{bmatrix}$$

Multiply row 1 by -2 and add it to row 2.

$$\begin{bmatrix} 6 & 9 & | & 1 & 0 \\ 0 & 0 & | & -2 & 3 \end{bmatrix}$$

We cannot obtain the identity matrix on the left since the second row contains only zeros to the left of the vertical line. Thus, $\mathbf{A}^{-1}$ does not exist.

9. $\mathbf{A} = \begin{bmatrix} 4 & -3 \\ 1 & -2 \end{bmatrix}$

Write the augmented matrix.

$\begin{bmatrix} 4 & -3 & | & 1 & 0 \\ 1 & -2 & | & 0 & 1 \end{bmatrix}$

Interchange the rows.

$\begin{bmatrix} 1 & -2 & | & 0 & 1 \\ 4 & -3 & | & 1 & 0 \end{bmatrix}$

Multiply row 1 by -4 and add it to row 2.

$\begin{bmatrix} 1 & -2 & | & 0 & 1 \\ 0 & 5 & | & 1 & -4 \end{bmatrix}$

Multiply row 2 by $\frac{1}{5}$.

$\begin{bmatrix} 1 & -2 & | & 0 & 1 \\ 0 & 1 & | & \frac{1}{5} & -\frac{4}{5} \end{bmatrix}$

Multiply row 2 by 2 and add it to row 1.

$\begin{bmatrix} 1 & 0 & | & \frac{2}{5} & -\frac{3}{5} \\ 0 & 1 & | & \frac{1}{5} & -\frac{4}{5} \end{bmatrix}$

Then $\mathbf{A}^{-1} = \begin{bmatrix} \frac{2}{5} & -\frac{3}{5} \\ \frac{1}{5} & -\frac{4}{5} \end{bmatrix}$, or $\begin{bmatrix} 0.4 & -0.6 \\ 0.2 & -0.8 \end{bmatrix}$.

11. $\mathbf{A} = \begin{bmatrix} 3 & 1 & 0 \\ 1 & 1 & 1 \\ 1 & -1 & 2 \end{bmatrix}$

Write the augmented matrix.

$\begin{bmatrix} 3 & 1 & 0 & | & 1 & 0 & 0 \\ 1 & 1 & 1 & | & 0 & 1 & 0 \\ 1 & -1 & 2 & | & 0 & 0 & 1 \end{bmatrix}$

Interchange the first two rows.

$\begin{bmatrix} 1 & 1 & 1 & | & 0 & 1 & 0 \\ 3 & 1 & 0 & | & 1 & 0 & 0 \\ 1 & -1 & 2 & | & 0 & 0 & 1 \end{bmatrix}$

Multiply row 1 by -3 and add it to row 2. Also, multiply row 1 by -1 and add it to row 3.

$\begin{bmatrix} 1 & 1 & 1 & | & 0 & 1 & 0 \\ 0 & -2 & -3 & | & 1 & -3 & 0 \\ 0 & -2 & 1 & | & 0 & -1 & 1 \end{bmatrix}$

Multiply row 2 by $-\frac{1}{2}$.

$\begin{bmatrix} 1 & 1 & 1 & | & 0 & 1 & 0 \\ 0 & 1 & \frac{3}{2} & | & -\frac{1}{2} & \frac{3}{2} & 0 \\ 0 & -2 & 1 & | & 0 & -1 & 1 \end{bmatrix}$

Multiply row 2 by -1 and add it to row 1. Also, multiply row 2 by 2 and add it to row 3.

$\begin{bmatrix} 1 & 0 & -\frac{1}{2} & | & \frac{1}{2} & -\frac{1}{2} & 0 \\ 0 & 1 & \frac{3}{2} & | & -\frac{1}{2} & \frac{3}{2} & 0 \\ 0 & 0 & 4 & | & -1 & 2 & 1 \end{bmatrix}$

Multiply row 3 by $\frac{1}{4}$.

$\begin{bmatrix} 1 & 0 & -\frac{1}{2} & | & \frac{1}{2} & -\frac{1}{2} & 0 \\ 0 & 1 & \frac{3}{2} & | & -\frac{1}{2} & \frac{3}{2} & 0 \\ 0 & 0 & 1 & | & -\frac{1}{4} & \frac{1}{2} & \frac{1}{4} \end{bmatrix}$

Multiply row 3 by $\frac{1}{2}$ and add it to row 1. Also, multiply row 3 by $-\frac{3}{2}$ and add it to row 2.

$\begin{bmatrix} 1 & 0 & 0 & | & \frac{3}{8} & -\frac{1}{4} & \frac{1}{8} \\ 0 & 1 & 0 & | & -\frac{1}{8} & \frac{3}{4} & -\frac{3}{8} \\ 0 & 0 & 1 & | & -\frac{1}{4} & \frac{1}{2} & \frac{1}{4} \end{bmatrix}$

Then $\mathbf{A}^{-1} = \begin{bmatrix} \frac{3}{8} & -\frac{1}{4} & \frac{1}{8} \\ -\frac{1}{8} & \frac{3}{4} & -\frac{3}{8} \\ -\frac{1}{4} & \frac{1}{2} & \frac{1}{4} \end{bmatrix}$.

13. $\mathbf{A} = \begin{bmatrix} 1 & -4 & 8 \\ 1 & -3 & 2 \\ 2 & -7 & 10 \end{bmatrix}$

Write the augmented matrix.

$\begin{bmatrix} 1 & -4 & 8 & | & 1 & 0 & 0 \\ 1 & -3 & 2 & | & 0 & 1 & 0 \\ 2 & -7 & 10 & | & 0 & 0 & 1 \end{bmatrix}$

Multiply row 1 by -1 and add it to row 2. Also, multiply row 1 by -2 and add it to row 3.

$\begin{bmatrix} 1 & -4 & 8 & | & 1 & 0 & 0 \\ 0 & 1 & -6 & | & -1 & 1 & 0 \\ 0 & 1 & -6 & | & -2 & 0 & 1 \end{bmatrix}$

Since the second and third rows are identical left of the vertical line, it will not be possible to obtain the identity matrix on the left side. Thus, $\mathbf{A}^{-1}$ does not exist.

15. $\mathbf{A} = \begin{bmatrix} 2 & 3 & 2 \\ 3 & 3 & 4 \\ -1 & -1 & -1 \end{bmatrix}$

Write the augmented matrix.

$\begin{bmatrix} 2 & 3 & 2 & | & 1 & 0 & 0 \\ 3 & 3 & 4 & | & 0 & 1 & 0 \\ -1 & -1 & -1 & | & 0 & 0 & 1 \end{bmatrix}$

Interchange rows 1 and 3.

$\begin{bmatrix} -1 & -1 & -1 & | & 0 & 0 & 1 \\ 3 & 3 & 4 & | & 0 & 1 & 0 \\ 2 & 3 & 2 & | & 1 & 0 & 0 \end{bmatrix}$

Multiply row 1 by 3 and add it to row 2. Also, multiply row 1 by 2 and add it to row 3.

$\begin{bmatrix} -1 & -1 & -1 & | & 0 & 0 & 1 \\ 0 & 0 & 1 & | & 0 & 1 & 3 \\ 0 & 1 & 0 & | & 1 & 0 & 2 \end{bmatrix}$

Multiply row 1 by -1.

$$\begin{bmatrix} 1 & 1 & 1 & | & 0 & 0 & -1 \\ 0 & 0 & 1 & | & 0 & 1 & 3 \\ 0 & 1 & 0 & | & 1 & 0 & 2 \end{bmatrix}$$

Interchange rows 2 and 3.

$$\begin{bmatrix} 1 & 1 & 1 & | & 0 & 0 & -1 \\ 0 & 1 & 0 & | & 1 & 0 & 2 \\ 0 & 0 & 1 & | & 0 & 1 & 3 \end{bmatrix}$$

Multiply row 2 by -1 and add it to row 1.

$$\begin{bmatrix} 1 & 0 & 1 & | & -1 & 0 & -3 \\ 0 & 1 & 0 & | & 1 & 0 & 2 \\ 0 & 0 & 1 & | & 0 & 1 & 3 \end{bmatrix}$$

Multiply row 3 by -1 and add it to row 1.

$$\begin{bmatrix} 1 & 0 & 0 & | & -1 & -1 & -6 \\ 0 & 1 & 0 & | & 1 & 0 & 2 \\ 0 & 0 & 1 & | & 0 & 1 & 3 \end{bmatrix}$$

Then $\mathbf{A}^{-1} = \begin{bmatrix} -1 & -1 & -6 \\ 1 & 0 & 2 \\ 0 & 1 & 3 \end{bmatrix}$.

17. $\mathbf{A} = \begin{bmatrix} 1 & 2 & -1 \\ -2 & 0 & 1 \\ 1 & -1 & 0 \end{bmatrix}$

Write the augmented matrix.

$$\begin{bmatrix} 1 & 2 & -1 & | & 1 & 0 & 0 \\ -2 & 0 & 1 & | & 0 & 1 & 0 \\ 1 & -1 & 0 & | & 0 & 0 & 1 \end{bmatrix}$$

Multiply row 1 by 2 and add it to row 2. Also, multiply row 1 by -1 and add it to row 3.

$$\begin{bmatrix} 1 & 2 & -1 & | & 1 & 0 & 0 \\ 0 & 4 & -1 & | & 2 & 1 & 0 \\ 0 & -3 & 1 & | & -1 & 0 & 1 \end{bmatrix}$$

Add row 3 to row 1 and also to row 2.

$$\begin{bmatrix} 1 & -1 & 0 & | & 0 & 0 & 1 \\ 0 & 1 & 0 & | & 1 & 1 & 1 \\ 0 & -3 & 1 & | & -1 & 0 & 1 \end{bmatrix}$$

Add row 2 to row 1. Also, multiply row 2 by 3 and add it to row 3.

$$\begin{bmatrix} 1 & 0 & 0 & | & 1 & 1 & 2 \\ 0 & 1 & 0 & | & 1 & 1 & 1 \\ 0 & 0 & 1 & | & 2 & 3 & 4 \end{bmatrix}$$

Then $\mathbf{A}^{-1} = \begin{bmatrix} 1 & 1 & 2 \\ 1 & 1 & 1 \\ 2 & 3 & 4 \end{bmatrix}$.

19. $\mathbf{A} = \begin{bmatrix} 1 & 3 & -1 \\ 0 & 2 & -1 \\ 1 & 1 & 0 \end{bmatrix}$

Write the augmented matrix.

$$\begin{bmatrix} 1 & 3 & -1 & | & 1 & 0 & 0 \\ 0 & 2 & -1 & | & 0 & 1 & 0 \\ 1 & 1 & 0 & | & 0 & 0 & 1 \end{bmatrix}$$

Multiply row 1 by -1 and add it to row 3.

$$\begin{bmatrix} 1 & 3 & -1 & | & 1 & 0 & 0 \\ 0 & 2 & -1 & | & 0 & 1 & 0 \\ 0 & -2 & 1 & | & -1 & 0 & 1 \end{bmatrix}$$

Add row 3 to row 1 and also to row 2.

$$\begin{bmatrix} 1 & 1 & 0 & | & 0 & 0 & 0 \\ 0 & 0 & 0 & | & -1 & 1 & 1 \\ 0 & -2 & 1 & | & -1 & 0 & 1 \end{bmatrix}$$

Since the second row consists only of zeros to the left of the vertical line, it will not be possible to obtain the identity matrix on the left side. Thus, $\mathbf{A}^{-1}$ does not exist. A graphing calculator will return an error message when we try to find $\mathbf{A}^{-1}$.

21. $\mathbf{A} = \begin{bmatrix} 1 & 2 & 3 & 4 \\ 0 & 1 & 3 & -5 \\ 0 & 0 & 1 & -2 \\ 0 & 0 & 0 & -1 \end{bmatrix}$

Write the augmented matrix.

$$\begin{bmatrix} 1 & 2 & 3 & 4 & | & 1 & 0 & 0 & 0 \\ 0 & 1 & 3 & -5 & | & 0 & 1 & 0 & 0 \\ 0 & 0 & 1 & -2 & | & 0 & 0 & 1 & 0 \\ 0 & 0 & 0 & -1 & | & 0 & 0 & 0 & 1 \end{bmatrix}$$

Multiply row 4 by -1.

$$\begin{bmatrix} 1 & 2 & 3 & 4 & | & 1 & 0 & 0 & 0 \\ 0 & 1 & 3 & -5 & | & 0 & 1 & 0 & 0 \\ 0 & 0 & 1 & -2 & | & 0 & 0 & 1 & 0 \\ 0 & 0 & 0 & 1 & | & 0 & 0 & 0 & -1 \end{bmatrix}$$

Multiply row 4 by -4 and add it to row 1. Multiply row 4 by 5 and add it to row 2. Also, multiply row 4 by 2 and add it to row 3.

$$\begin{bmatrix} 1 & 2 & 3 & 0 & | & 1 & 0 & 0 & 4 \\ 0 & 1 & 3 & 0 & | & 0 & 1 & 0 & -5 \\ 0 & 0 & 1 & 0 & | & 0 & 0 & 1 & -2 \\ 0 & 0 & 0 & 1 & | & 0 & 0 & 0 & -1 \end{bmatrix}$$

Multiply row 3 by -3 and add it to row 1 and to row 2.

$$\begin{bmatrix} 1 & 2 & 0 & 0 & | & 1 & 0 & -3 & 10 \\ 0 & 1 & 0 & 0 & | & 0 & 1 & -3 & 1 \\ 0 & 0 & 1 & 0 & | & 0 & 0 & 1 & -2 \\ 0 & 0 & 0 & 1 & | & 0 & 0 & 0 & -1 \end{bmatrix}$$

Multiply row 2 by -2 and add it to row 1.

$$\begin{bmatrix} 1 & 0 & 0 & 0 & | & 1 & -2 & 3 & 8 \\ 0 & 1 & 0 & 0 & | & 0 & 1 & -3 & 1 \\ 0 & 0 & 1 & 0 & | & 0 & 0 & 1 & -2 \\ 0 & 0 & 0 & 1 & | & 0 & 0 & 0 & -1 \end{bmatrix}$$

Then $\mathbf{A}^{-1} = \begin{bmatrix} 1 & -2 & 3 & 8 \\ 0 & 1 & -3 & 1 \\ 0 & 0 & 1 & -2 \\ 0 & 0 & 0 & -1 \end{bmatrix}$.

23. $\mathbf{A} = \begin{bmatrix} 1 & -14 & 7 & 38 \\ -1 & 2 & 1 & -2 \\ 1 & 2 & -1 & -6 \\ 1 & -2 & 3 & 6 \end{bmatrix}$

Write the augmented matrix.

$$\begin{bmatrix} 1 & -14 & 7 & 38 & | & 1 & 0 & 0 & 0 \\ -1 & 2 & 1 & -2 & | & 0 & 1 & 0 & 0 \\ 1 & 2 & -1 & -6 & | & 0 & 0 & 1 & 0 \\ 1 & -2 & 3 & 6 & | & 0 & 0 & 0 & 1 \end{bmatrix}$$

Add row 1 to row 2. Also, multiply row 1 by -1 and add it to row 3 and to row 4.

$$\left[\begin{array}{cccc|cccc} 1 & -14 & 7 & 38 & 1 & 0 & 0 & 0 \\ 0 & -12 & 8 & 36 & 1 & 1 & 0 & 0 \\ 0 & 16 & -8 & -44 & -1 & 0 & 1 & 0 \\ 0 & 12 & -4 & -32 & -1 & 0 & 0 & 1 \end{array}\right]$$

Add row 2 to row 4.

$$\left[\begin{array}{cccc|cccc} 1 & -14 & 7 & 38 & 1 & 0 & 0 & 0 \\ 0 & -12 & 8 & 36 & 1 & 1 & 0 & 0 \\ 0 & 16 & -8 & -44 & -1 & 0 & 1 & 0 \\ 0 & 0 & 4 & 4 & 0 & 1 & 0 & 1 \end{array}\right]$$

Multiply row 4 by $\frac{1}{4}$.

$$\left[\begin{array}{cccc|cccc} 1 & -14 & 7 & 38 & 1 & 0 & 0 & 0 \\ 0 & -12 & 8 & 36 & 1 & 1 & 0 & 0 \\ 0 & 16 & -8 & -44 & -1 & 0 & 1 & 0 \\ 0 & 0 & 1 & 1 & 0 & \frac{1}{4} & 0 & \frac{1}{4} \end{array}\right]$$

Multiply row 4 by -38 and add it to row 1. Multiply row 4 by -36 and add it to row 2. Also, multiply row 4 by 44 and add it to row 3.

$$\left[\begin{array}{cccc|cccc} 1 & -14 & -31 & 0 & 1 & -\frac{19}{2} & 0 & -\frac{19}{2} \\ 0 & -12 & -28 & 0 & 1 & -8 & 0 & -9 \\ 0 & 16 & 36 & 0 & -1 & 11 & 1 & 11 \\ 0 & 0 & 1 & 1 & 0 & \frac{1}{4} & 0 & \frac{1}{4} \end{array}\right]$$

Multiply row 3 by $\frac{1}{36}$.

$$\left[\begin{array}{cccc|cccc} 1 & -14 & -31 & 0 & 1 & -\frac{19}{2} & 0 & -\frac{19}{2} \\ 0 & -12 & -28 & 0 & 1 & -8 & 0 & -9 \\ 0 & \frac{4}{9} & 1 & 0 & -\frac{1}{36} & \frac{11}{36} & \frac{1}{36} & \frac{11}{36} \\ 0 & 0 & 1 & 1 & 0 & \frac{1}{4} & 0 & \frac{1}{4} \end{array}\right]$$

Multiply row 3 by 31 and add it to row 1. Multiply row 3 by 28 and add it to row 2. Also, multiply row 3 by -1 and add it to row 4.

$$\left[\begin{array}{cccc|cccc} 1 & -\frac{2}{9} & 0 & 0 & \frac{5}{36} & -\frac{1}{36} & \frac{31}{36} & -\frac{1}{36} \\ 0 & \frac{4}{9} & 0 & 0 & \frac{2}{9} & \frac{5}{9} & \frac{7}{9} & -\frac{4}{9} \\ 0 & \frac{4}{9} & 1 & 0 & -\frac{1}{36} & \frac{11}{36} & \frac{1}{36} & \frac{11}{36} \\ 0 & -\frac{4}{9} & 0 & 1 & \frac{1}{36} & -\frac{1}{18} & -\frac{1}{36} & -\frac{1}{18} \end{array}\right]$$

Multiply row 2 by $\frac{1}{2}$ and add it to row 1. Also, multiply row 2 by -1 and add it to row 3. Add row 2 to row 4.

$$\left[\begin{array}{cccc|cccc} 1 & 0 & 0 & 0 & \frac{1}{4} & \frac{1}{4} & \frac{5}{4} & -\frac{1}{4} \\ 0 & \frac{4}{9} & 0 & 0 & \frac{2}{9} & \frac{5}{9} & \frac{7}{9} & -\frac{4}{9} \\ 0 & 0 & 1 & 0 & -\frac{1}{4} & -\frac{1}{4} & -\frac{3}{4} & \frac{3}{4} \\ 0 & 0 & 0 & 1 & \frac{1}{4} & \frac{1}{2} & \frac{3}{4} & -\frac{1}{2} \end{array}\right]$$

Multiply row 2 by $\frac{9}{4}$.

$$\left[\begin{array}{cccc|cccc} 1 & 0 & 0 & 0 & \frac{1}{4} & \frac{1}{4} & \frac{5}{4} & -\frac{1}{4} \\ 0 & 1 & 0 & 0 & \frac{1}{2} & \frac{5}{4} & \frac{7}{4} & -1 \\ 0 & 0 & 1 & 0 & -\frac{1}{4} & -\frac{1}{4} & -\frac{3}{4} & \frac{3}{4} \\ 0 & 0 & 0 & 1 & \frac{1}{4} & \frac{1}{2} & \frac{3}{4} & -\frac{1}{2} \end{array}\right]$$

Then $\mathbf{A}^{-1} = \left[\begin{array}{cccc} \frac{1}{4} & \frac{1}{4} & \frac{5}{4} & -\frac{1}{4} \\ \frac{1}{2} & \frac{5}{4} & \frac{7}{4} & -1 \\ -\frac{1}{4} & -\frac{1}{4} & -\frac{3}{4} & \frac{3}{4} \\ \frac{1}{4} & \frac{1}{2} & \frac{3}{4} & -\frac{1}{2} \end{array}\right]$, or

$$\left[\begin{array}{cccc} 0.25 & 0.25 & 1.25 & -0.25 \\ 0.5 & 1.25 & 1.75 & -1 \\ -0.25 & -0.25 & -0.75 & 0.75 \\ 0.25 & 0.5 & 0.75 & -0.5 \end{array}\right].$$

25. Write an equivalent matrix equation, $\mathbf{AX} = \mathbf{B}$.

$$\left[\begin{array}{cc} 11 & 3 \\ 7 & 2 \end{array}\right]\left[\begin{array}{c} x \\ y \end{array}\right] = \left[\begin{array}{c} -4 \\ 5 \end{array}\right]$$

Then we have $\mathbf{X} = \mathbf{A}^{-1}\mathbf{B}$.

$$\left[\begin{array}{c} x \\ y \end{array}\right] = \left[\begin{array}{cc} 2 & -3 \\ -7 & 11 \end{array}\right]\left[\begin{array}{c} -4 \\ 5 \end{array}\right] = \left[\begin{array}{c} -23 \\ 83 \end{array}\right]$$

The solution is $(-23, 83)$.

27. Write an equivalent matrix equation, $\mathbf{AX} = \mathbf{B}$.

$$\left[\begin{array}{ccc} 3 & 1 & 0 \\ 2 & -1 & 2 \\ 1 & 1 & 1 \end{array}\right]\left[\begin{array}{c} x \\ y \\ z \end{array}\right] = \left[\begin{array}{c} 2 \\ -5 \\ 5 \end{array}\right]$$

Then we have $\mathbf{X} = \mathbf{A}^{-1}\mathbf{B}$.

$$\left[\begin{array}{c} x \\ y \\ z \end{array}\right] = \frac{1}{9}\left[\begin{array}{ccc} 3 & 1 & -2 \\ 0 & -3 & 6 \\ -3 & 2 & 5 \end{array}\right]\left[\begin{array}{c} 2 \\ -5 \\ 5 \end{array}\right] = \frac{1}{9}\left[\begin{array}{c} -9 \\ 45 \\ 9 \end{array}\right] =$$

$$\left[\begin{array}{c} -1 \\ 5 \\ 1 \end{array}\right]$$

The solution is $(-1, 5, 1)$.

29. $4x + 3y = 2,$
$\quad x - 2y = 6$

Write an equivalent matrix equation, $\mathbf{AX} = \mathbf{B}$.

$$\left[\begin{array}{cc} 4 & 3 \\ 1 & -2 \end{array}\right]\left[\begin{array}{c} x \\ y \end{array}\right] = \left[\begin{array}{c} 2 \\ 6 \end{array}\right]$$

Then $\mathbf{X} = \mathbf{A}^{-1}\mathbf{B} = \left[\begin{array}{cc} \frac{2}{11} & \frac{3}{11} \\ \frac{1}{11} & -\frac{4}{11} \end{array}\right]\left[\begin{array}{c} 2 \\ 6 \end{array}\right] = \left[\begin{array}{c} 2 \\ -2 \end{array}\right].$

The solution is $(2, -2)$.

31. $5x + y = 2,$
$\quad 3x - 2y = -4$

Write an equivalent matrix equation, $\mathbf{AX} = \mathbf{B}$.

$$\begin{bmatrix} 5 & 1 \\ 3 & -2 \end{bmatrix}\begin{bmatrix} x \\ y \end{bmatrix} = \begin{bmatrix} \dfrac{2}{13} & \dfrac{1}{13} \\ \dfrac{3}{13} & -\dfrac{5}{13} \end{bmatrix}\begin{bmatrix} 2 \\ -4 \end{bmatrix} =$$

$$\begin{bmatrix} 2 \\ -4 \end{bmatrix}$$

Then $\mathbf{X} = \mathbf{A}^{-1}\mathbf{B} = \begin{bmatrix} 0 \\ 2 \end{bmatrix}.$

The solution is $(0, 2)$.

33. $x \quad\ \ + z = 1,$
$\quad 2x + y \quad\ = 3,$
$\quad x - y + z = 4$

Write an equivalent matrix equation, $\mathbf{AX} = \mathbf{B}$.

$$\begin{bmatrix} 1 & 0 & 1 \\ 2 & 1 & 0 \\ 1 & -1 & 1 \end{bmatrix}\begin{bmatrix} x \\ y \\ z \end{bmatrix} = \begin{bmatrix} 1 \\ 3 \\ 4 \end{bmatrix}$$

Then $\mathbf{X} = \mathbf{A}^{-1}\mathbf{B} = \begin{bmatrix} -\dfrac{1}{2} & \dfrac{1}{2} & \dfrac{1}{2} \\ 1 & 0 & -1 \\ \dfrac{3}{2} & -\dfrac{1}{2} & -\dfrac{1}{2} \end{bmatrix}\begin{bmatrix} 1 \\ 3 \\ 4 \end{bmatrix} = \begin{bmatrix} 3 \\ -3 \\ -2 \end{bmatrix}.$

The solution is $(3, -3, -2)$.

35. $2x + 3y + 4z = 2,$
$\quad x - 4y + 3z = 2,$
$\quad 5x + y + z = -4$

Write an equivalent matrix equation, $\mathbf{AX} = \mathbf{B}$.

$$\begin{bmatrix} 2 & 3 & 4 \\ 1 & -4 & 3 \\ 5 & 1 & 1 \end{bmatrix}\begin{bmatrix} x \\ y \\ z \end{bmatrix} = \begin{bmatrix} 2 \\ 2 \\ -4 \end{bmatrix}$$

Then $\mathbf{X} = \mathbf{A}^{-1}\mathbf{B} = \begin{bmatrix} -\dfrac{1}{16} & \dfrac{1}{112} & \dfrac{25}{112} \\ \dfrac{1}{8} & -\dfrac{9}{56} & -\dfrac{1}{56} \\ \dfrac{3}{16} & \dfrac{13}{112} & -\dfrac{11}{112} \end{bmatrix}\begin{bmatrix} 2 \\ 2 \\ -4 \end{bmatrix} = \begin{bmatrix} -1 \\ 0 \\ 1 \end{bmatrix}.$

The solution is $(-1, 0, 1)$.

37. $2w - 3x + 4y - 5z = 0,$
$\quad 3w - 2x + 7y - 3z = 2,$
$\quad w + x - y + z = 1,$
$\quad -w - 3x - 6y + 4z = 6$

Write an equivalent matrix equation, $\mathbf{AX} = \mathbf{B}$.

$$\begin{bmatrix} 2 & -3 & 4 & -5 \\ 3 & -2 & 7 & -3 \\ 1 & 1 & -1 & 1 \\ -1 & -3 & -6 & 4 \end{bmatrix}\begin{bmatrix} w \\ x \\ y \\ z \end{bmatrix} = \begin{bmatrix} 0 \\ 2 \\ 1 \\ 6 \end{bmatrix}$$

Then $\mathbf{X} = \mathbf{A}^{-1}\mathbf{B} = \dfrac{1}{203}\begin{bmatrix} 26 & 11 & 127 & 9 \\ -8 & -19 & 39 & -34 \\ -37 & 39 & -48 & -5 \\ -55 & 47 & -11 & 20 \end{bmatrix}\begin{bmatrix} 0 \\ 2 \\ 1 \\ 6 \end{bmatrix} =$

$$\begin{bmatrix} 1 \\ -1 \\ 0 \\ 1 \end{bmatrix}.$$

The solution is $(1, -1, 0, 1)$.

39. *Familiarize*. Let x and y represent the number of pounds of cranberries grown in Wisconsin and Massachusetts, respectively, in millions.

Translate. A total of 660 million lb of cranberries were grown in the two states.

$$x + y = 660$$

Wisconsin grew 240 million lb more than Massachusetts.

$$x = y + 240$$

We have a system of equations.

$$\begin{array}{ll} x + y = 660, & x + y = 660, \\ x = y + 240, & \text{or} \quad x - y = 240. \end{array}$$

Carry out. Write an equivalent matrix equation, $\mathbf{AX} = \mathbf{B}$.

$$\begin{bmatrix} 1 & 1 \\ 1 & -1 \end{bmatrix}\begin{bmatrix} x \\ y \end{bmatrix} = \begin{bmatrix} 660 \\ 240 \end{bmatrix}$$

Then $\mathbf{X} = \mathbf{A}^{-1}\mathbf{B} = \begin{bmatrix} \dfrac{1}{2} & \dfrac{1}{2} \\ \dfrac{1}{2} & -\dfrac{1}{2} \end{bmatrix}\begin{bmatrix} 660 \\ 240 \end{bmatrix} = \begin{bmatrix} 450 \\ 210 \end{bmatrix},$

so the solution is $(450, 210)$.

Check. The total production is $450 + 210$, or 660 million lb. Also, 450 million lb is 240 million lb more than 210 million lb. The answer checks.

State. In Wisconsin, 450 million lb of cranberries were grown, and 210 million lb were grown in Massachusetts.

41. *Familiarize*. Let x, y, and z represent the prices of one ton of topsoil, mulch, and pea gravel, respectively.

Translate.

Four tons of topsoil, 3 tons of mulch, and 6 tons of pea gravel costs \$2825.

$$4x + 3y + 6z = 2825$$

Five tons of topsoil, 2 tons of mulch, and 5 tons of pea gravel costs \$2663.

$$5x + 2y + 5z = 2663$$

Pea gravel costs \$17 less per ton than topsoil.

$$z = x - 17$$

We have a system of equations.

$$4x + 3y + 6z = 2825,$$
$$5x + 2y + 5z = 2663,$$
$$z = x - 17, \text{ or}$$

$$4x + 3y + 6z = 2825,$$
$$5x + 2y + 5z = 2663,$$
$$x \quad - z = 17$$

Carry out. Write an equivalent matrix equation, $\mathbf{AX} = \mathbf{B}$.

$$\begin{bmatrix} 4 & 3 & 6 \\ 5 & 2 & 5 \\ 1 & 0 & -1 \end{bmatrix} \begin{bmatrix} x \\ y \\ z \end{bmatrix} = \begin{bmatrix} 2825 \\ 2663 \\ 17 \end{bmatrix}$$

Then $\mathbf{X} = \mathbf{A}^{-1}\mathbf{B} = \begin{bmatrix} -\dfrac{1}{5} & \dfrac{3}{10} & \dfrac{3}{10} \\ 1 & -1 & 1 \\ -\dfrac{1}{5} & \dfrac{3}{10} & -\dfrac{7}{10} \end{bmatrix} \begin{bmatrix} 2825 \\ 2663 \\ 17 \end{bmatrix} =$

$\begin{bmatrix} 239 \\ 179 \\ 222 \end{bmatrix}$, so the solution is $(239, 179, 222)$.

Check. Four tons of topsoil, 3 tons of mulch, and 6 tons of pea gravel costs $4 \cdot \$239 + 3 \cdot \$179 + 6 \cdot \$222$, or $\$956 + \$537 + \$1332$, or $\$2825$. Five tons of topsoil, 2 tons of mulch, and 5 tons of pea gravel costs $5 \cdot \$239 + 2 \cdot \$179 + 5 \cdot \$222$, or $\$1195 + \$358 + \$1110$, or $\$2663$. The price of pea gravel, $\$222$, is $\$17$ less than the price of topsoil, $\$239$. The solution checks.

State. The price of topsoil is $\$239$ per ton, of mulch is $\$179$ per ton, and of pea gravel is $\$222$ per ton.

43.
$$\begin{array}{r|rrrr} -2 & 1 & -6 & 4 & -8 \\ & & -2 & 16 & -40 \\ \hline & 1 & -8 & 20 & -48 \end{array}$$

$$f(-2) = -48$$

45.
$$2x^2 + x = 7$$
$$2x^2 + x - 7 = 0$$
$$a = 2, \ b = 1, \ c = -7$$
$$x = \frac{-b \pm \sqrt{b^2 - 4ac}}{2a}$$
$$= \frac{-1 \pm \sqrt{1^2 - 4 \cdot 2 \cdot (-7)}}{2 \cdot 2} = \frac{-1 \pm \sqrt{1 + 56}}{4}$$
$$= \frac{-1 \pm \sqrt{57}}{4}$$

The solutions are $\dfrac{-1 + \sqrt{57}}{4}$ and $\dfrac{-1 - \sqrt{57}}{4}$, or $\dfrac{-1 \pm \sqrt{57}}{4}$.

47.
$$\sqrt{2x + 1} - 1 = \sqrt{2x - 4}$$
$$(\sqrt{2x+1} - 1)^2 = (\sqrt{2x-4})^2 \quad \text{Squaring both sides}$$
$$2x + 1 - 2\sqrt{2x+1} + 1 = 2x - 4$$
$$2x + 2 - 2\sqrt{2x+1} = 2x - 4$$
$$2 - 2\sqrt{2x+1} = -4 \quad \text{Subtracting } 2x$$
$$-2\sqrt{2x+1} = -6 \quad \text{Subtracting } 2$$
$$\sqrt{2x+1} = 3 \quad \text{Dividing by } -2$$
$$(\sqrt{2x+1})^2 = 3^2 \quad \text{Squaring both sides}$$
$$2x + 1 = 9$$
$$2x = 8$$
$$x = 4$$

The number 4 checks. It is the solution.

49. $f(x) = x^3 - 3x^2 - 6x + 8$

We use synthetic division to find one factor. We first try $x - 1$.

$$\begin{array}{r|rrrr} 1 & 1 & -3 & -6 & 8 \\ & & 1 & -2 & -8 \\ \hline & 1 & -2 & -8 & 0 \end{array}$$

Since $f(1) = 0$, $x - 1$ is a factor of $f(x)$. We have $f(x) = (x-1)(x^2 - 2x - 8)$. Factoring the trinomial we get $f(x) = (x-1)(x-4)(x+2)$.

51. $\mathbf{A} = [x]$

Write the augmented matrix.

$$\left[\, x \mid 1 \,\right]$$

Multiply by $\dfrac{1}{x}$.

$$\left[\, 1 \mid \dfrac{1}{x} \,\right]$$

Then $\mathbf{A}^{-1}$ exists if and only if $x \neq 0$. $\mathbf{A}^{-1} = \left[\dfrac{1}{x}\right]$.

53. $\mathbf{A} = \begin{bmatrix} 0 & 0 & x \\ 0 & y & 0 \\ z & 0 & 0 \end{bmatrix}$

Write the augmented matrix.

$$\left[\begin{array}{ccc|ccc} 0 & 0 & x & 1 & 0 & 0 \\ 0 & y & 0 & 0 & 1 & 0 \\ z & 0 & 0 & 0 & 0 & 1 \end{array}\right]$$

Interchange row 1 and row 3.

$$\left[\begin{array}{ccc|ccc} z & 0 & 0 & 0 & 0 & 1 \\ 0 & y & 0 & 0 & 1 & 0 \\ 0 & 0 & x & 1 & 0 & 0 \end{array}\right]$$

Multiply row 1 by $\dfrac{1}{z}$, row 2 by $\dfrac{1}{y}$, and row 3 by $\dfrac{1}{x}$.

$$\left[\begin{array}{ccc|ccc} 1 & 0 & 0 & 0 & 0 & \dfrac{1}{z} \\ 0 & 1 & 0 & 0 & \dfrac{1}{y} & 0 \\ 0 & 0 & 1 & \dfrac{1}{x} & 0 & 0 \end{array}\right]$$

Then $\mathbf{A}^{-1}$ exists if and only if $x \neq 0$ and $y \neq 0$ and $z \neq 0$, or if and only if $xyz \neq 0$.

$$\mathbf{A}^{-1} = \begin{bmatrix} 0 & 0 & \dfrac{1}{z} \\ 0 & \dfrac{1}{y} & 0 \\ \dfrac{1}{x} & 0 & 0 \end{bmatrix}$$

Exercise Set 6.6

1. $\begin{vmatrix} 5 & 3 \\ -2 & -4 \end{vmatrix} = 5(-4) - (-2) \cdot 3 = -20 + 6 = -14$

3. $\begin{vmatrix} 4 & -7 \\ -2 & 3 \end{vmatrix} = 4 \cdot 3 - (-2)(-7) = 12 - 14 = -2$

5. $\begin{vmatrix} -2 & -\sqrt{5} \\ -\sqrt{5} & 3 \end{vmatrix} = -2 \cdot 3 - (-\sqrt{5})(-\sqrt{5}) = -6 - 5 = -11$

7. $\begin{vmatrix} x & 4 \\ x & x^2 \end{vmatrix} = x \cdot x^2 - x \cdot 4 = x^3 - 4x$

9. $\mathbf{A} = \begin{bmatrix} 7 & -4 & -6 \\ 2 & 0 & -3 \\ 1 & 2 & -5 \end{bmatrix}$

M_{11} is the determinant of the matrix formed by deleting the first row and first column of $\mathbf{A}$:

$M_{11} = \begin{vmatrix} 0 & -3 \\ 2 & -5 \end{vmatrix} = 0(-5) - 2(-3) = 0 + 6 = 6$

M_{32} is the determinant of the matrix formed by deleting the third row and second column of $\mathbf{A}$:

$M_{32} = \begin{vmatrix} 7 & -6 \\ 2 & -3 \end{vmatrix} = 7(-3) - 2(-6) = -21 + 12 = -9$

M_{22} is the determinant of the matrix formed by deleting the second row and second column of $\mathbf{A}$:

$M_{22} = \begin{vmatrix} 7 & -6 \\ 1 & -5 \end{vmatrix} = 7(-5) - 1(-6) = -35 + 6 = -29$

11. In Exercise 9 we found that $M_{11} = 6$.

$A_{11} = (-1)^{1+1} M_{11} = 1 \cdot 6 = 6$

In Exercise 9 we found that $M_{32} = -9$.

$A_{32} = (-1)^{3+2} M_{32} = -1(-9) = 9$

In Exercise 9 we found that $M_{22} = -29$.

$A_{22} = (-1)^{2+2}(-29) = 1(-29) = -29$

13. $\mathbf{A} = \begin{bmatrix} 7 & -4 & -6 \\ 2 & 0 & -3 \\ 1 & 2 & -5 \end{bmatrix}$

$|\mathbf{A}|$

$= 2A_{21} + 0A_{22} + (-3)A_{23}$

$= 2(-1)^{2+1} \begin{vmatrix} -4 & -6 \\ 2 & -5 \end{vmatrix} + 0 + (-3)(-1)^{2+3} \begin{vmatrix} 7 & -4 \\ 1 & 2 \end{vmatrix}$

$= 2(-1)[-4(-5) - 2(-6)] + 0 + $
$\qquad (-3)(-1)[7 \cdot 2 - 1(-4)]$

$= -2(32) + 0 + 3(18) = -64 + 0 + 54$

$= -10$

15. $\mathbf{A} = \begin{bmatrix} 7 & -4 & -6 \\ 2 & 0 & -3 \\ 1 & 2 & -5 \end{bmatrix}$

$|\mathbf{A}|$

$= -6A_{13} + (-3)A_{23} + (-5)A_{33}$

$= -6(-1)^{1+3} \begin{vmatrix} 2 & 0 \\ 1 & 2 \end{vmatrix} + (-3)(-1)^{2+3} \begin{vmatrix} 7 & -4 \\ 1 & 2 \end{vmatrix} + $

$\qquad (-5)(-1)^{3+3} \begin{vmatrix} 7 & -4 \\ 2 & 0 \end{vmatrix}$

$= -6 \cdot 1(2 \cdot 2 - 1 \cdot 0) + (-3)(-1)[7 \cdot 2 - 1(-4)] + $
$\qquad -5 \cdot 1(7 \cdot 0 - 2(-4))$

$= -6(4) + 3(18) - 5(8) = -24 + 54 - 40$

$= -10$

17. $\mathbf{A} = \begin{bmatrix} 1 & 0 & 0 & -2 \\ 4 & 1 & 0 & 0 \\ 5 & 6 & 7 & 8 \\ -2 & -3 & -1 & 0 \end{bmatrix}$

$M_{12} = \begin{bmatrix} 4 & 0 & 0 \\ 5 & 7 & 8 \\ -2 & -1 & 0 \end{bmatrix}$

We will expand M_{12} across the first row.

$M_{12} = 4(-1)^{1+1} \begin{vmatrix} 7 & 8 \\ -1 & 0 \end{vmatrix} + 0(-1)^{1+2} \begin{vmatrix} 5 & 8 \\ -2 & 0 \end{vmatrix} + $

$\qquad 0(-1)^{1+3} \begin{vmatrix} 5 & 7 \\ -2 & -1 \end{vmatrix}$

$= 4 \cdot 1[7 \cdot 0 - (-1)8] + 0 + 0$

$= 4(8) = 32$

$M_{44} = \begin{bmatrix} 1 & 0 & 0 \\ 4 & 1 & 0 \\ 5 & 6 & 7 \end{bmatrix}$

We will expand M_{44} across the first row.

$M_{44} = 1(-1)^{1+1} \begin{vmatrix} 1 & 0 \\ 6 & 7 \end{vmatrix} + 0(-1)^{1+2} \begin{vmatrix} 4 & 0 \\ 5 & 7 \end{vmatrix} + $

$\qquad 0(-1)^{1+3} \begin{vmatrix} 4 & 1 \\ 5 & 6 \end{vmatrix}$

$= 1 \cdot 1(1 \cdot 7 - 6 \cdot 0) + 0 + 0$

$= 1(7) = 7$

19. $\mathbf{A} = \begin{bmatrix} 1 & 0 & 0 & -2 \\ 4 & 1 & 0 & 0 \\ 5 & 6 & 7 & 8 \\ -2 & -3 & -1 & 0 \end{bmatrix}$

$A_{22} = (-1)^{2+2} M_{22} = M_{22}$

$= \begin{vmatrix} 1 & 0 & -2 \\ 5 & 7 & 8 \\ -2 & -1 & 0 \end{vmatrix}$

We will expand across the first row.

$$\begin{vmatrix} 1 & 0 & -2 \\ 5 & 7 & 8 \\ -2 & -1 & 0 \end{vmatrix}$$

$$= 1(-1)^{1+1}\begin{vmatrix} 7 & 8 \\ -1 & 0 \end{vmatrix} + 0(-1)^{1+2}\begin{vmatrix} 5 & 8 \\ -2 & 0 \end{vmatrix} +$$

$$(-2)(-1)^{1+3}\begin{vmatrix} 5 & 7 \\ -2 & -1 \end{vmatrix}$$

$$= 1 \cdot 1[7 \cdot 0 - (-1)8] + 0 + (-2) \cdot 1[5(-1) - (-2)7]$$

$$= 1(8) + 0 - 2(9) = 8 + 0 - 18$$

$$= -10$$

$$A_{34} = (-1)^{3+4}M_{34} = -1 \cdot M_{34}$$

$$= -1 \cdot \begin{vmatrix} 1 & 0 & 0 \\ 4 & 1 & 0 \\ -2 & -3 & -1 \end{vmatrix}$$

We will expand across the first row.

$$-1 \cdot \begin{vmatrix} 1 & 0 & 0 \\ 4 & 1 & 0 \\ -2 & -3 & -1 \end{vmatrix}$$

$$= -1\left[1(-1)^{1+1}\begin{vmatrix} 1 & 0 \\ -3 & -1 \end{vmatrix} + 0(-1)^{1+2}\begin{vmatrix} 4 & 0 \\ -2 & -1 \end{vmatrix} +\right.$$

$$\left. 0(-1)^{1+3}\begin{vmatrix} 4 & 1 \\ -2 & -3 \end{vmatrix}\right]$$

$$= -1[1 \cdot 1(1(-1) - (-3) \cdot 0) + 0 + 0]$$

$$= -1[1(-1)] = 1$$

21. $\mathbf{A} = \begin{bmatrix} 1 & 0 & 0 & -2 \\ 4 & 1 & 0 & 0 \\ 5 & 6 & 7 & 8 \\ -2 & -3 & -1 & 0 \end{bmatrix}$

$|\mathbf{A}|$

$$= 1 \cdot A_{11} + 0 \cdot A_{12} + 0 \cdot A_{13} + (-2)A_{14}$$

$$= A_{11} + (-2)A_{14}$$

$$= (-1)^{1+1}\begin{vmatrix} 1 & 0 & 0 \\ 6 & 7 & 8 \\ -3 & -1 & 0 \end{vmatrix} +$$

$$(-2)(-1)^{1+4}\begin{vmatrix} 4 & 1 & 0 \\ 5 & 6 & 7 \\ -2 & -3 & -1 \end{vmatrix}$$

$$= \begin{vmatrix} 1 & 0 & 0 \\ 6 & 7 & 8 \\ -3 & -1 & 0 \end{vmatrix} + 2\begin{vmatrix} 4 & 1 & 0 \\ 5 & 6 & 7 \\ -2 & -3 & -1 \end{vmatrix}$$

We will expand each determinant across the first row. We have:

$$1(-1)^{1+1}\begin{vmatrix} 7 & 8 \\ -1 & 0 \end{vmatrix} + 0 + 0 +$$

$$2\left[4(-1)^{1+1}\begin{vmatrix} 6 & 7 \\ -3 & -1 \end{vmatrix} + 1(-1)^{1+2}\begin{vmatrix} 5 & 7 \\ -2 & -1 \end{vmatrix} + 0\right]$$

$$= 1 \cdot 1[7 \cdot 0 - (-1)8] + 2[4 \cdot 1[6(-1) - (-3) \cdot 7] +$$

$$1(-1)[5(-1) - (-2) \cdot 7]$$

$$= 1(8) + 2[4(15) - 1(9)] = 8 + 2(51)$$

$$= 8 + 102 = 110$$

23. We will expand across the first row. We could have chosen any other row or column just as well.

$$\begin{vmatrix} 3 & 1 & 2 \\ -2 & 3 & 1 \\ 3 & 4 & -6 \end{vmatrix}$$

$$= 3(-1)^{1+1}\begin{vmatrix} 3 & 1 \\ 4 & -6 \end{vmatrix} + 1 \cdot (-1)^{1+2}\begin{vmatrix} -2 & 1 \\ 3 & -6 \end{vmatrix} +$$

$$2(-1)^{1+3}\begin{vmatrix} -2 & 3 \\ 3 & 4 \end{vmatrix}$$

$$= 3 \cdot 1[3(-6) - 4 \cdot 1] + 1(-1)[-2(-6) - 3 \cdot 1] +$$

$$2 \cdot 1(-2 \cdot 4 - 3 \cdot 3)$$

$$= 3(-22) - (9) + 2(-17)$$

$$= -109$$

25. We will expand down the second column. We could have chosen any other row or column just as well.

$$\begin{vmatrix} x & 0 & -1 \\ 2 & x & x^2 \\ -3 & x & 1 \end{vmatrix}$$

$$= 0(-1)^{1+2}\begin{vmatrix} 2 & x^2 \\ -3 & 1 \end{vmatrix} + x(-1)^{2+2}\begin{vmatrix} x & -1 \\ -3 & 1 \end{vmatrix} +$$

$$x(-1)^{3+2}\begin{vmatrix} x & -1 \\ 2 & x^2 \end{vmatrix}$$

$$= 0(-1)[2 \cdot 1 - (-3)x^2] + x \cdot 1[x \cdot 1 - (-3)(-1)] +$$

$$x(-1)[x \cdot x^2 - 2(-1)]$$

$$= 0 + x(x - 3) - x(x^3 + 2)$$

$$= x^2 - 3x - x^4 - 2x = -x^4 + x^2 - 5x$$

27. $-2x + 4y = 3,$

$3x - 7y = 1$

$$D = \begin{vmatrix} -2 & 4 \\ 3 & -7 \end{vmatrix} = -2(-7) - 3 \cdot 4 = 14 - 12 = 2$$

$$D_x = \begin{vmatrix} 3 & 4 \\ 1 & -7 \end{vmatrix} = 3(-7) - 1 \cdot 4 = -21 - 4 = -25$$

$$D_y = \begin{vmatrix} -2 & 3 \\ 3 & 1 \end{vmatrix} = -2 \cdot 1 - 3 \cdot 3 = -2 - 9 = -11$$

$$x = \frac{D_x}{D} = \frac{-25}{2} = -\frac{25}{2}$$

$$y = \frac{D_y}{D} = \frac{-11}{2} = -\frac{11}{2}$$

The solution is $\left(-\frac{25}{2}, -\frac{11}{2}\right)$.

29. $2x - y = 5,$
$\quad x - 2y = 1$

$$D = \begin{vmatrix} 2 & -1 \\ 1 & -2 \end{vmatrix} = 2(-2) - 1(-1) = -4 + 1 = -3$$

$$D_x = \begin{vmatrix} 5 & -1 \\ 1 & -2 \end{vmatrix} = 5(-2) - 1(-1) = -10 + 1 = -9$$

$$D_y = \begin{vmatrix} 2 & 5 \\ 1 & 1 \end{vmatrix} = 2 \cdot 1 - 1 \cdot 5 = 2 - 5 = -3$$

$$x = \frac{D_x}{D} = \frac{-9}{-3} = 3$$

$$y = \frac{D_y}{D} = \frac{-3}{-3} = 1$$

The solution is $(3, 1)$.

31. $2x + 9y = -2,$
$\quad 4x - 3y = 3$

$$D = \begin{vmatrix} 2 & 9 \\ 4 & -3 \end{vmatrix} = 2(-3) - 4 \cdot 9 = -6 - 36 = -42$$

$$D_x = \begin{vmatrix} -2 & 9 \\ 3 & -3 \end{vmatrix} = -2(-3) - 3 \cdot 9 = 6 - 27 = -21$$

$$D_y = \begin{vmatrix} 2 & -2 \\ 4 & 3 \end{vmatrix} = 2 \cdot 3 - 4(-2) = 6 + 8 = 14$$

$$x = \frac{D_x}{D} = \frac{-21}{-42} = \frac{1}{2}$$

$$y = \frac{D_y}{D} = \frac{14}{-42} = -\frac{1}{3}$$

The solution is $\left(\frac{1}{2}, -\frac{1}{3}\right)$.

33. $2x + 5y = 7,$
$\quad 3x - 2y = 1$

$$D = \begin{vmatrix} 2 & 5 \\ 3 & -2 \end{vmatrix} = 2(-2) - 3 \cdot 5 = -4 - 15 = -19$$

$$D_x = \begin{vmatrix} 7 & 5 \\ 1 & -2 \end{vmatrix} = 7(-2) - 1 \cdot 5 = -14 - 5 = -19$$

$$D_y = \begin{vmatrix} 2 & 7 \\ 3 & 1 \end{vmatrix} = 2 \cdot 1 - 3 \cdot 7 = 2 - 21 = -19$$

$$x = \frac{D_x}{D} = \frac{-19}{-19} = 1$$

$$y = \frac{D_y}{D} = \frac{-19}{-19} = 1$$

The solution is $(1, 1)$.

35. $3x + 2y - z = 4,$
$\quad 3x - 2y + z = 5,$
$\quad 4x - 5y - z = -1$

$$D = \begin{vmatrix} 3 & 2 & -1 \\ 3 & -2 & 1 \\ 4 & -5 & -1 \end{vmatrix} = 42$$

$$D_x = \begin{vmatrix} 4 & 2 & -1 \\ 5 & -2 & 1 \\ -1 & -5 & -1 \end{vmatrix} = 63$$

$$D_y = \begin{vmatrix} 3 & 4 & -1 \\ 3 & 5 & 1 \\ 4 & -1 & -1 \end{vmatrix} = 39$$

$$D_z = \begin{vmatrix} 3 & 2 & 4 \\ 3 & -2 & 5 \\ 4 & -5 & -1 \end{vmatrix} = 99$$

$$x = \frac{D_x}{D} = \frac{63}{42} = \frac{3}{2}$$

$$y = \frac{D_y}{D} = \frac{39}{42} = \frac{13}{14}$$

$$z = \frac{D_z}{D} = \frac{99}{42} = \frac{33}{14}$$

The solution is $\left(\frac{3}{2}, \frac{13}{14}, \frac{33}{14}\right)$.

(Note that we could have used Cramer's rule to find only two of the values and then used substitution to find the remaining value.)

37. $3x + 5y - z = -2,$
$\quad x - 4y + 2z = 13,$
$\quad 2x + 4y + 3z = 1$

$$D = \begin{vmatrix} 3 & 5 & -1 \\ 1 & -4 & 2 \\ 2 & 4 & 3 \end{vmatrix} = -67$$

$$D_x = \begin{vmatrix} -2 & 5 & -1 \\ 13 & -4 & 2 \\ 1 & 4 & 3 \end{vmatrix} = -201$$

$$D_y = \begin{vmatrix} 3 & -2 & -1 \\ 1 & 13 & 2 \\ 2 & 1 & 3 \end{vmatrix} = 134$$

$$D_z = \begin{vmatrix} 3 & 5 & -2 \\ 1 & -4 & 13 \\ 2 & 4 & 1 \end{vmatrix} = -67$$

$$x = \frac{D_x}{D} = \frac{-201}{-67} = 3$$

$$y = \frac{D_y}{D} = \frac{134}{-67} = -2$$

$$z = \frac{D_z}{D} = \frac{-67}{-67} = 1$$

The solution is $(3, -2, 1)$.

(Note that we could have used Cramer's rule to find only two of the values and then used substitution to find the remaining value.)

39. $x - 3y - 7z = 6,$

$2x + 3y + z = 9,$

$4x + y = 7$

$$D = \begin{vmatrix} 1 & -3 & -7 \\ 2 & 3 & 1 \\ 4 & 1 & 0 \end{vmatrix} = 57$$

$$D_x = \begin{vmatrix} 6 & -3 & -7 \\ 9 & 3 & 1 \\ 7 & 1 & 0 \end{vmatrix} = 57$$

$$D_y = \begin{vmatrix} 1 & 6 & -7 \\ 2 & 9 & 1 \\ 4 & 7 & 0 \end{vmatrix} = 171$$

$$D_z = \begin{vmatrix} 1 & -3 & 6 \\ 2 & 3 & 9 \\ 4 & 1 & 7 \end{vmatrix} = -114$$

$$x = \frac{D_x}{D} = \frac{57}{57} = 1$$

$$y = \frac{D_y}{D} = \frac{171}{57} = 3$$

$$z = \frac{D_z}{D} = \frac{-114}{57} = -2$$

The solution is $(1, 3, -2)$.

(Note that we could have used Cramer's rule to find only two of the values and then used substitution to find the remaining value.)

41. $6y + 6z = -1,$

$8x + 6z = -1,$

$4x + 9y = 8$

$$D = \begin{vmatrix} 0 & 6 & 6 \\ 8 & 0 & 6 \\ 4 & 9 & 0 \end{vmatrix} = 576$$

$$D_x = \begin{vmatrix} -1 & 6 & 6 \\ -1 & 0 & 6 \\ 8 & 9 & 0 \end{vmatrix} = 288$$

$$D_y = \begin{vmatrix} 0 & -1 & 6 \\ 8 & -1 & 6 \\ 4 & 8 & 0 \end{vmatrix} = 384$$

$$D_z = \begin{vmatrix} 0 & 6 & -1 \\ 8 & 0 & -1 \\ 4 & 9 & 8 \end{vmatrix} = -480$$

$$x = \frac{D_x}{D} = \frac{288}{576} = \frac{1}{2}$$

$$y = \frac{D_y}{D} = \frac{384}{576} = \frac{2}{3}$$

$$z = \frac{D_z}{D} = \frac{-480}{576} = -\frac{5}{6}$$

The solution is $\left(\frac{1}{2}, \frac{2}{3}, -\frac{5}{6}\right)$.

(Note that we could have used Cramer's rule to find only two of the values and then used substitution to find the remaining value.)

43. The graph of $f(x) = 3x + 2$ is shown below. Since it passes the horizontal-line test, the function is one-to-one.

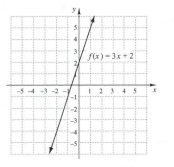

We find a formula for $f^{-1}(x)$.

Replace $f(x)$ with y: $y = 3x + 2$

Interchange x and y: $x = 3y + 2$

Solve for y: $y = \frac{x - 2}{3}$

Replace y with $f^{-1}(x)$: $f^{-1}(x) = \frac{x - 2}{3}$

45. The graph of $f(x) = |x| + 3$ is shown below. It fails the horizontal-line test, so it is not one-to-one.

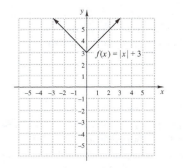

47. $(3 - 4i) - (-2 - i) = 3 - 4i + 2 + i =$

$(3 + 2) + (-4 + 1)i = 5 - 3i$

49. $(1 - 2i)(6 + 2i) = 6 + 2i - 12i - 4i^2 =$

$6 + 2i - 12i + 4 = 10 - 10i$

51. $$\begin{vmatrix} y & 2 \\ 3 & y \end{vmatrix} = y$$

$$y^2 - 6 = y$$

$$y^2 - y - 6 = 0$$

$$(y - 3)(y + 2) = 0$$

$$y - 3 = 0 \ or \ y + 2 = 0$$

$$y = 3 \ or \ y = -2$$

The solutions are 3 and -2.

53. $\begin{vmatrix} 2 & x & 1 \\ 1 & 2 & -1 \\ 3 & 4 & -2 \end{vmatrix} = -6$

$$-x - 2 = -6 \quad \text{Evaluating the}$$
$$\text{determinant}$$
$$-x = -4$$
$$x = 4$$

The solution is 4.

55. Answers may vary.

$\begin{vmatrix} a & b \\ -b & a \end{vmatrix}$

57. Answers may vary.

$\begin{vmatrix} 2\pi r & 2\pi r \\ -h & r \end{vmatrix}$

Exercise Set 6.7

1. Graph (f) is the graph of $y > x$.

3. Graph (h) is the graph of $y \le x - 3$.

5. Graph (g) is the graph of $2x + y < 4$.

7. Graph (b) is the graph of $2x - 5y > 10$.

9. Graph: $y > 2x$

   1. We first graph the related equation $y = 2x$. We draw the line dashed since the inequality symbol is $>$.

   2. To determine which half-plane to shade, test a point not on the line. We try $(1, 1)$ and substitute:

$$\frac{y > 2x}{1 \ ? \ 2 \cdot 1}$$
$$1 \ \bigm| \ 2 \qquad \text{FALSE}$$

   Since $1 > 2$ is false, $(1, 1)$ is not a solution, nor are any points in the half-plane containing $(1, 1)$. The points in the opposite half-plane are solutions, so we shade that half-plane and obtain the graph.

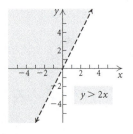

11. Graph: $y + x \ge 0$

   1. First graph the related equation $y + x = 0$. Draw the line solid since the inequality is $\ge$.

   2. Next determine which half-plane to shade by testing a point not on the line. Here we use $(2, 2)$ as a check.

$$\frac{y + x \ge 0}{2 + 2 \ ? \ 0}$$
$$4 \ \bigm| \ 0 \qquad \text{TRUE}$$

Since $4 \ge 0$ is true, $(2, 2)$ is a solution. Thus shade the half-plane containing $(2, 2)$.

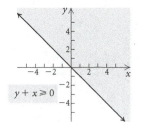

13. Graph: $y > x - 3$

   1. We first graph the related equation $y = x - 3$. Draw the line dashed since the inequality symbol is $>$.

   2. To determine which half-plane to shade, test a point not on the line. We try $(0, 0)$.

$$\frac{y > x - 3}{0 \ ? \ 0 - 3}$$
$$0 \ \bigm| \ -3 \qquad \text{TRUE}$$

Since $0 > -3$ is true, $(0, 0)$ is a solution. Thus we shade the half-plane containing $(0, 0)$.

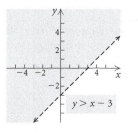

15. Graph: $x + y < 4$

   1. First graph the related equation $x + y = 4$. Draw the line dashed since the inequality is $<$.

   2. To determine which half-plane to shade, test a point not on the line. We try $(0, 0)$.

$$\frac{x + y < 4}{0 + 0 \ ? \ 4}$$
$$0 \ \bigm| \ 4 \qquad \text{TRUE}$$

Since $0 < 4$ is true, $(0, 0)$ is a solution. Thus shade the half-plane containing $(0, 0)$.

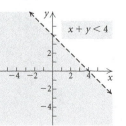

17. Graph: $3x - 2y \leq 6$

   1. First graph the related equation $3x - 2y = 6$. Draw the line solid since the inequality is $\leq$.

   2. To determine which half-plane to shade, test a point not on the line. We try $(0, 0)$.

$$\frac{3x - 2y \leq 6}{3(0) - 2(0) \ ? \ 6}$$
$$0 \ \bigm| \ 6 \ \text{TRUE}$$

Since $0 \leq 6$ is true, $(0, 0)$ is a solution. Thus shade the half-plane containing $(0, 0)$.

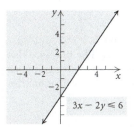

19. Graph: $3y + 2x \geq 6$

   1. First graph the related equation $3y + 2x = 6$. Draw the line solid since the inequality is $\geq$.

   2. To determine which half-plane to shade, test a point not on the line. We try $(0, 0)$.

$$\frac{3y + 2x \geq 6}{3 \cdot 0 + 2 \cdot 0 \ ? \ 6}$$
$$0 \ \bigm| \ 6 \ \text{FALSE}$$

Since $0 \geq 6$ is false, $(0, 0)$ is not a solution. We shade the half-plane which does not contain $(0, 0)$.

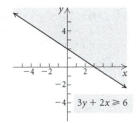

21. Graph: $3x - 2 \leq 5x + y$

$$-2 \leq 2x + y \quad \text{Adding } -3x$$

   1. First graph the related equation $2x + y = -2$. Draw the line solid since the inequality is $\leq$.

   2. To determine which half-plane to shade, test a point not on the line. We try $(0, 0)$.

$$\frac{2x + y \geq -2}{2(0) + 0 \ ? \ -2}$$
$$0 \ \bigm| \ -2 \ \text{TRUE}$$

Since $0 \geq -2$ is true, $(0, 0)$ is a solution. Thus shade the half-plane containing the origin.

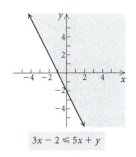

$3x - 2 \leq 5x + y$

23. Graph: $x < -4$

   1. We first graph the related equation $x = -4$. Draw the line dashed since the inequality is $<$.

   2. To determine which half-plane to shade, test a point not on the line. We try $(0, 0)$.

$$\frac{x < -4}{0 \ ? \ -4 \ \text{FALSE}}$$

Since $0 < -4$ is false, $(0, 0)$ is not a solution. Thus, we shade the half-plane which does not contain the origin.

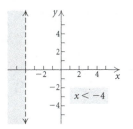

25. Graph: $y \geq 5$

   1. First we graph the related equation $y = 5$. Draw the line solid since the inequality is $\geq$.

   2. To determine which half-plane to shade we test a point not on the line. We try $(0, 0)$.

$$\frac{y \geq 5}{0 \ ? \ 5 \ \text{FALSE}}$$

Since $0 \geq 5$ is false, $(0, 0)$ is not a solution. We shade the half-plane that does not contain $(0, 0)$.

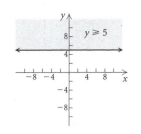

27. Graph: $-4 < y < -1$

This is a conjunction of two inequalities

$$-4 < y \quad \text{and} \quad y < -1.$$

We can graph $-4 < y$ and $y < -1$ separately and then graph the intersection, or region in both solution sets.

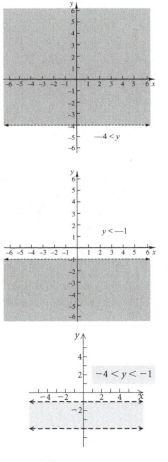

$-4 < y$

$y < -1$

$-4 < y < -1$

29. Graph: $y \geq |x|$

    1. Graph the related equation $y = |x|$. Draw the line solid since the inequality symbol is $\geq$.

    2. To determine the region to shade, observe that the solution set consists of all ordered pairs (x, y) where the second coordinate is greater than or equal to the absolute value of the first coordinate. We see that the solutions are the points on or above the graph of $y = |x|$.

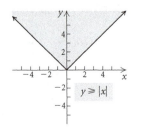

$y \geq |x|$

31. Graph (f) is the correct graph.

33. Graph (a) is the correct graph.

35. Graph (b) is the correct graph.

37. First we find the related equations. One line goes through $(0, 4)$ and $(4, 0)$. We find its slope:

$$m = \frac{0 - 4}{4 - 0} = \frac{-4}{4} = -1$$

This line has slope -1 and y-intercept $(0, 4)$, so its equation is $y = -x + 4$.

The other line goes through $(0, 0)$ and $(1, 3)$. We find the slope.

$$m = \frac{3 - 0}{1 - 0} = 3$$

This line has slope 3 and y-intercept $(0, 0)$, so its equation is $y = 3x + 0$, or $y = 3x$.

Observing the shading on the graph and the fact that the lines are solid, we can write the system of inqualities as

$$y \leq -x + 4,$$
$$y \leq 3x.$$

Answers may vary.

39. The equation of the vertical line is $x = 2$ and the equation of the horizontal line is $y = -1$. The lines are dashed and the shaded area is to the left of the vertical line and above the horizontal line, so the system of inequalities can be written

$$x < 2,$$
$$y > -1.$$

41. First we find the related equations. One line goes through $(0, 3)$ and $(3, 0)$. We find its slope:

$$m = \frac{0 - 3}{3 - 0} = \frac{-3}{3} = -1$$

This line has slope -1 and y-intercept $(0, 3)$, so its equation is $y = -x + 3$.

The other line goes through $(0, 1)$ and $(1, 2)$. We find its slope:

$$m = \frac{2 - 1}{1 - 0} = \frac{1}{1} = 1$$

This line has slope 1 and y-intercept $(0, 1)$, so its equation is $y = x + 1$.

Observe that both lines are solid and that the shading lies below both lines, to the right of the y-axis, and above the x-axis. We can write this system of inequalities as

$$y \leq -x + 3,$$
$$y \leq x + 1,$$
$$x \geq 0,$$
$$y \geq 0.$$

43. Graph: $y \leq x,$
             $y \geq 3 - x$

We graph the related equations $y = x$ and $y = 3 - x$ using solid lines and determine the solution set for each inequality. Then we shade the region common to both solution sets.

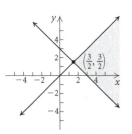

$\left(\frac{3}{2}, \frac{3}{2}\right)$

We find the vertex $\left(\dfrac{3}{2}, \dfrac{3}{2}\right)$ by solving the system

$$y = x,$$
$$y = 3 - x.$$

45. Graph: $y \geq x,$
$\qquad\quad y \leq 4 - x$

We graph the related equations $y = x$ and $y = 4 - x$ using solid lines and determine the solution set for each inequality. Then we shade the region common to both solution sets.

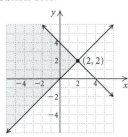

47. Graph: $y \geq -3,$
$\qquad\quad x \geq 1$

We graph the related equations $y = -3$ and $x = 1$ using solid lines and determine the solution set for each inequality. Then we shade the region common to both solution sets.

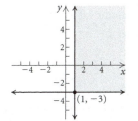

We find the vertex $(1, -3)$ by solving the system

$$y = -3,$$
$$x = 1.$$

49. Graph: $x \leq 3,$
$\qquad\quad y \geq 2 - 3x$

We graph the related equations $x = 3$ and $y = 2 - 3x$ using solid lines and determine the half-plane containing the solution set for each inequality. Then we shade the region common to both solution sets.

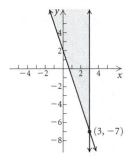

We find the vertex $(3, -7)$ by solving the system

$$x = 3,$$
$$y = 2 - 3x.$$

51. Graph: $x + y \leq 1,$
$\qquad\quad x - y \leq 2$

We graph the related equations $x + y = 1$ and $x - y = 2$ using solid lines and determine the half-plane containing the solution set for each inequality. Then we shade the region common to both solution sets.

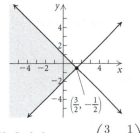

We find the vertex $\left(\dfrac{3}{2}, -\dfrac{1}{2}\right)$ by solving the system

$$x + y = 1,$$
$$x - y = 2.$$

53. Graph: $2y - x \leq 2,$
$\qquad\quad y + 3x \geq -1$

We graph the related equations $2y - x = 2$ and $y + 3x = -1$ using solid lines and determine the half-plane containing the solution set for each inequality. Then we shade the region common to both solution sets.

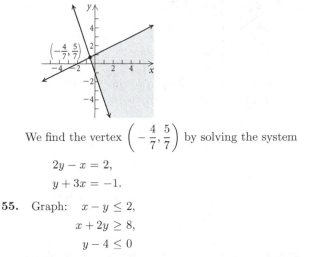

We find the vertex $\left(-\dfrac{4}{7}, \dfrac{5}{7}\right)$ by solving the system

$$2y - x = 2,$$
$$y + 3x = -1.$$

55. Graph: $x - y \leq 2,$
$\qquad\quad x + 2y \geq 8,$
$\qquad\quad y - 4 \leq 0$

We graph the related equations $x - y = 2$, $x + 2y = 8$, and $y - 4 = 0$ using solid lines and determine the half-plane containing the solution set for each inequality. Then we shade the region common to all three solution sets.

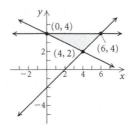

We find the vertex $(0, 4)$ by solving the system

$$x + 2y = 8,$$
$$y = 4.$$

We find the vertex $(6, 4)$ by solving the system

$$x - y = 2,$$
$$y = 4.$$

We find the vertex $(4, 2)$ by solving the system

$$x - y = 2,$$
$$x + 2y = 8.$$

57. Graph: $4x - 3y \geq -12,$

$$4x + 3y \geq -36,$$
$$y \leq 0,$$
$$x \leq 0$$

Shade the intersection of the graphs of the four inequalities.

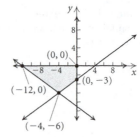

We find the vertex $(-12, 0)$ by solving the system

$$4y + 3x = -36,$$
$$y = 0.$$

We find the vertex $(0, 0)$ by solving the system

$$y = 0,$$
$$x = 0.$$

We find the vertex $(0, -3)$ by solving the system

$$4y - 3x = -12,$$
$$x = 0.$$

We find the vertex $(-4, -6)$ by solving the system

$$4y - 3x = -12,$$
$$4y + 3x = -36.$$

59. Graph: $3x + 4y \geq 12,$

$$5x + 6y \leq 30,$$
$$1 \leq x \leq 3$$

Shade the intersection of the graphs of the given inequalities.

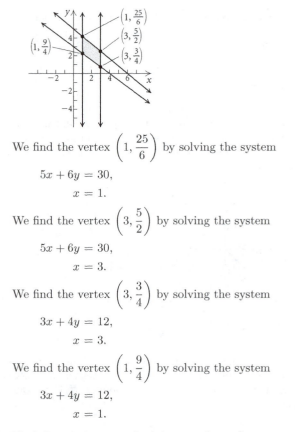

We find the vertex $\left(1, \dfrac{25}{6}\right)$ by solving the system

$$5x + 6y = 30,$$
$$x = 1.$$

We find the vertex $\left(3, \dfrac{5}{2}\right)$ by solving the system

$$5x + 6y = 30,$$
$$x = 3.$$

We find the vertex $\left(3, \dfrac{3}{4}\right)$ by solving the system

$$3x + 4y = 12,$$
$$x = 3.$$

We find the vertex $\left(1, \dfrac{9}{4}\right)$ by solving the system

$$3x + 4y = 12,$$
$$x = 1.$$

61. Find the maximum and minimum values of

$$P = 17x - 3y + 60, \text{ subject to}$$
$$6x + 8y \leq 48,$$
$$0 \leq y \leq 4,$$
$$0 \leq x \leq 7.$$

Graph the system of inequalities and determine the vertices.

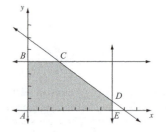

Vertex A: $(0, 0)$

Vertex B:

We solve the system $x = 0$ and $y = 4$. The coordinates of point B are $(0, 4)$.

Vertex C:

We solve the system $6x + 8y = 48$ and $y = 4$. The coordinates of point C are $\left(\dfrac{8}{3}, 4\right)$.

Vertex D:

We solve the system $6x + 8y = 48$ and $x = 7$. The coordinates of point D are $\left(7, \dfrac{3}{4}\right)$.

Vertex E:

  We solve the system $x = 7$ and $y = 0$. The coordinates of point E are $(7, 0)$.

Evaluate the objective function P at each vertex.

Vertex	$P = 17x - 3y + 60$
$A(0,0)$	$17 \cdot 0 - 3 \cdot 0 + 60 = 60$
$B(0,4)$	$17 \cdot 0 - 3 \cdot 4 + 60 = 48$
$C\left(\dfrac{8}{3}, 4\right)$	$17 \cdot \dfrac{8}{3} - 3 \cdot 4 + 60 = 66\dfrac{2}{3}$
$D\left(7, \dfrac{3}{4}\right)$	$17 \cdot 7 - 3 \cdot \dfrac{3}{4} + 60 = 176\dfrac{3}{4}$
$E(7,0)$	$17 \cdot 7 - 3 \cdot 0 + 60 = 179$

The maximum value of P is 179 when $x = 7$ and $y = 0$.

The minimum value of P is 48 when $x = 0$ and $y = 4$.

63. Find the maximum and minimum values of

$F = 5x + 36y$, subject to

$$5x + 3y \le 34,$$
$$3x + 5y \le 30,$$
$$x \ge 0,$$
$$y \ge 0.$$

Graph the system of inequalities and find the vertices.

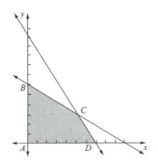

Vertex A: $(0, 0)$

Vertex B:

  We solve the system $3x + 5y = 30$ and $x = 0$. The coordinates of point B are $(0, 6)$.

Vertex C:

  We solve the system $5x + 3y = 34$ and $3x + 5y = 30$. The coordinates of point C are $(5, 3)$.

Vertex D:

  We solve the system $5x + 3y = 34$ and $y = 0$. The coordinates of point D are $\left(\dfrac{34}{5}, 0\right)$.

Evaluate the objective function F at each vertex.

Vertex	$F = 5x + 36y$
$A(0,0)$	$5 \cdot 0 + 36 \cdot 0+ = 0$
$B(0,6)$	$5 \cdot 0 + 36 \cdot 6 = 216$
$C(5,3)$	$5 \cdot 5 + 39 \cdot 3 = 133$
$D\left(\dfrac{34}{5}, 0\right)$	$5 \cdot \dfrac{34}{5} + 36 \cdot 0 = 34$

The maximum value of F is 216 when $x = 0$ and $y = 6$.

The minimum value of F is 0 when $x = 0$ and $y = 0$.

65. Let $x = $ the number of gallons the pickup truck uses and $y = $ the number of gallons the moped uses. The number of miles M that can be driven is given by

$$M = 20x + 100y$$

subject to the constraints

$$x + y \;\le\; 12,$$
$$0 \le x \le 10,$$
$$0 \le y \;\le 3.$$

We graph the system of inequalities, determine the vertices, and find the value of M at each vertex.

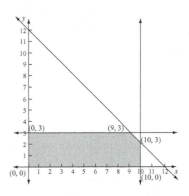

Vertex	$M = 20x + 100y$
$(0,0)$	$20(0) + 100(0) = 0$
$(0,3)$	$20(0) + 100(3) = 300$
$(9,3)$	$20(9) + 100(3) = 480$
$(10,2)$	$20(10) + 100(2) = 400$
$(10,0)$	$20(10) + 100(0) = 200$

The maximum number of miles is 480 when the pickup truck uses 9 gal and the moped uses 3 gal.

67. Let $x = $ the corn acreage and $y = $ the soybean acreage. The profit P is given by

$$P = 325x + 180y$$

subject to the constraints

$$x + y \le 240,$$
$$2x + y \le 320,$$
$$x \ge 0,$$
$$y \ge 0.$$

We graph the system of equations, determine the vertices, and find the value of P at each vertex.

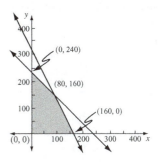

Vertex	$P = 325x + 180y$
$(0, 0)$	$325(0) + 180(0) = 0$
$(0, 240)$	$325(0) + 180(240) = 43,200$
$(80, 160)$	$325(80) + 180(160) = 54,800$
$(160, 0)$	$325(160) + 180(0) = 52,000$

The maximum profit of $54,800 occurs when 80 acres of corn and 160 acres of soybeans are planted.

69. Let $x =$ the number of sacks of soybean meal to be used and $y =$ the number of sacks of oats. The minimum cost is given by

$$C = 20x + 8y$$

subject to the constraints

$$50x + 15y \geq 120,$$
$$8x + 5y \geq 24,$$
$$5x + y \geq 10,$$
$$x \geq 0,$$
$$y \geq 0.$$

Graph the system of inequalities, determine the vertices, and find the value of C at each vertex.

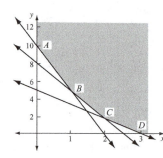

Vertex	$C = 20x + 8y$
$A(0, 10)$	$20 \cdot 0 + 8 \cdot 10 = 80$
$B\left(\dfrac{6}{5}, 4\right)$	$20 \cdot \dfrac{6}{5} + 8 \cdot 4 = 56$
$C\left(\dfrac{24}{13}, \dfrac{24}{13}\right)$	$20 \cdot \dfrac{24}{13} + 8 \cdot \dfrac{24}{13} = 51\dfrac{9}{13}$
$D(3, 0)$	$20 \cdot 3 + 8 \cdot 0 = 60$

The minimum cost of $\$51\dfrac{9}{13}$ is achieved by using $\dfrac{24}{13}$, or $1\dfrac{11}{13}$ sacks of soybean meal and $\dfrac{24}{13}$, or $1\dfrac{11}{13}$ sacks of oats.

71. Let $x =$ the amount invested in corporate bonds and $y =$ the amount invested in municipal bonds. The income I is given by

$$I = 0.03x + 0.0425y$$

subject to the constraints

$$x + y \leq 40,000,$$
$$6000 \leq x \leq 22,000,$$
$$0 \leq y \leq 30,000.$$

We graph the system of inequalities, determine the vertices, and find the value of I at each vertex.

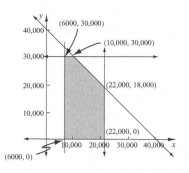

Vertex	$I = 0.03x + 0.0425y$
$(6000, 0)$	180
$(6000, 30,000)$	1455
$(10,000, 30,000)$	1575
$(22,000, 18,000)$	1425
$(22,000, 0)$	660

The maximum income of $1575 occurs when $10,000 is invested in corporate bonds and $30,000 is invested in municipal bonds.

73. Let $x =$ the number of P_1 airplanes and $y =$ the number of P_2 airplanes to be used. The operating cost C, in thousands of dollars, is given by

$$C = 12x + 10y$$

subject to the constraints

$$40x + 80y \geq 2000,$$
$$40x + 30y \geq 1500,$$
$$120x + 40y \geq 2400,$$
$$x \geq 0,$$
$$y \geq 0.$$

Graph the system of inequalities, determine the vertices, and find the value of C at each vertex.

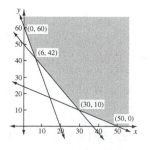

Vertex	$C = 12x + 10y$
$(0, 60)$	$12 \cdot 0 + 10 \cdot 60 = 600$
$(6, 42)$	$12 \cdot 6 + 10 \cdot 42 = 492$
$(30, 10)$	$12 \cdot 30 + 10 \cdot 10 = 460$
$(50, 0)$	$12 \cdot 50 + 10 \cdot 0 = 600$

The minimum cost of $460 thousand is achieved using 30 P_1's and 10 P_2's.

75. Let $x =$ the number of silk organza bridal dresses made and $y =$ the number of lace sheath bridal dresses made. The profit is given by

$$P = 320x + 305y$$

subject to

$$3x + 6y \leq 27,$$
$$6x + 3y \leq 36,$$
$$x \geq 0,$$
$$y \geq 0.$$

Graph the system of inequalities, determine the vertices, and find the value of P at each vertex.

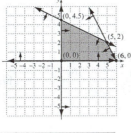

Vertex	$P = 320x + 305y$
$(0,0)$	0
$(0,4.5)$	1372.50
$(5,2)$	2210
$(6,0)$	1920

The maximum profit of $2210 occurs when 5 silk organza dresses and 2 lace dresses are made.

77. Let $x =$ the number of pounds of meat and $y =$ the number of pounds of cheese in the diet in a week. The cost is given by

$$C = 3.50x + 4.60y$$

subject to

$$2x + 3y \geq 12,$$
$$2x + y \geq 6,$$
$$x \geq 0,$$
$$y \geq 0.$$

Graph the system of inequalities, determine the vertices, and find the value of C at each vertex.

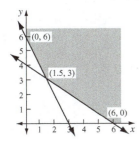

Vertex	$C = 3.50x + 4.60y$
$(0,6)$	$3.50(0) + 4.60(6) = 27.60$
$(1.5,3)$	$3.50(1.5) + 4.60(3) = 19.05$
$(6,0)$	$3.50(6) + 4.60(0) = 21.00$

The minimum weekly cost of $19.05 is achieved when 1.5 lb of meat and 3 lb of cheese are used.

79. Let $x =$ the number of animal A and $y =$ the number of animal B. The total number of animals is given by

$$T = x + y$$

subject to

$$x + 0.2y \leq 600,$$
$$0.5x + y \leq 525,$$
$$x \geq 0,$$
$$y \geq 0.$$

Graph the system of inequalities, determine the vertices, and find the value of T at each vertex.

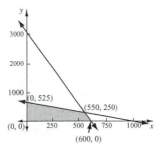

Vertex	$T = x + y$
$(0,0)$	$0 + 0 = 0$
$(0,525)$	$0 + 525 = 525$
$(550,250)$	$550 + 250 = 800$
$(600,0)$	$600 + 0 = 600$

The maximum total number of 800 is achieved when there are 550 of A and 250 of B.

81. $-5 \leq x + 2 < 4$

$-7 \leq x < 2$ Subtracting 2

The solution set is $\{x | -7 \leq x < 2\}$, or $[-7, 2)$.

83. $x^2 - 2x \leq 3$ Polynomial inequality

$x^2 - 2x - 3 \leq 0$

$x^2 - 2x - 3 = 0$ Related equation

$(x + 1)(x - 3) = 0$ Factoring

Using the principle of zero products or by observing the graph of $y = x^2 - 2x - 3$, we see that the solutions of the related equation are -1 and 3. These numbers divide the x-axis into the intervals $(-\infty, -1)$, $(-1, 3)$, and $(3, \infty)$. We let $f(x) = x^2 - 2x - 3$ and test a value in each interval.

$(-\infty, -1)$: $f(-2) = 5 > 0$

$(-1, 3)$: $f(0) = -3 < 0$

$(3, \infty)$: $f(4) = 5 > 0$

Function values are negative on $(-1, 3)$. This can also be determined from the graph of $y = x^2 - 2x - 3$. Since the inequality symbol is $\leq$, the endpoints of the interval must be included in the solution set. It is $\{x | -1 \leq x \leq 3\}$ or $[-1, 3]$.

85. Graph: $y \geq x^2 - 2$,

$\qquad y \leq 2 - x^2$

First graph the related equations $y = x^2 - 2$ and $y = 2 - x^2$ using solid lines. The solution set consists of the region above the graph of $y = x^2 - 2$ and below the graph of $y = 2 - x^2$.

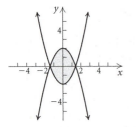

87.

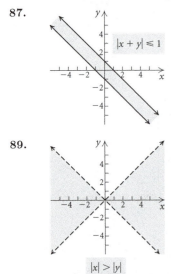

$|x + y| \leq 1$

89.

$|x| > |y|$

91. Let $x =$ the number of chairs and $y =$ the number of sofas produced. The maximum income is given by

$\qquad I = 200x + 750y$

subject to

$\qquad 20x + 100y \leq 1900,$

$\qquad x + 50y \leq 500,$

$\qquad 2x + 20y \leq 240,$

$\qquad x \geq 0,$

$\qquad y \geq 0.$

Graph the system of inequalities, determine the vertices, and find the value of I at each vertex.

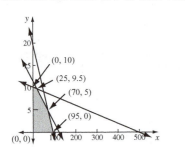

Vertex	$I = 200x + 750y$
$(0, 0)$	0
$(0, 10)$	7500
$(25, 9.5)$	$12,125$
$(70, 5)$	$17,750$
$(95, 0)$	$19,000$

The maximum income of $19,000 is achieved by making 95 chairs and 0 sofas.

Exercise Set 6.8

1. $\dfrac{x + 7}{(x - 3)(x + 2)} = \dfrac{A}{x - 3} + \dfrac{B}{x + 2}$

$\dfrac{x + 7}{(x - 3)(x + 2)} = \dfrac{A(x + 2) + B(x - 3)}{(x - 3)(x + 2)}$ Adding

Equate the numerators:

$x + 7 = A(x + 2) + B(x - 3)$

Let $x + 2 = 0$, or $x = -2$. Then we get

$\qquad -2 + 7 = 0 + B(-2 - 3)$

$\qquad\qquad 5 = -5B$

$\qquad\qquad -1 = B$

Next let $x - 3 = 0$, or $x = 3$. Then we get

$\qquad 3 + 7 = A(3 + 2) + 0$

$\qquad\qquad 10 = 5A$

$\qquad\qquad 2 = A$

The decomposition is as follows:

$\dfrac{2}{x - 3} - \dfrac{1}{x + 2}$

3. $\dfrac{7x - 1}{6x^2 - 5x + 1}$

$= \dfrac{7x - 1}{(3x - 1)(2x - 1)}$ Factoring the denominator

$= \dfrac{A}{3x - 1} + \dfrac{B}{2x - 1}$

$= \dfrac{A(2x - 1) + B(3x - 1)}{(3x - 1)(2x - 1)}$ Adding

Equate the numerators:

$7x - 1 = A(2x - 1) + B(3x - 1)$

Let $2x - 1 = 0$, or $x = \dfrac{1}{2}$. Then we get

$7\left(\dfrac{1}{2}\right) - 1 = 0 + B\left(3 \cdot \dfrac{1}{2} - 1\right)$

$\qquad\qquad \dfrac{5}{2} = \dfrac{1}{2}B$

$\qquad\qquad 5 = B$

Next let $3x - 1 = 0$, or $x = \dfrac{1}{3}$. We get

$$7\left(\frac{1}{3}\right) - 1 = A\left(2 \cdot \frac{1}{3} - 1\right) + 0$$

$$\frac{7}{3} - 1 = A\left(\frac{2}{3} - 1\right)$$

$$\frac{4}{3} = -\frac{1}{3}A$$

$$-4 = A$$

The decomposition is as follows:

$$-\frac{4}{3x-1} + \frac{5}{2x-1}$$

5. $\dfrac{3x^2 - 11x - 26}{(x^2 - 4)(x + 1)}$

$$= \frac{3x^2 - 11x - 26}{(x + 2)(x - 2)(x + 1)} \quad \text{Factoring the denominator}$$

$$= \frac{A}{x + 2} + \frac{B}{x - 2} + \frac{C}{x + 1}$$

$$= \frac{A(x-2)(x+1) + B(x+2)(x+1) + C(x+2)(x-2)}{(x+2)(x-2)(x+1)}$$

$$\text{Adding}$$

Equate the numerators:

$$3x^2 - 11x - 26 = A(x - 2)(x + 1) +$$
$$B(x + 2)(x + 1) + C(x + 2)(x - 2)$$

Let $x + 2 = 0$ or $x = -2$. Then we get

$$3(-2)^2 - 11(-2) - 26 = A(-2-2)(-2+1) + 0 + 0$$

$$12 + 22 - 26 = A(-4)(-1)$$

$$8 = 4A$$

$$2 = A$$

Next let $x - 2 = 0$, or $x = 2$. Then, we get

$$3 \cdot 2^2 - 11 \cdot 2 - 26 = 0 + B(2 + 2)(2 + 1) + 0$$

$$12 - 22 - 26 = B \cdot 4 \cdot 3$$

$$-36 = 12B$$

$$-3 = B$$

Finally let $x + 1 = 0$, or $x = -1$. We get

$$3(-1)^2 - 11(-1) - 26 = 0 + 0 + C(-1 + 2)(-1 - 2)$$

$$3 + 11 - 26 = C(1)(-3)$$

$$-12 = -3C$$

$$4 = C$$

The decomposition is as follows:

$$\frac{2}{x + 2} - \frac{3}{x - 2} + \frac{4}{x + 1}$$

7. $\dfrac{9}{(x + 2)^2(x - 1)}$

$$= \frac{A}{x + 2} + \frac{B}{(x + 2)^2} + \frac{C}{x - 1}$$

$$= \frac{A(x + 2)(x - 1) + B(x - 1) + C(x + 2)^2}{(x + 2)^2(x - 1)}$$

$$\text{Adding}$$

Equate the numerators:

$$9 = A(x + 2)(x - 1) + B(x - 1) + C(x + 2)^2 \qquad (1)$$

Let $x - 1 = 0$, or $x = 1$. Then, we get

$$9 = 0 + 0 + C(1 + 2)^2$$

$$9 = 9C$$

$$1 = C$$

Next let $x + 2 = 0$, or $x = -2$. Then, we get

$$9 = 0 + B(-2 - 1) + 0$$

$$9 = -3B$$

$$-3 = B$$

To find A we first simplify equation (1).

$$9 = A(x^2 + x - 2) + B(x - 1) + C(x^2 + 4x + 4)$$

$$= Ax^2 + Ax - 2A + Bx - B + Cx^2 + 4Cx + 4C$$

$$= (A + C)x^2 + (A + B + 4C)x + (-2A - B + 4C)$$

Then we equate the coefficients of x^2.

$$0 = A + C$$

$$0 = A + 1 \quad \text{Substituting 1 for } C$$

$$-1 = A$$

The decomposition is as follows:

$$-\frac{1}{x + 2} - \frac{3}{(x + 2)^2} + \frac{1}{x - 1}$$

9. $\dfrac{2x^2 + 3x + 1}{(x^2 - 1)(2x - 1)}$

$$= \frac{2x^2 + 3x + 1}{(x + 1)(x - 1)(2x - 1)} \quad \text{Factoring the denominator}$$

$$= \frac{A}{x + 1} + \frac{B}{x - 1} + \frac{C}{2x - 1}$$

$$= \frac{A(x-1)(2x-1) + B(x+1)(2x-1) + C(x+1)(x-1)}{(x + 1)(x - 1)(2x - 1)}$$

$$\text{Adding}$$

Equate the numerators:

$$2x^2 + 3x + 1 = A(x - 1)(2x - 1) +$$
$$B(x + 1)(2x - 1) + C(x + 1)(x - 1)$$

Let $x + 1 = 0$, or $x = -1$. Then, we get

$$2(-1)^2 + 3(-1) + 1 = A(-1 - 1)[2(-1) - 1] + 0 + 0$$

$$2 - 3 + 1 = A(-2)(-3)$$

$$0 = 6A$$

$$0 = A$$

Next let $x - 1 = 0$, or $x = 1$. Then, we get

$$2 \cdot 1^2 + 3 \cdot 1 + 1 = 0 + B(1 + 1)(2 \cdot 1 - 1) + 0$$

$$2 + 3 + 1 = B \cdot 2 \cdot 1$$

$$6 = 2B$$

$$3 = B$$

Finally we let $2x - 1 = 0$, or $x = \dfrac{1}{2}$. We get

$$2\left(\frac{1}{2}\right)^2 + 3\left(\frac{1}{2}\right) + 1 = 0 + 0 + C\left(\frac{1}{2} + 1\right)\left(\frac{1}{2} - 1\right)$$

$$\frac{1}{2} + \frac{3}{2} + 1 = C \cdot \frac{3}{2} \cdot \left(-\frac{1}{2}\right)$$

$$3 = -\frac{3}{4}C$$

$$-4 = C$$

The decomposition is as follows:

$$\frac{3}{x-1} - \frac{4}{2x-1}$$

11. $\dfrac{x^4 - 3x^3 - 3x^2 + 10}{(x+1)^2(x-3)}$

$= \dfrac{x^4 - 3x^3 - 3x^2 + 10}{x^3 - x^2 - 5x - 3}$ Multiplying the denominator

Since the degree of the numerator is greater than the degree of the denominator, we divide.

$$
\begin{array}{r}
x- \ \ 2 \\
x^3 - x^2 - 5x - 3 \overline{)\ x^4 - 3x^3 - 3x^2 +\ 0x + 10} \\
\underline{x^4 -\ x^3 - 5x^2 -\ 3x} \\
-2x^3 + 2x^2 +\ 3x + 10 \\
\underline{-2x^3 + 2x^2 + 10x +\ 6} \\
-\ 7x +\ 4
\end{array}
$$

The original expression is thus equivalent to the following:

$x - 2 + \dfrac{-7x+4}{x^3 - x^2 - 5x - 3}$

We proceed to decompose the fraction.

$\dfrac{-7x+4}{(x+1)^2(x-3)}$

$= \dfrac{A}{x+1} + \dfrac{B}{(x+1)^2} + \dfrac{C}{x-3}$

$= \dfrac{A(x+1)(x-3) + B(x-3) + C(x+1)^2}{(x+1)^2(x-3)}$

Adding

Equate the numerators:

$-7x + 4 = A(x+1)(x-3) + B(x-3) +$
$\qquad\qquad C(x+1)^2 \qquad\qquad (1)$

Let $x - 3 = 0$, or $x = 3$. Then, we get

$-7 \cdot 3 + 4 = 0 + 0 + C(3+1)^2$

$\qquad -17 = 16C$

$\qquad -\dfrac{17}{16} = C$

Let $x + 1 = 0$, or $x = -1$. Then, we get

$-7(-1) + 4 = 0 + B(-1-3) + 0$

$\qquad 11 = -4B$

$\qquad -\dfrac{11}{4} = B$

To find A we first simplify equation (1).

$-7x + 4$

$= A(x^2 - 2x - 3) + B(x - 3) + C(x^2 + 2x + 1)$

$= Ax^2 - 2Ax - 3A + Bx - 3B + Cx^2 - 2Cx + C$

$= (A+C)x^2 + (-2A+B-2C)x + (-3A-3B+C)$

Then equate the coefficients of x^2.

$\qquad 0 = A + C$

Substituting $-\dfrac{17}{16}$ for C, we get $A = \dfrac{17}{16}$.

The decomposition is as follows:

$$\frac{17/16}{x+1} - \frac{11/4}{(x+1)^2} - \frac{17/16}{x-3}$$

The original expression is equivalent to the following:

$$x - 2 + \frac{17/16}{x+1} - \frac{11/4}{(x+1)^2} - \frac{17/16}{x-3}$$

13. $\dfrac{-x^2 + 2x - 13}{(x^2+2)(x-1)}$

$= \dfrac{Ax+B}{x^2+2} + \dfrac{C}{x-1}$

$= \dfrac{(Ax+B)(x-1) + C(x^2+2)}{(x^2+2)(x-1)}$

Adding

Equate the numerators:

$-x^2 + 2x - 13 = (Ax+B)(x-1) + C(x^2+2) \quad (1)$

Let $x - 1 = 0$, or $x = 1$. Then we get

$-1^2 + 2 \cdot 1 - 13 = 0 + C(1^2+2)$

$\qquad -1 + 2 - 13 = C(1+2)$

$\qquad\qquad -12 = 3C$

$\qquad\qquad -4 = C$

To find A and B we first simplify equation (1).

$-x^2 + 2x - 13$

$= Ax^2 - Ax + Bx - B + Cx^2 + 2C$

$= (A+C)x^2 + (-A+B)x + (-B+2C)$

Equate the coefficients of x^2:

$-1 = A + C$

Substituting -4 for C, we get $A = 3$.

Equate the constant terms:

$-13 = -B + 2C$

Substituting -4 for C, we get $B = 5$.

The decomposition is as follows:

$$\frac{3x+5}{x^2+2} - \frac{4}{x-1}$$

15. $\dfrac{6 + 26x - x^2}{(2x-1)(x+2)^2}$

$= \dfrac{A}{2x-1} + \dfrac{B}{x+2} + \dfrac{C}{(x+2)^2}$

$= \dfrac{A(x+2)^2 + B(2x-1)(x+2) + C(2x-1)}{(2x-1)(x+2)^2}$

Adding

Equate the numerators:

$6 + 26x - x^2 = A(x+2)^2 + B(2x-1)(x+2) +$
$\qquad\qquad\qquad C(2x-1) \qquad\qquad (1)$

Let $2x - 1 = 0$, or $x = \dfrac{1}{2}$. Then, we get

$6 + 26 \cdot \dfrac{1}{2} - \left(\dfrac{1}{2}\right)^2 = A\left(\dfrac{1}{2} + 2\right)^2 + 0 + 0$

$\qquad 6 + 13 - \dfrac{1}{4} = A\left(\dfrac{5}{2}\right)^2$

$\qquad\qquad \dfrac{75}{4} = \dfrac{25}{4}A$

$\qquad\qquad 3 = A$

Let $x + 2 = 0$, or $x = -2$. We get

$$6 + 26(-2) - (-2)^2 = 0 + 0 + C[2(-2) - 1]$$
$$6 - 52 - 4 = -5C$$
$$-50 = -5C$$
$$10 = C$$

To find B we first simplify equation (1).

$$6 + 26x - x^2$$
$$= A(x^2 + 4x + 4) + B(2x^2 + 3x - 2) + C(2x - 1)$$
$$= Ax^2 + 4Ax + 4A + 2Bx^2 + 3Bx - 2B + 2Cx - C$$
$$= (A + 2B)x^2 + (4A + 3B + 2C)x + (4A - 2B - C)$$

Equate the coefficients of x^2:

$$-1 = A + 2B$$

Substituting 3 for A, we obtain $B = -2$.

The decomposition is as follows:

$$\frac{3}{2x - 1} - \frac{2}{x + 2} + \frac{10}{(x + 2)^2}$$

17. $\dfrac{6x^3 + 5x^2 + 6x - 2}{2x^2 + x - 1}$

Since the degree of the numerator is greater than the degree of the denominator, we divide.

$$\begin{array}{r} 3x + 1 \\ 2x^2 + x - 1 \overline{\smash{\big)}\, 6x^3 + 5x^2 + 6x - 2} \\ \underline{6x^3 + 3x^2 - 3x } \\ 2x^2 + 9x - 2 \\ \underline{2x^2 + x - 1} \\ 8x - 1 \end{array}$$

The original expression is equivalent to

$$3x + 1 + \frac{8x - 1}{2x^2 + x - 1}$$

We proceed to decompose the fraction.

$$\frac{8x - 1}{2x^2 + x - 1} = \frac{8x - 1}{(2x - 1)(x + 1)} \quad \text{Factoring the denominator}$$

$$= \frac{A}{2x - 1} + \frac{B}{x + 1}$$

$$= \frac{A(x + 1) + B(2x - 1)}{(2x - 1)(x + 1)} \quad \text{Adding}$$

Equate the numerators:

$$8x - 1 = A(x + 1) + B(2x - 1)$$

Let $x + 1 = 0$, or $x = -1$. Then we get

$$8(-1) - 1 = 0 + B[2(-1) - 1]$$
$$-8 - 1 = B(-2 - 1)$$
$$-9 = -3B$$
$$3 = B$$

Next let $2x - 1 = 0$, or $x = \dfrac{1}{2}$. We get

$$8\left(\frac{1}{2}\right) - 1 = A\left(\frac{1}{2} + 1\right) + 0$$

$$4 - 1 = A\left(\frac{3}{2}\right)$$

$$3 = \frac{3}{2}A$$

$$2 = A$$

The decomposition is

$$\frac{2}{2x - 1} + \frac{3}{x + 1}.$$

The original expression is equivalent to

$$3x + 1 + \frac{2}{2x - 1} + \frac{3}{x + 1}.$$

19. $\dfrac{2x^2 - 11x + 5}{(x - 3)(x^2 + 2x - 5)}$

$$= \frac{A}{x - 3} + \frac{Bx + C}{x^2 + 2x - 5}$$

$$= \frac{A(x^2 + 2x - 5) + (Bx + C)(x - 3)}{(x - 3)(x^2 + 2x - 5)} \quad \text{Adding}$$

Equate the numerators:

$$2x^2 - 11x + 5 = A(x^2 + 2x - 5) +$$
$$(Bx + C)(x - 3) \quad (1)$$

Let $x - 3 = 0$, or $x = 3$. Then, we get

$$2 \cdot 3^2 - 11 \cdot 3 + 5 = A(3^2 + 2 \cdot 3 - 5) + 0$$
$$18 - 33 + 5 = A(9 + 6 - 5)$$
$$-10 = 10A$$
$$-1 = A$$

To find B and C, we first simplify equation (1).

$$2x^2 - 11x + 5 = Ax^2 + 2Ax - 5A + Bx^2 - 3Bx +$$
$$Cx - 3C$$
$$= (A + B)x^2 + (2A - 3B + C)x +$$
$$(-5A - 3C)$$

Equate the coefficients of x^2:

$$2 = A + B$$

Substituting -1 for A, we get $B = 3$.

Equate the constant terms:

$$5 = -5A - 3C$$

Substituting -1 for A, we get $C = 0$.

The decomposition is as follows:

$$-\frac{1}{x - 3} + \frac{3x}{x^2 + 2x - 5}$$

21. $\dfrac{-4x^2 - 2x + 10}{(3x + 5)(x + 1)^2}$

The decomposition looks like

$$\frac{A}{3x + 5} + \frac{B}{x + 1} + \frac{C}{(x + 1)^2}.$$

Add and equate the numerators.

$$-4x^2 - 2x + 10$$
$$= A(x + 1)^2 + B(3x + 5)(x + 1) + C(3x + 5)$$
$$= A(x^2 + 2x + 1) + B(3x^2 + 8x + 5) + C(3x + 5)$$

or

$$-4x^2 - 2x + 10$$
$$= (A + 3B)x^2 + (2A + 8B + 3C)x + (A + 5B + 5C)$$

Then equate corresponding coefficients.

$-4 = A + 3B$ Coefficients of x^2-terms

$-2 = 2A + 8B + 3C$ Coefficients of x-terms

$10 = A + 5B + 5C$ Constant terms

We solve this system of three equations and find $A = 5$, $B = -3$, $C = 4$.

The decomposition is

$$\frac{5}{3x + 5} - \frac{3}{x + 1} + \frac{4}{(x + 1)^2}.$$

23. $\dfrac{36x + 1}{12x^2 - 7x - 10} = \dfrac{36x + 1}{(4x - 5)(3x + 2)}$

The decomposition looks like

$$\frac{A}{4x - 5} + \frac{B}{3x + 2}.$$

Add and equate the numerators.

$36x + 1 = A(3x + 2) + B(4x - 5)$

or $36x + 1 = (3A + 4B)x + (2A - 5B)$

Then equate corresponding coefficients.

$36 = 3A + 4B$ Coefficients of x-terms

$1 = 2A - 5B$ Constant terms

We solve this system of equations and find $A = 8$ and $B = 3$.

The decomposition is

$$\frac{8}{4x - 5} + \frac{3}{3x + 2}.$$

25. $\dfrac{-4x^2 - 9x + 8}{(3x^2 + 1)(x - 2)}$

The decomposition looks like

$$\frac{Ax + B}{3x^2 + 1} + \frac{C}{x - 2}.$$

Add and equate the numerators.

$-4x^2 - 9x + 8$

$= (Ax + B)(x - 2) + C(3x^2 + 1)$

$= Ax^2 - 2Ax + Bx - 2B + 3Cx^2 + C$

or

$-4x^2 - 9x + 8$

$= (A + 3C)x^2 + (-2A + B)x + (-2B + C)$

Then equate corresponding coefficients.

$-4 = A + 3C$ Coefficients of x^2-terms

$-9 = -2A + B$ Coefficients of x-terms

$8 = -2B + C$ Constant terms

We solve this system of equations and find $A = 2$, $B = -5$, $C = -2$.

The decomposition is

$$\frac{2x - 5}{3x^2 + 1} - \frac{2}{x - 2}.$$

27. $x^3 + x^2 + 9x + 9 = 0$

$x^2(x + 1) + 9(x + 1) = 0$

$(x + 1)(x^2 + 9) = 0$

$x + 1 = 0$ or $x^2 + 9 = 0$

$x = -1$ or $x^2 = -9$

$x = -1$ or $x = \pm 3i$

The solutions are -1, $3i$, and $-3i$.

29. $f(x) = x^3 + x^2 - 3x - 2$

We use synthetic division to factor the polynomial. Using the possibilities found by the rational zeros theorem we find that $x + 2$ is a factor:

$$\begin{array}{r|rrrr} -2 & 1 & 1 & -3 & -2 \\ & & -2 & 2 & 2 \\ \hline & 1 & -1 & -1 & 0 \end{array}$$

We have $x^3 + x^2 - 3x - 2 = (x + 2)(x^2 - x - 1)$.

$x^3 + x^2 - 3x - 2 = 0$

$(x + 2)(x^2 - x - 1) = 0$

$x + 2 = 0$ or $x^2 - x - 1 = 0$

The solution of the first equation is -2. We use the quadratic formula to solve the second equation.

$$x = \frac{-b \pm \sqrt{b^2 - 4ac}}{2a}$$

$$= \frac{-(-1) \pm \sqrt{(-1)^2 - 4 \cdot 1 \cdot (-1)}}{2 \cdot 1}$$

$$= \frac{1 \pm \sqrt{5}}{2}$$

The solutions are -2, $\dfrac{1 + \sqrt{5}}{2}$ and $\dfrac{1 - \sqrt{5}}{2}$.

31. $f(x) = x^3 + 5x^2 + 5x - 3$

$$\begin{array}{r|rrrr} -3 & 1 & 5 & 5 & -3 \\ & & -3 & -6 & 3 \\ \hline & 1 & 2 & -1 & 0 \end{array}$$

$x^3 + 5x^2 + 5x - 3 = 0$

$(x + 3)(x^2 + 2x - 1) = 0$

$x + 3 = 0$ or $x^2 + 2x - 1 = 0$

The solution of the first equation is -3. We use the quadratic formula to solve the second equation.

$$x = \frac{-b \pm \sqrt{b^2 - 4ac}}{2a}$$

$$= \frac{-2 \pm \sqrt{2^2 - 4 \cdot 1 \cdot (-1)}}{2 \cdot 1} = \frac{-2 \pm \sqrt{8}}{2}$$

$$= \frac{-2 \pm 2\sqrt{2}}{2} = \frac{2(-1 \pm \sqrt{2})}{2}$$

$$= -1 \pm \sqrt{2}$$

The solutions are -3, $-1 + \sqrt{2}$, and $-1 - \sqrt{2}$.

33. $\dfrac{x}{x^4 - a^4}$

$= \dfrac{x}{(x^2 + a^2)(x + a)(x - a)}$ Factoring the denominator

$= \dfrac{Ax + B}{x^2 + a^2} + \dfrac{C}{x + a} + \dfrac{D}{x - a}$

$= [(Ax + B)(x + a)(x - a) + C(x^2 + a^2)(x - a) +$

$\quad D(x^2 + a^2)(x + a)]/[(x^2 + a^2)(x + a)(x - a)]$

Equate the numerators:
$$x = (Ax + B)(x + a)(x - a) + C(x^2 + a^2)(x - a) +$$
$$D(x^2 + a^2)(x + a)$$

Let $x - a = 0$, or $x = a$. Then, we get
$$a = 0 + 0 + D(a^2 + a^2)(a + a)$$
$$a = D(2a^2)(2a)$$
$$a = 4a^3 D$$
$$\frac{1}{4a^2} = D$$

Let $x + a = 0$, or $x = -a$. We get
$$-a = 0 + C[(-a)^2 + a^2](-a - a) + 0$$
$$-a = C(2a^2)(-2a)$$
$$-a = -4a^3 C$$
$$\frac{1}{4a^2} = C$$

Equate the coefficients of x^3:
$$0 = A + C + D$$

Substituting $\frac{1}{4a^2}$ for C and for D, we get
$$A = -\frac{1}{2a^2}.$$

Equate the constant terms:
$$0 = -Ba^2 - Ca^3 + Da^3$$

Substitute $\frac{1}{4a^2}$ for C and for D. Then solve for B.
$$0 = -Ba^2 - \frac{1}{4a^2} \cdot a^3 + \frac{1}{4a^2} \cdot a^3$$
$$0 = -Ba^2$$
$$0 = B$$

The decomposition is as follows:
$$-\frac{\frac{1}{2a^2}x}{x^2 + a^2} + \frac{\frac{1}{4a^2}}{x + a} + \frac{\frac{1}{4a^2}}{x - a}$$

35. $\dfrac{1 + \ln x^2}{(\ln x + 2)(\ln x - 3)^2} = \dfrac{1 + 2\ln x}{(\ln x + 2)(\ln x - 3)^2}$

Let $u = \ln x$. Then we have:
$$\frac{1 + 2u}{(u + 2)(u - 3)^2}$$
$$= \frac{A}{u + 2} + \frac{B}{u - 3} + \frac{C}{(u - 3)^2}$$
$$= \frac{A(u - 3)^2 + B(u + 2)(u - 3) + C(u + 2)}{(u + 2)(u - 3)^2}$$

Equate the numerators:
$$1 + 2u = A(u - 3)^2 + B(u + 2)(u - 3) + C(u + 2)$$

Let $u - 3 = 0$, or $u = 3$.
$$1 + 2 \cdot 3 = 0 + 0 + C(5)$$
$$7 = 5C$$
$$\frac{7}{5} = C$$

Let $u + 2 = 0$, or $u = -2$.

$$1 + 2(-2) = A(-2 - 3)^2 + 0 + 0$$
$$-3 = 25A$$
$$-\frac{3}{25} = A$$

To find B, we equate the coefficients of u^2:
$$0 = A + B$$

Substituting $-\dfrac{3}{25}$ for A and solving for B, we get $B = \dfrac{3}{25}$.

The decomposition of $\dfrac{1 + 2u}{(u + 2)(u - 3)^2}$ is as follows:
$$-\frac{3}{25(u + 2)} + \frac{3}{25(u - 3)} + \frac{7}{5(u - 3)^2}$$

Substituting $\ln x$ for u we get
$$-\frac{3}{25(\ln x + 2)} + \frac{3}{25(\ln x - 3)} + \frac{7}{5(\ln x - 3)^2}.$$

Chapter 6 Review Exercises

1. The statement is true. See page 395 in the text.

3. The statement is true. See page 428 in the text.

5. (a)

7. (h)

9. (b)

11. (c)

13. $\quad 5x - 3y = -4, \quad (1)$
$\quad\quad 3x - y = -4 \quad (2)$

Multiply equation (2) by -3 and add.
$$5x - 3y = -4$$
$$\underline{-9x + 3y = 12}$$
$$-4x \quad\quad = 8$$
$$x = -2$$

Back-substitute to find y.
$$3(-2) - y = -4 \quad \text{Using equation (2)}$$
$$-6 - y = -4$$
$$-y = 2$$
$$y = -2$$

The solution is $(-2, -2)$.

15. $\quad x + 5y = 12, \quad (1)$
$\quad\quad 5x + 25y = 12 \quad (2)$

Solve equation (1) for x.
$$x = -5y + 12$$

Substitute in equation (2) and solve for y.
$$5(-5y + 12) + 25y = 12$$
$$-25y + 60 + 25y = 12$$
$$60 = 12$$

We get a false equation, so there is no solution.

17. $x + 5y - 3z = 4,$ (1)

$3x - 2y + 4z = 3,$ (2)

$2x + 3y - z = 5$ (3)

Multiply equation (1) by -3 and add it to equation (2).

Multiply equation (1) by -2 and add it to equation (3).

$x + 5y + 3z = 4$ (1)

$-17y + 13z = -9$ (4)

$-7y + 5z = -3$ (5)

Multiply equation (5) by 17.

$x + 5y + 3z = 4$ (1)

$-17y + 13z = -9$ (4)

$-119y + 85z = -51$ (6)

Multiply equation (4) by -7 and add it to equation (6).

$x + 5y + 3z = 4$ (1)

$-17y + 13z = -9$ (4)

$-6z = 12$ (7)

Now we solve equation (7) for z.

$-6z = 12$

$z = -2$

Back-substitute -2 for z in equation (4) and solve for y.

$-17y + 13(-2) = -9$

$-17y - 26 = -9$

$-17y = 17$

$y = -1$

Finally, we back-substitute -1 for y and -2 for z in equation (1) and solve for x.

$x + 5(-1) - 3(-2) = 4$

$x - 5 + 6 = 4$

$x + 1 = 4$

$x = 3$

The solution is $(3, -1, -2)$.

19. $x - y = 5,$ (1)

$y - z = 6,$ (2)

$z - w = 7,$ (3)

$x + w = 8$ (4)

Multiply equation (1) by -1 and add it to equation (4).

$x - y = 5$ (1)

$y - z = 6$ (2)

$z - w = 7$ (3)

$y + w = 3$ (5)

Multiply equation (2) by -1 and add it to equation (5).

$x - y = 5$ (1)

$y - z = 6$ (2)

$z - w = 7$ (3)

$z + w = -3$ (6)

Multiply equation (3) by -1 and add it to equation (6).

$x - y = 5$ (1)

$y - z = 6$ (2)

$z - w = 7$ (3)

$2w = -10$ (7)

Solve equation (7) for w.

$2w = -10$

$w = -5$

Back-substitute -5 for w in equation (3) and solve for z.

$z - w = 7$

$z - (-5) = 7$

$z + 5 = 7$

$z = 2$

Back-substitute 2 for z in equation (2) and solve for y.

$y - z = 6$

$y - 2 = 6$

$y = 8$

Back-substitute 8 for y in equation (1) and solve for x.

$x - y = 5$

$x - 8 = 5$

$x = 13$

Writing the solution as (w, x, y, z), we have $(-5, 13, 8, 2)$.

21. Systems 13, 14, 15, 17, 18, and 19 each have either no solution or exactly one solution, so the equations in those systems are independent. System 16 has infinitely many solutions, so the equations in that system are dependent.

23. $3x + 4y + 2z = 3$

$5x - 2y - 13z = 3$

$4x + 3y - 3z = 6$

Write the augmented matrix. We will use Gaussian elimination.

$$\begin{bmatrix} 3 & 4 & 2 & | & 3 \\ 5 & -2 & -13 & | & 3 \\ 4 & 3 & -3 & | & 6 \end{bmatrix}$$

Multiply row 2 and row 3 by 3.

$$\begin{bmatrix} 3 & 4 & 2 & | & 3 \\ 15 & -6 & -39 & | & 9 \\ 12 & 9 & -9 & | & 18 \end{bmatrix}$$

Multiply row 1 by -5 and add it to row 2.

Multiply row 1 by -4 and add it to row 3.

$$\begin{bmatrix} 3 & 4 & 2 & | & 3 \\ 0 & -26 & -49 & | & -6 \\ 0 & -7 & -17 & | & 6 \end{bmatrix}$$

Multiply row 3 by 26.

$$\begin{bmatrix} 3 & 4 & 2 & | & 3 \\ 0 & -26 & -49 & | & -6 \\ 0 & -182 & -442 & | & 156 \end{bmatrix}$$

Multiply row 2 by -7 and add it to row 3.

$$\begin{bmatrix} 3 & 4 & 2 & | & 3 \\ 0 & -26 & -49 & | & -6 \\ 0 & 0 & -99 & | & 198 \end{bmatrix}$$

Multiply row 1 by $\frac{1}{3}$, row 2 by $-\frac{1}{26}$, and row 3 by $-\frac{1}{99}$.

$$\begin{bmatrix} 1 & \frac{4}{3} & \frac{2}{3} & | & 1 \\ 0 & 1 & \frac{49}{26} & | & \frac{3}{13} \\ 0 & 0 & 1 & | & -2 \end{bmatrix}$$

$$x + \frac{4}{3}y + \frac{2}{3}z = 1 \quad (1)$$
$$y + \frac{49}{26}z = \frac{3}{13} \quad (2)$$
$$z = -2$$

Back-substitute in equation (2) and solve for y.

$$y + \frac{49}{26}(-2) = \frac{3}{13}$$
$$y - \frac{49}{13} = \frac{3}{13}$$
$$y = \frac{52}{13} = 4$$

Back-substitute in equation (1) and solve for x.

$$x + \frac{4}{3}(4) + \frac{2}{3}(-2) = 1$$
$$x + \frac{16}{3} - \frac{4}{3} = 1$$
$$x = -3$$

The solution is $(-3, 4, -2)$.

25.
$$w + x + y + z = -2,$$
$$-3w - 2x + 3y + 2z = 10,$$
$$2w + 3x + 2y - z = -12,$$
$$2w + 4x - y + z = 1$$

Write the augmented matrix. We will use Gauss-Jordan elimination.

$$\begin{bmatrix} 1 & 1 & 1 & 1 & | & -2 \\ -3 & -2 & 3 & 2 & | & 10 \\ 2 & 3 & 2 & -1 & | & -12 \\ 2 & 4 & -1 & 1 & | & 1 \end{bmatrix}$$

Multiply row 1 by 3 and add it to row 2.
Multiply row 1 by -2 and add it to row 3.
Multiply row 1 by -2 and add it to row 4.

$$\begin{bmatrix} 1 & 1 & 1 & 1 & | & -2 \\ 0 & 1 & 6 & 5 & | & 4 \\ 0 & 1 & 0 & -3 & | & -8 \\ 0 & 2 & -3 & -1 & | & 5 \end{bmatrix}$$

Multiply row 2 by -1 and add it to row 1.
Multiply row 2 by -1 and add it to row 3.
Multiply row 2 by -2 and add it to row 4.

$$\begin{bmatrix} 1 & 0 & -5 & -4 & | & -6 \\ 0 & 1 & 6 & 5 & | & 4 \\ 0 & 0 & -6 & -8 & | & -12 \\ 0 & 0 & -15 & -11 & | & -3 \end{bmatrix}$$

Multiply row 1 by 3.
Multiply row 3 by $-\frac{1}{2}$.

$$\begin{bmatrix} 3 & 0 & -15 & -12 & | & -18 \\ 0 & 1 & 6 & 5 & | & 4 \\ 0 & 0 & 3 & 4 & | & 6 \\ 0 & 0 & -15 & -11 & | & -3 \end{bmatrix}$$

Multiply row 3 by 5 and add it to row 1.
Multiply row 3 by -2 and add it to row 2.
Multiply row 3 by 5 and add it to row 4.

$$\begin{bmatrix} 3 & 0 & 0 & 8 & | & 12 \\ 0 & 1 & 0 & -3 & | & -8 \\ 0 & 0 & 3 & 4 & | & 6 \\ 0 & 0 & 0 & 9 & | & 27 \end{bmatrix}$$

Multiply row 4 by $\frac{1}{9}$.

$$\begin{bmatrix} 3 & 0 & 0 & 8 & | & 12 \\ 0 & 1 & 0 & -3 & | & -8 \\ 0 & 0 & 3 & 4 & | & 6 \\ 0 & 0 & 0 & 1 & | & 3 \end{bmatrix}$$

Multiply row 4 by -8 and add it to row 1.
Multiply row 4 by 3 and add it to row 2.
Multiply row 4 by -4 and add it to row 3.

$$\begin{bmatrix} 3 & 0 & 0 & 0 & | & -12 \\ 0 & 1 & 0 & 0 & | & 1 \\ 0 & 0 & 3 & 0 & | & -6 \\ 0 & 0 & 0 & 1 & | & 3 \end{bmatrix}$$

Multiply rows 1 and 3 by $\frac{1}{3}$.

$$\begin{bmatrix} 1 & 0 & 0 & 0 & | & -4 \\ 0 & 1 & 0 & 0 & | & 1 \\ 0 & 0 & 1 & 0 & | & -2 \\ 0 & 0 & 0 & 1 & | & 3 \end{bmatrix}$$

The solution is $(-4, 1, -2, 3)$.

27. ***Familiarize.*** Let $x =$ the amount invested at 3% and $y =$ the amount invested at 3.5%. Then the interest from the investments is 3%x and 3.5%y, or $0.03x$ and $0.035y$.

Translate.

The total investment is \$5000.

$$x + y = 5000$$

The total interest is \$167.

$$0.03x + 0.035y = 167$$

We have a system of equations.

$$x + y = 5000,$$
$$0.03x + 0.035y = 167$$

Multiplying the second equation by 1000 to clear the decimals, we have:

$$x + y = 5000, \qquad (1)$$
$$30x + 35y = 167,000. \qquad (2)$$

Carry out. We begin by multiplying equation (1) by -30 and adding.

$$\begin{aligned} -30x - 30y &= -150,000 \\ \underline{30x + 35y} &= \underline{167,000} \\ 5y &= 17,000 \\ y &= 3400 \end{aligned}$$

Back-substitute to find x.

$$x + 3400 = 5000 \quad \text{Using equation (1)}$$
$$x = 1600$$

Check. The total investment is \$1600 + \$3400, or \$5000. The total interest is 0.03(\$1600) + 0.035(\$3400), or \$48 + \$119, or \$167. The solution checks.

State. \$1600 was invested at 3% and \$3400 was invested at 3.5%.

29. ***Familiarize.*** Let x, y, and z represent the scores on the first, second, and third tests, respectively.

Translate.

The total score on the three tests is 226.

$$x + y + z = 226$$

The sum of the scores on the first and second tests exceeds the score on the third test by 62.

$$x + y = z + 62$$

The first score exceeds the second by 6.

$$x = y + 6$$

We have a system of equations.

$$x + y + z = 226,$$
$$x + y = z + 62,$$
$$x = y + 6$$

or $\quad x + y + z = 226,$
$$x + y - z = 62,$$
$$x - y = 6,$$

Carry out. Solving the system of equations, we get $(75, 69, 82)$.

Check. The sum of the scores is $75 + 69 + 82$, or 226. The sum of the scores on the first two tests is $75 + 69$, or 144. This exceeds the score on the third test, 82, by 62. The score on the first test, 75, exceeds the score on the second test, 69, by 6. The solution checks.

State. The scores on the first, second, and third tests were 75, 69, and 82, respectively.

31. $\mathbf{A} + \mathbf{B} = \begin{bmatrix} 1 & -1 & 0 \\ 2 & 3 & -2 \\ -2 & 0 & 1 \end{bmatrix} + \begin{bmatrix} -1 & 0 & 6 \\ 1 & -2 & 0 \\ 0 & 1 & -3 \end{bmatrix}$

$\phantom{\mathbf{A} + \mathbf{B}} = \begin{bmatrix} 1 + (-1) & -1 + 0 & 0 + 6 \\ 2 + 1 & 3 + (-2) & -2 + 0 \\ -2 + 0 & 0 + 1 & 1 + (-3) \end{bmatrix}$

$\phantom{\mathbf{A} + \mathbf{B}} = \begin{bmatrix} 0 & -1 & 6 \\ 3 & 1 & -2 \\ -2 & 1 & -2 \end{bmatrix}$

33. $-\mathbf{A} = -1 \begin{bmatrix} 1 & -1 & 0 \\ 2 & 3 & -2 \\ -2 & 0 & 1 \end{bmatrix} = \begin{bmatrix} -1 & 1 & 0 \\ -2 & -3 & 2 \\ 2 & 0 & -1 \end{bmatrix}$

35. $\mathbf{B}$ and $\mathbf{C}$ do not have the same order, so it is not possible to find $\mathbf{B} + \mathbf{C}$.

37. $\mathbf{BA} = \begin{bmatrix} -1 & 0 & 6 \\ 1 & -2 & 0 \\ 0 & 1 & -3 \end{bmatrix} \cdot \begin{bmatrix} 1 & -1 & 0 \\ 2 & 3 & -2 \\ -2 & 0 & 1 \end{bmatrix}$

$\phantom{\mathbf{BA}} = \begin{bmatrix} -1 + 0 - 12 & 1 + 0 + 0 & 0 + 0 + 6 \\ 1 - 4 + 0 & -1 - 6 + 0 & 0 + 4 + 0 \\ 0 + 2 + 6 & 0 + 3 + 0 & 0 - 2 - 3 \end{bmatrix}$

$\phantom{\mathbf{BA}} = \begin{bmatrix} -13 & 1 & 6 \\ -3 & -7 & 4 \\ 8 & 3 & -5 \end{bmatrix}$

39. a) $\mathbf{M} = \begin{bmatrix} 2.25 & 0.38 & 0.55 & 0.33 & 0.85 \\ 3.09 & 0.42 & 0.46 & 0.48 & 0.51 \\ 2.40 & 0.31 & 0.59 & 0.36 & 0.64 \\ 1.80 & 0.29 & 0.34 & 0.55 & 0.52 \end{bmatrix}$

b) $\mathbf{N} = \begin{bmatrix} 41 & 18 & 39 & 36 \end{bmatrix}$

c) $\mathbf{NM} = \begin{bmatrix} 306.27 & 45.67 & 66.08 & 56.01 & 87.71 \end{bmatrix}$

d) The entries of $\mathbf{NM}$ represent the total cost, in dollars, for each item for the day's meal.

41. $\mathbf{A} = \begin{bmatrix} 0 & 0 & 3 \\ 0 & -2 & 0 \\ 4 & 0 & 0 \end{bmatrix}$

Write the augmented matrix.

$$\left[\begin{array}{ccc|ccc} 0 & 0 & 3 & 1 & 0 & 0 \\ 0 & -2 & 0 & 0 & 1 & 0 \\ 4 & 0 & 0 & 0 & 0 & 1 \end{array} \right]$$

Interchange rows 1 and 3.

$$\left[\begin{array}{ccc|ccc} 4 & 0 & 0 & 0 & 0 & 1 \\ 0 & -2 & 0 & 0 & 1 & 0 \\ 0 & 0 & 3 & 1 & 0 & 0 \end{array} \right]$$

Multiply row 1 by $\frac{1}{4}$, row 2 by $-\frac{1}{2}$, and row 3 by $\frac{1}{3}$.

$$\left[\begin{array}{ccc|ccc} 1 & 0 & 0 & 0 & 0 & \frac{1}{4} \\ 0 & 1 & 0 & 0 & -\frac{1}{2} & 0 \\ 0 & 0 & 1 & \frac{1}{3} & 0 & 0 \end{array}\right]$$

$$\mathbf{A}^{-1} = \left[\begin{array}{ccc} 0 & 0 & \frac{1}{4} \\ 0 & -\frac{1}{2} & 0 \\ \frac{1}{3} & 0 & 0 \end{array}\right]$$

43. $3x - 2y + 4z = 13,$
$\quad x + 5y - 3z = 7,$
$\quad 2x - 3y + 7z = -8$

Write the coefficients on the left in a matrix. Then write the product of that matrix and the column matrix containing the variables, and set the result equal to the column matrix containing the constants on the right.

$$\left[\begin{array}{ccc} 3 & -2 & 4 \\ 1 & 5 & -3 \\ 2 & -3 & 7 \end{array}\right] \left[\begin{array}{c} x \\ y \\ z \end{array}\right] = \left[\begin{array}{c} 13 \\ 7 \\ -8 \end{array}\right]$$

45. $5x - y + 2z = 17,$
$\quad 3x + 2y - 3z = -16,$
$\quad 4x - 3y - z = 5$

Write an equivalent matrix equation, $\mathbf{AX} = \mathbf{B}$.

$$\left[\begin{array}{ccc} 5 & -1 & 2 \\ 3 & 2 & -3 \\ 4 & -3 & -1 \end{array}\right] \left[\begin{array}{c} x \\ y \\ z \end{array}\right] = \left[\begin{array}{c} 17 \\ -16 \\ 5 \end{array}\right]$$

Then,

$$\mathbf{X} = \mathbf{A}^{-1}\mathbf{B} = \left[\begin{array}{ccc} \frac{11}{80} & \frac{7}{80} & \frac{1}{80} \\ \frac{9}{80} & \frac{13}{80} & -\frac{21}{80} \\ \frac{17}{80} & -\frac{11}{80} & -\frac{13}{80} \end{array}\right] \left[\begin{array}{c} 17 \\ -16 \\ 5 \end{array}\right] = \left[\begin{array}{c} 1 \\ -2 \\ 5 \end{array}\right]$$

The solution is $(1, -2, 5)$.

47. $\begin{vmatrix} 1 & -2 \\ 3 & 4 \end{vmatrix} = 1 \cdot 4 - 3(-2) = 4 + 6 = 10$

49. We will expand across the first row.

$$\begin{vmatrix} 2 & -1 & 1 \\ 1 & 2 & -1 \\ 3 & 4 & -3 \end{vmatrix}$$

$$= 2(-1)^{1+1}\begin{vmatrix} 2 & -1 \\ 4 & -3 \end{vmatrix} + (-1)(-1)^{1+2}\begin{vmatrix} 1 & -1 \\ 3 & -3 \end{vmatrix} +$$

$$1(-1)^{1+3}\begin{vmatrix} 1 & 2 \\ 3 & 4 \end{vmatrix}$$

$$= 2 \cdot 1[2(-3) - 4(-1)] + (-1)(-1)[1(-3) - 3(-1)] +$$
$$1 \cdot 1[1(4) - 3(2)]$$

$$= 2(-2) + 1(0) + 1(-2)$$

$$= -6$$

51. $5x - 2y = 19,$
$\quad 7x + 3y = 15$

$$D = \begin{vmatrix} 5 & -2 \\ 7 & 3 \end{vmatrix} = 5(3) - 7(-2) = 29$$

$$D_x = \begin{vmatrix} 19 & -2 \\ 15 & 3 \end{vmatrix} = 19(3) - 15(-2) = 87$$

$$D_y = \begin{vmatrix} 5 & 19 \\ 7 & 15 \end{vmatrix} = 5(15) - 7(19) = -58$$

$$x = \frac{D_x}{D} = \frac{87}{29} = 3$$

$$y = \frac{D_y}{D} = \frac{-58}{29} = -2$$

The solution is $(3, -2)$.

53. $3x - 2y + z = 5,$
$\quad 4x - 5y - z = -1,$
$\quad 3x + 2y - z = 4$

$$D = \begin{vmatrix} 3 & -2 & 1 \\ 4 & -5 & -1 \\ 3 & 2 & -1 \end{vmatrix} = 42$$

$$D_x = \begin{vmatrix} 5 & -2 & 1 \\ -1 & -5 & -1 \\ 4 & 2 & -1 \end{vmatrix} = 63$$

$$D_y = \begin{vmatrix} 3 & 5 & 1 \\ 4 & -1 & -1 \\ 3 & 4 & -1 \end{vmatrix} = 39$$

$$D_z = \begin{vmatrix} 3 & -2 & 5 \\ 4 & -5 & -1 \\ 3 & 2 & 4 \end{vmatrix} = 99$$

$$x = \frac{D_x}{D} = \frac{63}{42} = \frac{3}{2}$$

$$y = \frac{D_y}{D} = \frac{39}{42} = \frac{13}{14}$$

$$z = \frac{D_z}{D} = \frac{99}{42} = \frac{33}{14}$$

The solution is $\left(\frac{3}{2}, \frac{13}{14}, \frac{33}{14}\right)$.

55.

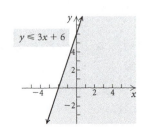

$y \leq 3x + 6$

57. Graph: $2x + y \geq 9$,

$4x + 3y \geq 23$,

$x + 3y \geq 8$,

$x \geq 0$,

$y \geq 0$

Shade the intersection of the graphs of the given inequalities.

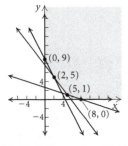

We find the vertex $(0, 9)$ by solving the system

$2x + y = 9$,

$x = 0$.

We find the vertex $(2, 5)$ by solving the system

$2x + y = 9$,

$4x + 3y = 23$.

We find the vertex $(5, 1)$ by solving the system

$4x + 3y = 23$,

$x + 3y = 8$.

We find the vertex $(8, 0)$ by solving the system

$x + 3y = 8$,

$y = 0$.

59. Let $x =$ the number of questions answered from group A and $y =$ the number of questions answered from group B. Find the maximum value of $S = 7x + 12y$ subject to

$x + y \geq 8$,

$8x + 10y \leq 80$,

$x \geq 0$,

$y \geq 0$

Graph the system of inequalities, determine the vertices, and find the value of T at each vertex.

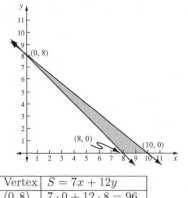

Vertex	$S = 7x + 12y$
$(0, 8)$	$7 \cdot 0 + 12 \cdot 8 = 96$
$(8, 0)$	$7 \cdot 8 + 12 \cdot 0 = 56$
$(10, 0)$	$7 \cdot 10 + 12 \cdot 0 = 70$

The maximum score of 96 occurs when 0 questions from group A and 8 questions from group B are answered correctly.

61. $\dfrac{-8x + 23}{2x^2 + 5x - 12} = \dfrac{-8x + 23}{(2x - 3)(x + 4)}$

$= \dfrac{A}{2x - 3} + \dfrac{B}{x + 4}$

$= \dfrac{A(x + 4) + B(2x - 3)}{(2x - 3)(x + 4)}$

Equate the numerators.

$-8x + 23 = A(x + 4) + B(2x - 3)$

Let $x = \dfrac{3}{2}$: $-8\left(\dfrac{3}{2}\right) + 23 = A\left(\dfrac{3}{2} + 4\right) + 0$

$-12 + 23 = \dfrac{11}{2}A$

$11 = \dfrac{11}{2}A$

$2 = A$

Let $x = -4$: $-8(-4) + 23 = 0 + B[2(-4) - 3]$

$32 + 23 = -11B$

$55 = -11B$

$-5 = B$

The decomposition is $\dfrac{2}{2x - 3} - \dfrac{5}{x + 4}$.

63. Interchanging columns of a matrix is not a row-equivalent operation, so answer A is correct. (See page 419 in the text.)

65. Let x, y, and z represent the amounts invested at 4%, 5%, and $5\dfrac{1}{2}$%, respectively.

Solve: $x + y + z = 40,000$,

$0.04x + 0.05y + 0.055z = 1990$,

$0.055z = 0.04x + 590$

$x = \$10,000$, $y = \$12,000$, $z = \$18,000$

67. $\dfrac{3}{x} - \dfrac{4}{y} + \dfrac{1}{z} = -2, \quad (1)$

$\dfrac{5}{x} + \dfrac{1}{y} - \dfrac{2}{z} = 1, \quad (2)$

$\dfrac{7}{x} + \dfrac{3}{y} + \dfrac{2}{z} = 19 \quad (3)$

Multiply equations (2) and (3) by 3.

$\dfrac{3}{x} - \dfrac{4}{y} + \dfrac{1}{z} = -2 \quad (1)$

$\dfrac{15}{x} + \dfrac{3}{y} - \dfrac{6}{z} = 3 \quad (4)$

$\dfrac{21}{x} + \dfrac{9}{y} + \dfrac{6}{z} = 57 \quad (5)$

Multiply equation (1) by -5 and add it to equation (2).

Multiply equation (1) by -7 and add it to equation (3).

$\dfrac{3}{x} - \dfrac{4}{y} + \dfrac{1}{z} = -2 \quad (1)$

$\dfrac{23}{y} - \dfrac{11}{z} = 13 \quad (6)$

$\dfrac{37}{y} - \dfrac{1}{z} = 71 \quad (7)$

Multiply equation (7) by 23.

$\dfrac{3}{x} - \dfrac{4}{y} + \dfrac{1}{z} = -2 \quad (1)$

$\dfrac{23}{y} - \dfrac{11}{z} = 13 \quad (6)$

$\dfrac{851}{y} - \dfrac{23}{z} = 1633 \quad (8)$

Multiply equation (6) by -37 and add it to equation (8).

$\dfrac{3}{x} - \dfrac{4}{y} + \dfrac{1}{z} = -2 \quad (1)$

$\dfrac{23}{y} - \dfrac{11}{z} = 13 \quad (6)$

$\dfrac{384}{z} = 1152 \quad (9)$

Complete the solution.

$\dfrac{384}{z} = 1152$

$\dfrac{1}{3} = z$

$\dfrac{23}{y} - \dfrac{11}{1/3} = 13$

$\dfrac{23}{y} - 33 = 13$

$\dfrac{23}{y} = 46$

$\dfrac{1}{2} = y$

$\dfrac{3}{x} - \dfrac{4}{1/2} + \dfrac{1}{1/3} = -2$

$\dfrac{3}{x} - 8 + 3 = -2$

$\dfrac{3}{x} - 5 = -2$

$\dfrac{3}{x} = 3$

$1 = x$

The solution is $\left(1, \dfrac{1}{2}, \dfrac{1}{3}\right)$.

(We could also have solved this system of equations by first substituting a for $\dfrac{1}{x}$, b for $\dfrac{1}{y}$, and c for $\dfrac{1}{z}$ and proceeding as we did in Exercise 66 above.)

69.

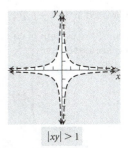

$|xy| > 1$

71. In general, $(\mathbf{AB})^2 \neq \mathbf{A}^2\mathbf{B}^2$. $(\mathbf{AB})^2 = \mathbf{ABAB}$ and $\mathbf{A}^2\mathbf{B}^2 = \mathbf{AABB}$. Since matrix multiplication is not commutative, $\mathbf{BA} \neq \mathbf{AB}$, so $(\mathbf{AB})^2 \neq \mathbf{A}^2\mathbf{B}^2$.

73. If $a_1 x + b_1 y = c_1$ and $a_2 x + b_2 y = c_2$ are parallel lines, then $a_1 = ka_2$, $b_1 = kb_2$, and $c_1 \neq kc_2$, for some number k. Then $\begin{vmatrix} a_1 & b_1 \\ a_2 & b_2 \end{vmatrix} = 0$, $\begin{vmatrix} c_1 & b_1 \\ c_2 & b_2 \end{vmatrix} \neq 0$, and $\begin{vmatrix} a_1 & c_1 \\ a_2 & c_2 \end{vmatrix} \neq 0$.

75. The denominator of the second fraction, $x^2 - 5x + 6$, can be factored into linear factors with real coefficients: $(x - 3)(x - 2)$. Thus, the given expression is not a partial fraction decomposition.

Chapter 6 Test

1. $3x + 2y = 1, \quad (1)$

$2x - y = -11 \quad (2)$

Multiply equation (2) by 2 and add.

$3x + 2y = 1$

$\underline{4x - 2y = -22}$

$7x \qquad = -21$

$x = -3$

Back-substitute to find y.

$2(-3) - y = -11 \quad \text{Using equation (2).}$

$-6 - y = -11$

$-y = -5$

$y = 5$

The solution is $(-3, 5)$. Since the system of equations has exactly one solution, it is consistent and the equations are independent.

2. $2x - y = 3$,   (1)

$\quad\quad 2y = 4x - 6$ (2)

Solve equation (1) for y.

$\quad y = 2x - 3$

Subsitute in equation (2) and solve for x.

$\quad 2(2x - 3) = 4x - 6$

$\quad\quad 4x - 6 = 4x - 6$

$\quad\quad\quad\quad 0 = 0$

The equation $0 = 0$ is true for all values of x and y. Thus the system of equations has infinitely many solutions. Solving either equation for y, we get $y = 2x - 3$, so the solutions are ordered pairs of the form $(x, 2x-3)$. Equivalently, if we solve either equation for x, we get $x = \dfrac{y+3}{2}$, so the solutions can also be expressed as $\left(\dfrac{y+3}{2}, y\right)$. Since there are infinitely many solutions, the system of equations is consistent and the equations are dependent.

3. $x - y = 4$,   (1)

$\quad\quad 3y = 3x - 8$ (2)

Solve equation (1) for x.

$\quad x = y + 4$

Subsitute in equation (2) and solve for y.

$\quad 3y = 3(y + 4) - 8$

$\quad 3y = 3y + 12 - 8$

$\quad\quad 0 = 4$

We get a false equation so there is no solution. Since there is no solution the system of equations is inconsistent and the equations are independent.

4. $2x - 3y = 8$,   (1)

$\quad 5x - 2y = 9$ (2)

Multiply equation (1) by 5 and equation (2) by -2 and add.

$\quad\quad 10x - 15y = 40$

$\quad\underline{-10x + 4y = -18}$

$\quad\quad\quad\ -11y = 22$

$\quad\quad\quad\quad\quad y = -2$

Back-substitute to find x.

$\quad 2x - 3(-2) = 8$

$\quad\quad\ 2x + 6 = 8$

$\quad\quad\quad\ 2x = 2$

$\quad\quad\quad\quad x = 1$

The solution is $(1, -2)$. Since the system of equations has exactly one solution, it is consistent and the equations are independent.

5. $4x + 2y + z = 4$,   (1)

$\quad 3x - y + 5z = 4$,   (2)

$\quad 5x + 3y - 3z = -2$ (3)

Multiply equations (2) and (3) by 4.

$\quad\quad 4x + 2y + z = 4$ (1)

$\quad\quad 12x - 4y + 20z = 16$ (4)

$\quad\quad 20x + 12y - 12z = -8$ (5)

Multiply equation (1) by -3 and add it to equation (4).

Multiply equation (1) by -5 and add it to equation (5).

$\quad\quad 4x + 2y + z = 4$ (1)

$\quad\quad\quad -10y + 17z = 4$ (6)

$\quad\quad\quad\ 2y - 17z = -28$ (7)

Interchange equations (6) and (7).

$\quad\quad 4x + 2y + z = 4$ (1)

$\quad\quad\quad 2y - 17z = -28$ (7)

$\quad\quad\quad -10y + 17z = 4$ (6)

Multiply equation (7) by 5 and add it to equation (6).

$\quad\quad 4x + 2y + z = 4$ (1)

$\quad\quad\quad 2y - 17z = -28$ (7)

$\quad\quad\quad\quad -68z = -136$ (8)

Solve equation (8) for z.

$\quad\quad -68z = -136$

$\quad\quad\quad\quad z = 2$

Back-substitute 2 for z in equation (7) and solve for y.

$\quad 2y - 17 \cdot 2 = -28$

$\quad\quad 2y - 34 = -28$

$\quad\quad\quad\ 2y = 6$

$\quad\quad\quad\quad y = 3$

Back-substitute 3 for y and 2 for z in equation (1) and solve for x.

$\quad 4x + 2 \cdot 3 + 2 = 4$

$\quad\quad\ 4x + 8 = 4$

$\quad\quad\quad\ 4x = -4$

$\quad\quad\quad\quad x = -1$

The solution is $(-1, 3, 2)$.

6. *Familiarize*. Let x and y represent the number of student and nonstudent tickets sold, respectively. Then the receipts from the student tickets were $8x$ and the receipts from the nonstudent tickets were $12y$.

***Translate*.** One equation comes from the fact that 620 tickets were sold.

$\quad x + y = 620$

A second equation comes from the fact that the total receipts were \$5592.

$\quad 8x + 12y = 5592$

***Carry out*.** We solve the system of equations.

$\quad\quad x + y = 620$,   (1)

$\quad 8x + 12y = 5592$ (2)

Multiply equation (1) by -8 and add.

$$-8x - 8y = -4960$$

$$\underline{8x + 12y = 5592}$$

$$4y = 632$$

$$y = 158$$

Substitute 158 for y in equation (1) and solve for x.

$$x + 158 = 620$$

$$x = 462$$

Check. The number of tickets sold was $462 + 158$, or 620. The total receipts were $\$8 \cdot 462 + \$12 \cdot 158 = \$3696 + \$1896 = \$5592$. The solution checks.

State. 462 student tickets and 158 nonstudent tickets were sold.

7. Familiarize. Let x, y, and z represent the number of orders that can be processed per day by Hui, Ashlyn, and Sherriann, respectively.

Translate.

Hui, Ashlyn, and Sherriann can process 352 orders per day.

$$x + y + z = 352$$

Hui and Ashlyn together can process 224 orders per day.

$$x + y = 224$$

Hui and Sherriann together can process 248 orders per day.

$$x + z = 248$$

We have a system of equations:

$$x + y + z = 352,$$
$$x + y \quad\quad = 224,$$
$$x \quad\quad + z = 248.$$

Carry out. Solving the system of equations, we get $(120, 104, 128)$.

Check. Hui, Ashlyn, and Sherriann can process $120 + 104 + 128$, or 352, orders per day. Together, Hui and Ashlyn can process $120 + 104$, or 224, orders per day. Together, Hui and Sheriann can process $120 + 128$, or 248, orders per day. The solution checks.

State. Hui can process 120 orders per day, Ashlyn can process 104 orders per day, and Sheriann can process 128 orders per day.

8. $\mathbf{B} + \mathbf{C} = \begin{bmatrix} -5 & 1 \\ -2 & 4 \end{bmatrix} + \begin{bmatrix} 3 & -4 \\ -1 & 0 \end{bmatrix}$

$= \begin{bmatrix} -5 + 3 & 1 + (-4) \\ -2 + (-1) & 4 + 0 \end{bmatrix}$

$= \begin{bmatrix} -2 & -3 \\ -3 & 4 \end{bmatrix}$

9. $\mathbf{A}$ and $\mathbf{C}$ do not have the same order, so it is not possible to find $\mathbf{A} - \mathbf{C}$.

10. $\mathbf{CB} = \begin{bmatrix} 3 & -4 \\ -1 & 0 \end{bmatrix} \begin{bmatrix} -5 & 1 \\ -2 & 4 \end{bmatrix}$

$= \begin{bmatrix} 3(-5) + (-4)(-2) & 3(1) + (-4)(4) \\ -1(-5) + 0(-2) & -1(1) + 0(4) \end{bmatrix}$

$= \begin{bmatrix} -7 & -13 \\ 5 & -1 \end{bmatrix}$

11. The product $\mathbf{AB}$ is not defined because the number of columns of $\mathbf{A}$, 3, is not equal to the number of rows of $\mathbf{B}$, 2.

12. $2\mathbf{A} = 2\begin{bmatrix} 1 & -1 & 3 \\ -2 & 5 & 2 \end{bmatrix} = \begin{bmatrix} 2 & -2 & 6 \\ -4 & 10 & 4 \end{bmatrix}$

13. $\mathbf{C} = \begin{bmatrix} 3 & -4 \\ -1 & 0 \end{bmatrix}$

Write the augmented matrix.

$$\left[\begin{array}{rr|rr} 3 & -4 & 1 & 0 \\ -1 & 0 & 0 & 1 \end{array}\right]$$

Interchange rows.

$$\left[\begin{array}{rr|rr} -1 & 0 & 0 & 1 \\ 3 & -4 & 1 & 0 \end{array}\right]$$

Multiply row 1 by 3 and add it to row 2.

$$\left[\begin{array}{rr|rr} -1 & 0 & 0 & 1 \\ 0 & -4 & 1 & 3 \end{array}\right]$$

Multiply row 1 by -1 and row 2 by $-\dfrac{1}{4}$.

$$\left[\begin{array}{rr|rr} 1 & 0 & 0 & -1 \\ 0 & 1 & -\dfrac{1}{4} & -\dfrac{3}{4} \end{array}\right]$$

$$\mathbf{C}^{-1} = \begin{bmatrix} 0 & -1 \\ -\dfrac{1}{4} & -\dfrac{3}{4} \end{bmatrix}$$

14. a) $\mathbf{M} = \begin{bmatrix} 1.55 & 1.00 & 0.99 \\ 1.70 & 0.95 & 1.01 \\ 1.65 & 0.99 & 0.96 \end{bmatrix}$

b) $\mathbf{N} = \begin{bmatrix} 26 & 18 & 23 \end{bmatrix}$

c) $\mathbf{NM} = \begin{bmatrix} 108.85 & 65.87 & 66.00 \end{bmatrix}$

d) The entries of $\mathbf{NM}$ represent the total cost, in dollars, for each type of menu item served on the given day.

15. $\begin{bmatrix} 3 & -4 & 2 \\ 2 & 3 & 1 \\ 1 & -5 & -3 \end{bmatrix} \begin{bmatrix} x \\ y \\ z \end{bmatrix} = \begin{bmatrix} -8 \\ 7 \\ 3 \end{bmatrix}$

16. $3x + 2y + 6z = 2,$
$\quad x + y + 2z = 1,$
$2x + 2y + 5z = 3$

Write an equivalent matrix equation, $\mathbf{AX} = \mathbf{B}$.

$$\begin{bmatrix} 3 & 2 & 6 \\ 1 & 1 & 2 \\ 2 & 2 & 5 \end{bmatrix} \begin{bmatrix} x \\ y \\ z \end{bmatrix} = \begin{bmatrix} 2 \\ 1 \\ 3 \end{bmatrix}$$

Then,

$$\mathbf{X} = \mathbf{A}^{-1}\mathbf{B} = \begin{bmatrix} 1 & 2 & -2 \\ -1 & 3 & 0 \\ 0 & -2 & 1 \end{bmatrix} \begin{bmatrix} 2 \\ 1 \\ 3 \end{bmatrix} = \begin{bmatrix} -2 \\ 1 \\ 1 \end{bmatrix}$$

The solution is $(-2, 1, 1)$.

17. $\begin{vmatrix} 3 & -5 \\ 8 & 7 \end{vmatrix} = 3 \cdot 7 - 8(-5) = 21 + 40 = 61$

18. We will expand across the first row.

$$\begin{vmatrix} 2 & -1 & 4 \\ -3 & 1 & -2 \\ 5 & 3 & -1 \end{vmatrix}$$

$$= 2(-1)^{1+1} \begin{vmatrix} 1 & -2 \\ 3 & -1 \end{vmatrix} + (-1)(-1)^{1+2} \begin{vmatrix} -3 & -2 \\ 5 & -1 \end{vmatrix} +$$

$$4(-1)^{1+3} \begin{vmatrix} -3 & 1 \\ 5 & 3 \end{vmatrix}$$

$$= 2 \cdot 1[1(-1) - 3(-2)] + (-1)(-1)[-3(-1) - 5(-2)] +$$
$$4 \cdot 1[-3(3) - 5(1)]$$

$$= 2(5) + 1(13) + 4(-14)$$

$$= -33$$

19. $5x + 2y = -1,$
$7x + 6y = 1$

$$D = \begin{vmatrix} 5 & 2 \\ 7 & 6 \end{vmatrix} = 5(6) - 7(2) = 16$$

$$D_x = \begin{vmatrix} -1 & 2 \\ 1 & 6 \end{vmatrix} = -1(6) - (1)(2) = -8$$

$$D_y = \begin{vmatrix} 5 & -1 \\ 7 & 1 \end{vmatrix} = 5(1) - 7(-1) = 12$$

$$x = \frac{D_x}{D} = \frac{-8}{16} = -\frac{1}{2}$$

$$y = \frac{D_y}{D} = \frac{12}{16} = \frac{3}{4}$$

The solution is $\left(-\frac{1}{2}, \frac{3}{4}\right)$.

20.

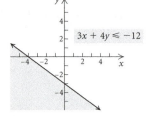

21. Find the maximum value and the minimum value of

$Q = 2x + 3y$ subject to

$$x + y \geq 6,$$
$$2x - 3y \geq -3,$$
$$x \geq 1,$$
$$y \geq 0.$$

Graph the system of inequalities and determine the vertices.

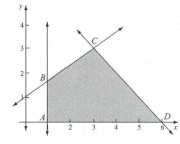

Vertex A:
We solve the system $x = 1$ and $y = 0$. The coordinates of point A are $(1, 0)$.

Vertex B:
We solve the system $2x - 3y = -3$ and $x = 1$. The coordinates of point B are $\left(1, \frac{5}{3}\right)$.

Vertex C:
We solve the system $x + y = 6$ and $2x - 3y = -3$. The coordinates of point C are $(3, 3)$.

Vertex D:
We solve the system $x + y = 6$ and $y = 0$. The coordinates of point D are $(6, 0)$.

Evaluate the objective function Q at each vertex.

Vertex	$Q = 2x + 3y$
$(1, 0)$	$2 \cdot 1 + 3 \cdot 0 = 2$
$\left(1, \frac{5}{3}\right)$	$2 \cdot 1 + 3 \cdot \frac{5}{3} = 7$
$(3, 3)$	$2 \cdot 3 + 3 \cdot 3 = 15$
$(6, 0)$	$2 \cdot 6 + 3 \cdot 0 = 12$

The maximum value of Q is 15 when $x = 3$ and $y = 3$.

The minimum value of Q is 2 when $x = 1$ and $y = 0$.

22. Let $x =$ the number of pound cakes prepared and $y =$ the number of carrot cakes. Find the maximum value of $P = 6x + 8y$ subject to

$$x + y \leq 100,$$
$$x \geq 25,$$
$$y \geq 15$$

Graph the system of inequalities, determine the vertices, and find the value of P at each vertex.

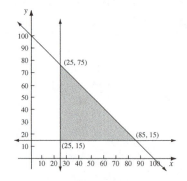

Vertex	$P = 6x + 8y$
$(25, 15)$	$6 \cdot 25 + 8 \cdot 15 = 270$
$(25, 75)$	$6 \cdot 25 + 8 \cdot 75 = 750$
$(85, 15)$	$6 \cdot 85 + 8 \cdot 15 = 630$

The maximum profit of \$750 occurs when 25 pound cakes and 75 carrot cakes are prepared.

23.
$$\frac{3x - 11}{x^2 + 2x - 3} = \frac{3x - 11}{(x - 1)(x + 3)}$$
$$= \frac{A}{x - 1} + \frac{B}{x + 3}$$
$$= \frac{A(x + 3) + B(x - 1)}{(x - 1)(x + 3)}$$

Equate the numerators.

$$3x - 11 = A(x + 3) + B(x - 1)$$

Let $x = -3 : 3(-3) - 11 = 0 + B(-3 - 1)$
$$-20 = -4B$$
$$5 = B$$

Let $x = 1 : 3(1) - 11 = A(1 + 3) + 0$
$$-8 = 4A$$
$$-2 = A$$

The decomposition is $-\dfrac{2}{x - 1} + \dfrac{5}{x + 3}$.

24. Graph the system of inequalities. We see that D is the correct graph.

25. Solve:
$$A(2) - B(-2) = C(2) - 8$$
$$A(-3) - B(-1) = C(1) - 8$$
$$A(4) - B(2) = C(9) - 8$$
or
$$2A + 2B - 2C = -8$$
$$-3A + B - C = -8$$
$$4A - 2B - 9C = -8$$

The solution is $(1, -3, 2)$, so $A = 1$, $B = -3$, and $C = 2$.

Chapter 7

Conic Sections

Exercise Set 7.1

1. Graph (f) is the graph of $x^2 = 8y$.

3. Graph (b) is the graph of $(y-2)^2 = -3(x+4)$.

5. Graph (d) is the graph of $13x^2 - 8y - 9 = 0$.

7. $x^2 = 20y$

$x^2 = 4 \cdot 5 \cdot y$ Writing $x^2 = 4py$

Vertex: $(0,0)$

Focus: $(0,5)$ $[(0,p)]$

Directrix: $y = -5$ $(y = -p)$

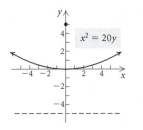

9. $y^2 = -6x$

$y^2 = 4\left(-\dfrac{3}{2}\right)x$ Writing $y^2 = 4px$

Vertex: $(0,0)$

Focus: $\left(-\dfrac{3}{2}, 0\right)$ $[(p,0)]$

Directrix: $x = -\left(-\dfrac{3}{2}\right) = \dfrac{3}{2}$ $(x = -p)$

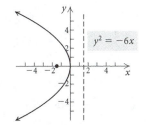

11. $x^2 - 4y = 0$

$x^2 = 4y$

$x^2 = 4 \cdot 1 \cdot y$ Writing $x^2 = 4py$

Vertex: $(0,0)$

Focus: $(0,1)$ $[(0,p)]$

Directrix: $y = -1$ $(y = -p)$

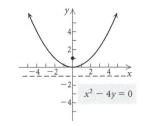

13. $x = 2y^2$

$y^2 = \dfrac{1}{2}x$

$y^2 = 4 \cdot \dfrac{1}{8} \cdot x$ Writing $y^2 = 4px$

Vertex: $(0,0)$

Focus: $\left(\dfrac{1}{8}, 0\right)$

Directrix: $x = -\dfrac{1}{8}$

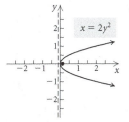

15. The vertex is at the origin and the focus is $(-3,0)$, so the equation is of the form $y^2 = 4px$ where $p = -3$.

$y^2 = 4px$

$y^2 = 4(-3)x$

$y^2 = -12x$

17. Since the directrix, $x = -4$, is a vertical line, the equation is of the form $(y-k)^2 = 4p(x-h)$. The focus, $(7,0)$, is on the x-axis so the axis of symmetry is the x-axis and $p = 7$. The vertex, (h,k), is the point on the x-axis midway between the directrix and the focus. Thus, it is $(0,0)$. We have

$(y-k)^2 = 4p(x-h)$

$(y-0)^2 = 4 \cdot 7(x-0)$ Substituting

$y^2 = 28x.$

19. Since the directrix, $y = \pi$, is a horizontal line, the equation is of the form $(x-h)^2 = 4p(y-k)$. The focus, $(0,-\pi)$, is on the y-axis so the axis of symmetry is the y-axis and $p = -\pi$. The vertex (h,k) is the point on the y-axis midway between the directrix and the focus. Thus, it is $(0,0)$. We have

$$(x - h)^2 = 4p(y - k)$$
$$(x - 0)^2 = 4(-\pi)(y - 0) \quad \text{Substituting}$$
$$x^2 = -4\pi y$$

21. Since the directrix, $x = -4$, is a vertical line, the equation is of the form $(y - k)^2 = 4p(x - h)$. The focus, $(3, 2)$, is on the horizontal line $y = 2$, so the axis of symmetry is $y = 2$. The vertex is the point on the line $y = 2$ that is midway between the directrix and the focus. That is, it is the midpoint of the segment from $(-4, 2)$ to $(3, 2)$: $\left(\dfrac{-4 + 3}{2}, \dfrac{2 + 2}{2}\right)$, or $\left(-\dfrac{1}{2}, 2\right)$. Then $h = -\dfrac{1}{2}$ and the directrix is $x = h - p$, so we have

$$x = h - p$$
$$-4 = -\frac{1}{2} - p$$
$$-\frac{7}{2} = -p$$
$$\frac{7}{2} = p.$$

Now we find the equation of the parabola.

$$(y - k)^2 = 4p(x - h)$$
$$(y - 2)^2 = 4\left(\frac{7}{2}\right)\left[x - \left(-\frac{1}{2}\right)\right]$$
$$(y - 2)^2 = 14\left(x + \frac{1}{2}\right)$$

23.
$$(x + 2)^2 = -6(y - 1)$$
$$[x - (-2)]^2 = 4\left(-\frac{3}{2}\right)(y - 1) \quad [(x - h)^2 = 4p(y - k)]$$

Vertex: $(-2, 1)$ $\qquad [(h, k)]$

Focus: $\left(-2, 1 + \left(-\frac{3}{2}\right)\right)$, or $\left(-2, -\frac{1}{2}\right)$

$\qquad\qquad\qquad [(h, k + p)]$

Directrix: $y = 1 - \left(-\frac{3}{2}\right) = \frac{5}{2}$ $\quad (y = k - p)$

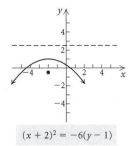

$$(x + 2)^2 = -6(y - 1)$$

25. $x^2 + 2x + 2y + 7 = 0$
$$x^2 + 2x = -2y - 7$$
$$(x^2 + 2x + 1) = -2y - 7 + 1 = -2y - 6$$
$$(x + 1)^2 = -2(y + 3)$$
$$[x - (-1)]^2 = 4\left(-\frac{1}{2}\right)[y - (-3)]$$
$$[(x - h)^2 = 4p(y - k)]$$

Vertex: $(-1, -3)$ $\qquad [(h, k)]$

Focus: $\left(-1, -3 + \left(-\frac{1}{2}\right)\right)$, or $\left(-1, -\frac{7}{2}\right)$

$\qquad\qquad\qquad\qquad\qquad [(h, k+p)]$

Directrix: $y = -3 - \left(-\frac{1}{2}\right) = -\frac{5}{2}$ $\quad (y = k - p)$

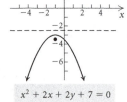

$$x^2 + 2x + 2y + 7 = 0$$

27. $x^2 - y - 2 = 0$
$$x^2 = y + 2$$
$$(x - 0)^2 = 4 \cdot \frac{1}{4} \cdot [y - (-2)]$$
$$[(x - h)^2 = 4p(y - k)]$$

Vertex: $(0, -2)$ $\qquad [(h, k)]$

Focus: $\left(0, -2 + \frac{1}{4}\right)$, or $\left(0, -\frac{7}{4}\right)$ $\;[(h, k + p)]$

Directrix: $y = -2 - \frac{1}{4} = -\frac{9}{4}$ $\qquad (y = k - p)$

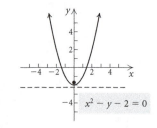

$$x^2 - y - 2 = 0$$

29.
$$y = x^2 + 4x + 3$$
$$y - 3 = x^2 + 4x$$
$$y - 3 + 4 = x^2 + 4x + 4$$
$$y + 1 = (x + 2)^2$$
$$4 \cdot \frac{1}{4} \cdot [y - (-1)] = [x - (-2)]^2$$
$$[(x - h)^2 = 4p(y - k)]$$

Vertex: $(-2, -1)$ $\qquad [(h, k)]$

Focus: $\left(-2, -1 + \frac{1}{4}\right)$, or $\left(-2, -\frac{3}{4}\right)$ $\;[(h, k + p)]$

Directrix: $y = -1 - \frac{1}{4} = -\frac{5}{4}$ $\quad (y = k - p)$

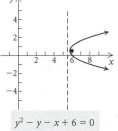

$$y = x^2 + 4x + 3$$

31. $y^2 - y - x + 6 = 0$

$$y^2 - y = x - 6$$

$$y^2 - y + \frac{1}{4} = x - 6 + \frac{1}{4}$$

$$\left(y - \frac{1}{2}\right)^2 = x - \frac{23}{4}$$

$$\left(y - \frac{1}{2}\right)^2 = 4 \cdot \frac{1}{4}\left(x - \frac{23}{4}\right)$$

$$\qquad\qquad [(y-k)^2 = 4p(x-h)]$$

Vertex: $\left(\dfrac{23}{4}, \dfrac{1}{2}\right) \qquad [(h,k)]$

Focus: $\left(\dfrac{23}{4} + \dfrac{1}{4}, \dfrac{1}{2}\right)$, or $\left(6, \dfrac{1}{2}\right) \qquad [(h+p,k)]$

Directrix: $x = \dfrac{23}{4} - \dfrac{1}{4} = \dfrac{22}{4}$ or $\dfrac{11}{2} \quad (x = h - p)$

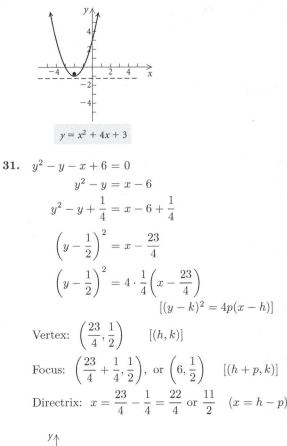

$$y^2 - y - x + 6 = 0$$

33. a) The vertex is $(0,0)$. The focus is $(4,0)$, so $p = 4$. The parabola has a horizontal axis of symmetry so the equation is of the form $y^2 = 4px$. We have

$$y^2 = 4px$$
$$y^2 = 4 \cdot 4 \cdot x$$
$$y^2 = 16x$$

b) We make a drawing.

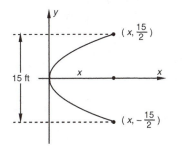

The depth of the satellite dish at the vertex is x where $\left(x, \dfrac{15}{2}\right)$ is a point on the parabola.

$$y^2 = 16x$$
$$\left(\frac{15}{2}\right)^2 = 16x \qquad \text{Substituting } \frac{15}{2} \text{ for } y$$
$$\frac{225}{4} = 16x$$
$$\frac{225}{64} = x, \text{ or}$$
$$3\frac{33}{64} = x$$

The depth of the satellite dish at the vertex is $3\dfrac{33}{64}$ ft.

35. We position a coordinate system with the origin at the vertex and the x-axis on the parabola's axis of symmetry.

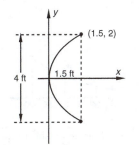

The parabola is of the form $y^2 = 4px$ and a point on the parabola is $\left(1.5, \dfrac{4}{2}\right)$, or $(1.5, 2)$.

$$y^2 = 4px$$
$$2^2 = 4 \cdot p \cdot (1.5) \quad \text{Substituting}$$
$$4 = 6p$$
$$\frac{4}{6} = p, \text{ or}$$
$$\frac{2}{3} = p$$

Since the focus is at $(p, 0)$, or $\left(\dfrac{2}{3}, 0\right)$, the focus is $\dfrac{2}{3}$ ft, or 8 in., from the vertex.

37. When we let $y = 0$ and solve for x, the only equation for which $x = \dfrac{2}{3}$ is (h), so only equation (h) has x-intercept $\left(\dfrac{2}{3}, 0\right)$.

39. Note that equation (g) is equivalent to $y = 2x - \dfrac{7}{4}$ and equation (h) is equivalent to $y = -\dfrac{1}{2}x + \dfrac{1}{3}$. When we look at the equations in the form $y = mx + b$, we see that $m > 0$ for (a), (b), (f), and (g) so these equations have positive slope, or slant up front left to right.

41. When we look at the equations in the form $y = mx + b$ (See Exercise 39.), only (b) has $m = \dfrac{1}{3}$ so only (b) has slope $\dfrac{1}{3}$.

43. Parallel lines have the same slope and different y-intercepts. When we look at the equations in the form $y = mx + b$ (See Exercise 39.), we see that (a) and (g) represent parallel lines.

45. A parabola with a vertical axis of symmetry has an equation of the type $(x - h)^2 = 4p(y - k)$.

Solve for p substituting $(-1, 2)$ for (h, k) and $(-3, 1)$ for (x, y).

$$[-3 - (-1)]^2 = 4p(1 - 2)$$
$$4 = -4p$$
$$-1 = p$$

The equation of the parabola is

$$[x - (-1)]^2 = 4(-1)(y - 2), \text{ or}$$
$$(x + 1)^2 = -4(y - 2).$$

47. Position a coordinate system as shown below with the y-axis on the parabola's axis of symmetry.

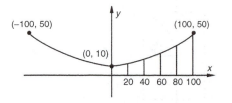

The equation of the parabola is of the form $(x - h)^2 = 4p(y - k)$. Substitute 100 for x, 50 for y, 0 for h, and 10 for k and solve for p.

$$(x - h)^2 = 4p(y - k)$$
$$(100 - 0)^2 = 4p(50 - 10)$$
$$10,000 = 160p$$
$$\frac{250}{4} = p$$

Then the equation is

$$x^2 = 4\left(\frac{250}{4}\right)(y - 10), \text{ or}$$
$$x^2 = 250(y - 10).$$

To find the lengths of the vertical cables, find y when $x = 0$, 20, 40, 60, 80, and 100.

When $x = 0$: $0^2 = 250(y - 10)$
$$0 = y - 10$$
$$10 = y$$

When $x = 20$: $20^2 = 250(y - 10)$
$$400 = 250(y - 10)$$
$$1.6 = y - 10$$
$$11.6 = y$$

When $x = 40$: $40^2 = 250(y - 10)$
$$1600 = 250(y - 10)$$
$$6.4 = y - 10$$
$$16.4 = y$$

When $x = 60$: $60^2 = 250(y - 10)$
$$3600 = 250(y - 10)$$
$$14.4 = y - 10$$
$$24.4 = y$$

When $x = 80$: $80^2 = 250(y - 10)$
$$6400 = 250(y - 10)$$
$$25.6 = y - 10$$
$$35.6 = y$$

When $x = 100$, we know from the given information that $y = 50$.

The lengths of the vertical cables are 10 ft, 11.6 ft, 16.4 ft, 24.4 ft, 35.6 ft, and 50 ft.

Exercise Set 7.2

1. Graph (b) is the graph of $x^2 + y^2 = 5$.

3. Graph (d) is the graph of $x^2 + y^2 - 6x + 2y = 6$.

5. Graph (a) is the graph of $x^2 + y^2 - 5x + 3y = 0$.

7. Complete the square twice.

$$x^2 + y^2 - 14x + 4y = 11$$
$$x^2 - 14x + y^2 + 4y = 11$$
$$x^2 - 14x + 49 + y^2 + 4y + 4 = 11 + 49 + 4$$
$$(x - 7)^2 + (y + 2)^2 = 64$$
$$(x - 7)^2 + [y - (-2)]^2 = 8^2$$

Center: $(7, -2)$

Radius: 8

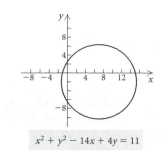

$x^2 + y^2 - 14x + 4y = 11$

9. Complete the square twice.

$$x^2 + y^2 + 6x - 2y = 6$$
$$x^2 + 6x + y^2 - 2y = 6$$
$$x^2 + 6x + 9 + y^2 - 2y + 1 = 6 + 9 + 1$$
$$(x + 3)^2 + (y - 1)^2 = 16$$
$$[x - (-3)]^2 + (y - 1)^2 = 4^2$$

Center: $(-3, 1)$

Radius: 4

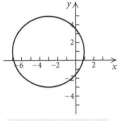

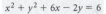

$$x^2 + y^2 + 6x - 2y = 6$$

11. Complete the square twice.

$$x^2 + y^2 + 4x - 6y - 12 = 0$$
$$x^2 + 4x + y^2 - 6y = 12$$
$$x^2 + 4x + 4 + y^2 - 6y + 9 = 12 + 4 + 9$$
$$(x + 2)^2 + (y - 3)^2 = 25$$
$$[x - (-2)]^2 + (y - 3)^2 = 5^2$$

Center: $(-2, 3)$

Radius: 5

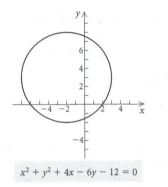

$$x^2 + y^2 + 4x - 6y - 12 = 0$$

13. Complete the square twice.

$$x^2 + y^2 - 6x - 8y + 16 = 0$$
$$x^2 - 6x + y^2 - 8y = -16$$
$$x^2 - 6x + 9 + y^2 - 8y + 16 = -16 + 9 + 16$$
$$(x - 3)^2 + (y - 4)^2 = 9$$
$$(x - 3)^2 + (y - 4)^2 = 3^2$$

Center: $(3, 4)$

Radius: 3

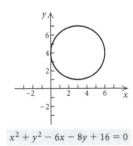

$$x^2 + y^2 - 6x - 8y + 16 = 0$$

15. Complete the square twice.

$$x^2 + y^2 + 6x - 10y = 0$$
$$x^2 + 6x + y^2 - 10y = 0$$
$$x^2 + 6x + 9 + y^2 - 10y + 25 = 0 + 9 + 25$$
$$(x + 3)^2 + (y - 5)^2 = 34$$
$$[x - (-3)]^2 + (y - 5)^2 = (\sqrt{34})^2$$

Center: $(-3, 5)$

Radius: $\sqrt{34}$

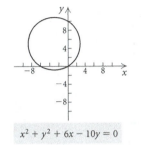

$$x^2 + y^2 + 6x - 10y = 0$$

17. Complete the square twice.

$$x^2 + y^2 - 9x = 7 - 4y$$
$$x^2 - 9x + y^2 + 4y = 7$$
$$x^2 - 9x + \frac{81}{4} + y^2 + 4y + 4 = 7 + \frac{81}{4} + 4$$
$$\left(x - \frac{9}{2}\right)^2 + (y + 2)^2 = \frac{125}{4}$$
$$\left(x - \frac{9}{2}\right)^2 + [y - (-2)]^2 = \left(\frac{5\sqrt{5}}{2}\right)^2$$

Center: $\left(\dfrac{9}{2}, -2\right)$

Radius: $\dfrac{5\sqrt{5}}{2}$

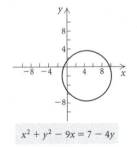

$$x^2 + y^2 - 9x = 7 - 4y$$

19. Graph (c) is the graph of $16x^2 + 4y^2 = 64$.

21. Graph (d) is the graph of $x^2 + 9y^2 - 6x + 90y = -225$.

23. $\dfrac{x^2}{4} + \dfrac{y^2}{1} = 1$

$\dfrac{x^2}{2^2} + \dfrac{y^2}{1^2} = 1$ Standard form

$a = 2$, $b = 1$

The major axis is horizontal, so the vertices are $(-2, 0)$ and $(2, 0)$. Since we know that $c^2 = a^2 - b^2$, we have

$c^2 = 4 - 1 = 3$, so $c = \sqrt{3}$ and the foci are $(-\sqrt{3}, 0)$ and $(\sqrt{3}, 0)$.

To graph the ellipse, plot the vertices. Note also that since $b = 1$, the y-intercepts are $(0, -1)$ and $(0, 1)$. Plot these points as well and connect the four plotted points with a smooth curve.

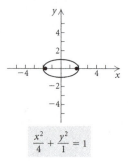

$$\frac{x^2}{4} + \frac{y^2}{1} = 1$$

25. $16x^2 + 9y^2 = 144$

$$\frac{x^2}{9} + \frac{y^2}{16} = 1 \qquad \text{Dividing by 144}$$

$$\frac{x^2}{3^2} + \frac{y^2}{4^2} = 1 \qquad \text{Standard form}$$

$a = 4$, $b = 3$

The major axis is vertical, so the vertices are $(0, -4)$ and $(0, 4)$. Since $c^2 = a^2 - b^2$, we have $c^2 = 16 - 9 = 7$, so $c = \sqrt{7}$ and the foci are $(0, -\sqrt{7})$ and $(0, \sqrt{7})$.

To graph the ellipse, plot the vertices. Note also that since $b = 3$, the x-intercepts are $(-3, 0)$ and $(3, 0)$. Plot these points as well and connect the four plotted points with a smooth curve.

$16x^2 + 9y^2 = 144$

27. $2x^2 + 3y^2 = 6$

$$\frac{x^2}{3} + \frac{y^2}{2} = 1$$

$$\frac{x^2}{(\sqrt{3})^2} + \frac{y^2}{(\sqrt{2})^2} = 1$$

$a = \sqrt{3}$, $b = \sqrt{2}$

The major axis is horizontal, so the vertices are $(-\sqrt{3}, 0)$ and $(\sqrt{3}, 0)$. Since $c^2 = a^2 - b^2$, we have $c^2 = 3 - 2 = 1$, so $c = 1$ and the foci are $(-1, 0)$ and $(1, 0)$.

To graph the ellipse, plot the vertices. Note also that since $b = \sqrt{2}$, the y-intercepts are $(0, -\sqrt{2})$ and $(0, \sqrt{2})$. Plot these points as well and connect the four plotted points with a smooth curve.

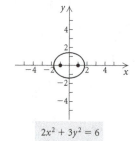

$2x^2 + 3y^2 = 6$

29. $4x^2 + 9y^2 = 1$

$$\frac{x^2}{\frac{1}{4}} + \frac{y^2}{\frac{1}{9}} = 1$$

$$\frac{x^2}{\left(\frac{1}{2}\right)^2} + \frac{y^2}{\left(\frac{1}{3}\right)^2} = 1$$

$a = \dfrac{1}{2}$, $b = \dfrac{1}{3}$

The major axis is horizontal, so the vertices are $\left(-\dfrac{1}{2}, 0\right)$ and $\left(\dfrac{1}{2}, 0\right)$. Since $c^2 = a^2 - b^2$, we have $c^2 = \dfrac{1}{4} - \dfrac{1}{9} = \dfrac{5}{36}$, so $c = \dfrac{\sqrt{5}}{6}$ and the foci are $\left(-\dfrac{\sqrt{5}}{6}, 0\right)$ and $\left(\dfrac{\sqrt{5}}{6}, 0\right)$.

To graph the ellipse, plot the vertices. Note also that since $b = \dfrac{1}{3}$, the y-intercepts are $\left(0, -\dfrac{1}{3}\right)$ and $\left(0, \dfrac{1}{3}\right)$. Plot these points as well and connect the four plotted points with a smooth curve.

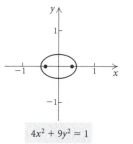

$4x^2 + 9y^2 = 1$

31. The vertices are on the x-axis, so the major axis is horizontal. We have $a = 7$ and $c = 3$, so we can find b^2:

$$c^2 = a^2 - b^2$$
$$3^2 = 7^2 - b^2$$
$$b^2 = 49 - 9 = 40$$

Write the equation:

$$\frac{x^2}{a^2} + \frac{y^2}{b^2} = 1$$

$$\frac{x^2}{49} + \frac{y^2}{40} = 1$$

33. The vertices, $(0, -8)$ and $(0, 8)$, are on the y-axis, so the major axis is vertical and $a = 8$. Since the vertices are equidistant from the origin, the center of the ellipse is at the origin. The length of the minor axis is 10, so $b = 10/2$, or 5.

Write the equation:

$$\frac{x^2}{b^2} + \frac{y^2}{a^2} = 1$$

$$\frac{x^2}{5^2} + \frac{y^2}{8^2} = 1$$

$$\frac{x^2}{25} + \frac{y^2}{64} = 1$$

35. The foci, $(-2, 0)$ and $(2, 0)$ are on the x-axis, so the major axis is horizontal and $c = 2$. Since the foci are equidistant from the origin, the center of the ellipse is at the origin. The length of the major axis is 6, so $a = 6/2$, or 3. Now we find b^2:

$$c^2 = a^2 - b^2$$

$$2^2 = 3^2 - b^2$$

$$4 = 9 - b^2$$

$$b^2 = 5$$

Write the equation:

$$\frac{x^2}{a^2} + \frac{y^2}{b^2} = 1$$

$$\frac{x^2}{9} + \frac{y^2}{5} = 1$$

37. $\dfrac{(x-1)^2}{9} + \dfrac{(y-2)^2}{4} = 1$

$$\frac{(x-1)^2}{3^2} + \frac{(y-2)^2}{2^2} = 1 \quad \text{Standard form}$$

The center is $(1, 2)$. Note that $a = 3$ and $b = 2$. The major axis is horizontal so the vertices are 3 units left and right of the center:

$(1 - 3, 2)$ and $(1 + 3, 2)$, or $(-2, 2)$ and $(4, 2)$.

We know that $c^2 = a^2 - b^2$, so $c^2 = 9 - 4 = 5$ and $c = \sqrt{5}$. Then the foci are $\sqrt{5}$ units left and right of the center:

$(1 - \sqrt{5}, 2)$ and $(1 + \sqrt{5}, 2)$.

To graph the ellipse, plot the vertices. Since $b = 2$, two other points on the graph are 2 units below and above the center:

$(1, 2 - 2)$ and $(1, 2 + 2)$ or $(1, 0)$ and $(1, 4)$

Plot these points also and connect the four plotted points with a smooth curve.

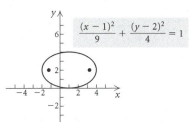

39. $\dfrac{(x+3)^2}{25} + \dfrac{(y-5)^2}{36} = 1$

$$\frac{[x-(-3)]^2}{5^2} + \frac{(y-5)^2}{6^2} = 1 \quad \text{Standard form}$$

The center is $(-3, 5)$. Note that $a = 6$ and $b = 5$. The major axis is vertical so the vertices are 6 units below and above the center:

$(-3, 5 - 6)$ and $(-3, 5 + 6)$, or $(-3, -1)$ and $(-3, 11)$.

We know that $c^2 = a^2 - b^2$, so $c^2 = 36 - 25 = 11$ and $c = \sqrt{11}$. Then the foci are $\sqrt{11}$ units below and above the vertex:

$(-3, 5 - \sqrt{11})$ and $(-3, 5 + \sqrt{11})$.

To graph the ellipse, plot the vertices. Since $b = 5$, two other points on the graph are 5 units left and right of the center:

$(-3 - 5, 5)$ and $(-3 + 5, 5)$, or $(-8, 5)$ and $(2, 5)$

Plot these points also and connect the four plotted points with a smooth curve.

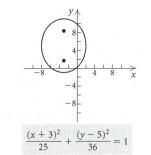

41. $3(x+2)^2 + 4(y-1)^2 = 192$

$$\frac{(x+2)^2}{64} + \frac{(y-1)^2}{48} = 1 \quad \text{Dividing by 192}$$

$$\frac{[x-(-2)]^2}{8^2} + \frac{(y-1)^2}{(\sqrt{48})^2} = 1 \quad \text{Standard form}$$

The center is $(-2, 1)$. Note that $a = 8$ and $b = \sqrt{48}$, or $4\sqrt{3}$. The major axis is horizontal so the vertices are 8 units left and right of the center:

$(-2 - 8, 1)$ and $(-2 + 8, 1)$, or $(-10, 1)$ and $(6, 1)$.

We know that $c^2 = a^2 - b^2$, so $c^2 = 64 - 48 = 16$ and $c = 4$. Then the foci are 4 units left and right of the center:

$(-2 - 4, 1)$ and $(-2 + 4, 1)$ or $(-6, 1)$ and $(2, 1)$.

To graph the ellipse, plot the vertices. Since $b = 4\sqrt{3} \approx 6.928$, two other points on the graph are about 6.928 units below and above the center:

$(-2, 1 - 6.928)$ and $(-2, 1 + 6.928)$, or

$(-2, -5.928)$ and $(-2, 7.928)$.

Plot these points also and connect the four plotted points with a smooth curve.

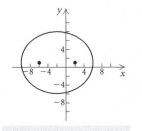

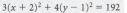

$$3(x+2)^2 + 4(y-1)^2 = 192$$

43. Begin by completing the square twice.
$$4x^2 + 9y^2 - 16x + 18y - 11 = 0$$
$$4x^2 - 16x + 9y^2 + 18y = 11$$
$$4(x^2 - 4x) + 9(y^2 + 2y) = 11$$
$$4(x^2 - 4x + 4) + 9(y^2 + 2y + 1) = 11 + 4 \cdot 4 + 9 \cdot 1$$
$$4(x-2)^2 + 9(y+1)^2 = 36$$
$$\frac{(x-2)^2}{9} + \frac{(y+1)^2}{4} = 1$$
$$\frac{(x-2)^2}{3^2} + \frac{[y-(-1)]^2}{2^2} = 1$$

The center is $(2, -1)$. Note that $a = 3$ and $b = 2$. The major axis is horizontal so the vertices are 3 units left and right of the center:

$(2-3, -1)$ and $(2+3, -1)$, or $(-1, -1)$ and $(5, -1)$.

We know that $c^2 = a^2 - b^2$, so $c^2 = 9 - 4 = 5$ and $c = \sqrt{5}$. Then the foci are $\sqrt{5}$ units left and right of the center:

$(2 - \sqrt{5}, -1)$ and $(2 + \sqrt{5}, -1)$.

To graph the ellipse, plot the vertices. Since $b = 2$, two other points on the graph are 2 units below and above the center:

$(2, -1 - 2)$ and $(2, -1 + 2)$, or $(2, -3)$ and $(2, 1)$.

Plot these points also and connect the four plotted points with a smooth curve.

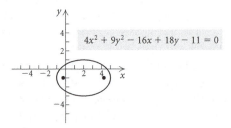

45. Begin by completing the square twice.
$$4x^2 + y^2 - 8x - 2y + 1 = 0$$
$$4x^2 - 8x + y^2 - 2y = -1$$
$$4(x^2 - 2x) + y^2 - 2y = -1$$
$$4(x^2 - 2x + 1) + y^2 - 2y + 1 = -1 + 4 \cdot 1 + 1$$
$$4(x-1)^2 + (y-1)^2 = 4$$
$$\frac{(x-1)^2}{1} + \frac{(y-1)^2}{4} = 1$$
$$\frac{(x-1)^2}{1^2} + \frac{(y-1)^2}{2^2} = 1$$

The center is $(1, 1)$. Note that $a = 2$ and $b = 1$. The major axis is vertical so the vertices are 2 units below and above the center:

$(1, 1-2)$ and $(1, 1+2)$, or $(1, -1)$ and $(1, 3)$.

We know that $c^2 = a^2 - b^2$, so $c^2 = 4 - 1 = 3$ and $c = \sqrt{3}$. Then the foci are $\sqrt{3}$ units below and above the center:

$(1, 1 - \sqrt{3})$ and $(1, 1 + \sqrt{3})$.

To graph the ellipse, plot the vertices. Since $b = 1$, two other points on the graph are 1 unit left and right of the center:

$(1 - 1, 1)$ and $(1 + 1, 1)$ or $(0, 1)$ and $(2, 1)$.

Plot these points also and connect the four plotted points with a smooth curve.

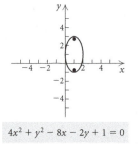

$$4x^2 + y^2 - 8x - 2y + 1 = 0$$

47. The ellipse in Example 4 is flatter than the one in Example 2, so the ellipse in Example 2 has the smaller eccentricity.

We compute the eccentricities: In Example 2, $c = 3$ and $a = 5$, so $e = c/a = 3/5 = 0.6$. In Example 4, $c = 2\sqrt{3}$ and $a = 4$, so $e = c/a = 2\sqrt{3}/4 \approx 0.866$. These computations confirm that the ellipse in Example 2 has the smaller eccentricity.

49. Since the vertices, $(0, -4)$ and $(0, 4)$ are on the y-axis and are equidistant from the origin, we know that the major axis of the ellipse is vertical, its center is at the origin, and $a = 4$. Use the information that $e = 1/4$ to find c:
$$e = \frac{c}{a}$$
$$\frac{1}{4} = \frac{c}{4} \quad \text{Substituting}$$
$$c = 1$$

Now $c^2 = a^2 - b^2$, so we can find b^2:
$$1^2 = 4^2 - b^2$$
$$1 = 16 - b^2$$
$$b^2 = 15$$

Write the equation of the ellipse:
$$\frac{x^2}{b^2} + \frac{y^2}{a^2} = 1$$
$$\frac{x^2}{15} + \frac{y^2}{16} = 1$$

51. From the figure in the text we see that the center of the ellipse is $(0, 0)$, the major axis is horizontal, the vertices are $(-50, 0)$ and $(50, 0)$, and one y-intercept is $(0, 12)$. Then $a = 50$ and $b = 12$. The equation is

$$\frac{x^2}{a^2} + \frac{y^2}{b^2} = 1$$

$$\frac{x^2}{50^2} + \frac{y}{12^2} = 1.$$

53. Positioning a coordinate system with the origin at the center of the ellipse and with the major axis horizontal, the vertices are $(-45, 0)$ and $(45, 0)$ and one y-intercept is $(0, 30)$. Thus we have $a = 45$ and $b = 30$. We find c.

$$c^2 = a^2 - b^2$$

$$c^2 = 45^2 - 30^2 = 1125$$

$$c \approx 33.5$$

The foci are about 33.5 ft from the center of the ellipse.

55. Position a coordinate system as shown below where 1 unit $= 10^7$ mi.

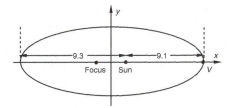

The length of the major axis is $9.3 + 9.1$, or 18.4. Then the distance from the center of the ellipse (the origin) to V is $18.4/2$, or 9.2. Since the distance from the sun to V is 9.1, the distance from the sun to the center is $9.2 - 9.1$, or 0.1. Then the distance from the sun to the other focus is twice this distance:

$$2(0.1 \times 10^7 \text{ mi}) = 0.2 \times 10^7 \text{ mi}$$

$$= 2 \times 10^6 \text{ mi}$$

57. midpoint

59. y-intercept

61. remainder

63. parabola

65. The center of the ellipse is the midpoint of the segment connecting the vertices:

$$\left(\frac{3+3}{2}, \frac{-4+6}{2}\right), \text{ or } (3, 1).$$

Now a is the distance from the origin to a vertex. We use the vertex $(3, 6)$.

$$a = \sqrt{(3-3)^2 + (6-1)^2} = 5$$

Also b is one-half the length of the minor axis.

$$b = \frac{\sqrt{(5-1)^2 + (1-1)^2}}{2} = \frac{4}{2} = 2$$

The vertices lie on the vertical line $x = 3$, so the major axis is vertical. We write the equation of the ellipse.

$$\frac{(x-h)^2}{b^2} + \frac{(y-k)^2}{a^2} = 1$$

$$\frac{(x-3)^2}{4} + \frac{(y-1)^2}{25} = 1$$

67. The center is the midpoint of the segment connecting the vertices:

$$\left(\frac{-3+3}{2}, \frac{0+0}{2}\right), \text{ or } (0, 0).$$

Then $a = 3$ and since the vertices are on the x-axis, the major axis is horizontal. The equation is of the form $\frac{x^2}{a^2} + \frac{y^2}{b^2} = 1$.

Substitute 3 for a, 2 for x, and $\frac{22}{3}$ for y and solve for b^2.

$$\frac{4}{9} + \frac{\frac{484}{9}}{b^2} = 1$$

$$\frac{4}{9} + \frac{484}{9b^2} = 1$$

$$4b^2 + 484 = 9b^2$$

$$484 = 5b^2$$

$$\frac{484}{5} = b^2$$

Then the equation is $\frac{x^2}{9} + \frac{y^2}{484/5} = 1.$

69. Position a coordinate system as shown.

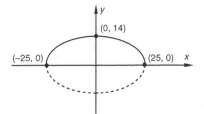

The equation of the ellipse is

$$\frac{x^2}{25^2} + \frac{y^2}{14^2} = 1$$

$$\frac{x^2}{625} + \frac{y^2}{196} = 1.$$

A point 6 ft from the riverbank corresponds to $(25 - 6, 0)$, or $(19, 0)$ or to $(-25 + 6, 0)$, or $(-19, 0)$. Substitute either 19 or -19 for x and solve for y, the clearance.

$$\frac{19^2}{625} + \frac{y^2}{196} = 1$$

$$\frac{y^2}{196} = 1 - \frac{361}{625}$$

$$y^2 = 196\left(1 - \frac{361}{625}\right)$$

$$y \approx 9.1$$

The clearance 6 ft from the riverbank is about 9.1 ft.

Chapter 7 Mid-Chapter Mixed Review

1. The equation $(x+3)^2 = 8(y-2)$ is equivalent to the equation $[x - (-3)]^2 = 4 \cdot 2(y-2)$, so the given statement is true. See page 481 in the text.

3. The equation $(x-4)^2 + (y+1)^2 = 9$ is equivalent to the equation $(x-4)^2 + [y-(-1)]^2 = 3^2$. This is the equation of a circle with center $(4,-1)$ and radius 3, so the given statement is false.

5. Graph (c) is the graph of $x^2 = -4y$.

7. Graph (d) is the graph of $16x^2 + 9y^2 = 144$.

9. Graph (b) is the graph of $(x-1)^2 = 2(y+3)$.

11. Graph (g) is the graph of $(x-2)^2 + (y+3)^2 = 4$.

13. $y^2 = 12x$

$y^2 = 4 \cdot 3 \cdot x$

Vertex: $(0,0)$

Focus: $(3,0)$

Directrix: $x = -3$

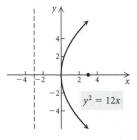

15. Since the directrix, $y = 1$, is a horizontal line, the equation is of the form $(x-h)^2 = 4p(y-k)$. The focus, $(0,3)$, is on the y-axis so the axis of symmetry is the y-axis. The vertex is the point on the y-axis that is midway between the directrix and the focus. That is, it is the midpoint of the segment from $(0,1)$ to $(0,3)$: $\left(\dfrac{0+0}{2}, \dfrac{1+3}{2}\right)$, or $(0,2)$. Then $k = 2$ and the directrix is $y = k - p$, so we have

$y = k - p$

$1 = 2 - p$

$p = 1.$

Now we find the equation of the parabola.

$(x-h)^2 = 4p(y-k)$

$(x-0)^2 = 4 \cdot 1 \cdot (y-2)$

$x^2 = 4(y-2)$

17. $\qquad x^2 + y^2 + 4x - 8y = 5$

$\qquad x^2 + 4x + y^2 - 8y = 5$

$x^2 + 4x + 4 + y^2 - 8y + 16 = 5 + 4 + 16$

$\qquad (x+2)^2 + (y-4)^2 = 25$

$\qquad [x-(-2)]^2 + (y-4)^2 = 5^2$

Center: $(-2,4)$; radius: 5

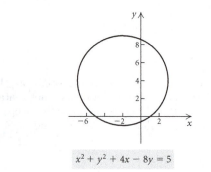

$$x^2 + y^2 + 4x - 8y = 5$$

19. $\dfrac{x^2}{1} + \dfrac{y^2}{9} = 1$

$\dfrac{x^2}{1^2} + \dfrac{y^2}{3^2} = 1$

$a = 3,\ b = 1$

The major axis is vertical, so the vertices are $(0,-3)$ and $(0,3)$. Since $c^2 = a^2 - b^2$ we have $c^2 = 9 - 1 = 8$, so $c = \sqrt{8}$, or $2\sqrt{2}$, and the foci are $(0,-2\sqrt{2})$ and $(0,2\sqrt{2})$.

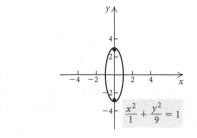

$$\frac{x^2}{1} + \frac{y^2}{9} = 1$$

21. $\dfrac{(x-2)^2}{16} + \dfrac{(y+1)^2}{4} = 1$

$\dfrac{(x-2)^2}{4^2} + \dfrac{[y-(-1)]^2}{2^2} = 1$

The center is $(2,-1)$. Note that $a = 4$ and $b = 2$. The major axis is horizontal, so the vertices are 4 units to the left and right of the center:

$(2-4,-1)$ and $(2+4,-1)$, or $(-2,-1)$ and $(6,-1)$.

We know that $c^2 = a^2 - b^2$ so $c^2 = 16 - 4 = 12$ and $c = \sqrt{12}$, or $2\sqrt{3}$. Then the foci are $2\sqrt{3}$ units to the left and right of the center:

$(2-2\sqrt{3},-1)$ and $(2+2\sqrt{3},-1)$.

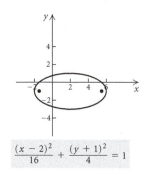

$$\frac{(x-2)^2}{16} + \frac{(y+1)^2}{4} = 1$$

23. The vertices, $(-5,0)$ and $(5,0)$ are on the x-axis, so the major axis is horizontal and $a = 5$. Since the vertices are equidistant from the origin, the center of the ellipse is the origin. The foci are $(-2, 0)$ and $(2, 0)$, so we know that $c = 2$. Then we have

$$c^2 = a^2 - b^2$$
$$4 = 25 - b^2$$
$$b^2 = 21.$$

Now we write the equation of the ellipse.

$$\frac{x^2}{25} + \frac{y^2}{21} = 1$$

25. The foci, $(-3, 0)$ and $(3, 0)$ are on the x-axis, so the major axis is horizontal. The foci are equidistant from the origin, so the center of the ellipse is the origin. The length of the major axis is 8, so $a = 8/2$, or 4. Now we find b^2.

$$c^2 = a^2 - b^2$$
$$9 = 16 - b^2$$
$$b^2 = 7$$

We write the equation of the ellipse.

$$\frac{x^2}{16} + \frac{y^2}{7} = 1$$

27. See page 479 of the text.

29. No, the center of an ellipse is not part of the graph of the ellipse. Its coordinates do not satisfy the equation of the ellipse.

Exercise Set 7.3

1. Graph (b) is the graph of $\dfrac{x^2}{25} - \dfrac{y^2}{9} = 1$.

3. Graph (c) is the graph of $\dfrac{(y-1)^2}{16} - \dfrac{(x+3)^2}{1} = 1$.

5. Graph (a) is the graph of $25x^2 - 16y^2 = 400$.

7. The vertices are equidistant from the origin and are on the y-axis, so the center is at the origin and the transverse axis is vertical. Since $c^2 = a^2 + b^2$, we have $5^2 = 3^2 + b^2$ so $b^2 = 16$.

The equation is of the form $\dfrac{y^2}{a^2} - \dfrac{x^2}{b^2} = 1$, so we have $\dfrac{y^2}{9} - \dfrac{x^2}{16} = 1$.

9. The asymptotes pass through the origin, so the center is the origin. The given vertex is on the x-axis, so the transverse axis is horizontal. Since $\dfrac{b}{a}x = \dfrac{3}{2}x$ and $a = 2$, we have $b = 3$. The equation is of the form $\dfrac{x^2}{a^2} - \dfrac{y^2}{b^2} = 1$, so we have $\dfrac{x^2}{2^2} - \dfrac{y^2}{3^2} = 1$, or $\dfrac{x^2}{4} - \dfrac{y^2}{9} = 1$.

11. $\dfrac{x^2}{4} - \dfrac{y^2}{4} = 1$

$\dfrac{x^2}{2^2} - \dfrac{y^2}{2^2} = 1$ Standard form

The center is $(0, 0)$; $a = 2$ and $b = 2$. The transverse axis is horizontal so the vertices are $(-2, 0)$ and $(2, 0)$. Since

$c^2 = a^2 + b^2$, we have $c^2 = 4 + 4 = 8$ and $c = \sqrt{8}$, or $2\sqrt{2}$. Then the foci are $(-2\sqrt{2}, 0)$ and $(2\sqrt{2}, 0)$.

Find the asymptotes:

$$y = \frac{b}{a}x \quad \text{and} \quad y = -\frac{b}{a}x$$
$$y = \frac{2}{2}x \quad \text{and} \quad y = -\frac{2}{2}x$$
$$y = x \quad \text{and} \quad y = -x$$

To draw the graph sketch the asymptotes, plot the vertices, and draw the branches of the hyperbola outward from the vertices toward the asymptotes.

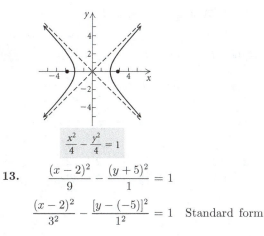

$$\boxed{\dfrac{x^2}{4} - \dfrac{y^2}{4} = 1}$$

13. $\dfrac{(x-2)^2}{9} - \dfrac{(y+5)^2}{1} = 1$

$\dfrac{(x-2)^2}{3^2} - \dfrac{[y-(-5)]^2}{1^2} = 1$ Standard form

The center is $(2, -5)$; $a = 3$ and $b = 1$. The transverse axis is horizontal, so the vertices are 3 units left and right of the center:

$(2 - 3, -5)$ and $(2 + 3, -5)$, or $(-1, -5)$ and $(5, -5)$.

Since $c^2 = a^2 + b^2$, we have $c^2 = 9 + 1 = 10$ and $c = \sqrt{10}$. Then the foci are $\sqrt{10}$ units left and right of the center:

$$(2 - \sqrt{10}, -5) \text{ and } (2 + \sqrt{10}, -5).$$

Find the asymptotes:

$$y - k = \frac{b}{a}(x - h) \quad \text{and} \quad y - k = -\frac{b}{a}(x - h)$$
$$y - (-5) = \frac{1}{3}(x - 2) \quad \text{and} \quad y - (-5) = -\frac{1}{3}(x - 2)$$
$$y + 5 = \frac{1}{3}(x - 2) \quad \text{and} \quad y + 5 = -\frac{1}{3}(x - 2), \text{ or}$$
$$y = \frac{1}{3}x - \frac{17}{3} \quad \text{and} \quad y = -\frac{1}{3}x - \frac{13}{3}$$

Sketch the asymptotes, plot the vertices, and draw the graph.

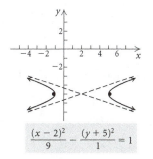

$$\boxed{\dfrac{(x-2)^2}{9} - \dfrac{(y+5)^2}{1} = 1}$$

15.
$$\frac{(y+3)^2}{4} - \frac{(x+1)^2}{16} = 1$$

$$\frac{[y-(-3)]^2}{2^2} - \frac{[x-(-1)]^2}{4^2} = 1 \quad \text{Standard form}$$

The center is $(-1,-3)$; $a = 2$ and $b = 4$. The transverse axis is vertical, so the vertices are 2 units below and above the center:

$(-1, -3-2)$ and $(1, -3+2)$, or $(-1,-5)$ and $(-1,-1)$.

Since $c^2 = a^2 + b^2$, we have $c^2 = 4 + 16 = 20$ and $c = \sqrt{20}$, or $2\sqrt{5}$. Then the foci are $2\sqrt{5}$ units below and above of the center:

$$(-1, -3-2\sqrt{5}) \text{ and } (-1, -3+2\sqrt{5}).$$

Find the asymptotes:

$$y - k = \frac{a}{b}(x-h) \quad \text{and} \quad y-k = -\frac{a}{b}(x-h)$$

$$y-(-3) = \frac{2}{4}(x-(-1)) \text{ and } y-(-3) = -\frac{2}{4}(x-(-1))$$

$$y+3 = \frac{1}{2}(x+1) \quad \text{and} \quad y+3 = -\frac{1}{2}(x+1), \text{ or}$$

$$y = \frac{1}{2}x - \frac{5}{2} \quad \text{and} \quad y = -\frac{1}{2}x - \frac{7}{2}$$

Sketch the asymptotes, plot the vertices, and draw the graph.

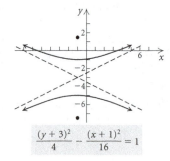

$$\frac{(y+3)^2}{4} - \frac{(x+1)^2}{16} = 1$$

17. $x^2 - 4y^2 = 4$

$$\frac{x^2}{4} - \frac{y^2}{1} = 1$$

$$\frac{x^2}{2^2} - \frac{y^2}{1^2} = 1 \quad \text{Standard form}$$

The center is $(0,0)$; $a = 2$ and $b = 1$. The transverse axis is horizontal, so the vertices are $(-2,0)$ and $(2,0)$. Since $c^2 = a^2 + b^2$, we have $c^2 = 4 + 1 = 5$ and $c = \sqrt{5}$. Then the foci are $(-\sqrt{5},0)$ and $(\sqrt{5},0)$.

Find the asymptotes:

$$y = \frac{b}{a}x \quad \text{and} \quad y = -\frac{b}{a}x$$

$$y = \frac{1}{2}x \quad \text{and} \quad y = -\frac{1}{2}x$$

Sketch the asymptotes, plot the vertices, and draw the graph.

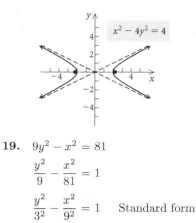

19. $9y^2 - x^2 = 81$

$$\frac{y^2}{9} - \frac{x^2}{81} = 1$$

$$\frac{y^2}{3^2} - \frac{x^2}{9^2} = 1 \quad \text{Standard form}$$

The center is $(0,0)$; $a = 3$ and $b = 9$. The transverse axis is vertical, so the vertices are $(0,-3)$ and $(0,3)$. Since $c^2 = a^2 + b^2$, we have $c^2 = 9 + 81 = 90$ and $c = \sqrt{90}$, or $3\sqrt{10}$. Then the foci are $(0, -3\sqrt{10})$ and $(0, 3\sqrt{10})$.

Find the asymptotes:

$$y = \frac{a}{b}x \quad \text{and} \quad y = -\frac{a}{b}x$$

$$y = \frac{3}{9}x \quad \text{and} \quad y = -\frac{3}{9}x$$

$$y = \frac{1}{3}x \quad \text{and} \quad y = -\frac{1}{3}x$$

Sketch the asymptotes, plot the vertices, and draw the graph.

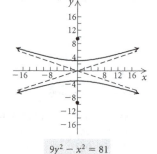

$$9y^2 - x^2 = 81$$

21.
$$x^2 - y^2 = 2$$

$$\frac{x^2}{2} - \frac{y^2}{2} = 1$$

$$\frac{x^2}{(\sqrt{2})^2} - \frac{y^2}{(\sqrt{2})^2} = 1 \quad \text{Standard form}$$

The center is $(0,0)$; $a = \sqrt{2}$ and $b = \sqrt{2}$. The transverse axis is horizontal, so the vertices are $(-\sqrt{2},0)$ and $(\sqrt{2},0)$. Since $c^2 = a^2 + b^2$, we have $c^2 = 2 + 2 = 4$ and $c = 2$. Then the foci are $(-2,0)$ and $(2,0)$.

Find the asymptotes:

$$y = \frac{b}{a}x \quad \text{and} \quad y = -\frac{b}{a}x$$

$$y = \frac{\sqrt{2}}{\sqrt{2}}x \quad \text{and} \quad y = -\frac{\sqrt{2}}{\sqrt{2}}x$$

$$y = x \qquad \text{and} \quad y = -x$$

Sketch the asymptotes, plot the vertices, and draw the graph.

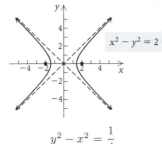

23. $y^2 - x^2 = \dfrac{1}{4}$

$$\dfrac{y^2}{1/4} - \dfrac{x^2}{1/4} = 1$$

$$\dfrac{y^2}{(1/2)^2} - \dfrac{x^2}{(1/2)^2} = 1 \quad \text{Standard form}$$

The center is $(0,0)$; $a = \dfrac{1}{2}$ and $b = \dfrac{1}{2}$. The transverse axis is vertical, so the vertices are $\left(0, -\dfrac{1}{2}\right)$ and $\left(0, \dfrac{1}{2}\right)$. Since $c^2 = a^2 + b^2$, we have $c^2 = \dfrac{1}{4} + \dfrac{1}{4} = \dfrac{1}{2}$ and $c = \sqrt{\dfrac{1}{2}}$, or $\dfrac{\sqrt{2}}{2}$. Then the foci are $\left(0, -\dfrac{\sqrt{2}}{2}\right)$ and $\left(0, \dfrac{\sqrt{2}}{2}\right)$.

Find the asymptotes:

$$y = \dfrac{a}{b}x \quad \text{and} \quad y = -\dfrac{a}{b}x$$

$$y = \dfrac{1/2}{1/2}x \quad \text{and} \quad y = -\dfrac{1/2}{1/2}x$$

$$y = x \quad \text{and} \quad y = -x$$

Sketch the asymptotes, plot the vertices, and draw the graph.

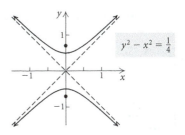

25. Begin by completing the square twice.

$$x^2 - y^2 - 2x - 4y - 4 = 0$$

$$(x^2 - 2x) - (y^2 + 4y) = 4$$

$$(x^2 - 2x + 1) - (y^2 + 4y + 4) = 4 + 1 - 1 \cdot 4$$

$$(x - 1)^2 - (y + 2)^2 = 1$$

$$\dfrac{(x - 1)^2}{1^2} - \dfrac{[y - (-2)]^2}{1^2} = 1 \quad \text{Standard form}$$

The center is $(1, -2)$; $a = 1$ and $b = 1$. The transverse axis is horizontal, so the vertices are 1 unit left and right of the center:

$(1 - 1, -2)$ and $(1 + 1, -2)$ or $(0, -2)$ and $(2, -2)$

Since $c^2 = a^2 + b^2$, we have $c^2 = 1 + 1 = 2$ and $c = \sqrt{2}$. Then the foci are $\sqrt{2}$ units left and right of the center:

$(1 - \sqrt{2}, -2)$ and $(1 + \sqrt{2}, -2)$.

Find the asymptotes:

$$y - k = \dfrac{b}{a}(x - h) \quad \text{and} \quad y - k = -\dfrac{b}{a}(x - h)$$

$$y - (-2) = \dfrac{1}{1}(x - 1) \quad \text{and} \quad y - (-2) = -\dfrac{1}{1}(x - 1)$$

$$y + 2 = x - 1 \quad \text{and} \quad y + 2 = -(x - 1), \text{ or}$$

$$y = x - 3 \quad \text{and} \quad y = -x - 1$$

Sketch the asymptotes, plot the vertices, and draw the graph.

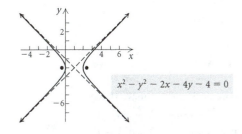

27. Begin by completing the square twice.

$$36x^2 - y^2 - 24x + 6y - 41 = 0$$

$$(36x^2 - 24x) - (y^2 - 6y) = 41$$

$$36\left(x^2 - \dfrac{2}{3}x\right) - (y^2 - 6y) = 41$$

$$36\left(x^2 - \dfrac{2}{3}x + \dfrac{1}{9}\right) - (y^2 - 6y + 9) = 41 + 36 \cdot \dfrac{1}{9} - 1 \cdot 9$$

$$36\left(x - \dfrac{1}{3}\right)^2 - (y - 3)^2 = 36$$

$$\dfrac{\left(x - \dfrac{1}{3}\right)^2}{1} - \dfrac{(y - 3)^2}{36} = 1$$

$$\dfrac{\left(x - \dfrac{1}{3}\right)^2}{1^2} - \dfrac{(y - 3)^2}{6^2} = 1 \quad \text{Standard form}$$

The center is $\left(\dfrac{1}{3}, 3\right)$; $a = 1$ and $b = 6$. The transverse axis is horizontal, so the vertices are 1 unit left and right of the center:

$\left(\dfrac{1}{3} - 1, 3\right)$ and $\left(\dfrac{1}{3} + 1, 3\right)$ or $\left(-\dfrac{2}{3}, 3\right)$ and $\left(\dfrac{4}{3}, 3\right)$.

Since $c^2 = a^2 + b^2$, we have $c^2 = 1 + 36 = 37$ and $c = \sqrt{37}$. Then the foci are $\sqrt{37}$ units left and right of the center:

$\left(\dfrac{1}{3} - \sqrt{37}, 3\right)$ and $\left(\dfrac{1}{3} + \sqrt{37}, 3\right)$.

Find the asymptotes:

$$y - k = \frac{b}{a}(x - h) \quad \text{and} \quad y - k = -\frac{b}{a}(x - h)$$

$$y - 3 = \frac{6}{1}\left(x - \frac{1}{3}\right) \quad \text{and} \quad y - 3 = -\frac{6}{1}\left(x - \frac{1}{3}\right)$$

$$y - 3 = 6\left(x - \frac{1}{3}\right) \quad \text{and} \quad y - 3 = -6\left(x - \frac{1}{3}\right), \text{ or}$$

$$y = 6x + 1 \quad \text{and} \quad y = -6x + 5$$

Sketch the asymptotes, plot the vertices, and draw the graph.

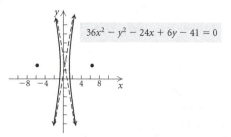

29. Begin by completing the square twice.

$$9y^2 - 4x^2 - 18y + 24x - 63 = 0$$

$$9(y^2 - 2y) - 4(x^2 - 6x) = 63$$

$$9(y^2 - 2y + 1) - 4(x^2 - 6x + 9) = 63 + 9\cdot1 - 4\cdot9$$

$$9(y - 1)^2 - 4(x - 3)^2 = 36$$

$$\frac{(y - 1)^2}{4} - \frac{(x - 3)^2}{9} = 1$$

$$\frac{(y - 1)^2}{2^2} - \frac{(x - 3)^2}{3^2} = 1 \quad \text{Standard form}$$

The center is $(3, 1)$; $a = 2$ and $b = 3$. The transverse axis is vertical, so the vertices are 2 units below and above the center:

$(3, 1 - 2)$ and $(3, 1 + 2)$, or $(3, -1)$ and $(3, 3)$.

Since $c^2 = a^2 + b^2$, we have $c^2 = 4 + 9 = 13$ and $c = \sqrt{13}$. Then the foci are $\sqrt{13}$ units below and above the center:

$$(3, 1 - \sqrt{13}) \text{ and } (3, 1 + \sqrt{13}).$$

Find the asymptotes:

$$y - k = \frac{a}{b}(x - h) \text{ and } y - k = -\frac{a}{b}(x - h)$$

$$y - 1 = \frac{2}{3}(x - 3) \text{ and } y - 1 = -\frac{2}{3}(x - 3), \text{ or}$$

$$y = \frac{2}{3}x - 1 \quad \text{and} \quad y = -\frac{2}{3}x + 3$$

Sketch the asymptotes, plot the vertices, and draw the graph.

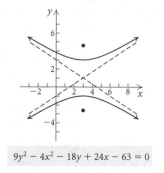

31. Begin by completing the square twice.

$$x^2 - y^2 - 2x - 4y = 4$$

$$(x^2 - 2x + 1) - (y^2 + 4y + 4) = 4 + 1 - 4$$

$$(x - 1)^2 - (y + 2)^2 = 1$$

$$\frac{(x - 1)^2}{1^2} - \frac{[y - (-2)]^2}{1^2} = 1 \quad \text{Standard form}$$

The center is $(1, -2)$; $a = 1$ and $b = 1$. The transverse axis is horizontal, so the vertices are 1 unit left and right of the center:

$(1 - 1, -2)$ and $(1 + 1, -2)$, or $(0, -2)$ and $(2, -2)$.

Since $c^2 = a^2 + b^2$, we have $c^2 = 1 + 1 = 2$ and $c = \sqrt{2}$. Then the foci are $\sqrt{2}$ units left and right of the center:

$$(1 - \sqrt{2}, -2) \text{ and } (1 + \sqrt{2}, -2).$$

Find the asymptotes:

$$y - k = \frac{b}{a}(x - h) \text{ and } \quad y - k = -\frac{b}{a}(x - h)$$

$$y - (-2) = \frac{1}{1}(x - 1) \text{ and } y - (-2) = -\frac{1}{1}(x - 1)$$

$$y + 2 = x - 1 \quad \text{and} \quad y + 2 = -(x - 1), \text{ or}$$

$$y = x - 3 \quad \text{and} \quad y = -x - 1$$

Sketch the asymptotes, plot the vertices, and draw the graph.

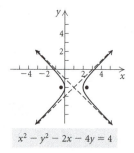

33. Begin by completing the square twice.

$$y^2 - x^2 - 6x - 8y - 29 = 0$$

$$(y^2 - 8y + 16) - (x^2 + 6x + 9) = 29 + 16 - 9$$

$$(y - 4)^2 - (x + 3)^2 = 36$$

$$\frac{(y - 4)^2}{36} - \frac{(x + 3)^2}{36} = 1$$

$$\frac{(y - 4)^2}{6^2} - \frac{[x - (-3)]^2}{6^2} = 1 \quad \text{Standard form}$$

The center is $(-3, 4)$; $a = 6$ and $b = 6$. The transverse axis is vertical, so the vertices are 6 units below and above the center:

$(-3, 4 - 6)$ and $(-3, 4 + 6)$, or $(-3, -2)$ and $(-3, 10)$.

Since $c^2 = a^2 + b^2$, we have $c^2 = 36 + 36 = 72$ and $c = \sqrt{72}$, or $6\sqrt{2}$. Then the foci are $6\sqrt{2}$ units below and above the center:

$(-3, 4 - 6\sqrt{2})$ and $(-3, 4 + 6\sqrt{2})$.

Find the asymptotes:

$y - k = \dfrac{a}{b}(x - h)$ and $y - k = -\dfrac{a}{b}(x - h)$

$y - 4 = \dfrac{6}{6}(x - (-3))$ and $y - 4 = -\dfrac{6}{6}(x - (-3))$

$y - 4 = x + 3$ and $y - 4 = -(x + 3)$, or

$y = x + 7$ and $y = -x + 1$

Sketch the asymptotes, plot the vertices, and draw the graph.

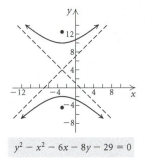

$$y^2 - x^2 - 6x - 8y - 29 = 0$$

35. The hyperbola in Example 3 is wider than the one in Example 2, so the hyperbola in Example 3 has the larger eccentricity.

Compute the eccentricities: In Example 2, $c = 5$ and $a = 4$, so $e = 5/4$, or 1.25. In Example 3, $c = \sqrt{5}$ and $a = 1$, so $e = \sqrt{5}/1 \approx 2.24$. These computations confirm that the hyperbola in Example 3 has the larger eccentricity.

37. The center is the midpoint of the segment connecting the vertices:

$\left(\dfrac{3 - 3}{2}, \dfrac{7 + 7}{2} \right)$, or $(0, 7)$.

The vertices are on the horizontal line $y = 7$, so the transverse axis is horizontal. Since the vertices are 3 units left and right of the center, $a = 3$.

Find c:

$e = \dfrac{c}{a} = \dfrac{5}{3}$

$\dfrac{c}{3} = \dfrac{5}{3}$ Substituting 3 for a

$c = 5$

Now find b^2:

$c^2 = a^2 + b^2$

$5^2 = 3^2 + b^2$

$16 = b^2$

Write the equation:

$\dfrac{(x - h)^2}{a^2} - \dfrac{(y - k)^2}{b^2} = 1$

$\dfrac{x^2}{9} - \dfrac{(y - 7)^2}{16} = 1$

39.

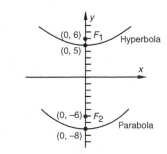

One focus is 6 units above the center of the hyperbola, so $c = 6$. One vertex is 5 units above the center, so $a = 5$. Find b^2:

$c^2 = a^2 + b^2$

$6^2 = 5^2 + b^2$

$11 = b^2$

Write the equation:

$\dfrac{y^2}{a^2} - \dfrac{x^2}{b^2} = 1$

$\dfrac{y^2}{25} - \dfrac{x^2}{11} = 1$

41. a) The graph of $f(x) = 2x - 3$ is shown below.

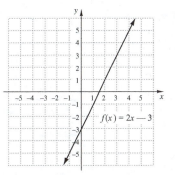

Since there is no horizontal line that crosses the graph more than once, the function is one-to-one.

b) Replace $f(x)$ with y: $y = 2x - 3$

Interchange x and y: $x = 2y - 3$

Solve for y: $x + 3 = 2y$

$\dfrac{x + 3}{2} = y$

Replace y with $f^{-1}(x)$: $f^{-1}(x) = \dfrac{x + 3}{2}$

43. a) The graph of $f(x) = \dfrac{5}{x-1}$ is shown below.

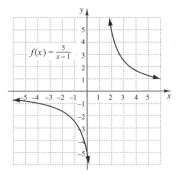

$f(x) = \dfrac{5}{x-1}$

Since there is no horizontal line that crosses the graph more than once, the function is one-to-one.

b) Replace $f(x)$ with y: $y = \dfrac{5}{x-1}$

Interchange x and y: $x = \dfrac{5}{y-1}$

Solve for y: $x(y-1) = 5$

$$y - 1 = \dfrac{5}{x}$$

$$y = \dfrac{5}{x} + 1$$

Replace y with $f^{-1}(x)$: $f^{-1} = \dfrac{5}{x} + 1$, or $\dfrac{5+x}{x}$

45. $x + y = 5$,   (1)

$\dfrac{x - y = 7}{2x \qquad = 12}$ (2)

$2x = 12$ Adding

$x = 6$

Back-substitute in either equation (1) or (2) and solve for y. We use equation (1).

$$6 + y = 5$$

$$y = -1$$

The solution is $(6, -1)$.

47. $2x - 3y = 7$,   (1)

$3x + 5y = 1$ (2)

Multiply equation (1) by 5 and equation (2) by 3 and add to eliminate y.

$10x - 15y = 35$

$\dfrac{9x + 15y = 3}{19x \qquad = 38}$

$19x = 38$

$x = 2$

Back-substitute and solve for y.

$3 \cdot 2 + 5y = 1$ Using equation (2)

$5y = -5$

$y = -1$

The solution is $(2, -1)$.

49. The center is the midpoint of the segment connecting $(3, -8)$ and $(3, -2)$:

$$\left(\dfrac{3+3}{2}, \dfrac{-8-2}{2} \right), \text{ or } (3, -5).$$

The vertices are on the vertical line $x = 3$ and are 3 units above and below the center so the transverse axis is vertical and $a = 3$. Use the equation of an asymptote to find b:

$$y - k = \dfrac{a}{b}(x - h)$$

$$y + 5 = \dfrac{3}{b}(x - 3)$$

$$y = \dfrac{3}{b}x - \dfrac{9}{b} - 5$$

This equation corresponds to the asymptote $y = 3x - 14$, so $\dfrac{3}{b} = 3$ and $b = 1$.

Write the equation of the hyperbola:

$$\dfrac{(y-k)^2}{a^2} - \dfrac{(x-h)^2}{b^2} = 1$$

$$\dfrac{(y+5)^2}{9} - \dfrac{(x-3)^2}{1} = 1$$

51. S and T are the foci of the hyperbola, so $c = 300/2 = 150$.

$200 \text{ microseconds} \cdot \dfrac{0.186 \text{ mi}}{1 \text{ microsecond}} = 37.2 \text{ mi}$, the difference of the ships' distances from the foci. That is, $2a = 37.2$, so $a = 18.6$.

Find b^2:

$$c^2 = a^2 + b^2$$

$$150^2 = 18.6^2 + b^2$$

$$22,154.04 = b^2$$

Then the equation of the hyperbola is

$$\dfrac{x^2}{18.6^2} - \dfrac{y^2}{22,154.04} = 1, \text{ or } \dfrac{x^2}{345.96} - \dfrac{y^2}{22,154.04} = 1.$$

Exercise Set 7.4

1. The correct graph is (e).

3. The correct graph is (c).

5. The correct graph is (b).

7. $x^2 + y^2 = 25$,   (1)

$y - x = 1$ (2)

First solve equation (2) for y.

$y = x + 1$ (3)

Then substitute $x + 1$ for y in equation (1) and solve for x.

$$x^2 + y^2 = 25$$

$$x^2 + (x+1)^2 = 25$$

$$x^2 + x^2 + 2x + 1 = 25$$

$$2x^2 + 2x - 24 = 0$$

$$x^2 + x - 12 = 0 \quad \text{Multiplying by } \dfrac{1}{2}$$

$$(x+4)(x-3) = 0 \quad \text{Factoring}$$

$x + 4 = 0 \quad \text{or} \quad x - 3 = 0$ \quad Principle of zero products

$x = -4 \quad \text{or} \qquad x = 3$

Now substitute these numbers into equation (3) and solve for y.

$$y = -4 + 1 = -3$$
$$y = 3 + 1 = 4$$

The pairs $(-4, -3)$ and $(3, 4)$ check, so they are the solutions.

9. $4x^2 + 9y^2 = 36$,   (1)

    $3y + 2x = 6$ (2)

First solve equation (2) for y.

$$3y = -2x + 6$$
$$y = -\frac{2}{3}x + 2 \qquad (3)$$

Then substitute $-\frac{2}{3}x + 2$ for y in equation (1) and solve for x.

$$4x^2 + 9y^2 = 36$$
$$4x^2 + 9\left(-\frac{2}{3}x + 2\right)^2 = 36$$
$$4x^2 + 9\left(\frac{4}{9}x^2 - \frac{8}{3}x + 4\right) = 36$$
$$4x^2 + 4x^2 - 24x + 36 = 36$$
$$8x^2 - 24x = 0$$
$$x^2 - 3x = 0$$
$$x(x - 3) = 0$$

$x = 0$ or $x = 3$

Now substitute these numbers in equation (3) and solve for y.

$$y = -\frac{2}{3} \cdot 0 + 2 = 2$$
$$y = -\frac{2}{3} \cdot 3 + 2 = 0$$

The pairs $(0, 2)$ and $(3, 0)$ check, so they are the solutions.

11. $x^2 + y^2 = 25$,   (1)

    $y^2 = x + 5$ (2)

We substitute $x + 5$ for y^2 in equation (1) and solve for x.

$$x^2 + y^2 = 25$$
$$x^2 + (x + 5) = 25$$
$$x^2 + x - 20 = 0$$
$$(x + 5)(x - 4) = 0$$

$x + 5 = 0$ or $x - 4 = 0$

    $x = -5$ or $x = 4$

We substitute these numbers for x in either equation (1) or equation (2) and solve for y. Here we use equation (2).

$y^2 = -5 + 5 = 0$ and $y = 0$.

$y^2 = 4 + 5 = 9$ and $y = \pm 3$.

The pairs $(-5, 0)$, $(4, 3)$ and $(4, -3)$ check. They are the solutions.

13. $x^2 + y^2 = 9$,   (1)

    $x^2 - y^2 = 9$ (2)

Here we use the elimination method.

$$
\begin{array}{lll}
x^2 + y^2 = & 9 & \quad (1) \\
\underline{x^2 - y^2 = \;\; 9} & & \quad (2) \\
2x^2 \qquad\;\; = & 18 & \quad \text{Adding} \\
x^2 = & 9 & \\
x = & \pm 3 &
\end{array}
$$

If $x = 3$, $x^2 = 9$, and if $x = -3$, $x^2 = 9$, so substituting 3 or -3 in equation (1) gives us

$$x^2 + y^2 = 9$$
$$9 + y^2 = 9$$
$$y^2 = 0$$
$$y = 0.$$

The pairs $(3, 0)$ and $(-3, 0)$ check. They are the solutions.

15. $y^2 - x^2 = 9$ (1)

    $2x - 3 = y$ (2)

Substitute $2x - 3$ for y in equation (1) and solve for x.

$$y^2 - x^2 = 9$$
$$(2x - 3)^2 - x^2 = 9$$
$$4x^2 - 12x + 9 - x^2 = 9$$
$$3x^2 - 12x = 0$$
$$x^2 - 4x = 0$$
$$x(x - 4) = 0$$

$x = 0$ or $x = 4$

Now substitute these numbers into equation (2) and solve for y.

If $x = 0$, $y = 2 \cdot 0 - 3 = -3$.

If $x = 4$, $y = 2 \cdot 4 - 3 = 5$.

The pairs $(0, -3)$ and $(4, 5)$ check. They are the solutions.

17. $y^2 = x + 3$,   (1)

    $2y = x + 4$ (2)

First solve equation (2) for x.

    $2y - 4 = x$ (3)

Then substitute $2y - 4$ for x in equation (1) and solve for y.

$$y^2 = x + 3$$
$$y^2 = (2y - 4) + 3$$
$$y^2 = 2y - 1$$
$$y^2 - 2y + 1 = 0$$
$$(y - 1)(y - 1) = 0$$

$y - 1 = 0$ or $y - 1 = 0$

    $y = 1$ or $y = 1$

Now substitute 1 for y in equation (3) and solve for x.

$$2 \cdot 1 - 4 = x$$
$$-2 = x$$

The pair $(-2, 1)$ checks. It is the solution.

19. $x^2 + y^2 = 25$,   (1)

    $xy = 12$ (2)

First we solve equation (2) for y.

$$xy = 12$$

$$y = \frac{12}{x}$$

Then we substitute $\dfrac{12}{x}$ for y in equation (1) and solve for x.

$$x^2 + y^2 = 25$$

$$x^2 + \left(\frac{12}{x}\right)^2 = 25$$

$$x^2 + \frac{144}{x^2} = 25$$

$$x^4 + 144 = 25x^2 \quad \text{Multiplying by } x^2$$

$$x^4 - 25x^2 + 144 = 0$$

$$u^2 - 25u + 144 = 0 \qquad \text{Letting } u = x^2$$

$$(u - 9)(u - 16) = 0$$

$$u = 9 \quad \text{or} \quad u = 16$$

We now substitute x^2 for u and solve for x.

$$x^2 = 9 \quad \text{or} \quad x^2 = 16$$

$$x = \pm 3 \quad \text{or} \quad x = \pm 4$$

Since $y = 12/x$, if $x = 3$, $y = 4$; if $x = -3$, $y = -4$; if $x = 4$, $y = 3$; and if $x = -4$, $y = -3$. The pairs $(3, 4)$, $(-3, -4)$, $(4, 3)$, and $(-4, -3)$ check. They are the solutions.

21. $x^2 + y^2 = 4$,   (1)

    $16x^2 + 9y^2 = 144$ (2)

$$\begin{aligned} -9x^2 - 9y^2 &= -36 \quad \text{Multiplying (1) by } -9 \\ \underline{16x^2 + 9y^2} &= \underline{144} \\ 7x^2 &= 108 \quad \text{Adding} \end{aligned}$$

$$x^2 = \frac{108}{7}$$

$$x = \pm\sqrt{\frac{108}{7}} = \pm 6\sqrt{\frac{3}{7}}$$

$$x = \pm\frac{6\sqrt{21}}{7} \quad \begin{array}{l}\text{Rationalizing the de-}\\ \text{nominator}\end{array}$$

Substituting $\dfrac{6\sqrt{21}}{7}$ or $-\dfrac{6\sqrt{21}}{7}$ for x in equation (1) gives us

$$\frac{36 \cdot 21}{49} + y^2 = 4$$

$$y^2 = 4 - \frac{108}{7}$$

$$y^2 = -\frac{80}{7}$$

$$y = \pm\sqrt{-\frac{80}{7}} = \pm 4i\sqrt{\frac{5}{7}}$$

$$y = \pm\frac{4i\sqrt{35}}{7}. \quad \begin{array}{l}\text{Rationalizing the}\\ \text{denominator}\end{array}$$

The pairs $\left(\dfrac{6\sqrt{21}}{7}, \dfrac{4i\sqrt{35}}{7}\right)$,

$\left(\dfrac{6\sqrt{21}}{7}, -\dfrac{4i\sqrt{35}}{7}\right)$, $\left(-\dfrac{6\sqrt{21}}{7}, \dfrac{4i\sqrt{35}}{7}\right)$, and

$\left(-\dfrac{6\sqrt{21}}{7}, -\dfrac{4i\sqrt{35}}{7}\right)$ check. They are the solutions.

23. $x^2 + 4y^2 = 25$,   (1)

    $x + 2y = 7$ (2)

First solve equation (2) for x.

$$x = -2y + 7 \qquad (3)$$

Then substitute $-2y + 7$ for x in equation (1) and solve for y.

$$x^2 + 4y^2 = 25$$

$$(-2y + 7)^2 + 4y^2 = 25$$

$$4y^2 - 28y + 49 + 4y^2 = 25$$

$$8y^2 - 28y + 24 = 0$$

$$2y^2 - 7y + 6 = 0$$

$$(2y - 3)(y - 2) = 0$$

$$y = \frac{3}{2} \quad \text{or} \quad y = 2$$

Now substitute these numbers in equation (3) and solve for x.

$$x = -2 \cdot \frac{3}{2} + 7 = 4$$

$$x = -2 \cdot 2 + 7 = 3$$

The pairs $\left(4, \dfrac{3}{2}\right)$ and $(3, 2)$ check, so they are the solutions.

25. $x^2 - xy + 3y^2 = 27$,   (1)

    $x - y = 2$ (2)

First solve equation (2) for y.

$$x - 2 = y \qquad (3)$$

Then substitute $x - 2$ for y in equation (1) and solve for x.

$$x^2 - xy + 3y^2 = 27$$

$$x^2 - x(x - 2) + 3(x - 2)^2 = 27$$

$$x^2 - x^2 + 2x + 3x^2 - 12x + 12 = 27$$

$$3x^2 - 10x - 15 = 0$$

$$x = \frac{-(-10) \pm \sqrt{(-10)^2 - 4(3)(-15)}}{2 \cdot 3}$$

$$x = \frac{10 \pm \sqrt{100 + 180}}{6} = \frac{10 \pm \sqrt{280}}{6}$$

$$x = \frac{10 \pm 2\sqrt{70}}{6} = \frac{5 \pm \sqrt{70}}{3}$$

Now substitute these numbers in equation (3) and solve for y.

$$y = \frac{5 + \sqrt{70}}{3} - 2 = \frac{-1 + \sqrt{70}}{3}$$

$$y = \frac{5 - \sqrt{70}}{3} - 2 = \frac{-1 - \sqrt{70}}{3}$$

The pairs $\left(\dfrac{5 + \sqrt{70}}{3}, \dfrac{-1 + \sqrt{70}}{3}\right)$ and

$\left(\dfrac{5 - \sqrt{70}}{3}, \dfrac{-1 - \sqrt{70}}{3}\right)$ check, so they are the solutions.

27. $x^2 + y^2 = 16,$ $x^2 + y^2 = 16,$ (1)

           or

$y^2 - 2x^2 = 10$ $-2x^2 + y^2 = 10$ (2)

Here we use the elimination method.

$$\begin{aligned} 2x^2 + 2y^2 &= 32 \quad \text{Multiplying (1) by 2} \\ -2x^2 + y^2 &= 10 \\ \hline 3y^2 &= 42 \quad \text{Adding} \\ y^2 &= 14 \\ y &= \pm\sqrt{14} \end{aligned}$$

Substituting $\sqrt{14}$ or $-\sqrt{14}$ for y in equation (1) gives us

$$x^2 + 14 = 16$$
$$x^2 = 2$$
$$x = \pm\sqrt{2}$$

The pairs $(-\sqrt{2}, -\sqrt{14})$, $(-\sqrt{2}, \sqrt{14})$, $(\sqrt{2}, -\sqrt{14})$, and $(\sqrt{2}, \sqrt{14})$ check. They are the solutions.

29. $x^2 + y^2 = 5,$ (1)

    $xy = 2$ (2)

First we solve equation (2) for y.

$$xy = 2$$
$$y = \frac{2}{x}$$

Then we substitute $\dfrac{2}{x}$ for y in equation (1) and solve for x.

$$x^2 + y^2 = 5$$
$$x^2 + \left(\frac{2}{x}\right)^2 = 5$$
$$x^2 + \frac{4}{x^2} = 5$$
$$x^4 + 4 = 5x^2 \quad \text{Multiplying by } x^2$$
$$x^4 - 5x^2 + 4 = 0$$
$$u^2 - 5u + 4 = 0 \quad \text{Letting } u = x^2$$
$$(u - 4)(u - 1) = 0$$
$$u = 4 \quad \text{or} \quad u = 1$$

We now substitute x^2 for u and solve for x.

$$x^2 = 4 \quad \text{or} \quad x^2 = 1$$
$$x = \pm 2 \qquad x = \pm 1$$

Since $y = 2/x$, if $x = 2$, $y = 1$; if $x = -2$, $y = -1$; if $x = 1$, $y = 2$; and if $x = -1$, $y = -2$. The pairs $(2, 1)$, $(-2, -1)$, $(1, 2)$, and $(-1, -2)$ check. They are the solutions.

31. $3x + y = 7$ (1)

    $4x^2 + 5y = 56$ (2)

First solve equation (1) for y.

$$3x + y = 7$$
$$y = 7 - 3x \quad (3)$$

Next substitute $7 - 3x$ for y in equation (2) and solve for x.

$$4x^2 + 5y = 56$$
$$4x^2 + 5(7 - 3x) = 56$$
$$4x^2 + 35 - 15x = 56$$
$$4x^2 - 15x - 21 = 0$$

Using the quadratic formula, we find that

$$x = \frac{15 - \sqrt{561}}{8} \quad \text{or} \quad x = \frac{15 + \sqrt{561}}{8}.$$

Now substitute these numbers into equation (3) and solve for y.

If $x = \dfrac{15 - \sqrt{561}}{8}$, $y = 7 - 3\left(\dfrac{15 - \sqrt{561}}{8}\right)$, or

$$\frac{11 + 3\sqrt{561}}{8}.$$

If $x = \dfrac{15 + \sqrt{561}}{8}$, $y = 7 - 3\left(\dfrac{15 + \sqrt{561}}{8}\right)$, or

$$\frac{11 - 3\sqrt{561}}{8}.$$

The pairs $\left(\dfrac{15 - \sqrt{561}}{8}, \dfrac{11 + 3\sqrt{561}}{8}\right)$ and

$\left(\dfrac{15 + \sqrt{561}}{8}, \dfrac{11 - 3\sqrt{561}}{8}\right)$ check and are the solutions.

33. $a + b = 7,$ (1)

    $ab = 4$ (2)

First solve equation (1) for a.

$$a = -b + 7 \quad (3)$$

Then substitute $-b + 7$ for a in equation (2) and solve for b.

$$(-b + 7)b = 4$$
$$-b^2 + 7b = 4$$
$$0 = b^2 - 7b + 4$$
$$b = \frac{-(-7) \pm \sqrt{(-7)^2 - 4 \cdot 1 \cdot 4}}{2 \cdot 1}$$
$$b = \frac{7 \pm \sqrt{33}}{2}$$

Now substitute these numbers in equation (3) and solve for a.

$$a = -\left(\frac{7 + \sqrt{33}}{2}\right) + 7 = \frac{7 - \sqrt{33}}{2}$$
$$a = -\left(\frac{7 - \sqrt{33}}{2}\right) + 7 = \frac{7 + \sqrt{33}}{2}$$

The pairs $\left(\dfrac{7 - \sqrt{33}}{2}, \dfrac{7 + \sqrt{33}}{2}\right)$ and

$\left(\dfrac{7 + \sqrt{33}}{2}, \dfrac{7 - \sqrt{33}}{2}\right)$ check, so they are the solutions.

35. $x^2 + y^2 = 13,$ (1)

    $xy = 6$ (2)

First we solve Equation (2) for y.

$$xy = 6$$
$$y = \frac{6}{x}$$

Then we substitute $\dfrac{6}{x}$ for y in equation (1) and solve for x.

$$x^2 + y^2 = 13$$

$$x^2 + \left(\frac{6}{x}\right)^2 = 13$$

$$x^2 + \frac{36}{x^2} = 13$$

$$x^4 + 36 = 13x^2 \qquad \text{Multiplying by } x^2$$

$$x^4 - 13x^2 + 36 = 0$$

$$u^2 - 13u + 36 = 0 \qquad \text{Letting } u = x^2$$

$$(u - 9)(u - 4) = 0$$

$$u = 9 \quad \text{or} \quad u = 4$$

We now substitute x^2 for u and solve for x.

$$x^2 = 9 \quad \text{or} \quad x^2 = 4$$

$$x = \pm 3 \quad \text{or} \quad x = \pm 2$$

Since $y = 6/x$, if $x = 3$, $y = 2$; if $x = -3$, $y = -2$; if $x = 2$, $y = 3$; and if $x = -2$, $y = -3$. The pairs $(3, 2)$, $(-3, -2)$, $(2, 3)$, and $(-2, -3)$ check. They are the solutions.

37. $\quad x^2 + y^2 + 6y + 5 = 0 \qquad (1)$

$\quad\quad x^2 + y^2 - 2x - 8 = 0 \qquad (2)$

Using the elimination method, multiply equation (2) by -1 and add the result to equation (1).

$$\begin{array}{ll} x^2 + y^2 + 6y + 5 = 0 & (1) \\ -x^2 - y^2 + 2x + 8 = 0 & (2) \\ \hline 2x + 6y + 13 = 0 & (3) \end{array}$$

Solve equation (3) for x.

$$2x + 6y + 13 = 0$$

$$2x = -6y - 13$$

$$x = \frac{-6y - 13}{2}$$

Substitute $\dfrac{-6y - 13}{2}$ for x in equation (1) and solve for y.

$$x^2 + y^2 + 6y + 5 = 0$$

$$\left(\frac{-6y - 13}{2}\right)^2 + y^2 + 6y + 5 = 0$$

$$\frac{36y^2 + 156y + 169}{4} + y^2 + 6y + 5 = 0$$

$$36y^2 + 156y + 169 + 4y^2 + 24y + 20 = 0$$

$$40y^2 + 180y + 189 = 0$$

Using the quadratic formula, we find that

$y = \dfrac{-45 \pm 3\sqrt{15}}{20}$. Substitute $\dfrac{-45 \pm 3\sqrt{15}}{20}$ for y in

$x = \dfrac{-6y - 13}{2}$ and solve for x.

If $y = \dfrac{-45 + 3\sqrt{15}}{20}$, then

$$x = \frac{-6\left(\dfrac{-45 + 3\sqrt{15}}{20}\right) - 13}{2} = \frac{5 - 9\sqrt{15}}{20}.$$

If $y = \dfrac{-45 - 3\sqrt{15}}{20}$, then

$$x = \frac{-6\left(\dfrac{-45 - 3\sqrt{15}}{20}\right) - 13}{2} = \frac{5 + 9\sqrt{15}}{20}.$$

The pairs $\left(\dfrac{5 + 9\sqrt{15}}{20}, \dfrac{-45 - 3\sqrt{15}}{20}\right)$ and

$\left(\dfrac{5 - 9\sqrt{15}}{20}, \dfrac{-45 + 3\sqrt{15}}{20}\right)$ check and are the solutions.

39. $\quad 2a + b = 1, \qquad (1)$

$\quad\quad b = 4 - a^2 \qquad (2)$

Equation (2) is already solved for b. Substitute $4 - a^2$ for b in equation (1) and solve for a.

$$2a + 4 - a^2 = 1$$

$$0 = a^2 - 2a - 3$$

$$0 = (a - 3)(a + 1)$$

$$a = 3 \quad \text{or} \quad a = -1$$

Substitute these numbers in equation (2) and solve for b.

$$b = 4 - 3^2 = -5$$

$$b = 4 - (-1)^2 = 3$$

The pairs $(3, -5)$ and $(-1, 3)$ check. They are the solutions.

41. $\quad a^2 + b^2 = 89, \qquad (1)$

$\quad\quad a - b = 3 \qquad (2)$

First solve equation (2) for a.

$$a = b + 3 \qquad (3)$$

Then substitute $b + 3$ for a in equation (1) and solve for b.

$$(b + 3)^2 + b^2 = 89$$

$$b^2 + 6b + 9 + b^2 = 89$$

$$2b^2 + 6b - 80 = 0$$

$$b^2 + 3b - 40 = 0$$

$$(b + 8)(b - 5) = 0$$

$$b = -8 \text{ or } b = 5$$

Substitute these numbers in equation (3) and solve for a.

$$a = -8 + 3 = -5$$

$$a = 5 + 3 = 8$$

The pairs $(-5, -8)$ and $(8, 5)$ check. They are the solutions.

43. $\quad xy - y^2 = 2, \qquad (1)$

$\quad\quad 2xy - 3y^2 = 0 \qquad (2)$

$$\begin{array}{ll} -2xy + 2y^2 = -4 & \text{Multiplying (1) by } -2 \\ 2xy - 3y^2 = 0 & \\ \hline -y^2 = -4 & \text{Adding} \\ y^2 = 4 & \\ y = \pm 2 & \end{array}$$

We substitute for y in equation (1) and solve for x.

When $y = 2$: $\quad x \cdot 2 - 2^2 = 2$

$$2x - 4 = 2$$

$$2x = 6$$

$$x = 3$$

When $y = -2$: $\quad x(-2) - (-2)^2 = 2$

$$-2x - 4 = 2$$
$$-2x = 6$$
$$x = -3$$

The pairs $(3, 2)$ and $(-3, -2)$ check. They are the solutions.

45. $\quad m^2 - 3mn + n^2 + 1 = 0, \quad (1)$

$\quad 3m^2 - mn + 3n^2 \quad\quad = 13 \quad (2)$

$\quad m^2 - 3mn + n^2 = -1 \quad (3) \quad$ Rewriting (1)

$\quad 3m^2 - mn + 3n^2 = 13 \quad (2)$

$\quad -3m^2 + 9mn - 3n^2 = 3 \quad$ Multiplying (3) by -3

$\quad \underline{3m^2 - mn + 3n^2 = 13}$

$$8mn = 16$$
$$mn = 2$$
$$n = \frac{2}{m} \quad (4)$$

Substitute $\dfrac{2}{m}$ for n in equation (1) and solve for m.

$$m^2 - 3m\left(\frac{2}{m}\right) + \left(\frac{2}{m}\right)^2 + 1 = 0$$

$$m^2 - 6 + \frac{4}{m^2} + 1 = 0$$

$$m^2 - 5 + \frac{4}{m^2} = 0$$

$$m^4 - 5m^2 + 4 = 0 \quad \text{Multiplying} \atop \text{by } m^2$$

Substitute u for m^2.

$$u^2 - 5u + 4 = 0$$
$$(u - 4)(u - 1) = 0$$
$$u = 4 \quad or \quad u = 1$$
$$m^2 = 4 \quad or \quad m^2 = 1$$
$$m = \pm 2 \quad or \quad m = \pm 1$$

Substitute for m in equation (4) and solve for n.

When $m = 2$, $n = \dfrac{2}{2} = 1$.

When $m = -2$, $n = \dfrac{2}{-2} = -1$.

When $m = 1$, $n = \dfrac{2}{1} = 2$.

When $m = -1$, $n = \dfrac{2}{-1} = -2$.

The pairs $(2, 1)$, $(-2, -1)$, $(1, 2)$, and $(-1, -2)$ check. They are the solutions.

47. $\quad x^2 + y^2 = 5, \quad (1)$

$\quad x - y = 8 \quad\quad (2)$

First solve equation (2) for x.

$x = y + 8 \quad (3)$

Then substitute $y + 8$ for x in equation (1) and solve for y.

$$(y + 8)^2 + y^2 = 5$$
$$y^2 + 16y + 64 + y^2 = 5$$
$$2y^2 + 16y + 59 = 0$$

$$y = \frac{-16 \pm \sqrt{(16)^2 - 4(2)(59)}}{2 \cdot 2}$$

$$y = \frac{-16 \pm \sqrt{-216}}{4}$$

$$y = \frac{-16 \pm 6i\sqrt{6}}{4}$$

$$y = -4 \pm \frac{3}{2}i\sqrt{6}$$

Now substitute these numbers in equation (3) and solve for x.

$$x = -4 + \frac{3}{2}i\sqrt{6} + 8 = 4 + \frac{3}{2}i\sqrt{6}$$

$$x = -4 - \frac{3}{2}i\sqrt{6} + 8 = 4 - \frac{3}{2}i\sqrt{6}$$

The pairs $\left(4 + \dfrac{3}{2}i\sqrt{6}, -4 + \dfrac{3}{2}i\sqrt{6}\right)$ and

$\left(4 - \dfrac{3}{2}i\sqrt{6}, -4 - \dfrac{3}{2}i\sqrt{6}\right)$ check. They are the solutions.

49. $\quad a^2 + b^2 = 14, \quad (1)$

$\quad ab = 3\sqrt{5} \quad\quad (2)$

Solve equation (2) for b.

$$b = \frac{3\sqrt{5}}{a}$$

Substitute $\dfrac{3\sqrt{5}}{a}$ for b in equation (1) and solve for a.

$$a^2 + \left(\frac{3\sqrt{5}}{a}\right)^2 = 14$$

$$a^2 + \frac{45}{a^2} = 14$$

$$a^4 + 45 = 14a^2$$

$$a^4 - 14a^2 + 45 = 0$$

$$u^2 - 14u + 45 = 0 \quad\quad \text{Letting } u = a^2$$

$$(u - 9)(u - 5) = 0$$

$$u = 9 \quad or \quad u = 5$$

$$a^2 = 9 \quad or \quad a^2 = 5$$

$$a = \pm 3 \quad or \quad a = \pm\sqrt{5}$$

Since $b = 3\sqrt{5}/a$, if $a = 3$, $b = \sqrt{5}$; if $a = -3$, $b = -\sqrt{5}$; if $a = \sqrt{5}$, $b = 3$; and if $a = -\sqrt{5}$, $b = -3$. The pairs $(3, \sqrt{5})$, $(-3, -\sqrt{5})$, $(\sqrt{5}, 3)$, $(-\sqrt{5}, -3)$ check. They are the solutions.

51. $\quad x^2 + y^2 = 25, \quad\quad (1)$

$\quad 9x^2 + 4y^2 = 36 \quad (2)$

$-4x^2 - 4y^2 = -100$ Multiplying (1) by -4

$$\frac{9x^2 + 4y^2 = 36}{5x^2 \qquad = -64}$$

$$x^2 = -\frac{64}{5}$$

$$x = \pm\sqrt{\frac{-64}{5}} = \pm\frac{8i}{\sqrt{5}}$$

$$x = \pm\frac{8i\sqrt{5}}{5} \qquad \text{Rationalizing the}$$
denominator

Substituting $\frac{8i\sqrt{5}}{5}$ or $-\frac{8i\sqrt{5}}{5}$ for x in equation (1) and solving for y gives us

$$-\frac{64}{5} + y^2 = 25$$

$$y^2 = \frac{189}{5}$$

$$y = \pm\sqrt{\frac{189}{5}} = \pm3\sqrt{\frac{21}{5}}$$

$$y = \pm\frac{3\sqrt{105}}{5}. \qquad \text{Rationalizing the}$$
denominator

The pairs $\left(\dfrac{8i\sqrt{5}}{5}, \dfrac{3\sqrt{105}}{5}\right)$, $\left(-\dfrac{8i\sqrt{5}}{5}, \dfrac{3\sqrt{105}}{5}\right)$,

$\left(\dfrac{8i\sqrt{5}}{5}, -\dfrac{3\sqrt{105}}{5}\right)$, and $\left(-\dfrac{8i\sqrt{5}}{5}, -\dfrac{3\sqrt{105}}{5}\right)$
check.

They are the solutions.

53. $5y^2 - x^2 = 1,$ (1)

$xy = 2$ (2)

Solve equation (2) for x.

$$x = \frac{2}{y}$$

Substitute $\dfrac{2}{y}$ for x in equation (1) and solve for y.

$$5y^2 - \left(\frac{2}{y}\right)^2 = 1$$

$$5y^2 - \frac{4}{y^2} = 1$$

$$5y^4 - 4 = y^2$$

$$5y^4 - y^2 - 4 = 0$$

$$5u^2 - u - 4 = 0 \qquad \text{Letting } u = y^2$$

$$(5u + 4)(u - 1) = 0$$

$5u + 4 = 0 \qquad\qquad or \quad u - 1 = 0$

$u = -\dfrac{4}{5} \qquad or \qquad u = 1$

$y^2 = -\dfrac{4}{5} \qquad or \qquad y^2 = 1$

$y = \pm\dfrac{2i}{\sqrt{5}} \qquad or \qquad y = \pm 1$

$y = \pm\dfrac{2i\sqrt{5}}{5} \qquad or \qquad y = \pm 1$

Since $x = 2/y$, if $y = \dfrac{2i\sqrt{5}}{5}$, $x = \dfrac{2}{\dfrac{2i\sqrt{5}}{5}} = \dfrac{5}{i\sqrt{5}} =$

$\dfrac{5}{i\sqrt{5}} \cdot \dfrac{-i\sqrt{5}}{-i\sqrt{5}} = -i\sqrt{5}$; if $y = -\dfrac{2i\sqrt{5}}{5}$,

$x = \dfrac{2}{-\dfrac{2i\sqrt{5}}{5}} = i\sqrt{5}$;

if $y = 1$, $x = 2/1 = 2$; if $y = -1$, $x = 2/-1 = -2$.

The pairs $\left(-i\sqrt{5}, \dfrac{2i\sqrt{5}}{5}\right)$, $\left(i\sqrt{5}, -\dfrac{2i\sqrt{5}}{5}\right)$, $(2, 1)$ and $(-2, -1)$ check. They are the solutions.

55. The statement is true. See Example 4, for instance.

57. The statement is true because a line and a circle can intersect in at most two points.

59. *Familiarize.* We first make a drawing. We let l and w represent the length and width, respectively.

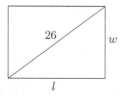

Translate. The perimeter is 68 in.

$2l + 2w = 68$, or $l + w = 34$

Using the Pythagorean theorem we have another equation.

$l^2 + w^2 = 26^2$, or $l^2 + w^2 = 676$

Carry out. We solve the system:

$l + w = 34,$ (1)

$l^2 + w^2 = 676$ (2)

First solve equation (1) for w.

$w = 34 - l$ (3)

Then substitute $34 - l$ for w in equation (2) and solve for l.

$$l^2 + w^2 = 676$$

$$l^2 + (34 - l)^2 = 676$$

$$l^2 + 1156 - 68l + l^2 = 676$$

$$2l^2 - 68l + 480 = 0$$

$$l^2 - 34l + 240 = 0$$

$$(l - 10)(l - 24) = 0$$

$l = 10$ or $l = 24$

If $l = 10$, then $w = 34 - 10$, or 24. If $l = 24$, then $w = 34 - 24$, or 10. Since the length is usually considered to be longer than the width, we have the solution $l = 24$ and $w = 10$, or $(24, 10)$.

Check. If $l = 24$ and $w = 10$, then the perimeter is $2 \cdot 24 + 2 \cdot 10$, or 68. The length of a diagonal is $\sqrt{24^2 + 10^2}$, or $\sqrt{676}$, or 26. The numbers check.

State. The length is 24 in., and the width is 10 in.

61. Familiarize. We first make a drawing. Let l = the length and w = the width of the brochure.

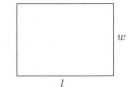

Translate.

Area: $lw = 20$

Perimeter: $2l + 2w = 18$, or $l + w = 9$

Carry out. We solve the system:

Solve the second equation for l: $l = 9 - w$

Substitute $9 - w$ for l in the first equation and solve for w.

$$(9 - w)w = 20$$
$$9w - w^2 = 20$$
$$0 = w^2 - 9w + 20$$
$$0 = (w - 5)(w - 4)$$
$$w = 5 \text{ or } w = 4$$

If $w = 5$, then $l = 9 - w$, or 4. If $w = 4$, then $l = 9 - 4$, or 5. Since length is usually considered to be longer than width, we have the solution $l = 5$ and $w = 4$, or $(5, 4)$.

Check. If $l = 5$ and $w = 4$, the area is $5 \cdot 4$, or 20. The perimeter is $2 \cdot 5 + 2 \cdot 4$, or 18. The numbers check.

State. The length of the brochure is 5 in. and the width is 4 in.

63. Familiarize. We make a drawing of the dog run. Let l = the length and w = the width.

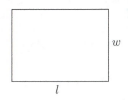

Since it takes 210 yd of fencing to enclose the run, we know that the perimeter is 210 yd.

Translate.

Perimeter: $2l + 2w = 210$, or $l + w = 105$

Area: $lw = 2250$

Carry out. We solve the system:

Solve the first equation for l: $l = 105 - w$

Substitute $105 - w$ for l in the second equation and solve for w.

$$(105 - w)w = 2250$$
$$105w - w^2 = 2250$$
$$0 = w^2 - 105w + 2250$$
$$0 = (w - 30)(w - 75)$$
$$w = 30 \text{ or } w = 75$$

If $w = 30$, then $l = 105 - 30$, or 75. If $w = 75$, then $l = 105 - 75$, or 30. Since length is usually considered to be longer than width, we have the solution $l = 75$ and $w = 30$, or $(75, 30)$.

Check. If $l = 75$ and $w = 30$, the perimeter is $2 \cdot 75 + 2 \cdot 30$, or 210. The area is $75(30)$, or 2250. The numbers check.

State. The length is 75 yd and the width is 30 yd.

65. Familiarize. We first make a drawing. Let l = the length and w = the width.

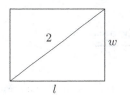

Translate.

Area: $lw = \sqrt{3}$ (1)

From the Pythagorean theorem: $l^2 + w^2 = 2^2$ (2)

Carry out. We solve the system of equations.

We first solve equation (1) for w.

$$lw = \sqrt{3}$$
$$w = \frac{\sqrt{3}}{l}$$

Then we substitute $\dfrac{\sqrt{3}}{l}$ for w in equation 2 and solve for l.

$$l^2 + \left(\frac{\sqrt{3}}{l}\right)^2 = 4$$
$$l^2 + \frac{3}{l^2} = 4$$
$$l^4 + 3 = 4l^2$$
$$l^4 - 4l^2 + 3 = 0$$
$$u^2 - 4u + 3 = 0 \quad \text{Letting } u = l^2$$
$$(u - 3)(u - 1) = 0$$
$$u = 3 \text{ or } u = 1$$

We now substitute l^2 for u and solve for l.

$$l^2 = 3 \qquad \text{or} \quad l^2 = 1$$
$$l = \pm\sqrt{3} \text{ or } \quad l = \pm 1$$

Measurements cannot be negative, so we only need to consider $l = \sqrt{3}$ and $l = 1$. Since $w = \sqrt{3}/l$, if $l = \sqrt{3}$, $w = 1$ and if $l = 1$, $w = \sqrt{3}$. Length is usually considered to be longer than width, so we have the solution $l = \sqrt{3}$ and $w = 1$, or $(\sqrt{3}, 1)$.

Check. If $l = \sqrt{3}$ and $w = 1$, the area is $\sqrt{3} \cdot 1 = \sqrt{3}$. Also $(\sqrt{3})^2 + 1^2 = 3 + 1 = 4 = 2^2$. The numbers check.

State. The length is $\sqrt{3}$ m, and the width is 1 m.

67. Familiarize. We let x = the length of a side of one test plot and y = the length of a side of the other plot. Make a drawing.

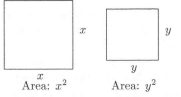

x

y

Area: x^2 Area: y^2

Translate.

The sum of the areas is 832 ft^2.

$\underbrace{\hspace{3cm}}$ $\underbrace{\hspace{0.5cm}}$ $\underbrace{\hspace{0.8cm}}$

$\downarrow$ $\downarrow$ $\downarrow$

$x^2 + y^2$ $=$ 832

The difference of the areas is 320 ft^2.

$\underbrace{\hspace{3cm}}$ $\underbrace{\hspace{0.5cm}}$ $\underbrace{\hspace{0.8cm}}$

$\downarrow$ $\downarrow$ $\downarrow$

$x^2 - y^2$ $=$ 320

Carry out. We solve the system of equations.

$$x^2 + y^2 = 832$$
$$\underline{x^2 - y^2 = 320}$$
$$2x^2 = 1152 \quad \text{Adding}$$
$$x^2 = 576$$
$$x = \pm 24$$

Since measurements cannot be negative, we consider only $x = 24$. Substitute 24 for x in the first equation and solve for y.

$$24^2 + y^2 = 832$$
$$576 + y^2 = 832$$
$$y^2 = 256$$
$$y = \pm 16$$

Again, we consider only the positive value, 16. The possible solution is $(24, 16)$.

Check. The areas of the test plots are 24^2, or 576, and 16^2, or 256. The sum of the areas is $576 + 256$, or 832. The difference of the areas is $576 - 256$, or 320. The values check.

State. The lengths of the test plots are 24 ft and 16 ft.

69. The correct graph is (b).

71. The correct graph is (d).

73. The correct graph is (a).

75. Graph: $x^2 + y^2 \leq 16$,

$y < x$

The solution set of $x^2 + y^2 \leq 16$ is the circle $x^2 + y^2 = 16$ and the region inside it. The solution set of $y < x$ is the half-plane below the line $y = x$. We shade the region common to the two solution sets.

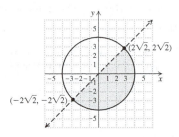

To find the points of intersection of the graphs we solve the system of equations

$$x^2 + y^2 = 16,$$
$$y = x.$$

The points of intersection are $(-2\sqrt{2}, -2\sqrt{2})$ and $(2\sqrt{2}, 2\sqrt{2})$.

77. Graph: $x^2 \leq y$,

$x + y \geq 2$

The solution set of $x^2 \leq y$ is the parabola $x^2 = y$ and the region inside it. The solution set of $x + y \geq 2$ is the line $x + y = 2$ and the half-plane above the line. We shade the region common to the two solution sets.

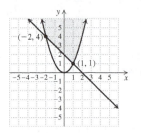

To find the points of intersection of the graphs we solve the system of equations

$$x^2 = y,$$
$$x + y = 2.$$

The points of intersection are $(-2, 4)$ and $(1, 1)$.

79. Graph: $x^2 + y^2 \leq 25$,

$x - y > 5$

The solution set of $x^2 + y^2 \leq 25$ is the circle $x^2 + y^2 = 25$ and the region inside it. The solution set of $x - y > 5$ is the half-plane below the line $x - y = 5$. We shade the region common to the two solution sets.

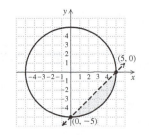

To find the points of intersection of the graphs we solve the system of equations

$$x^2 + y^2 = 25,$$
$$x - y = 5.$$

The points of intersection are $(0, -5)$ and $(5, 0)$.

81. Graph: $y \geq x^2 - 3$,

$y \leq 2x$

The solution set of $y \geq x^2 - 3$ is the parabola $y = x^2 - 3$ and the region inside it. The solution set of $y \leq 2x$ is the line $y = 2x$ and the half-plane below it. We shade the region common to the two solution sets.

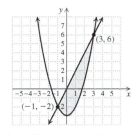

To find the points of intersection of the graphs we solve the system of equations

$$y = x^2 - 3,$$

$$y = 2x.$$

The points of intersection are $(-1, -2)$ and $(3, 6)$.

83. Graph: $y \geq x^2$,

$\qquad y < x + 2$

The solution set of $y \geq x^2$ is the parabola $y = x^2$ and the region inside it. The solution set of $y < x + 2$ is the half-plane below the line $y = x + 2$. We shade the region common to the two solution sets.

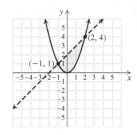

To find the points of intersection of the graphs we solve the system of equations

$$y = x^2,$$

$$y = x + 2.$$

The points of intersection are $(-1, 1)$ and $(2, 4)$.

85. $2^{3x} = 64$

$\qquad 2^{3x} = 2^6$

$\qquad 3x = 6$

$\qquad x = 2$

The solution is 2.

87. $\log_3 x = 4$

$\qquad x = 3^4$

$\qquad x = 81$

The solution is 81.

89. $(x - h)^2 + (y - k)^2 = r^2$

If $(2, 4)$ is a point on the circle, then

$(2 - h)^2 + (4 - k)^2 = r^2$.

If $(3, 3)$ is a point on the circle, then

$(3 - h)^2 + (3 - k)^2 = r^2$.

Thus

$$(2 - h)^2 + (4 - k)^2 = (3 - h)^2 + (3 - k)^2$$

$$4 - 4h + h^2 + 16 - 8k + k^2 =$$

$$9 - 6h + h^2 + 9 - 6k + k^2$$

$$-4h - 8k + 20 = -6h - 6k + 18$$

$$2h - 2k = -2$$

$$h - k = -1$$

If the center (h, k) is on the line $3x - y = 3$, then $3h - k = 3$. Solving the system

$\qquad h - k = -1,$

$\qquad 3h - k = 3$

we find that $(h, k) = (2, 3)$.

Find r^2, substituting $(2, 3)$ for (h, k) and $(2, 4)$ for (x, y). We could also use $(3, 3)$ for (x, y).

$$(x - h)^2 + (y - k)^2 = r^2$$

$$(2 - 2)^2 + (4 - 3)^2 = r^2$$

$$0 + 1 = r^2$$

$$1 = r^2$$

The equation of the circle is $(x - 2)^2 + (y - 3)^2 = 1$.

91. The equation of the ellipse is of the form $\dfrac{x^2}{a^2} + \dfrac{y^2}{b^2} = 1$. Substitute $\left(1, \dfrac{\sqrt{3}}{2}\right)$ and $\left(\sqrt{3}, \dfrac{1}{2}\right)$ for (x, y) to get two equations.

$$\frac{1^2}{a^2} + \frac{\left(\frac{\sqrt{3}}{2}\right)^2}{b^2} = 1, \ or \ \frac{1}{a^2} + \frac{3}{4b^2} = 1$$

$$\frac{(\sqrt{3})^2}{a^2} + \frac{\left(\frac{1}{2}\right)^2}{b^2} = 1, \ or \ \frac{3}{a^2} + \frac{1}{4b^2} = 1$$

Substitute u for $\dfrac{1}{a^2}$ and v for $\dfrac{1}{b^2}$.

$$u + \frac{3}{4}v = 1, \qquad\qquad 4u + 3v = 4,$$

$$or$$

$$3u + \frac{1}{4}v = 1 \qquad\qquad 12u + v = 4$$

Solving for u and v, we get $u = \dfrac{1}{4}$, $v = 1$. Then

$u = \dfrac{1}{a^2} = \dfrac{1}{4}$, so $a^2 = 4$; $v = \dfrac{1}{b^2} = 1$, so $b^2 = 1$.

Then the equation of the ellipse is

$$\frac{x^2}{4} + \frac{y^2}{1} = 1, \ or \ \frac{x^2}{4} + y^2 = 1.$$

93. See the answer section in the text.

95. See the answer section in the text.

97. $x^3 + y^3 = 72,$ $\quad$ (1)

$\qquad x + y = 6$ $\qquad$ (2)

Solve equation (2) for y: $y = 6 - x$

Substitute for y in equation (1) and solve for x.

$$x^3 + (6-x)^3 = 72$$
$$x^3 + 216 - 108x + 18x^2 - x^3 = 72$$
$$18x^2 - 108x + 144 = 0$$
$$x^2 - 6x + 8 = 0$$

Multiplying by $\frac{1}{18}$

$$(x-4)(x-2) = 0$$

$x = 4$ or $x = 2$

If $x = 4$, then $y = 6 - 4 = 2$.

If $x = 2$, then $y = 6 - 2 = 4$.

The pairs $(4, 2)$ and $(2, 4)$ check.

99. $p^2 + q^2 = 13$,   (1)

$\dfrac{1}{pq} = -\dfrac{1}{6}$ (2)

Solve equation (2) for p.

$$\frac{1}{q} = -\frac{p}{6}$$
$$-\frac{6}{q} = p$$

Substitute $-6/q$ for p in equation (1) and solve for q.

$$\left(-\frac{6}{q}\right)^2 + q^2 = 13$$
$$\frac{36}{q^2} + q^2 = 13$$
$$36 + q^4 = 13q^2$$
$$q^4 - 13q^2 + 36 = 0$$
$$u^2 - 13u + 36 = 0 \qquad \text{Letting } u = q^2$$
$$(u-9)(u-4) = 0$$
$$u = 9 \quad \text{or} \quad u = 4$$
$$x^2 = 9 \quad \text{or} \quad x^2 = 4$$
$$x = \pm 3 \text{ or} \quad x = \pm 2$$

Since $p = -6/q$, if $q = 3$, $p = -2$; if $q = -3$, $p = 2$; if $q = 2$, $p = -3$; and if $q = -2$, $p = 3$. The pairs $(-2, 3)$, $(2, -3)$, $(-3, 2)$, and $(3, -2)$ check. They are the solutions.

Chapter 7 Review Exercises

1. $x + y^2 = 1$

$$y^2 = x - 1$$
$$(y - 0)^2 = 4 \cdot \frac{1}{4}(x - 1)$$

This parabola has a horizontal axis of symmetry, the focus is $\left(\dfrac{5}{4}, 0\right)$, and the directrix is $x = \dfrac{3}{4}$. Thus it opens to the left and the statement is true.

3. The statement is true. See page 497 in the text.

5. The statement is false. See Example 4 on page 510 in the text.

7. Graph (a) is the graph of $y^2 = 9 - x^2$.

9. Graph (g) is the graph of $9y^2 - 4x^2 = 36$.

11. Graph (f) is the graph of $4x^2 + y^2 - 16x - 6y = 15$.

13. Graph (c) is the graph of $\dfrac{(x+3)^2}{16} - \dfrac{(y-1)^2}{25} = 1$.

15. $y^2 = -12x$

$$y^2 = 4(-3)x$$

F: $(-3, 0)$, V: $(0, 0)$, D: $x = 3$

17. Begin by completing the square twice.

$$16x^2 + 25y^2 - 64x + 50y - 311 = 0$$
$$16(x^2 - 4x) + 25(y^2 + 2y) = 311$$
$$16(x^2 - 4x + 4) + 25(y^2 + 2y + 1) = 311 + 16 \cdot 4 + 25 \cdot 1$$
$$16(x - 2)^2 + 25(y + 1)^2 = 400$$
$$\frac{(x-2)^2}{25} + \frac{[y - (-1)]^2}{16} = 1$$

The center is $(2, -1)$. Note that $a = 5$ and $b = 4$. The major axis is horizontal so the vertices are 5 units left and right of the center: $(2 - 5, -1)$ and $(2 + 5, -1)$, or $(-3, -1)$ and $(7, -1)$. We know that $c^2 = a^2 - b^2 = 25 - 16 = 9$ and $c = \sqrt{9} = 3$. Then the foci are 3 units left and right of the center: $(2 - 3, -1)$ and $(2 + 3, -1)$, or $(-1, -1)$ and $(5, -1)$.

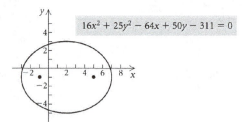

19. Begin by completing the square twice.

$$x^2 - 2y^2 + 4x + y - \frac{1}{8} = 0$$
$$(x^2 + 4x) - 2\left(y^2 - \frac{1}{2}y\right) = \frac{1}{8}$$
$$(x^2 + 4x + 4) - 2\left(y^2 - \frac{1}{2}y + \frac{1}{16}\right) = \frac{1}{8} + 4 - 2 \cdot \frac{1}{16}$$
$$(x + 2)^2 - 2\left(y - \frac{1}{4}\right)^2 = 4$$
$$\frac{[x - (-2)]^2}{4} - \frac{\left(y - \frac{1}{4}\right)^2}{2} = 1$$

The center is $\left(-2, \dfrac{1}{4}\right)$. The transverse axis is horizontal, so the vertices are 2 units left and right of the center: $\left(-2 - 2, \dfrac{1}{4}\right)$ and $\left(-2 + 2, \dfrac{1}{4}\right)$, or $\left(-4, \dfrac{1}{4}\right)$ and $\left(0, \dfrac{1}{4}\right)$. Since $c^2 = a^2 + b^2$, we have $c^2 = 4 + 2 = 6$ and $c = \sqrt{6}$. Then the foci are $\sqrt{6}$ units left and right of the center: $\left(-2 - \sqrt{6}, \dfrac{1}{4}\right)$ and $\left(-2 + \sqrt{6}, \dfrac{1}{4}\right)$.

Find the asymptotes:

$$y - k = \frac{b}{a}(x - h) \quad \text{and} \quad y - k = -\frac{b}{a}(x - h)$$

$$y - \frac{1}{4} = \frac{\sqrt{2}}{2}(x + 2) \quad \text{and} \quad y - \frac{1}{4} = -\frac{\sqrt{2}}{2}(x + 2)$$

21. $x^2 - 16y = 0, \quad (1)$

$x^2 - y^2 = 64 \quad (2)$

From equation (1) we have $x^2 = 16y$. Substitute in equation (2).

$$16y - y^2 = 64$$
$$0 = y^2 - 16y + 64$$
$$0 = (y - 8)^2$$
$$0 = y - 8$$
$$8 = y$$
$$x^2 - (8)^2 = 64 \qquad \text{Substituting in equation (2)}$$
$$x^2 = 128$$
$$x = \pm\sqrt{128} = \pm 8\sqrt{2}$$

The pairs $(-8\sqrt{2}, 8)$ and $(8\sqrt{2}, 8)$ check.

23. $x^2 - y^2 = 33, \qquad (1)$

$x + y = 11 \qquad (2)$

$$y = -x + 11$$
$$x^2 - (-x + 11)^2 = 33 \qquad \text{Substituting in (1)}$$
$$x^2 - (x^2 - 22x + 121) = 33$$
$$x^2 - x^2 + 22x - 121 = 33$$
$$22x = 154$$
$$x = 7$$

$y = -7 + 11 = 4$

The pair $(7, 4)$ checks.

25. $x^2 - y = 3, \quad (1)$

$2x - y = 3 \quad (2)$

From equation (1) we have $y = x^2 - 3$. Substitute in equation (2).

$$2x - (x^2 - 3) = 3$$
$$2x - x^2 + 3 = 3$$
$$0 = x^2 - 2x$$
$$0 = x(x - 2)$$

$x = 0 \quad or \quad x = 2$

$y = 0^2 - 3 = -3$

$y = 2^2 - 3 = 1$

The pairs $(0, -3)$ and $(2, 1)$ check.

27. $x^2 - y^2 = 3, \quad (1)$

$y = x^2 - 3 \quad (2)$

From equation (2) we have $x^2 = y + 3$. Substitute in equation (1).

$$y + 3 - y^2 = 3$$
$$0 = y^2 - y$$
$$0 = y(y - 1)$$

$y = 0 \quad or \quad y = 1$

$$x^2 = 0 + 3$$
$$x^2 = 3$$
$$x = \pm\sqrt{3}$$

$$x^2 = 1 + 3$$
$$x^2 = 4$$
$$x = \pm 2$$

The pairs $(\sqrt{3}, 0)$, $(-\sqrt{3}, 0)$, $(2, 1)$, and $(-2, 1)$ check.

29. $\quad x^2 + y^2 = 100, \quad (1)$

$2x^2 - 3y^2 = -120 \quad (2)$

$$\begin{array}{l} 3x^2 + 3y^2 = 300 \qquad \text{Multiplying (1) by 3} \\ \underline{2x^2 - 3y^2 = -120} \\ 5x^2 \qquad\quad = 180 \qquad \text{Adding} \\ \quad x^2 = 36 \\ \quad x = \pm 6 \end{array}$$

$$(\pm 6)^2 + y^2 = 100$$
$$y^2 = 64$$
$$y = \pm 8$$

The pairs $(6, 8)$, $(-6, 8)$, $(6, -8)$, and $(-6, -8)$ check.

31. *Familiarize*. Let x and y represent the numbers.

***Translate*.** The sum of the numbers is 11.

$$x + y = 11$$

The sum of the squares of the numbers is 65.

$$x^2 + y^2 = 65$$

***Carry out*.** We solve the system of equations.

$$x + y = 11, \quad (1)$$
$$x^2 + y^2 = 65 \quad (2)$$

First we solve equation (1) for y.

$$y = 11 - x$$

Then substitute $11 - x$ for y in equation (2) and solve for x.

$$x^2 + (11 - x)^2 = 65$$
$$x^2 + 121 - 22x + x^2 = 65$$
$$2x^2 - 22x + 121 = 65$$
$$2x^2 - 22x + 56 = 0$$
$$x^2 - 11x + 28 = 0 \qquad \text{Dividing by 2}$$
$$(x - 4)(x - 7) = 0$$
$$x - 4 = 0 \quad or \quad x - 7 = 0$$
$$x = 4 \quad or \qquad x = 7$$

If $x = 4$, then $y = 11 - 4 = 7$.

If $x = 7$, then $y = 11 - 7 = 4$.

In either case, the possible numbers are 4 and 7.

***Check*.** $4 + 7 = 11$ and $4^2 + 7^2 = 16 + 49 = 65$. The answer checks.

***State*.** The numbers are 4 and 7.

33. *Familiarize.* Let x and y represent the positive integers.

Translate. The sum of the numbers is 12.

$$x + y = 12$$

The sum of the reciprocals is $\dfrac{3}{8}$.

$$\frac{1}{x} + \frac{1}{y} = \frac{3}{8}$$

Carry out. We solve the system of equations.

$$x + y = 12, \quad (1)$$
$$\frac{1}{x} + \frac{1}{y} = \frac{3}{8} \quad (2)$$

First solve equation (1) for y.

$$y = 12 - x$$

Then substitute $12 - x$ for y in equation (2) and solve for x.

$$\frac{1}{x} + \frac{1}{12 - x} = \frac{3}{8}, \text{ LCD is } 8x(12 - x)$$

$$8x(12 - x)\left(\frac{1}{x} + \frac{1}{12 - x}\right) = 8x(12 - x) \cdot \frac{3}{8}$$

$$8(12 - x) + 8x = x(12 - x) \cdot 3$$

$$96 - 8x + 8x = 36x - 3x^2$$

$$96 = 36x - 3x^2$$

$$3x^2 - 36x + 96 = 0$$

$$x^2 - 12x + 32 = 0 \quad \text{Dividing by 3}$$

$$(x - 4)(x - 8) = 0$$

$$x - 4 = 0 \ \ or \ \ x - 8 = 0$$

$$x = 4 \ \ or \ \ \ \ \ \ x = 8$$

If $x = 4$, $y = 12 - 4 = 8$.

If $x = 8$, $y = 12 - 8 = 4$.

In either case, the possible numbers are 4 and 8.

Check. $4 + 8 = 12$; $\dfrac{1}{4} + \dfrac{1}{8} = \dfrac{2}{8} + \dfrac{1}{8} = \dfrac{3}{8}$. The answer checks.

State. The numbers are 4 and 8.

35. *Familiarize.* Let $x =$ the radius of the larger circle and let $y =$ the radius of the smaller circle. We will use the formula for the area of a circle, $A = \pi r^2$.

Translate. The sum of the areas is 130π ft^2.

$$\pi x^2 + \pi y^2 = 130\pi$$

The difference of the areas is 112π ft^2.

$$\pi x^2 - \pi y^2 = 112\pi$$

We have a system of equations.

$$\pi x^2 + \pi y^2 = 130\pi, \quad (1)$$
$$\pi x^2 - \pi y^2 = 112\pi \quad (2)$$

Carry out. We add.

$$\pi x^2 + \pi y^2 = 130\pi$$
$$\underline{\pi x^2 - \pi y^2 = 112\pi}$$
$$2\pi x^2 \qquad\quad = 242\pi$$
$$x^2 = 121 \quad \text{Dividing by } 2\pi$$
$$x = \pm 11$$

Since the length of a radius cannot be negative, we consider only $x = 11$. Substitute 11 for x in equation (1) and solve for y.

$$\pi \cdot 11^2 + \pi y^2 = 130\pi$$
$$121\pi + \pi y^2 = 130\pi$$
$$\pi y^2 = 9\pi$$
$$y^2 = 9$$
$$y = \pm 3$$

Again, we consider only the positive solution.

Check. If the radii are 11 ft and 3 ft, the sum of the areas is $\pi \cdot 11^2 + \pi \cdot 3^2 = 121\pi + 9\pi = 130\pi$ ft^2. The difference of the areas is $121\pi - 9\pi = 112\pi$ ft^2. The answer checks.

State. The radius of the larger circle is 11 ft, and the radius of the smaller circle is 3 ft.

37. Graph: $x^2 + y^2 \leq 16$,

$$x + y < 4$$

The solution set of $x^2 + y^2 \leq 16$ is the circle $x^2 + y^2 = 16$ and the region inside it. The solution set of $x + y < 4$ is the half-plane below the line $x + y = 4$. We shade the region common to the two solution sets.

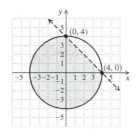

39. Graph: $x^2 + y^2 \leq 9$,

$$x \leq -1$$

The solution set of $x^2 + y^2 \leq 9$ is the circle $x^2 + y^2 = 9$ and the region inside it. The solution set of $x \leq -1$ is the line $x = -1$ and the half-plane to the left of it. We shade the region common to the two solution sets.

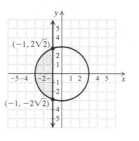

41. A straight line can intersect an ellipse at 0 points, 1 point, or 2 points but not at 4 points, so answer D is correct.

43. *Familiarize.* Let x and y represent the numbers.

Translate. The product of the numbers is 4.

$$xy = 4$$

The sum of the reciprocals is $\dfrac{65}{56}$.

$$\frac{1}{x} + \frac{1}{y} = \frac{65}{56}$$

Carry out. We solve the system of equations.

$$xy = 4, \qquad (1)$$

$$\frac{1}{x} + \frac{1}{y} = \frac{65}{56} \quad (2)$$

First solve equation (1) for y.

$$y = \frac{4}{x}$$

Then substitute $\frac{4}{x}$ for y in equation (2) and solve for x.

$$\frac{1}{x} + \frac{1}{4/x} = \frac{65}{56}$$

$$\frac{1}{x} + \frac{x}{4} = \frac{65}{56}$$

$$56x\left(\frac{1}{x} + \frac{x}{4}\right) = 56x \cdot \frac{65}{56}$$

$$56 + 14x^2 = 65x$$

$$14x^2 - 65x + 56 = 0$$

$$(2x - 7)(7x - 8) = 0$$

$$2x - 7 = 0 \quad or \quad 7x - 8 = 0$$

$$2x = 7 \quad or \qquad 7x = 8$$

$$x = \frac{7}{2} \quad or \qquad x = \frac{8}{7}$$

If $x = \frac{7}{2}$, $y = \frac{4}{7/2} = 4 \cdot \frac{2}{7} = \frac{8}{7}$.

If $x = \frac{8}{7}$, $y = \frac{4}{8/7} = 4 \cdot \frac{7}{8} = \frac{7}{2}$.

In either case the possible numbers are $\frac{7}{2}$ and $\frac{8}{7}$.

Check. $\frac{7}{2} \cdot \frac{8}{7} = 4$; $\frac{1}{7/2} + \frac{1}{8/7} = \frac{2}{7} + \frac{7}{8} = \frac{16}{56} + \frac{49}{56} = \frac{65}{56}$. The answer checks.

State. The numbers are $\frac{7}{2}$ and $\frac{8}{7}$.

45. The vertices are $(0, -3)$ and $(0, 3)$, so the center is $(0, 0)$, and the major axis is vertical.

$$\frac{x^2}{a^2} + \frac{y^2}{3^2} = 1$$

Substitute $\left(-\frac{1}{2}, \frac{3\sqrt{3}}{2}\right)$ and solve for a.

$$\frac{\left(-\frac{1}{2}\right)^2}{a^2} + \frac{\left(\frac{3\sqrt{3}}{2}\right)^2}{3^2} = 1$$

$$\frac{1}{4a^2} + \frac{3}{4} = 1$$

$$1 + 3a^2 = 4a^2 \quad \text{Multiplying by } 4a^2$$

$$1 = a^2$$

$$1 = a$$

The equation of the ellipse is $x^2 + \frac{y^2}{9} = 1$.

47. The equation of a circle can be written as

$$\frac{(x - h)^2}{a^2} + \frac{(y - k)^2}{b^2} = 1$$

where $a = b = r$, the radius of the circle. In an ellipse, $a > b$, so a circle is not a special type of ellipse.

49. Although we can always visualize the real-number solutions, we cannot visualize the imaginary-number solutions.

Chapter 7 Test

1. Graph (c) is the graph of $4x^2 - y^2 = 4$.

2. Graph (b) is the graph of $x^2 - 2x - 3y = 5$.

3. Graph (a) is the graph of $x^2 + 4x + y^2 - 2y - 4 = 0$.

4. Graph (d) is the graph of $9x^2 + 4y^2 = 36$.

5. $\quad x^2 = 12y$

$$x^2 = 4 \cdot 3y$$

$$V: \ (0, 0), \ F: \ (0, 3), \ D: \ y = -3$$

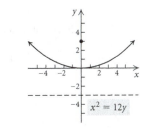

6. $\quad y^2 + 2y - 8x - 7 = 0$

$$y^2 + 2y = 8x + 7$$

$$y^2 + 2y + 1 = 8x + 7 + 1$$

$$(y + 1)^2 = 8x + 8$$

$$[y - (-1)]^2 = 4(2)[x - (-1)]$$

$$V: \ (-1, -1)$$

$$F: \ (-1 + 2, -1) \quad or \quad (1, -1)$$

$$D: \ x = -1 - 2 = -3$$

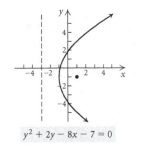

7. $\quad (x - h)^2 = 4p(y - k)$

$$(x - 0)^2 = 4 \cdot 2(y - 0)$$

$$x^2 = 8y$$

8. Begin by completing the square twice.

$$x^2 + y^2 + 2x - 6y - 15 = 0$$

$$x^2 + 2x + y^2 - 6y = 15$$

$$(x^2 + 2x + 1) + (y^2 - 6y + 9) = 15 + 1 + 9$$

$$(x + 1)^2 + (y - 3)^2 = 25$$

$$[x - (-1)]^2 + (y - 3)^2 = 5^2$$

Center: $(-1, 3)$, radius: 5

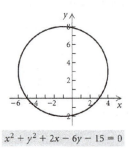

$$x^2 + y^2 + 2x - 6y - 15 = 0$$

9. $9x^2 + 16y^2 = 144$

$$\frac{x^2}{16} + \frac{y^2}{9} = 1$$

$$\frac{x^2}{4^2} + \frac{y^2}{3^2} = 1$$

$a = 4$, $b = 3$

The center is $(0, 0)$. The major axis is horizontal, so the vertices are $(-4, 0)$ and $(4, 0)$. Since $c^2 = a^2 - b^2$, we have $c^2 = 16 - 9 = 7$, so $c = \sqrt{7}$ and the foci are $(-\sqrt{7}, 0)$ and $(\sqrt{7}, 0)$.

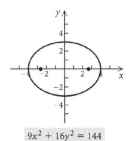

$$9x^2 + 16y^2 = 144$$

10. $\dfrac{(x+1)^2}{4} + \dfrac{(y-2)^2}{9} = 1$

$$\frac{[x - (-1)]^2}{2^2} + \frac{(y-2)^2}{3^2} = 1$$

The center is $(-1, 2)$. Note that $a = 3$ and $b = 2$. The major axis is vertical, so the vertices are 3 units below and above the center:

$(-1, 2 - 3)$ and $(-1, 2 + 3)$ or $(-1, -1)$ and $(-1, 5)$.

We know that $c^2 = a^2 - b^2$, so $c^2 = 9 - 4 = 5$ and $c = \sqrt{5}$. Then the foci are $\sqrt{5}$ units below and above the center:

$(-1, 2 - \sqrt{5})$ and $(-1, 2 + \sqrt{5})$.

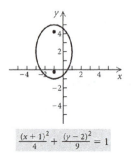

$$\frac{(x+1)^2}{4} + \frac{(y-2)^2}{9} = 1$$

11. The vertices $(0, -5)$ and $(0, 5)$ are on the y-axis, so the major axis is vertical and $a = 5$. Since the vertices are equidistant from the center, the center of the ellipse is at the origin. The length of the minor axis is 4, so $b = \dfrac{4}{2} = 2$.

The equation is $\dfrac{x^2}{4} + \dfrac{y^2}{25} = 1$.

12. $4x^2 - y^2 = 4$

$$\frac{x^2}{1} - \frac{y^2}{4} = 1$$

$$\frac{x^2}{1^2} - \frac{y^2}{2^2} = 1$$

The center is $(0, 0)$; $a = 1$ and $b = 2$.

The transverse axis is horizontal, so the vertices are $(-1, 0)$ and $(1, 0)$. Since $c^2 = a^2 + b^2$, we have $c^2 = 1 + 4 = 5$ and $c = \sqrt{5}$. Then the foci are $(-\sqrt{5}, 0)$ and $(\sqrt{5}, 0)$.

Find the asymptotes:

$$y = \frac{b}{a}x \quad \text{and} \quad y = -\frac{b}{a}x$$

$$y = \frac{2}{1}x \quad \text{and} \quad y = -\frac{2}{1}x$$

$$y = 2x \quad \text{and} \quad y = -2x$$

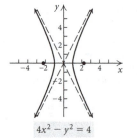

$$4x^2 - y^2 = 4$$

13. $\dfrac{(y-2)^2}{4} - \dfrac{(x+1)^2}{9} = 1$

$$\frac{(y-2)^2}{2^2} - \frac{[x - (-1)]^2}{3^2} = 1$$

The center is $(-1, 2)$; $a = 2$ and $b = 3$.

The transverse axis is vertical, so the vertices are 2 units below and above the center:

$(-1, 2 - 2)$ and $(-1, 2 + 2)$ or $(-1, 0)$ and $(-1, 4)$.

Since $c^2 = a^2 + b^2$, we have $c^2 = 4 + 9 = 13$ and $c = \sqrt{13}$. Then the foci are $\sqrt{13}$ units below and above the center:

$(-1, 2 - \sqrt{13})$ and $(-1, 2 + \sqrt{13})$.

Find the asymptotes:

$$y - k = \frac{a}{b}(x - h) \quad \text{and} \quad y - k = -\frac{a}{b}(x - h)$$

$$y - 2 = \frac{2}{3}(x - (-1)) \quad \text{and} \quad y - 2 = -\frac{2}{3}(x - (-1))$$

$$y - 2 = \frac{2}{3}(x + 1) \quad \text{and} \quad y - 2 = -\frac{2}{3}(x + 1)$$

$$y = \frac{2}{3}x + \frac{8}{3} \quad \text{and} \quad y = -\frac{2}{3}x + \frac{4}{3}$$

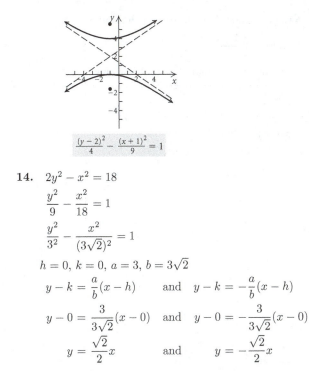

$$\frac{(y-2)^2}{4} - \frac{(x+1)^2}{9} = 1$$

14. $2y^2 - x^2 = 18$

$$\frac{y^2}{9} - \frac{x^2}{18} = 1$$

$$\frac{y^2}{3^2} - \frac{x^2}{(3\sqrt{2})^2} = 1$$

$h = 0$, $k = 0$, $a = 3$, $b = 3\sqrt{2}$

$y - k = \dfrac{a}{b}(x - h)$ and $y - k = -\dfrac{a}{b}(x - h)$

$y - 0 = \dfrac{3}{3\sqrt{2}}(x - 0)$ and $y - 0 = -\dfrac{3}{3\sqrt{2}}(x - 0)$

$y = \dfrac{\sqrt{2}}{2}x$ and $y = -\dfrac{\sqrt{2}}{2}x$

15. The parabola is of the form $y^2 = 4px$. A point on the parabola is $\left(6, \dfrac{18}{2}\right)$, or $(6, 9)$.

$$y^2 = 4px$$
$$9^2 = 4 \cdot p \cdot 6$$
$$81 = 24p$$
$$\frac{27}{8} = p$$

Since the focus is at $(p, 0) = \left(\dfrac{27}{8}, 0\right)$, the focus is $\dfrac{27}{8}$ in. from the vertex.

16. $2x^2 - 3y^2 = -10$,   (1)

$\quad\ \ x^2 + 2y^2 = 9$ (2)

$\quad\ \ 2x^2 - 3y^2 = -10$

$\underline{\ -2x^2 - 4y^2 = -18}\ $ Multiplying (2) by -2

$\qquad\ \ - 7y^2 = -28$ Adding

$\qquad\qquad y^2 = 4$

$\qquad\qquad\ y = \pm 2$

$x^2 + 2(\pm 2)^2 = 9$ Substituting into (2)

$\qquad x^2 + 8 = 9$

$\qquad\qquad x^2 = 1$

$\qquad\qquad\ x = \pm 1$

The pairs $(1, 2)$, $(1, -2)$, $(-1, 2)$ and $(-1, -2)$ check.

17. $x^2 + y^2 = 13$,   (1)

$\quad\ \ x + y = 1$ (2)

First solve equation (2) for y.

$\qquad y = 1 - x$

Then substitute $1 - x$ for y in equation (1) and solve for x.

$$x^2 + (1 - x)^2 = 13$$
$$x^2 + 1 - 2x + x^2 = 13$$
$$2x^2 - 2x - 12 = 0$$
$$2(x^2 - x - 6) = 0$$
$$2(x - 3)(x + 2) = 0$$
$$x = 3 \ \ or \ \ x = -2$$

If $x = 3$, $y = 1 - 3 = -2$. If $x = -2$, $y = 1 - (-2) = 3$.

The pairs $(3, -2)$ and $(-2, 3)$ check.

18. $x + y = 5$,   (1)

$\quad\ \ xy = 6$ (2)

First solve equation (1) for y.

$\qquad y = -x + 5$

Then substitute $-x + 5$ for y in equation (2) and solve for x.

$$x(-x + 5) = 6$$
$$-x^2 + 5x - 6 = 0$$
$$-1(x^2 - 5x + 6) = 0$$
$$-1(x - 2)(x - 3) = 0$$
$$x = 2 \ \ or \ \ x = 3$$

If $x = 2$, $y = -2 + 5 = 3$. If $x = 3$, $y = -3 + 5 = 2$.

The pairs $(2, 3)$ and $(3, 2)$ check.

19. *Familiarize*. Let l and w represent the length and width of the rectangle, in feet, respectively.

Translate. The perimeter is 18 ft.

$\qquad 2l + 2w = 18$ (1)

From the Pythagorean theorem, we have

$\qquad l^2 + w^2 = (\sqrt{41})^2$ (2)

Carry out. We solve the system of equations. We first solve equation (1) for w.

$$2l + 2w = 18$$
$$2w = 18 - 2l$$
$$w = 9 - l$$

Then substitute $9 - l$ for w in equation (2) and solve for l.

$$l^2 + (9 - l)^2 = (\sqrt{41})^2$$
$$l^2 + 81 - 18l + l^2 = 41$$
$$2l^2 - 18l + 40 = 0$$
$$2(l^2 - 9l + 20) = 0$$
$$2(l - 4)(l - 5) = 0$$
$$l = 4 \ \ or \ \ l = 5$$

If $l = 4$, then $w = 9 - 4 = 5$. If $l = 5$, then $w = 9 - 5 = 4$. Since length is usually considered to be longer than width, we have $l = 5$ and $w = 4$.

Check. The perimeter is $2 \cdot 5 + 2 \cdot 4$, or 18 ft. The length of a diagonal is $\sqrt{5^2 + 4^2}$, or $\sqrt{41}$ ft. The solution checks.

State. The dimensions of the garden are 5 ft by 4 ft.

20. *Familiarize*. Let l and w represent the length and width of the playground, in feet, respectively.

Translate.

Perimeter: $2l + 2w = 210$ (1)

Area: $lw = 2700$ (2)

Carry out. We solve the system of equations. First solve equation (2) for w.

$$w = \frac{2700}{l}$$

Then substitute $\dfrac{2700}{l}$ for w in equation (1) and solve for l.

$$2l + 2 \cdot \frac{2700}{l} = 210$$

$$2l + \frac{5400}{l} = 210$$

$$2l^2 + 5400 = 210l \quad \text{Multiplying by } l$$

$$2l^2 - 210l + 5400 = 0$$

$$2(l^2 - 105l + 2700) = 0$$

$$2(l - 45)(l - 60) = 0$$

$l = 45 \ \ or \ \ l = 60$

If $l = 45$, then $w = \dfrac{2700}{45} = 60$. If $l = 60$, then $w = \dfrac{2700}{60} = 45$. Since length is usually considered to be longer than width, we have $l = 60$ and $w = 45$.

Check. Perimeter: $2 \cdot 60 + 2 \cdot 45 = 210$ ft

Area: $60 \cdot 45 = 2700$ ft^2

The solution checks.

State. The dimensions of the playground are 60 ft by 45 ft.

21. Graph: $y \geq x^2 - 4$,

$\qquad\quad y < 2x - 1$

The solution set of $y \geq x^2 - 4$ is the parabola $y = x^2 - 4$ and the region inside it. The solution set of $y < 2x - 1$ is the half-plane below the line $y = 2x - 1$. We shade the region common to the two solution sets.

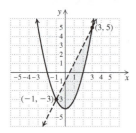

To find the points of intersection of the graphs of the related equations we solve the system of equations

$\qquad y = x^2 - 4$,

$\qquad y = 2x - 1$.

The points of intersection are $(-1, -3)$ and $(3, 5)$.

22. $(y-1)^2 = 4(x+1)$ represents a parabola with vertex $(-1, 1)$ that opens to the right. Thus the correct answer is A.

23. Use the midpoint formula to find the center.

$$(h, k) = \left(\frac{1 + 5}{2}, \frac{1 + (-3)}{2} \right) = (3, -1)$$

Use the distance formula to find the radius.

$$r = \frac{1}{2} \sqrt{(1 - 5)^2 + (1 - (-3))^2} = \frac{1}{2} \sqrt{(-4)^2 + (4)^2} = 2\sqrt{2}$$

Write the equation of the circle.

$$(x - h)^2 + (y - k)^2 = r^2$$

$$(x - 3)^2 + [y - (-1)]^2 = (2\sqrt{2})^2$$

$$(x - 3)^2 + (y + 1)^2 = 8$$

Chapter 8

Sequences, Series, and Combinatorics

1. $a_n = 4n - 1$

$a_1 = 4 \cdot 1 - 1 = 3,$

$a_2 = 4 \cdot 2 - 1 = 7,$

$a_3 = 4 \cdot 3 - 1 = 11,$

$a_4 = 4 \cdot 4 - 1 = 15;$

$a_{10} = 4 \cdot 10 - 1 = 39;$

$a_{15} = 4 \cdot 15 - 1 = 59$

3. $a_n = \dfrac{n}{n-1}, \; n \geq 2$

The first 4 terms are a_2, a_3, a_4, and a_5:

$a_2 = \dfrac{2}{2-1} = 2,$

$a_3 = \dfrac{3}{3-1} = \dfrac{3}{2},$

$a_4 = \dfrac{4}{4-1} = \dfrac{4}{3},$

$a_5 = \dfrac{5}{5-1} = \dfrac{5}{4};$

$a_{10} = \dfrac{10}{10-1} = \dfrac{10}{9};$

$a_{15} = \dfrac{15}{15-1} = \dfrac{15}{14}$

5. $a_n = \dfrac{n^2-1}{n^2+1},$

$a_1 = \dfrac{1^2-1}{1^2+1} = 0,$

$a_2 = \dfrac{2^2-1}{2^2+1} = \dfrac{3}{5},$

$a_3 = \dfrac{3^2-1}{3^2+1} = \dfrac{8}{10} = \dfrac{4}{5},$

$a_4 = \dfrac{4^2-1}{4^2+1} = \dfrac{15}{17};$

$a_{10} = \dfrac{10^2-1}{10^2+1} = \dfrac{99}{101};$

$a_{15} = \dfrac{15^2-1}{15^2+1} = \dfrac{224}{226} = \dfrac{112}{113}$

7. $a_n = (-1)^n n^2$

$a_1 = (-1)^1 1^2 = -1,$

$a_2 = (-1)^2 2^2 = 4,$

$a_3 = (-1)^3 3^2 = -9,$

$sa_4 = (-1)^4 4^2 = 16;$

$a_{10} = (-1)^{10} 10^2 = 100;$

$a_{15} = (-1)^{15} 15^2 = -225$

9. $a_n = 5 + \dfrac{(-2)^{n+1}}{2^n}$

$a_1 = 5 + \dfrac{(-2)^{1+1}}{2^1} = 5 + \dfrac{4}{2} = 7,$

$a_2 = 5 + \dfrac{(-2)^{2+1}}{2^2} = 5 + \dfrac{-8}{4} = 3,$

$a_3 = 5 + \dfrac{(-2)^{3+1}}{2^3} = 5 + \dfrac{16}{8} = 7,$

$a_4 = 5 + \dfrac{(-2)^{4+1}}{2^4} = 5 + \dfrac{-32}{16} = 3;$

$a_{10} = 5 + \dfrac{(-2)^{10+1}}{2^{10}} = 5 + \dfrac{-1 \cdot 2^{11}}{2^{10}} = 3;$

$a_{15} = 5 + \dfrac{(-2)^{15+1}}{2^{15}} = 5 + \dfrac{2^{16}}{2^{15}} = 7$

11. $a_n = 5n - 6$

$a_8 = 5 \cdot 8 - 6 = 40 - 6 = 34$

13. $a_n = (2n + 3)^2$

$a_6 = (2 \cdot 6 + 3)^2 = 225$

15. $a_n = 5n^2(4n - 100)$

$a_{11} = 5(11)^2(4 \cdot 11 - 100) = 5(121)(-56) =$
$-33,880$

17. $a_n = \ln e^n$

$a_{67} = \ln e^{67} = 67$

19. $2, 4, 6, 8, 10, \ldots$

These are the even integers, so the general term might be $2n$.

21. $-2, 6, -18, 54, \ldots$

We can see a pattern if we write the sequence as

$-1 \cdot 2 \cdot 1, \; 1 \cdot 2 \cdot 3, \; -1 \cdot 2 \cdot 9, \; 1 \cdot 2 \cdot 27, \ldots$

The general term might be $(-1)^n 2(3)^{n-1}$.

23. $\dfrac{2}{3}, \dfrac{3}{4}, \dfrac{4}{5}, \dfrac{5}{6}, \dfrac{6}{7}, \ldots$

These are fractions in which the denominator is 1 greater than the numerator. Also, each numerator is 1 greater than the preceding numerator. The general term might be $\dfrac{n+1}{n+2}$.

25. $1 \cdot 2, \; 2 \cdot 3, \; 3 \cdot 4, \; 4 \cdot 5, \ldots$

These are the products of pairs of consecutive natural numbers. The general term might be $n(n + 1)$.

27. $0, \log 10, \log 100, \log 1000, \ldots$

We can see a pattern if we write the sequence as

$\log 1, \log 10, \log 100, \log 1000, \ldots$

The general term might be $\log 10^{n-1}$. This is equivalent to $n - 1$.

29. $1, 2, 3, 4, 5, 6, 7, \ldots$

$S_3 = 1 + 2 + 3 = 6$

$S_7 = 1 + 2 + 3 + 4 + 5 + 6 + 7 = 28$

31. $2, 4, 6, 8, \ldots$

$S_4 = 2 + 4 + 6 + 8 = 20$

$S_5 = 2 + 4 + 6 + 8 + 10 = 30$

33. $\displaystyle\sum_{k=1}^{5} \frac{1}{2k} = \frac{1}{2 \cdot 1} + \frac{1}{2 \cdot 2} + \frac{1}{2 \cdot 3} + \frac{1}{2 \cdot 4} + \frac{1}{2 \cdot 5}$

$\qquad = \frac{1}{2} + \frac{1}{4} + \frac{1}{6} + \frac{1}{8} + \frac{1}{10}$

$\qquad = \frac{60}{120} + \frac{30}{120} + \frac{20}{120} + \frac{15}{120} + \frac{12}{120}$

$\qquad = \frac{137}{120}$

35. $\displaystyle\sum_{i=0}^{6} 2^i = 2^0 + 2^1 + 2^2 + 2^3 + 2^4 + 2^5 + 2^6$

$\qquad = 1 + 2 + 4 + 8 + 16 + 32 + 64$

$\qquad = 127$

37. $\displaystyle\sum_{k=7}^{10} \ln k = \ln 7 + \ln 8 + \ln 9 + \ln 10 =$

$\ln(7 \cdot 8 \cdot 9 \cdot 10) = \ln 5040 \approx 8.5252$

39. $\displaystyle\sum_{k=1}^{8} \frac{k}{k+1} = \frac{1}{1+1} + \frac{2}{2+1} + \frac{3}{3+1} + \frac{4}{4+1} +$

$\qquad \frac{5}{5+1} + \frac{6}{6+1} + \frac{7}{7+1} + \frac{8}{8+1}$

$\qquad = \frac{1}{2} + \frac{2}{3} + \frac{3}{4} + \frac{4}{5} + \frac{5}{6} + \frac{6}{7} + \frac{7}{8} + \frac{8}{9}$

$\qquad = \frac{15,551}{2520}$

41. $\displaystyle\sum_{i=1}^{5} (-1)^i$

$= (-1)^1 + (-1)^2 + (-1)^3 + (-1)^4 + (-1)^5$

$= -1 + 1 - 1 + 1 - 1$

$= -1$

43. $\displaystyle\sum_{k=1}^{8} (-1)^{k+1} 3k$

$= (-1)^2 3 \cdot 1 + (-1)^3 3 \cdot 2 + (-1)^4 3 \cdot 3 +$

$\quad (-1)^5 3 \cdot 4 + (-1)^6 3 \cdot 5 + (-1)^7 3 \cdot 6 +$

$\quad (-1)^8 3 \cdot 7 + (-1)^9 3 \cdot 8$

$= 3 - 6 + 9 - 12 + 15 - 18 + 21 - 24$

$= -12$

45. $\displaystyle\sum_{k=0}^{6} \frac{2}{k^2 + 1} = \frac{2}{0^2 + 1} + \frac{2}{1^2 + 1} + \frac{2}{2^2 + 1} + \frac{2}{3^2 + 1} +$

$\qquad \frac{2}{4^2 + 1} + \frac{2}{5^2 + 1} + \frac{2}{6^2 + 1}$

$\qquad = 2 + 1 + \frac{2}{5} + \frac{2}{10} + \frac{2}{17} + \frac{2}{26} + \frac{2}{37}$

$\qquad = 2 + 1 + \frac{2}{5} + \frac{1}{5} + \frac{2}{17} + \frac{1}{13} + \frac{2}{37}$

$\qquad = \frac{157,351}{40,885}$

47. $\displaystyle\sum_{k=0}^{5} (k^2 - 2k + 3)$

$= (0^2 - 2 \cdot 0 + 3) + (1^2 - 2 \cdot 1 + 3) +$

$\quad (2^2 - 2 \cdot 2 + 3) + (3^2 - 2 \cdot 3 + 3) +$

$\quad (4^2 - 2 \cdot 4 + 3) + (5^2 - 2 \cdot 5 + 3)$

$= 3 + 2 + 3 + 6 + 11 + 18$

$= 43$

49. $\displaystyle\sum_{i=0}^{10} \frac{2^i}{2^i + 1}$

$= \frac{2^0}{2^0 + 1} + \frac{2^1}{2^1 + 1} + \frac{2^2}{2^2 + 1} + \frac{2^3}{2^3 + 1} + \frac{2^4}{2^4 + 1} +$

$\quad \frac{2^5}{2^5 + 1} + \frac{2^6}{2^6 + 1} + \frac{2^7}{2^7 + 1} + \frac{2^8}{2^8 + 1} + \frac{2^9}{2^9 + 1} +$

$\quad \frac{2^{10}}{2^{10} + 1}$

$= \frac{1}{2} + \frac{2}{3} + \frac{4}{5} + \frac{8}{9} + \frac{16}{17} + \frac{32}{33} + \frac{64}{65} + \frac{128}{129} +$

$\quad \frac{256}{257} + \frac{512}{513} + \frac{1024}{1025}$

≈ 9.736

51. $5 + 10 + 15 + 20 + 25 + \ldots$

This is a sum of multiples of 5, and it is an infinite series. Sigma notation is

$$\sum_{k=1}^{\infty} 5k.$$

53. $2 - 4 + 8 - 16 + 32 - 64$

This is a sum of powers of 2 with alternating signs. Sigma notation is

$$\sum_{k=1}^{6} (-1)^{k+1} 2k, \text{ or } \sum_{k=1}^{6} (-1)^{k-1} 2k$$

55. $-\dfrac{1}{2} + \dfrac{2}{3} - \dfrac{3}{4} + \dfrac{4}{5} - \dfrac{5}{6} + \dfrac{6}{7}$

This is a sum of fractions in which the denominator is one greater than the numerator. Also, each numerator is 1 greater than the preceding numerator and the signs alternate. Sigma notation is

$$\sum_{k=1}^{6} (-1)^k \frac{k}{k+1}.$$

57. $4 - 9 + 16 - 25 + \ldots + (-1)^n n^2$

This is a sum of terms of the form $(-1)^k k^2$, beginning with $k = 2$ and continuing through $k = n$. Sigma notation is

$$\sum_{k=2}^{n} (-1)^k k^2.$$

59. $\dfrac{1}{1 \cdot 2} + \dfrac{1}{2 \cdot 3} + \dfrac{1}{3 \cdot 4} + \dfrac{1}{4 \cdot 5} + \ldots$

This is a sum of fractions in which the numerator is 1 and the denominator is a product of two consecutive integers. The larger integer in each product is the smaller integer in the succeeding product. It is an infinite series. Sigma notation is

$$\sum_{k=1}^{\infty} \frac{1}{k(k+1)}.$$

61. $a_1 = 4, \qquad a_{k+1} = 1 + \dfrac{1}{a_k}$

$a_2 = 1 + \dfrac{1}{4} = 1\dfrac{1}{4}$, or $\dfrac{5}{4}$

$a_3 = 1 + \dfrac{1}{\frac{5}{4}} = 1 + \dfrac{4}{5} = 1\dfrac{4}{5}$, or $\dfrac{9}{5}$

$a_4 = 1 + \dfrac{1}{\frac{9}{5}} = 1 + \dfrac{5}{9} = 1\dfrac{5}{9}$, or $\dfrac{14}{9}$

63. $a_1 = 6561, \qquad a_{k+1} = (-1)^k \sqrt{a_k}$

$a_2 = (-1)^1 \sqrt{6561} = -81$

$a_3 = (-1)^2 \sqrt{-81} = 9i$

$a_4 = (-1)^3 \sqrt{9i} = -3\sqrt{i}$

65. $a_1 = 2, \qquad a_{k+1} = a_k + a_{k-1}$

$a_2 = 3$

$a_3 = 3 + 2 = 5$

$a_4 = 5 + 3 = 8$

67. a) $a_1 = \$1000(1.062)^1 = \$1062$

$a_2 = \$1000(1.062)^2 \approx \$1127.84$

$a_3 = \$1000(1.062)^3 \approx \$1197.77$

$a_4 = \$1000(1.062)^4 \approx \$1272.03$

$a_5 = \$1000(1.062)^5 \approx \$1350.90$

$a_6 = \$1000(1.062)^6 \approx \$1434.65$

$a_7 = \$1000(1.062)^7 \approx \$1523.60$

$a_8 = \$1000(1.062)^8 \approx \$1618.07$

$a_9 = \$1000(1.062)^9 \approx \$1718.39$

$a_{10} = \$1000(1.062)^{10} \approx \$1824.93$

b) $a_{20} = \$1000(1.062)^{20} \approx \$3330.35$

69. Find each term by adding $1.10 to the preceding term.

$9.80, $10.90, $12.00, $13.10, $14.20, $15.30,

$16.40, $17.50, $18.60, $19.70

71. $a_1 = 1$ (Given)

$a_2 = 1$ (Given)

$a_3 = a_2 + a_1 = 1 + 1 = 2$

$a_4 = a_3 + a_2 = 2 + 1 = 3$

$a_5 = a_4 + a_3 = 3 + 2 = 5$

$a_6 = a_5 + a_4 = 5 + 3 = 8$

$a_7 = a_6 + a_5 = 8 + 5 = 13$

73. *Familiarize*. Let x and y represent the number of acres of pumpkins harvested in 2012 in Illinois and Ohio, respectively.

Translate. The total number of acres harvested was 23,400.

$x + y = 23,400$

The number of acres harvested in Ohio was 9000 fewer then the number of acres harvested in Illinois.

$y = x - 9000$

We have a system of equations.

$x + y = 23,400, \quad (1)$

$y = x - 9000 \qquad (2)$

Carry out. Substitute $x - 9000$ for y in equation (1) and solve for x.

$x + x - 9000 = 23,400$

$2x - 9000 = 23,400$

$2x = 32,400$

$x = 16,200$

Now substitute 16,200 for x in equation (2) and find y.

$y = 16,200 - 9000 = 7200$

Check. $16,200 + 7200 = 23,400$ and 7200 is 9000 less than 16,200 so the answer checks.

State. In 2012, 16,200 acres of pumpkins were harvested in Illinois, and 7200 acres were harvested in Ohio.

75. We complete the square twice.

$$x^2 + y^2 + 5x - 8y = 2$$
$$x^2 + 5x + y^2 - 8y = 2$$
$$x^2 + 5x + \frac{25}{4} + y^2 - 8y + 16 = 2 + \frac{25}{4} + 16$$
$$\left(x + \frac{5}{2}\right)^2 + (y - 4)^2 = \frac{97}{4}$$
$$\left[x - \left(-\frac{5}{2}\right)\right]^2 + (y - 4)^2 = \left(\frac{\sqrt{97}}{2}\right)^2$$

The center is $\left(-\dfrac{5}{2}, 4\right)$ and the radius is $\dfrac{\sqrt{97}}{2}$.

77. $a_n = i^n$

$a_1 = i$

$a_2 = i^2 = -1$

$a_3 = i^3 = -i$

$a_4 = i^4 = 1$

$a_5 = i^5 = i^4 \cdot i = i;$

$S_5 = i - 1 - i + 1 + i = i$

79. $S_n = \ln 1 + \ln 2 + \ln 3 + \cdots + \ln n$

$\qquad = \ln(1 \cdot 2 \cdot 3 \cdots n)$

Exercise Set 8.2

1. 3, 8, 13, 18, . . .

$\quad a_1 = 3$

$\quad d = 5 \quad (8 - 3 = 5, \; 13 - 8 = 5, \; 18 - 13 = 5)$

3. 9, 5, 1, −3, . . .

$\quad a_1 = 9$

$\quad d = -4 \quad (5-9 = -4, \; 1-5 = -4, \; -3-1 = -4)$

5. $\dfrac{3}{2}, \dfrac{9}{4}, 3, \dfrac{15}{4}, \ldots$

$\quad a_1 = \dfrac{3}{2}$

$\quad d = \dfrac{3}{4} \quad \left(\dfrac{9}{4} - \dfrac{3}{2} = \dfrac{3}{4}, 3 - \dfrac{9}{4} = \dfrac{3}{4} \right)$

7. $a_1 = \$316$

$\quad d = -\$3 \quad (\$313 - \$316 = -\$3,$
$\qquad\qquad\quad \$310 - \$313 = -\$3, \; \$307 - \$310 = -\$3)$

9. 2, 6, 10, . . .

$\quad a_1 = 2, \; d = 4, \text{ and } n = 12$

$\quad a_n = a_1 + (n-1)d$

$\quad a_{12} = 2 + (12 - 1)4 = 2 + 11 \cdot 4 = 2 + 44 = 46$

11. $3, \dfrac{7}{3}, \dfrac{5}{3}, \ldots$

$\quad a_1 = 3, \; d = -\dfrac{2}{3}, \text{ and } n = 14$

$\quad a_n = a_1 + (n-1)d$

$\quad a_{14} = 3 + (14 - 1)\left(-\dfrac{2}{3} \right) = 3 - \dfrac{26}{3} = -\dfrac{17}{3}$

13. \$2345.78, \$2967.54, \$3589.30, . . .

$\quad a_1 = \$2345.78, \; d = \$621.76, \text{ and } n = 10$

$\quad a_n = a_1 + (n-1)d$

$\quad a_{10} = \$2345.78 + (10 - 1)(\$621.76) = \$7941.62$

15. $a_1 = 2, \; d = 4$

$\quad a_n = a_1 + (n-1)d$

Let $a_n = 106$, and solve for n.

$\quad 106 = 2 + (n-1)(4)$

$\quad 106 = 2 + 4n - 4$

$\quad 106 = 4n - 2$

$\quad 108 = 4n$

$\quad 27 = n$

The 27th term is 106.

17. $a_1 = 3, \; d = -\dfrac{2}{3}$

$\quad a_n = a_1 + (n-1)d$

Let $a_n = -27$, and solve for n.

$\quad -27 = 3 + (n-1)\left(-\dfrac{2}{3} \right)$

$\quad -81 = 9 + (n-1)(-2)$

$\quad -81 = 9 - 2n + 2$

$\quad -92 = -2n$

$\quad 46 = n$

The 46th term is -27.

19. $a_n = a_1 + (n-1)d$

$\quad 33 = a_1 + (8-1)4 \qquad$ Substituting 33 for a_8,
$\qquad\qquad\qquad\qquad\qquad$ 8 for n, and 4 for d

$\quad 33 = a_1 + 28$

$\quad 5 = a_1$

(Note that this procedure is equivalent to subtracting d from a_8 seven times to get a_1: $33 - 7(4) = 33 - 28 = 5$)

21. $\quad a_n = a_1 + (n-1)d$

$\quad -507 = 25 + (n-1)(-14)$

$\quad -507 = 25 - 14n + 14$

$\quad -546 = -14n$

$\quad 39 = n$

23. $\dfrac{25}{3} + 15d = \dfrac{95}{6}$

$\qquad\quad 15d = \dfrac{45}{6}$

$\qquad\qquad d = \dfrac{1}{2}$

$\quad a_1 = \dfrac{25}{3} - 16\left(\dfrac{1}{2} \right) = \dfrac{25}{3} - 8 = \dfrac{1}{3}$

The first five terms of the sequence are $\dfrac{1}{3}, \dfrac{5}{6}, \dfrac{4}{3}, \dfrac{11}{6}, \dfrac{7}{3}$.

25. $5 + 8 + 11 + 14 + \ldots$

Note that $a_1 = 5$, $d = 3$, and $n = 20$. First we find a_{20}:

$\quad a_n = a_1 + (n-1)d$

$\quad a_{20} = 5 + (20-1)3$

$\qquad\; = 5 + 19 \cdot 3 = 62$

Then

$\quad S_n = \dfrac{n}{2}(a_1 + a_n)$

$\quad S_{20} = \dfrac{20}{2}(5 + 62)$

$\qquad\; = 10(67) = 670.$

27. The sum is $2 + 4 + 6 + \ldots + 798 + 800$. This is the sum of the arithmetic sequence for which $a_1 = 2$, $a_n = 800$, and $n = 400$.

$\quad S_n = \dfrac{n}{2}(a_1 + a_n)$

$\quad S_{400} = \dfrac{400}{2}(2 + 800) = 200(802) = 160,400$

29. The sum is $7 + 14 + 21 + \ldots + 91 + 98$. This is the sum of the arithmetic sequence for which $a_1 = 7$, $a_n = 98$, and $n = 14$.

$$S_n = \frac{n}{2}(a_1 + a_n)$$

$$S_{14} = \frac{14}{2}(7 + 98) = 7(105) = 735$$

31. First we find a_{20}:

$$a_n = a_1 + (n - 1)d$$

$$a_{20} = 2 + (20 - 1)5$$

$$= 2 + 19 \cdot 5 = 97$$

Then

$$S_n = \frac{n}{2}(a_1 + a_n)$$

$$S_{20} = \frac{20}{2}(2 + 97)$$

$$= 10(99) = 990.$$

33. $\displaystyle\sum_{k=1}^{40} (2k + 3)$

Write a few terms of the sum:

$$5 + 7 + 9 + \ldots + 83$$

This is a series coming from an arithmetic sequence with $a_1 = 5$, $n = 40$, and $a_{40} = 83$. Then

$$S_n = \frac{n}{2}(a_1 + a_n)$$

$$S_{40} = \frac{40}{2}(5 + 83)$$

$$= 20(88) = 1760$$

35. $\displaystyle\sum_{k=0}^{19} \frac{k - 3}{4}$

Write a few terms of the sum:

$$-\frac{3}{4} - \frac{1}{2} - \frac{1}{4} + 0 + \frac{1}{4} + \ldots + 4$$

Since k goes from 0 through 19, there are 20 terms. Thus, this is equivalent to a series coming from an arithmetic sequence with $a_1 = -\frac{3}{4}$, $n = 20$, and $a_{20} = 4$. Then

$$S_n = \frac{n}{2}(a_1 + a_n)$$

$$S_{20} = \frac{20}{2}\left(-\frac{3}{4} + 4\right)$$

$$= 10 \cdot \frac{13}{4} = \frac{65}{2}.$$

37. $\displaystyle\sum_{k=12}^{57} \frac{7 - 4k}{13}$

Write a few terms of the sum:

$$-\frac{41}{13} - \frac{45}{13} - \frac{49}{13} - \ldots - \frac{221}{13}$$

Since k goes from 12 through 57, there are 46 terms. Thus, this is equivalent to a series coming from an arithmetic sequence with $a_1 = -\frac{41}{13}$, $n = 46$, and $a_{46} = -\frac{221}{13}$. Then

$$S_n = \frac{n}{2}(a_1 + a_n)$$

$$S_{46} = \frac{46}{2}\left(-\frac{41}{13} - \frac{221}{13}\right)$$

$$= 23\left(-\frac{262}{13}\right) = -\frac{6026}{13}.$$

39. *Familiarize.* We have a sequence 10, 20, 30, ... It is an arithmetic sequence with $a_1 = 10$, $d = 10$, and $n = 31$.

Translate. We want to find $S_n = \frac{n}{2}(a_1 + a_n)$ where $a_n = a_1 + (n - 1)d$, $a_1 = 10$, $d = 10$, and $n = 31$.

Carry out. First we find a_{31}.

$$a_{31} = 10 + (31 - 1)10 = 10 + 30 \cdot 10 = 310$$

Then $S_{31} = \frac{31}{2}(10 + 310) = \frac{31}{2} \cdot 320 = 4960$.

Check. We can do the calculation again, or we can do the entire addition:

$$10 + 20 + 30 + \ldots + 310.$$

State. A total of 4960¢, or \$49.60 is saved.

41. *Familiarize.* We have arithmetic sequence with $a_1 = 28$, $d = 4$, and $n = 20$.

Translate. We want to find $S_n = \frac{n}{2}(a_1 + a_n)$ where $a_n = a_1 + (n - 1)d$, $a_1 = 28$, $d = 4$, and $n = 20$.

Carry out. First we find a_{20}.

$$a_{20} = 28 + (20 - 1)4 = 104$$

Then $S_{20} = \frac{20}{2}(28 + 104) = 10 \cdot 132 = 1320$.

Check. We can do the calculations again, or we can do the entire addition:

$$28 + 32 + 36 + \ldots 104.$$

State. There are 1320 seats in the first balcony.

43. Yes; $d = 48 - 16 = 80 - 48 = 112 - 80 = 144 - 112 = 32$.

$$a_{10} = 16 + (10 - 1)32 = 304$$

$$S_{10} = \frac{10}{2}(16 + 304) = 1600 \text{ ft}$$

45. We first find how many plants will be in the last row.

Familiarize. The sequence is 35, 31, 27, ... It is an arithmetic sequence with $a_1 = 35$ and $d = -4$. Since each row must contain a positive number of plants, we must determine how many times we can add -4 to 35 and still have a positive result.

Translate. We find the largest integer x for which $35 + x(-4) > 0$. Then we evaluate the expression $35 - 4x$ for that value of x.

Carry out. We solve the inequality.

$$35 - 4x > 0$$

$$35 > 4x$$

$$\frac{35}{4} > x$$

$$8\frac{3}{4} > x$$

The integer we are looking for is 8. Thus $35 - 4x = 35 - 4(8) = 3$.

Check. If we add -4 to 35 eight times we get 3, a positive number, but if we add -4 to 35 more than eight times we get a negative number.

State. There will be 3 plants in the last row.

Next we find how many plants there are altogether.

Familiarize. We want to find the sum $35+31+27+\ldots+3$. We know $a_1 = 35$ $a_n = 3$, and, since we add -4 to 35 eight times, $n = 9$. (There are 8 terms after a_1, for a total of 9 terms.) We will use the formula $S_n = \dfrac{n}{2}(a_1 + a_n)$.

Translate. We want to find the sum of the first 9 terms of an arithmetic sequence in which $a_1 = 35$ and $a_9 = 3$.

Carry out. Substituting into the formula, we have
$$S_9 = \frac{9}{2}(35 + 3)$$
$$= \frac{9}{2} \cdot 38 = 171$$

Check. We can check the calculations by doing them again. We could also do the entire addition:
$$35 + 31 + 27 + \ldots + 3.$$

State. There are 171 plants altogether.

47. Yes; $d = 6 - 3 = 9 - 6 = 3n - 3(n-1) = 3$.

49.
$$2x + y + 3z = 12$$
$$x - 3y - 2z = -1$$
$$5x + 2y - 4z = -4$$

We will use Gauss-Jordan elimination with matrices. First we write the augmented matrix.

$$\left[\begin{array}{ccc|c} 2 & 1 & 3 & 12 \\ 1 & -3 & -2 & -1 \\ 5 & 2 & -4 & -4 \end{array} \right]$$

Next we interchange the first two rows.

$$\left[\begin{array}{ccc|c} 1 & -3 & -2 & -1 \\ 2 & 1 & 3 & 12 \\ 5 & 2 & -4 & -4 \end{array} \right]$$

Now multiply the first row by -2 and add it to the second row. Also multiply the first row by -5 and add it to the third row.

$$\left[\begin{array}{ccc|c} 1 & -3 & -2 & -1 \\ 0 & 7 & 7 & 14 \\ 0 & 17 & 6 & 1 \end{array} \right]$$

Multiply the second row by $\dfrac{1}{7}$.

$$\left[\begin{array}{ccc|c} 1 & -3 & -2 & -1 \\ 0 & 1 & 1 & 2 \\ 0 & 17 & 6 & 1 \end{array} \right]$$

Multiply the second row by 3 and add it to the first row. Also multiply the second row by -17 and add it to the third row.

$$\left[\begin{array}{ccc|c} 1 & 0 & 1 & 5 \\ 0 & 1 & 1 & 2 \\ 0 & 0 & -11 & -33 \end{array} \right]$$

Multiply the third row by $-\dfrac{1}{11}$.

$$\left[\begin{array}{ccc|c} 1 & 0 & 1 & 5 \\ 0 & 1 & 1 & 2 \\ 0 & 0 & 1 & 3 \end{array} \right]$$

Multiply the third row by -1 and add it to the first row and also to the second row.

$$\left[\begin{array}{ccc|c} 1 & 0 & 0 & 2 \\ 0 & 1 & 0 & -1 \\ 0 & 0 & 1 & 3 \end{array} \right]$$

Now we can read the solution from the matrix. It is $(2, -1, 3)$.

51. The vertices are on the y-axis, so the transverse axis is vertical and $a = 5$. The length of the minor axis is 4, so $b = 4/2 = 2$. The equation is
$$\frac{x^2}{4} + \frac{y^2}{25} = 1.$$

53. $S_n = \dfrac{n}{2}(1 + 2n - 1) = n^2$

55. Let $d = $ the common difference. Then $a_4 = a_2 + 2d$, or
$$10p + q = 40 - 3q + 2d$$
$$10p + 4q - 40 = 2d$$
$$5p + 2q - 20 = d.$$
Also, $a_1 = a_2 - d$, so we have
$$a_1 = 40 - 3q - (5p + 2q - 20)$$
$$= 40 - 3q - 5p - 2q + 20$$
$$= 60 - 5p - 5q.$$

57. $4, m_1, m_2, m_3, m_4, 13$

We look for m_1, m_2, m_3, and m_4 such that $4, m_1, m_2, m_3, m_4, 13$ is an arithmetic sequence. In this case $a_1 = 4$, $n = 6$, and $a_6 = 13$. First we find d.
$$a_n = a_1 + (n-1)d$$
$$13 = 4 + (6 - 1)d$$
$$9 = 5d$$
$$1\frac{4}{5} = d$$
Then we have

$$m_1 = a_1 + d = 4 + 1\frac{4}{5} = 5\frac{4}{5},$$

$$m_2 = m_1 + d = 5\frac{4}{5} + 1\frac{4}{5} = 6\frac{8}{5} = 7\frac{3}{5},$$

$$m_3 = m_2 + d = 7\frac{3}{5} + 1\frac{4}{5} = 8\frac{7}{5} = 9\frac{2}{5},$$

$$m_4 = m_3 + d = 9\frac{2}{5} + 1\frac{4}{5} = 10\frac{6}{5} = 11\frac{1}{5}.$$

Exercise Set 8.3

1. $2, 4, 8, 16, \ldots$

$\frac{4}{2} = 2, \frac{8}{4} = 2, \frac{16}{8} = 2$

$r = 2$

3. $1, -1, 1, -1, \ldots$

$\frac{-1}{1} = -1, \frac{1}{-1} = -1, \frac{-1}{1} = -1$

$r = -1$

5. $\frac{2}{3}, -\frac{4}{3}, \frac{8}{3}, -\frac{16}{3}, \ldots$

$\dfrac{-\frac{4}{3}}{\frac{2}{3}} = -2, \dfrac{\frac{8}{3}}{-\frac{4}{3}} = -2, \dfrac{-\frac{16}{3}}{\frac{8}{3}} = -2$

$r = -2$

7. $\frac{0.6275}{6.275} = 0.1, \frac{0.06275}{0.6275} = 0.1$

$r = 0.1$

9. $\dfrac{\frac{5a}{2}}{5} = \frac{a}{2}, \dfrac{\frac{5a^2}{4}}{\frac{5a}{2}} = \frac{a}{2}, \dfrac{\frac{5a^3}{8}}{\frac{5a^2}{4}} = \frac{a}{2}$

$r = \frac{a}{2}$

11. $2, 4, 8, 16, \ldots$

$a_1 = 2$, $n = 7$, and $r = \frac{4}{2}$, or 2.

We use the formula $a_n = a_1 r^{n-1}$.

$a_7 = 2(2)^{7-1} = 2 \cdot 2^6 = 2 \cdot 64 = 128$

13. $2, 2\sqrt{3}, 6, \ldots$

$a_1 = 2$, $n = 9$, and $r = \frac{2\sqrt{3}}{2}$, or $\sqrt{3}$

$a_n = a_1 r^{n-1}$

$a_9 = 2(\sqrt{3})^{9-1} = 2(\sqrt{3})^8 = 2 \cdot 81 = 162$

15. $\frac{7}{625}, -\frac{7}{25}, \ldots$

$a_1 = \frac{7}{625}$, $n = 23$, and $r = \dfrac{-\frac{7}{25}}{\frac{7}{625}} = -25$.

$a_n = a_1 r^{n-1}$

$a_{23} = \frac{7}{625}(-25)^{23-1} = \frac{7}{625}(-25)^{22}$

$= \frac{7}{25^2} \cdot 25^2 \cdot 25^{20} = 7(25)^{20}$, or $7(5)^{40}$

17. $1, 3, 9, \ldots$

$a_1 = 1$ and $r = \frac{3}{1}$, or 3

$a_n = a_1 r^{n-1}$

$a_n = 1(3)^{n-1} = 3^{n-1}$

19. $1, -1, 1, -1, \ldots$

$a_1 = 1$ and $r = \frac{-1}{1} = -1$

$a_n = a_1 r^{n-1}$

$a_n = 1(-1)^{n-1} = (-1)^{n-1}$

21. $\frac{1}{x}, \frac{1}{x^2}, \frac{1}{x^3}, \ldots$

$a_1 = \frac{1}{x}$ and $r = \dfrac{\frac{1}{x^2}}{\frac{1}{x}} = \frac{1}{x}$

$a_n = a_1 r^{n-1}$

$a_n = \frac{1}{x}\left(\frac{1}{x}\right)^{n-1} = \frac{1}{x} \cdot \frac{1}{x^{n-1}} = \frac{1}{x^{1+n-1}} = \frac{1}{x^n}$

23. $6 + 12 + 24 + \ldots$

$a_1 = 6$, $n = 7$, and $r = \frac{12}{6}$, or 2

$S_n = \frac{a_1(1 - r^n)}{1 - r}$

$S_7 = \frac{6(1 - 2^7)}{1 - 2} = \frac{6(1 - 128)}{-1} = \frac{6(-127)}{-1} = 762$

25. $\frac{1}{18} - \frac{1}{6} + \frac{1}{2} - \ldots$

$a_1 = \frac{1}{18}$, $n = 9$, and $r = \dfrac{-\frac{1}{6}}{\frac{1}{18}} = -\frac{1}{6} \cdot \frac{18}{1} = -3$

$S_n = \frac{a_1(1 - r^n)}{1 - r}$

$S_9 = \dfrac{\frac{1}{18}\left[1 - (-3)^9\right]}{1 - (-3)} = \dfrac{\frac{1}{18}(1 + 19,683)}{4}$

$= \dfrac{\frac{1}{18}(19,684)}{4} = \frac{1}{18}(19,684)\left(\frac{1}{4}\right) = \frac{4921}{18}$

27. Multiplying each term of the sequence by $-\sqrt{2}$ produces the next term, so it is true that the sequence is geometric.

29. Since $\frac{2^{n+1}}{2^n} = 2$, the sequence has a common ratio so it is true that the sequence is geometric.

31. Since $|-0.75| < 1$, it is true that the series has a sum.

33. $4 + 2 + 1 + \ldots$

$|r| = \left|\frac{2}{4}\right| = \left|\frac{1}{2}\right| = \frac{1}{2}$, and since $|r| < 1$, the series does have a sum.

$S_\infty = \frac{a_1}{1 - r} = \frac{4}{1 - \frac{1}{2}} = \frac{4}{\frac{1}{2}} = 4 \cdot \frac{2}{1} = 8$

35. $25 + 20 + 16 + \ldots$

$|r| = \left|\dfrac{20}{25}\right| = \left|\dfrac{4}{5}\right| = \dfrac{4}{5}$, and since $|r| < 1$, the series does have a sum.

$S_\infty = \dfrac{a_1}{1 - r} = \dfrac{25}{1 - \dfrac{4}{5}} = \dfrac{25}{\dfrac{1}{5}} = 25 \cdot \dfrac{5}{1} = 125$

37. $8 + 40 + 200 + \ldots$

$|r| = \left|\dfrac{40}{8}\right| = |5| = 5$, and since $|r| > 1$ the series does not have a sum.

39. $0.6 + 0.06 + 0.006 + \ldots$

$|r| = \left|\dfrac{0.06}{0.6}\right| = |0.1| = 0.1$, and since $|r| < 1$, the series does have a sum.

$S_\infty = \dfrac{a_1}{1 - r} = \dfrac{0.6}{1 - 0.1} = \dfrac{0.6}{0.9} = \dfrac{6}{9} = \dfrac{2}{3}$

41. $\displaystyle\sum_{k=1}^{11} 15\left(\dfrac{2}{3}\right)^k$

$a_1 = 15 \cdot \dfrac{2}{3}$ or 10; $|r| = \left|\dfrac{2}{3}\right| = \dfrac{2}{3}$, $n = 11$

$S_{11} = \dfrac{10\left[1 - \left(\dfrac{2}{3}\right)^{11}\right]}{1 - \dfrac{2}{3}} = \dfrac{10\left[1 - \dfrac{2048}{177,147}\right]}{\dfrac{1}{3}}$

$= 10 \cdot \dfrac{175,099}{177,147} \cdot 3$

$= \dfrac{1,750,990}{59,049}$, or $29\dfrac{38,569}{59,049}$

≈ 29.65317

43. $\displaystyle\sum_{k=1}^{\infty} \left(\dfrac{1}{2}\right)^{k-1}$

$a_1 = 1$, $|r| = \left|\dfrac{1}{2}\right| = \dfrac{1}{2}$

$S_\infty = \dfrac{a_1}{1 - r} = \dfrac{1}{1 - \dfrac{1}{2}} = \dfrac{1}{\dfrac{1}{2}} = 2$

45. $\displaystyle\sum_{k=1}^{\infty} 12.5^k$

Since $|r| = 12.5 > 1$, the sum does not exist.

47. $\displaystyle\sum_{k=1}^{\infty} \$500(1.11)^{-k}$

$a_1 = \$500(1.11)^{-1}$, or $\dfrac{\$500}{1.11}$; $|r| = |1.11^{-1}| = \dfrac{1}{1.11}$

$S_\infty = \dfrac{a_1}{1 - r} = \dfrac{\dfrac{\$500}{1.11}}{1 - \dfrac{1}{1.11}} = \dfrac{\dfrac{\$500}{1.11}}{\dfrac{0.11}{1.11}} \approx \$4545.\overline{45}$

49. $\displaystyle\sum_{k=1}^{\infty} 16(0.1)^{k-1}$

$a_1 = 16$, $|r| = |0.1| = 0.1$

$S_\infty = \dfrac{a_1}{1 - r} = \dfrac{16}{1 - 0.1} = \dfrac{16}{0.9} = \dfrac{160}{9}$

51. $0.131313\ldots = 0.13 + 0.0013 + 0.000013 + \ldots$

This is an infinite geometric series with $a_1 = 0.13$.

$|r| = \left|\dfrac{0.0013}{0.13}\right| = |0.01| = 0.01 < 1$, so the series has a limit.

$S_\infty = \dfrac{a_1}{1 - r} = \dfrac{0.13}{1 - 0.01} = \dfrac{0.13}{0.99} = \dfrac{13}{99}$

53. We will find fraction notation for $0.999\overline{9}$ and then add 8.

$0.999\overline{9} = 0.9 + 0.09 + 0.009 + 0.0009 + \ldots$

This is an infinite geometric series with $a_1 = 0.9$.

$|r| = \left|\dfrac{0.09}{0.9}\right| = |0.1| = 0.1 < 1$, so the series has a limit.

$S_\infty = \dfrac{a_1}{1 - r} = \dfrac{0.9}{1 - 0.1} = \dfrac{0.9}{0.9} = 1$

Then $8.999\overline{9} = 8 + 1 = 9$.

55. $3.4125\overline{125} = 3.4 + 0.0125\overline{125}$

We will find fraction notation for $0.0125\overline{125}$ and then add 3.4, or $\dfrac{34}{10}$, or $\dfrac{17}{5}$.

$0.0125\overline{125} = 0.0125 + 0.0000125 + \ldots$

This is an infinite geometric series with $a_1 = 0.0125$.

$|r| = \left|\dfrac{0.0000125}{0.0125}\right| = |0.001| = 0.001 < 1$, so the series has a limit.

$S_\infty = \dfrac{a_1}{1 - r} = \dfrac{0.0125}{1 - 0.001} = \dfrac{0.0125}{0.999} = \dfrac{125}{9990}$

Then $\dfrac{17}{5} + \dfrac{125}{9990} = \dfrac{33,966}{9990} + \dfrac{125}{9990} = \dfrac{34,091}{9990}$

57. ***Familiarize.*** The total earnings are represented by the geometric series

$\$0.01 + \$0.01(2) + \$0.01(2)^2 + \ldots + \$0.01(2)^{27}$, where $a_1 = \$0.01$, $r = 2$, and $n = 28$.

Translate. Using the formula

$S_n = \dfrac{a_1(1 - r^n)}{1 - r}$

we have

$S_{28} = \dfrac{\$0.01(1 - 2^{28})}{1 - 2}$.

Carry out. We carry out the computation and get $\$2,684,354.55$.

Check. Repeat the calculation.

State. You would earn $\$2,684,354.55$.

59. a) *Familiarize*. The rebound distances form a geometric sequence:

$$0.6 \times 200, \ (0.6)^2 \times 200, \ (0.6)^3 \times 200, \ \ldots,$$

or $120, \ 0.6 \times 120, \ (0.6)^2 \times 120, \ \ldots$

The total rebound distance after 9 rebounds is the sum of the first 9 terms of this sequence.

Translate. We will use the formula

$$S_n = \frac{a_1(1 - r^n)}{1 - r} \text{ with } a_1 = 120, \ r = 0.6, \text{ and } n = 9.$$

Carry out.

$$S_9 = \frac{120[1 - (0.6)^9]}{1 - 0.6} \approx 297$$

Check. We repeat the calculation.

State. The bungee jumper has traveled about 297 ft upward after 9 rebounds.

b) $S_\infty = \dfrac{a_1}{1 - r} = \dfrac{120}{1 - 0.6} = 300$ ft

61. *Familiarize*. The amount of the annuity is the geometric series

$$\$3200 + \$3200(1.046) + \$3200(1.046)^2 + \ldots + \$3200(1.046)^9, \text{ where } a_1 = \$3200, \ r = 1.046, \text{ and } n = 10.$$

Translate. Using the formula

$$S_n = \frac{a_1(1 - r^n)}{1 - r}$$

we have

$$S_{10} = \frac{\$3200[1 - (1.046)^{10}]}{1 - 1.046}.$$

Carry out. We carry out the computation and get $S_{10} \approx \$39,505.71$.

Check. Repeat the calculations.

State. The amount of the annuity is \$39,505.71.

63. *Familiarize*. We have a sequence $0.01, \ 2(0.01), \ 2^2(0.01), \ 2^3(0.01), \ \ldots$. The thickness after 20 folds is given by the 21st term of the sequence.

Translate. Using the formula

$$a_n = a_1 r^{n-1},$$

where $a_1 = 0.01$, $r = 2$, and $n = 21$, we have

$$a_{21} = 0.01(2)^{21-1}.$$

Carry out. We carry out the computation and get 10,485.76.

Check. Repeat the calculations.

State. The result is 10,485.76 in. thick.

65. We use the formula

$$V = \frac{P\left[\left(1 + \dfrac{i}{n}\right)^{nN} - 1\right]}{i/n}$$

with $P = 300$, $i = 5.1\%$, or 0.051, $n = 4$, and $N = 12$.

$$V = \frac{300\left[\left(1 + \dfrac{0.051}{4}\right)^{4 \cdot 12} - 1\right]}{0.051/4}$$

$$V \approx \$19,694.01$$

The amount of the annuity is \$19,694.01.

67. *Familiarize*. The total effect on the economy is the sum of an infinite geometric series

$$\$13,000,000,000 + \$13,000,000,000(0.85) + \$13,000,000,000(0.85)^2 + \ldots$$

with $a_1 = \$13,000,000,000$ and $r = 0.85$.

Translate. Using the formula

$$S_\infty = \frac{a_1}{1 - r}$$

we have

$$S_\infty = \frac{\$13,000,000,000}{1 - 0.85}.$$

Carry out. Perform the calculation:

$$S_\infty \approx \$86,666,666,667.$$

Check. Repeat the calculation.

State. The total effect on the economy is \$86,666,666,667.

69. $f(x) = x^2$, $g(x) = 4x + 5$

$(f \circ g)(x) = f(g(x)) = f(4x + 5) = (4x + 5)^2 = 16x^2 + 40x + 25$

$(g \circ f)(x) = g(f(x)) = g(x^2) = 4x^2 + 5$

71.
$$5^x = 35$$
$$\ln 5^x = \ln 35$$
$$x \ln 5 = \ln 35$$
$$x = \frac{\ln 35}{\ln 5}$$
$$x \approx 2.209$$

73. See the answer section in the text.

75. a) If the sequence is arithmetic, then $a_2 - a_1 = a_3 - a_2$.

$$x + 7 - (x + 3) = 4x - 2 - (x + 7)$$
$$x = \frac{13}{3}$$

The three given terms are $\dfrac{13}{3} + 3 = \dfrac{22}{3}$, $\dfrac{13}{3} + 7 = \dfrac{34}{3}$, and $4 \cdot \dfrac{13}{3} - 2 = \dfrac{46}{3}$.

Then $d = \dfrac{12}{3}$, or 4, so the fourth term is

$$\frac{46}{3} + \frac{12}{3} = \frac{58}{3}.$$

b) If the sequence is geometric, then $a_2/a_1 = a_3/a_2$.

$$\frac{x + 7}{x + 3} = \frac{4x - 2}{x + 7}$$
$$x = -\frac{11}{3} \text{ or } x = 5$$

For $x = -\dfrac{11}{3}$: The three given terms are

$$-\frac{11}{3} + 3 = -\frac{2}{3}, \ -\frac{11}{3} + 7 = \frac{10}{3}, \text{ and}$$
$$4\left(-\frac{11}{3}\right) - 2 = -\frac{50}{3}.$$

Then $r = -5$, so the fourth term is

$$-\frac{50}{3}(-5) = \frac{250}{3}.$$

For $x = 5$: The three given terms are $5 + 3 = 8$, $5 + 7 = 12$, and $4 \cdot 5 - 2 = 18$. Then $r = \frac{3}{2}$, so the fourth term is $18 \cdot \frac{3}{2} = 27$.

77. $x^2 - x^3 + x^4 - x^5 + \ldots$

This is a geometric series with $a_1 = x^2$ and $r = -x$.

$$S_n = \frac{a_1(1 - r^n)}{1 - r} = \frac{x^2(1 - (-x)^n)}{1 - (-x)} = \frac{x^2(1 - (-x)^n)}{1 + x}$$

Exercise Set 8.4

1. $n^2 < n^3$

$1^2 < 1^3,\ 2^2 < 2^3,\ 3^2 < 3^3,\ 4^2 < 4^3,\ 5^2 < 5^3$

The first statement is false, and the others are true.

3. A polygon of n sides has $\dfrac{n(n-3)}{2}$ diagonals.

A polygon of 3 sides has $\dfrac{3(3-3)}{2}$ diagonals.

A polygon of 4 sides has $\dfrac{4(4-3)}{2}$ diagonals.

A polygon of 5 sides has $\dfrac{5(5-3)}{2}$ diagonals.

A polygon of 6 sides has $\dfrac{6(6-3)}{2}$ diagonals.

A polygon of 7 sides has $\dfrac{7(7-3)}{2}$ diagonals.

Each of these statements is true.

5. - 23. See the answer section in the text.

25. *Familiarize*. Let x, y, and z represent the amounts invested at 1.5%, 2%, and 3%, respectively.

***Translate*.** We know that simple interest for one year was $104. This gives us one equation:

$0.015x + 0.02y + 0.03z = 104$

The amount invested at 2% is twice the amount invested at 1.5%:

$y = 2x$, or $-2x + y = 0$

There is $400 more invested at 3% than at 2%:

$z = y + 400$, or $-y + z = 400$

We have a system of equations:

$0.015x + 0.02y + 0.03z = 104,$

$-2x + \quad y \qquad\qquad = \quad 0$

$- \quad y + \quad z = \ 400$

***Carry out*.** Solving the system of equations, we get $(800, 1600, 2000)$.

***Check*.** Simple interest for one year would be $0.015(\$800) + 0.02(\$1600) + 0.03(\$2000)$, or $\$12 + \$32 + \$60$, or $104. The amount invested at 2%, $1600, is twice $800, the amount invested at 1.5%. The amount invested at 3%,

$2000, is $400 more than $1600, the amount invested at 2%. The answer checks.

***State*.** Martin invested $800 at 1.5%, $1600 at 2%, and $2000 at 3%.

27. - 31. See the answer section in the text.

Chapter 8 Mid-Chapter Mixed Review

1. All of the terms of a sequence with general term $a_n = n$ are positive. Since the given sequence has negative terms, the given statement is false.

3. $a_2/a_1 = 7/3$; $a_3/a_2 = 3/-1 = -3$; since $7/3 \neq -3$, the sequence is not geometric. The given statement is false.

5. $a_n = 3n + 5$

$a_1 = 3 \cdot 1 + 5 = 8,$

$a_2 = 3 \cdot 2 + 5 = 11,$

$a_3 = 3 \cdot 3 + 5 = 14,$

$a_4 = 3 \cdot 4 + 5 = 17;$

$a_9 = 3 \cdot 9 + 5 = 32;$

$a_{14} = 3 \cdot 14 + 5 = 47$

7. $3, 6, 9, 12, 15, \ldots$

These are multiples of 3, so the general term could be $3n$.

9. $S_4 = 1 + \dfrac{1}{2} + \dfrac{1}{4} + \dfrac{1}{8} = 1\dfrac{7}{8}$, or $\dfrac{15}{8}$

11. $-4 + 8 - 12 + 16 - 20 + \ldots$

This is an infinite sum of multiples of 4 with alternating signs. Sigma notation is

$$\sum_{k=1}^{\infty} (-1)^k 4k.$$

13. $7 - 12 = -5$; $2 - 7 = -5$; $-3 - 2 = -5$

The common difference is -5.

15. In Exercise 14 we found that $d = 2$.

$a_n = a_1 + (n-1)d$

$44 = 4 + (n-1)2$

$44 = 4 + 2n - 2$

$44 = 2 + 2n$

$42 = 2n$

$21 = n$

The 21st term is 44.

17. $\dfrac{-8}{-16} = -\dfrac{1}{2}$; $\dfrac{4}{-8} = -\dfrac{1}{2}$; $\dfrac{-2}{4} = -\dfrac{1}{2}$

The common ratio is $-\dfrac{1}{2}$.

19. $|r| = \left|\dfrac{4}{-8}\right| = \left|-\dfrac{1}{2}\right| = \dfrac{1}{2} < 1$, so the series has a sum.

$$S_\infty = \frac{a_1}{1-r} = \frac{-8}{1 - \left(-\dfrac{1}{2}\right)} = \frac{-8}{\dfrac{3}{2}} = -8 \cdot \frac{2}{3} = -\frac{16}{3}$$

21. *Familiarize*. The number of plants is represented by the arithmetic series $36 + 30 + 24 + \ldots$ with $a_1 = 36$, $d = 30 - 36 = -6$, and $n = 6$.

Translate. We want to find $S_n = \dfrac{n}{2}(a_1 + a_n)$ where $a_n = a_1 + (n-1)d$.

Carry out.

$$a_6 = 36 + (6-1)(-6) = 36 + 5(-6) = 36 - 30 = 6$$

$$S_6 = \frac{6}{2}(36 + 6) = 3 \cdot 42 = 126$$

Check. We can do the calculations again or we can do the entire addition $36 + 30 + 24 + 18 + 12 + 6$. The answer checks.

State. In all, there will be 126 plants.

23. See the answer section in the text.

25.
$$1 + 2 + 3 + \ldots + 100$$
$$= (1 + 100) + (2 + 99) + (3 + 98) + \ldots + (50 + 51)$$
$$= \underbrace{101 + 101 + 101 + \ldots + 101}_{50 \text{ addends of } 101}$$
$$= 50 \cdot 101$$
$$= 5050$$

A formula for the first n natural numbers is $\dfrac{n}{2}(1 + n)$.

27. We can prove an infinite sequence of statements S_n by showing that a basis statement S_1 is true and then that for all natural numbers k, if S_k is true, then S_{k+1} is true.

Exercise Set 8.5

1. $_6P_6 = 6! = 6 \cdot 5 \cdot 4 \cdot 3 \cdot 2 \cdot 1 = 720$

3. Using formula (1), we have
$$_{10}P_7 = 10 \cdot 9 \cdot 8 \cdot 7 \cdot 6 \cdot 5 \cdot 4 = 604,800.$$
Using formula (2), we have
$$_{10}P_7 = \frac{10!}{(10-7)!} = \frac{10!}{3!} = \frac{10 \cdot 9 \cdot 8 \cdot 7 \cdot 6 \cdot 5 \cdot 4 \cdot 3!}{3!} =$$
$$604,800.$$

5. $5! = 5 \cdot 4 \cdot 3 \cdot 2 \cdot 1 = 120$

7. $0!$ is defined to be 1.

9. $\dfrac{9!}{5!} = \dfrac{9 \cdot 8 \cdot 7 \cdot 6 \cdot 5!}{5!} = 9 \cdot 8 \cdot 7 \cdot 6 = 3024$

11. $(8-3)! = 5! = 5 \cdot 4 \cdot 3 \cdot 2 \cdot 1 = 120$

13. $\dfrac{10!}{7!3!} = \dfrac{10 \cdot 9 \cdot 8 \cdot 7!}{7!3 \cdot 2 \cdot 1} = \dfrac{10 \cdot 3 \cdot 3 \cdot 4 \cdot 2}{3 \cdot 2 \cdot 1} = 10 \cdot 3 \cdot 4 = 120$

15. Using formula (2), we have
$$_8P_0 = \frac{8!}{(8-0)!} = \frac{8!}{8!} = 1.$$

17. Using a calculator, we find
$$_{52}P_4 = 6,497,400$$

19. Using formula (1), we have $_nP_3 = n(n-1)(n-2)$.

Using formula (2), we have
$$_nP_3 = \frac{n!}{(n-3)!} = \frac{n(n-1)(n-2)(n-3)!}{(n-3)!} =$$
$$n(n-1)(n-2).$$

21. Using formula (1), we have $_nP_1 = n$.

Using formula (2), we have
$$_nP_1 = \frac{n!}{(n-1)!} = \frac{n(n-1)!}{(n-1)!} = n.$$

23. $_6P_6 = 6! = 720$

25. $_9P_9 = 9! = 362,880$

27. $_9P_4 = 9 \cdot 8 \cdot 7 \cdot 6 = 3024$

29. Without repetition: $_5P_5 = 5! = 120$

With repetition: $5^5 = 3125$

31. There are $_5P_5$ choices for the order of the rock numbers and $_4P_4$ choices for the order of the speeches, so we have $_5P_5 \cdot_4 P_4 = 5!4! = 2880.$

33. The first number can be any of the eight digits other than 0 and 1. The remaining 6 numbers can each be any of the ten digits 0 through 9. We have
$$8 \cdot 10^6 = 8,000,000$$

Accordingly, there can be 8,000,000 telephone numbers within a given area code before the area needs to be split with a new area code.

35. $a^2b^3c^4 = a \cdot a \cdot b \cdot b \cdot b \cdot c \cdot c \cdot c \cdot c$

There are 2 a's, 3 b's, and 4 c's, for a total of 9. We have
$$\frac{9!}{2! \cdot 3! \cdot 4!}$$
$$= \frac{9 \cdot 8 \cdot 7 \cdot 6 \cdot 5 \cdot 4!}{2 \cdot 1 \cdot 3 \cdot 2 \cdot 1 \cdot 4!} = \frac{9 \cdot 8 \cdot 7 \cdot 6 \cdot 5}{2 \cdot 3 \cdot 2} = 1260.$$

37. a) $_6P_5 = 6 \cdot 5 \cdot 4 \cdot 3 \cdot 2 = 720$

b) $6^5 = 7776$

c) The first letter can only be D. The other four letters are chosen from A, B, C, E, F without repetition. We have
$$1 \cdot_5 P_4 = 1 \cdot 5 \cdot 4 \cdot 3 \cdot 2 = 120.$$

d) The first letter can only be D. The second letter can only be E. The other three letters are chosen from A, B, C, F without repetition. We have
$$1 \cdot 1 \cdot_4 P_3 = 1 \cdot 1 \cdot 4 \cdot 3 \cdot 2 = 24.$$

39. a) Since repetition is allowed, each of the 5 digits can be chosen in 10 ways. The number of zip-codes possible is $10 \cdot 10 \cdot 10 \cdot 10 \cdot 10$, or 100,000.

b) Since there are 100,000 possible zip-codes, there could be 100,000 post offices.

41. a) Since repetition is allowed, each digit can be chosen in 10 ways. There can be
$10 \cdot 10 \cdot 10 \cdot 10 \cdot 10 \cdot 10 \cdot 10 \cdot 10 \cdot 10$, or $1,000,000,000$ social security numbers.

b) Since more than 303 million social security numbers are possible, each person can have a social security number.

43.
$$x^2 + x - 6 = 0$$
$$(x + 3)(x - 2) = 0$$
$$x + 3 = 0 \quad or \quad x - 2 = 0$$
$$x = -3 \quad or \quad x = 2$$
The solutions are -3 and 2.

45. $f(x) = x^3 - 4x^2 - 7x + 10$

We use synthetic division to find one factor of the polynomial. We try $x - 1$.

$$\begin{array}{r|rrrr} 1 & 1 & -4 & -7 & 10 \\ & & 1 & -3 & -10 \\ \hline & 1 & -3 & -10 & 0 \end{array}$$

$$x^3 - 4x^2 - 7x + 10 = 0$$
$$(x - 1)(x^2 - 3x - 10) = 0$$
$$(x - 1)(x - 5)(x + 2) = 0$$
$$x - 1 = 0 \quad or \quad x - 5 = 0 \quad or \quad x + 2 = 0$$
$$x = 1 \quad or \quad x = 5 \quad or \quad x = -2$$
The solutions are -2, 1, and 5.

47.
$$_nP_4 = 8 \cdot {}_{n-1}P_3$$
$$\frac{n!}{(n-4)!} = 8 \cdot \frac{(n-1)!}{(n-1-3)!}$$
$$\frac{n!}{(n-4)!} = 8 \cdot \frac{(n-1)!}{(n-4)!}$$
$$n! = 8 \cdot (n-1)! \qquad \text{Multiplying by } (n-4)!$$
$$n(n-1)! = 8 \cdot (n-1)!$$
$$n = 8 \qquad \text{Dividing by } (n-1)!$$

49.
$$_nP_4 = 8 \cdot {}_nP_3$$
$$\frac{n!}{(n-4)!} = 8 \cdot \frac{n!}{(n-3)!}$$
$$(n-3)! = 8(n-4)! \qquad \begin{array}{l}\text{Multiplying by} \\ \frac{(n-4)!(n-3)!}{n!}\end{array}$$
$$(n-3)(n-4)! = 8(n-4)!$$
$$n - 3 = 8 \qquad \text{Dividing by } (n-4)!$$
$$n = 11$$

51. There is one losing team per game. In order to leave one tournament winner there must be $n-1$ losers produced in $n-1$ games.

Exercise Set 8.6

1. $_{13}C_2 = \dfrac{13!}{2!(13-2)!}$
$$= \frac{13!}{2!11!} = \frac{13 \cdot 12 \cdot 11!}{2 \cdot 1 \cdot 11!}$$
$$= \frac{13 \cdot 12}{2 \cdot 1} = \frac{13 \cdot 6 \cdot 2}{2 \cdot 1}$$
$$= 78$$

3. $\dbinom{13}{11} = \dfrac{13!}{11!(13-11)!}$
$$= \frac{13!}{11!2!}$$
$$= 78 \qquad \text{(See Exercise 1.)}$$

5. $\dbinom{7}{1} = \dfrac{7!}{1!(7-1)!}$
$$= \frac{7!}{1!6!} = \frac{7 \cdot 6!}{1 \cdot 6!}$$
$$= 7$$

7. $\dfrac{_5P_3}{3!} = \dfrac{5 \cdot 4 \cdot 3}{3!}$
$$= \frac{5 \cdot 4 \cdot 3}{3 \cdot 2 \cdot 1} = \frac{5 \cdot 2 \cdot 2 \cdot 3}{3 \cdot 2 \cdot 1}$$
$$= 5 \cdot 2 = 10$$

9. $\dbinom{6}{0} = \dfrac{6!}{0!(6-0)!}$
$$= \frac{6!}{0!6!} = \frac{6!}{6! \cdot 1}$$
$$= 1$$

11. $\dbinom{6}{2} = \dfrac{6 \cdot 5}{2 \cdot 1} = 15$

13. $\dbinom{n}{r} = \dbinom{n}{n-r}$, so
$$\dbinom{7}{0} + \dbinom{7}{1} + \dbinom{7}{2} + \dbinom{7}{3} + \dbinom{7}{4} +$$
$$\dbinom{7}{5} + \dbinom{7}{6} + \dbinom{7}{7}$$
$$= 2\left[\dbinom{7}{0} + \dbinom{7}{1} + \dbinom{7}{2} + \dbinom{7}{3}\right]$$
$$= 2\left[\frac{7!}{7!0!} + \frac{7!}{6!1!} + \frac{7!}{5!2!} + \frac{7!}{4!3!}\right]$$
$$= 2(1 + 7 + 21 + 35) = 2 \cdot 64 = 128$$

15. We will use form (1).
$$_{52}C_4 = \frac{52!}{4!(52-4)!}$$
$$= \frac{52 \cdot 51 \cdot 50 \cdot 49 \cdot 48!}{4 \cdot 3 \cdot 2 \cdot 1 \cdot 48!}$$
$$= \frac{52 \cdot 51 \cdot 50 \cdot 49}{4 \cdot 3 \cdot 2 \cdot 1}$$
$$= 270,725$$

17. We will use form (2).

$$\binom{27}{11}$$

$$= \frac{27 \cdot 26 \cdot 25 \cdot 24 \cdot 23 \cdot 22 \cdot 21 \cdot 20 \cdot 19 \cdot 18 \cdot 17}{11 \cdot 10 \cdot 9 \cdot 8 \cdot 7 \cdot 6 \cdot 5 \cdot 4 \cdot 3 \cdot 2 \cdot 1}$$

$$= 13,037,895$$

19. $\binom{n}{1} = \dfrac{n!}{1!(n-1)!} = \dfrac{n(n-1)!}{1!(n-1)!} = n$

21. $\binom{m}{m} = \dfrac{m!}{m!(m-m)!} = \dfrac{m!}{m!0!} = 1$

23. $_{36}C_4 = \dfrac{36!}{4!(36-4)!}$

$$= \frac{36!}{4!32!} = \frac{36 \cdot 35 \cdot 34 \cdot 33 \cdot 32!}{4 \cdot 3 \cdot 2 \cdot 1 \cdot 32!}$$

$$= \frac{36 \cdot 35 \cdot 34 \cdot 33}{4 \cdot 3 \cdot 2 \cdot 1}$$

$$= 58,905$$

25. $_{13}C_{10} = \dfrac{13!}{10!(13-10)!}$

$$= \frac{13!}{10!3!} = \frac{13 \cdot 12 \cdot 11 \cdot 10!}{10! \cdot 3 \cdot 2 \cdot 1}$$

$$= \frac{13 \cdot 12 \cdot 11}{3 \cdot 2 \cdot 1} = \frac{13 \cdot 3 \cdot 2 \cdot 2 \cdot 11}{3 \cdot 2 \cdot 1}$$

$$= 286$$

27. $_{10}C_7 \cdot _5 C_3 = \binom{10}{7} \cdot \binom{5}{3}$ Using the fundamental counting principle

$$= \frac{10!}{7!(10-7)!} \cdot \frac{5!}{3!(5-3)!}$$

$$= \frac{10 \cdot 9 \cdot 8 \cdot 7!}{7! \cdot 3!} \cdot \frac{5 \cdot 4 \cdot 3!}{3! \cdot 2!}$$

$$= \frac{10 \cdot 9 \cdot 8}{3 \cdot 2 \cdot 1} \cdot \frac{5 \cdot 4}{2 \cdot 1} = 120 \cdot 10 = 1200$$

29. $_{52}C_5 = 2,598,960$

31. a) $_{31}P_2 = 930$

b) $31 \cdot 31 = 961$

c) $_{31}C_2 = 465$

33. $3x - 7 = 5x + 10$

$$-7 = 2x + 10$$

$$-17 = 2x$$

$$-\frac{17}{2} = x$$

The solution is $-\dfrac{17}{2}$.

35. $x^2 + 5x + 1 = 0$

$$a = 1, \ b = 5, \ c = 1$$

$$x = \frac{-b \pm \sqrt{b^2 - 4ac}}{2a}$$

$$= \frac{-5 \pm \sqrt{5^2 - 4 \cdot 1 \cdot 1}}{2 \cdot 1}$$

$$= \frac{-5 \pm \sqrt{25 - 4}}{2} = \frac{-5 \pm \sqrt{21}}{2}$$

The solutions are $\dfrac{-5 + \sqrt{21}}{2}$ and $\dfrac{-5 - \sqrt{21}}{2}$, or

$$\frac{-5 \pm \sqrt{21}}{2}.$$

37. There are 13 diamonds, and we choose 5. We have $_{13}C_5 = 1287$.

39. Playing once: $_nC_2$

Playing twice: $2 \cdot _n C_2$

41. $$\binom{n}{n-2} = 6$$

$$\frac{n!}{(n-(n-2))!(n-2)!} = 6$$

$$\frac{n!}{2!(n-2)!} = 6$$

$$\frac{n(n-1)(n-2)!}{2 \cdot 1 \cdot (n-2)!} = 6$$

$$\frac{n(n-1)}{2} = 6$$

$$n(n-1) = 12$$

$$n^2 - n = 12$$

$$n^2 - n - 12 = 0$$

$$(n-4)(n+3) = 0$$

$$n = 4 \ \text{ or } \ n = -3$$

Only 4 checks. The solution is 4.

43. $$\binom{n+2}{4} = 6 \cdot \binom{n}{2}$$

$$\frac{(n+2)!}{(n+2-4)!4!} = 6 \cdot \frac{n!}{(n-2)!2!}$$

$$\frac{(n+2)!}{(n-2)!4!} = 6 \cdot \frac{n!}{(n-2)!2!}$$

$$\frac{(n+2)!}{4!} = 6 \cdot \frac{n!}{2!} \quad \text{Multiplying by } (n-2)!$$

$$4! \cdot \frac{(n+2)!}{4!} = 4! \cdot 6 \cdot \frac{n!}{2!}$$

$$(n+2)! = 72 \cdot n!$$

$$(n+2)(n+1)n! = 72 \cdot n!$$

$$(n+2)(n+1) = 72 \quad \text{Dividing by } n!$$

$$n^2 + 3n + 2 = 72$$

$$n^2 + 3n - 70 = 0$$

$$(n+10)(n-7) = 0$$

$$n = -10 \ \text{ or } \ n = 7$$

Only 7 checks. The solution is 7.

45. Line segments: $_nC_2 = \dfrac{n!}{2!(n-2)!} =$

$\dfrac{n(n-1)(n-2)!}{2 \cdot 1 \cdot (n-2)!} = \dfrac{n(n-1)}{2}$

Diagonals: The n line segments that form the sides of the n-agon are not diagonals. Thus, the number of diagonals is $_nC_2 - n = \dfrac{n(n-1)}{2} - n =$

$\dfrac{n^2 - n - 2n}{2} = \dfrac{n^2 - 3n}{2} = \dfrac{n(n-3)}{2}, \, n \geq 4.$

Let D_n be the number of diagonals on an n-agon. Prove the result above for diagonals using mathematical induction.

$S_n : \quad D_n = \dfrac{n(n-3)}{2}, \text{ for } n = 4, 5, 6, \ldots$

$S_4 : \quad D_4 = \dfrac{4 \cdot 1}{2}$

$S_k : \quad D_k = \dfrac{k(k-3)}{2}$

$S_{k+1} : \quad D_{k+1} = \dfrac{(k+1)(k-2)}{2}$

1) *Basis step:* S_4 is true (a quadrilateral has 2 diagonals).

2) *Induction step:* Assume S_k. Observe that when an additional vertex V_{k+1} is added to the k-gon, we gain k segments, 2 of which are sides of the $(k+1)$-gon, and a former side $\overline{V_1 V_k}$ becomes a diagonal. Thus the additional number of diagonals is $k - 2 + 1$, or $k - 1$. Then the new total of diagonals is $D_k + (k-1)$, or

$\begin{aligned} D_{k+1} &= D_k + (k-1) \\ &= \dfrac{k(k-3)}{2} + (k-1) \quad \text{(by } S_k) \\ &= \dfrac{(k+1)(k-2)}{2} \end{aligned}$

Exercise Set 8.7

1. Expand: $(x+5)^4$.

We have $a = x$, $b = 5$, and $n = 4$.

Pascal's triangle method: Use the fifth row of Pascal's triangle.

$\quad 1 \quad 4 \quad 6 \quad 4 \quad 1$

$(x+5)^4$

$= 1 \cdot x^4 + 4 \cdot x^3 \cdot 5 + 6 \cdot x^2 \cdot 5^2 +$

$\quad 4 \cdot x \cdot 5^3 + 1 \cdot 5^4$

$= x^4 + 20x^3 + 150x^2 + 500x + 625$

Factorial notation method:

$(x+5)^4$

$= \binom{4}{0} x^4 + \binom{4}{1} x^3 \cdot 5 + \binom{4}{2} x^2 \cdot 5^2 +$

$\quad \binom{4}{3} x \cdot 5^3 + \binom{4}{4} 5^4$

$= \dfrac{4!}{0!4!} x^4 + \dfrac{4!}{1!3!} x^3 \cdot 5 + \dfrac{4!}{2!2!} x^2 \cdot 5^2 +$

$\quad \dfrac{4!}{3!1!} x \cdot 5^3 + \dfrac{4!}{4!0!} 5^4$

$= x^4 + 20x^3 + 150x^2 + 500x + 625$

3. Expand: $(x-3)^5$.

We have $a = x$, $b = -3$, and $n = 5$.

Pascal's triangle method: Use the sixth row of Pascal's triangle.

$\quad 1 \quad 5 \quad 10 \quad 10 \quad 5 \quad 1$

$(x-3)^5$

$= 1 \cdot x^5 + 5x^4(-3) + 10x^3(-3)^2 + 10x^2(-3)^3 +$

$\quad 5x(-3)^4 + 1 \cdot (-3)^5$

$= x^5 - 15x^4 + 90x^3 - 270x^2 + 405x - 243$

Factorial notation method:

$(x-3)^5$

$= \binom{5}{0} x^5 + \binom{5}{1} x^4(-3) + \binom{5}{2} x^3(-3)^2 +$

$\quad \binom{5}{3} x^2(-3)^3 + \binom{5}{4} x(-3)^4 + \binom{5}{5}(-3)^5$

$= \dfrac{5!}{0!5!} x^5 + \dfrac{5!}{1!4!} x^4(-3) + \dfrac{5!}{2!3!} x^3(9) +$

$\quad \dfrac{5!}{3!2!} x^2(-27) + \dfrac{5!}{4!1!} x(81) + \dfrac{5!}{5!0!}(-243)$

$= x^5 - 15x^4 + 90x^3 - 270x^2 + 405x - 243$

5. Expand: $(x-y)^5$.

We have $a = x$, $b = -y$, and $n = 5$.

Pascal's triangle method: We use the sixth row of Pascal's triangle.

$\quad 1 \quad 5 \quad 10 \quad 10 \quad 5 \quad 1$

$(x-y)^5$

$= 1 \cdot x^5 + 5x^4(-y) + 10x^3(-y)^2 + 10x^2(-y)^3 +$

$\quad 5x(-y)^4 + 1 \cdot (-y)^5$

$= x^5 - 5x^4 y + 10x^3 y^2 - 10x^2 y^3 + 5xy^4 - y^5$

Factorial notation method:

$(x-y)^5$

$$= \binom{5}{0}x^5 + \binom{5}{1}x^4(-y) + \binom{5}{2}x^3(-y)^2 +$$

$$\binom{5}{3}x^2(-y)^3 + \binom{5}{4}x(-y)^4 + \binom{5}{5}(-y)^5$$

$$= \frac{5!}{0!5!}x^5 + \frac{5!}{1!4!}x^4(-y) + \frac{5!}{2!3!}x^3(y^2) +$$

$$\frac{5!}{3!2!}x^2(-y^3) + \frac{5!}{4!1!}x(y^4) + \frac{5!}{5!0!}(-y^5)$$

$$= x^5 - 5x^4y + 10x^3y^2 - 10x^2y^3 + 5xy^4 - y^5$$

7. Expand: $(5x+4y)^6$.

We have $a = 5x$, $b = 4y$, and $n = 6$.

Pascal's triangle method: Use the seventh row of Pascal's triangle.

$\quad 1 \quad 6 \quad 15 \quad 20 \quad 15 \quad 6 \quad 1$

$(5x+4y)^6$

$$= 1 \cdot (5x)^6 + 6 \cdot (5x)^5(4y) + 15(5x)^4(4y)^2 +$$

$$20(5x)^3(4y)^3 + 15(5x)^2(4y)^4 + 6(5x)(4y)^5 +$$

$$1 \cdot (4y)^6$$

$$= 15{,}625x^6 + 75{,}000x^5y + 150{,}000x^4y^2 +$$

$$160{,}000x^3y^3 + 96{,}000x^2y^4 + 30{,}720xy^5 + 4096y^6$$

Factorial notation method:

$(5x+4y)^6$

$$= \binom{6}{0}(5x)^6 + \binom{6}{1}(5x)^5(4y) +$$

$$\binom{6}{2}(5x)^4(4y)^2 + \binom{6}{3}(5x)^3(4y)^3 +$$

$$\binom{6}{4}(5x)^2(4y)^4 + \binom{6}{5}(5x)(4y)^5 + \binom{6}{6}(4y)^6$$

$$= \frac{6!}{0!6!}(15{,}625x^6) + \frac{6!}{1!5!}(3125x^5)(4y) +$$

$$\frac{6!}{2!4!}(625x^4)(16y^2) + \frac{6!}{3!3!}(125x^3)(64y^3) +$$

$$\frac{6!}{4!2!}(25x^2)(256y^4) + \frac{6!}{5!1!}(5x)(1024y^5) +$$

$$\frac{6!}{6!0!}(4096y^6)$$

$$= 15{,}625x^6 + 75{,}000x^5y + 150{,}000x^4y^2 +$$

$$160{,}000x^3y^3 + 96{,}000x^2y^4 + 30{,}720xy^5 +$$

$$4096y^6$$

9. Expand: $\left(2t + \dfrac{1}{t}\right)^7$.

We have $a = 2t$, $b = \dfrac{1}{t}$, and $n = 7$.

Pascal's triangle method: Use the eighth row of Pascal's triangle.

$\quad 1 \quad 7 \quad 21 \quad 35 \quad 35 \quad 21 \quad 7 \quad 1$

$$\left(2t + \frac{1}{t}\right)^7$$

$$= 1 \cdot (2t)^7 + 7(2t)^6\left(\frac{1}{t}\right) + 21(2t)^5\left(\frac{1}{t}\right)^2 +$$

$$35(2t)^4\left(\frac{1}{t}\right)^3 + 35(2t)^3\left(\frac{1}{t}\right)^4 + 21(2t)^2\left(\frac{1}{t}\right)^5 +$$

$$7(2t)\left(\frac{1}{t}\right)^6 + 1 \cdot \left(\frac{1}{t}\right)^7$$

$$= 128t^7 + 7 \cdot 64t^6 \cdot \frac{1}{t} + 21 \cdot 32t^5 \cdot \frac{1}{t^2} +$$

$$35 \cdot 16t^4 \cdot \frac{1}{t^3} + 35 \cdot 8t^3 \cdot \frac{1}{t^4} + 21 \cdot 4t^2 \cdot \frac{1}{t^5} +$$

$$7 \cdot 2t \cdot \frac{1}{t^6} + \frac{1}{t^7}$$

$$= 128t^7 + 448t^5 + 672t^3 + 560t + 280t^{-1} +$$

$$84t^{-3} + 14t^{-5} + t^{-7}$$

Factorial notation method:

$$\left(2t + \frac{1}{t}\right)^7$$

$$= \binom{7}{0}(2t)^7 + \binom{7}{1}(2t)^6\left(\frac{1}{t}\right) +$$

$$\binom{7}{2}(2t)^5\left(\frac{1}{t}\right)^2 + \binom{7}{3}(2t)^4\left(\frac{1}{t}\right)^3 +$$

$$\binom{7}{4}(2t)^3\left(\frac{1}{t}\right)^4 + \binom{7}{5}(2t)^2\left(\frac{1}{t}\right)^5 +$$

$$\binom{7}{6}(2t)\left(\frac{1}{t}\right)^6 + \binom{7}{7}\left(\frac{1}{t}\right)^7$$

$$= \frac{7!}{0!7!}(128t^7) + \frac{7!}{1!6!}(64t^6)\left(\frac{1}{t}\right) + \frac{7!}{2!5!}(32t^5)\left(\frac{1}{t^2}\right) +$$

$$\frac{7!}{3!4!}(16t^4)\left(\frac{1}{t^3}\right) + \frac{7!}{4!3!}(8t^3)\left(\frac{1}{t^4}\right) +$$

$$\frac{7!}{5!2!}(4t^2)\left(\frac{1}{t^5}\right) + \frac{7!}{6!1!}(2t)\left(\frac{1}{t^6}\right) + \frac{7!}{7!0!}\left(\frac{1}{t^7}\right)$$

$$= 128t^7 + 448t^5 + 672t^3 + 560t + 280t^{-1} +$$

$$84t^{-3} + 14t^{-5} + t^{-7}$$

11. Expand: $(x^2 - 1)^5$.

We have $a = x^2$, $b = -1$, and $n = 5$.

Pascal's triangle method: Use the sixth row of Pascal's triangle.

$$1 \quad 5 \quad 10 \quad 10 \quad 5 \quad 1$$

$$(x^2 - 1)^5$$

$$= 1 \cdot (x^2)^5 + 5(x^2)^4(-1) + 10(x^2)^3(-1)^2 +$$
$$\quad 10(x^2)^2(-1)^3 + 5(x^2)(-1)^4 + 1 \cdot (-1)^5$$

$$= x^{10} - 5x^8 + 10x^6 - 10x^4 + 5x^2 - 1$$

Factorial notation method:

$$(x^2 - 1)^5$$

$$= \binom{5}{0}(x^2)^5 + \binom{5}{1}(x^2)^4(-1) +$$

$$\binom{5}{2}(x^2)^3(-1)^2 + \binom{5}{3}(x^2)^2(-1)^3 +$$

$$\binom{5}{4}(x^2)(-1)^4 + \binom{5}{5}(-1)^5$$

$$= \frac{5!}{0!5!}(x^{10}) + \frac{5!}{1!4!}(x^8)(-1) + \frac{5!}{2!3!}(x^6)(1) +$$

$$\frac{5!}{3!2!}(x^4)(-1) + \frac{5!}{4!1!}(x^2)(1) + \frac{5!}{5!0!}(-1)$$

$$= x^{10} - 5x^8 + 10x^6 - 10x^4 + 5x^2 - 1$$

13. Expand: $(\sqrt{5} + t)^6$.

We have $a = \sqrt{5}$, $b = t$, and $n = 6$.

Pascal's triangle method: We use the seventh row of Pascal's triangle:

$$1 \quad 6 \quad 15 \quad 20 \quad 15 \quad 6 \quad 1$$

$$(\sqrt{5} + t)^6 = 1 \cdot (\sqrt{5})^6 + 6(\sqrt{5})^5(t) +$$
$$\quad 15(\sqrt{5})^4(t^2) + 20(\sqrt{5})^3(t^3) +$$
$$\quad 15(\sqrt{5})^2(t^4) + 6\sqrt{5}t^5 + 1 \cdot t^6$$

$$= 125 + 150\sqrt{5}\,t + 375t^2 + 100\sqrt{5}\,t^3 +$$
$$\quad 75t^4 + 6\sqrt{5}\,t^5 + t^6$$

Factorial notation method:

$$(\sqrt{5} + t)^6 = \binom{6}{0}(\sqrt{5})^6 + \binom{6}{1}(\sqrt{5})^5(t) +$$

$$\binom{6}{2}(\sqrt{5})^4(t^2) + \binom{6}{3}(\sqrt{5})^3(t^3) +$$

$$\binom{6}{4}(\sqrt{5})^2(t^4) + \binom{6}{5}(\sqrt{5})(t^5) +$$

$$\binom{6}{6}(t^6)$$

$$= \frac{6!}{0!6!}(125) + \frac{6!}{1!5!}(25\sqrt{5})t + \frac{6!}{2!4!}(25)(t^2) +$$

$$\frac{6!}{3!3!}(5\sqrt{5})(t^3) + \frac{6!}{4!2!}(5)(t^4) +$$

$$\frac{6!}{5!1!}(\sqrt{5})(t^5) + \frac{6!}{6!0!}(t^6)$$

$$= 125 + 150\sqrt{5}\,t + 375t^2 + 100\sqrt{5}\,t^3 +$$
$$\quad 75t^4 + 6\sqrt{5}\,t^5 + t^6$$

15. Expand: $\left(a - \dfrac{2}{a}\right)^9$.

We have $a = a$, $b = -\dfrac{2}{a}$, and $n = 9$.

Pascal's triangle method: Use the tenth row of Pascal's triangle.

$$1 \quad 9 \quad 36 \quad 84 \quad 126 \quad 126 \quad 84 \quad 36 \quad 9 \quad 1$$

$$\left(a - \frac{2}{a}\right)^9 = 1 \cdot a^9 + 9a^8\left(-\frac{2}{a}\right) + 36a^7\left(-\frac{2}{a}\right)^2 +$$

$$84a^6\left(-\frac{2}{a}\right)^3 + 126a^5\left(-\frac{2}{a}\right)^4 +$$

$$126a^4\left(-\frac{2}{a}\right)^5 + 84a^3\left(-\frac{2}{a}\right)^6 +$$

$$36a^2\left(-\frac{2}{a}\right)^7 + 9a\left(-\frac{2}{a}\right)^8 + 1 \cdot \left(-\frac{2}{a}\right)^9$$

$$= a^9 - 18a^7 + 144a^5 - 672a^3 + 2016a -$$
$$\quad 4032a^{-1} + 5376a^{-3} - 4608a^{-5} +$$
$$\quad 2304a^{-7} - 512a^{-9}$$

Factorial notation method:

$$\left(a - \frac{2}{a}\right)^9$$

$$= \binom{9}{0}a^9 + \binom{9}{1}a^8\left(-\frac{2}{a}\right) + \binom{9}{2}a^7\left(-\frac{2}{a}\right)^2 +$$

$$\binom{9}{3}a^6\left(-\frac{2}{a}\right)^3 + \binom{9}{4}a^5\left(-\frac{2}{a}\right)^4 +$$

$$\binom{9}{5}a^4\left(-\frac{2}{a}\right)^5 + \binom{9}{6}a^3\left(-\frac{2}{a}\right)^6 +$$

$$\binom{9}{7}a^2\left(-\frac{2}{a}\right)^7 + \binom{9}{8}a\left(-\frac{2}{a}\right)^8 +$$

$$\binom{9}{9}\left(-\frac{2}{a}\right)^9$$

$$= \frac{9!}{9!0!}a^9 + \frac{9!}{8!1!}a^8\left(-\frac{2}{a}\right) + \frac{9!}{7!2!}a^7\left(\frac{4}{a^2}\right) +$$

$$\frac{9!}{6!3!}a^6\left(-\frac{8}{a^3}\right) + \frac{9!}{5!4!}a^5\left(\frac{16}{a^4}\right) +$$

$$\frac{9!}{4!5!}a^4\left(-\frac{32}{a^5}\right) + \frac{9!}{3!6!}a^3\left(\frac{64}{a^6}\right) +$$

$$\frac{9!}{2!7!}a^2\left(-\frac{128}{a^7}\right) + \frac{9!}{1!8!}a\left(\frac{256}{a^8}\right) +$$

$$\frac{9!}{0!9!}\left(-\frac{512}{a^9}\right)$$

$$= a^9 - 9(2a^7) + 36(4a^5) - 84(8a^3) + 126(16a) -$$
$$\quad 126(32a^{-1}) + 84(64a^{-3}) - 36(128a^{-5}) +$$
$$\quad 9(256a^{-7}) - 512a^{-9}$$

$$= a^9 - 18a^7 + 144a^5 - 672a^3 + 2016a - 4032a^{-1} +$$
$$\quad 5376a^{-3} - 4608a^{-5} + 2304a^{-7} - 512a^{-9}$$

17. $(\sqrt{2}+1)^6 - (\sqrt{2}-1)^6$

First, expand $(\sqrt{2}+1)^6$.

$$(\sqrt{2}+1)^6 = \binom{6}{0}(\sqrt{2})^6 + \binom{6}{1}(\sqrt{2})^5(1)+$$

$$\binom{6}{2}(\sqrt{2})^4(1)^2 + \binom{6}{3}(\sqrt{2})^3(1)^3+$$

$$\binom{6}{4}(\sqrt{2})^2(1)^4 + \binom{6}{5}(\sqrt{2})(1)^5+$$

$$\binom{6}{6}(1)^6$$

$$= \frac{6!}{6!0!}\cdot 8 + \frac{6!}{5!1!}\cdot 4\sqrt{2} + \frac{6!}{4!2!}\cdot 4+$$

$$\frac{6!}{3!3!}\cdot 2\sqrt{2} + \frac{6!}{2!4!}\cdot 2 + \frac{6!}{1!5!}\cdot\sqrt{2}+\frac{6!}{0!6!}$$

$$= 8 + 24\sqrt{2} + 60 + 40\sqrt{2} + 30 + 6\sqrt{2} + 1$$

$$= 99 + 70\sqrt{2}$$

Next, expand $(\sqrt{2}-1)^6$.

$$(\sqrt{2}-1)^6$$

$$= \binom{6}{0}(\sqrt{2})^6 + \binom{6}{1}(\sqrt{2})^5(-1)+$$

$$\binom{6}{2}(\sqrt{2})^4(-1)^2 + \binom{6}{3}(\sqrt{2})^3(-1)^3+$$

$$\binom{6}{4}(\sqrt{2})^2(-1)^4 + \binom{6}{5}(\sqrt{2})(-1)^5+$$

$$\binom{6}{6}(-1)^6$$

$$= \frac{6!}{6!0!}\cdot 8 - \frac{6!}{5!1!}\cdot 4\sqrt{2} + \frac{6!}{4!2!}\cdot 4 - \frac{6!}{3!3!}\cdot 2\sqrt{2}+$$

$$\frac{6!}{2!4!}\cdot 2 - \frac{6!}{1!5!}\cdot\sqrt{2}+\frac{6!}{0!6!}$$

$$= 8 - 24\sqrt{2} + 60 - 40\sqrt{2} + 30 - 6\sqrt{2} + 1$$

$$= 99 - 70\sqrt{2}$$

$$(\sqrt{2}+1)^6 - (\sqrt{2}-1)^6$$

$$= (99 + 70\sqrt{2}) - (99 - 70\sqrt{2})$$

$$= 99 + 70\sqrt{2} - 99 + 70\sqrt{2}$$

$$= 140\sqrt{2}$$

19. Expand: $(x^{-2} + x^2)^4$.

We have $a = x^{-2}$, $b = x^2$, and $n = 4$.

Pascal's triangle method: Use the fifth row of Pascal's triangle.

$$1 \quad 4 \quad 6 \quad 4 \quad 1.$$

$$(x^{-2} + x^2)^4$$

$$= 1\cdot(x^{-2})^4 + 4(x^{-2})^3(x^2) + 6(x^{-2})^2(x^2)^2+$$

$$4(x^{-2})(x^2)^3 + 1\cdot(x^2)^4$$

$$= x^{-8} + 4x^{-4} + 6 + 4x^4 + x^8$$

Factorial notation method:

$$(x^{-2} + x^2)^4$$

$$= \binom{4}{0}(x^{-2})^4 + \binom{4}{1}(x^{-2})^3(x^2)+$$

$$\binom{4}{2}(x^{-2})^2(x^2)^2 + \binom{4}{3}(x^{-2})(x^2)^3+$$

$$\binom{4}{4}(x^2)^4$$

$$= \frac{4!}{4!0!}(x^{-8}) + \frac{4!}{3!1!}(x^{-6})(x^2) + \frac{4!}{2!2!}(x^{-4})(x^4)+$$

$$\frac{4!}{1!3!}(x^{-2})(x^6) + \frac{4!}{0!4!}(x^8)$$

$$= x^{-8} + 4x^{-4} + 6 + 4x^4 + x^8$$

21. Find the 3rd term of $(a + b)^7$.

First, we note that $3 = 2 + 1$, $a = a$, $b = b$, and $n = 7$. Then the 3rd term of the expansion of $(a + b)^7$ is

$$\binom{7}{2}a^{7-2}b^2, \text{ or } \frac{7!}{2!5!}a^5b^2, \text{ or } 21a^5b^2.$$

23. Find the 6th term of $(x - y)^{10}$.

First, we note that $6 = 5 + 1$, $a = x$, $b = -y$, and $n = 10$. Then the 6th term of the expansion of $(x - y)^{10}$ is

$$\binom{10}{5}x^5(-y)^5, \text{ or } -252x^5y^5.$$

25. Find the 12th term of $(a - 2)^{14}$.

First, we note that $12 = 11 + 1$, $a = a$, $b = -2$, and $n = 14$. Then the 12th term of the expansion of $(a - 2)^{14}$ is

$$\binom{14}{11}a^{14-11}\cdot(-2)^{11} = \frac{14!}{3!11!}a^3(-2048)$$

$$= 364a^3(-2048)$$

$$= -745,472a^3$$

27. Find the 5th term of $(2x^3 - \sqrt{y})^8$.

First, we note that $5 = 4 + 1$, $a = 2x^3$, $b = -\sqrt{y}$, and $n = 8$. Then the 5th term of the expansion of $(2x^3 - \sqrt{y})^8$ is

$$\binom{8}{4}(2x^3)^{8-4}(-\sqrt{y})^4$$

$$= \frac{8!}{4!4!}(2x^3)^4(-\sqrt{y})^4$$

$$= 70(16x^{12})(y^2)$$

$$= 1120x^{12}y^2$$

29. The expansion of $(2u - 3v^2)^{10}$ has 11 terms so the 6th term is the middle term. Note that $6 = 5 + 1$, $a = 2u$, $b = -3v^2$, and $n = 10$. Then the 6th term of the expansion of $(2u - 3v^2)^{10}$ is

$$\binom{10}{5}(2u)^{10-5}(-3v^2)^5$$

$$= \frac{10!}{5!5!}(2u)^5(-3v^2)^5$$

$$= 252(32u^5)(-243v^{10})$$

$$= -1,959,552u^5v^{10}$$

31. The number of subsets is 2^7, or 128

33. The number of subsets is 2^{24}, or 16,777,216.

35. The term of highest degree of $(x^5 + 3)^4$ is the first term, or

$$\binom{4}{0}(x^5)^{4-0}3^0 = \frac{4!}{4!0!}x^{20} = x^{20}.$$

Therefore, the degree of $(x^5 + 3)^4$ is 20.

37. We use factorial notation. Note that $a = 3$, $b = i$, and $n = 5$.

$$(3+i)^5$$
$$= \binom{5}{0}(3^5) + \binom{5}{1}(3^4)(i) + \binom{5}{2}(3^3)(i^2)+$$
$$\binom{5}{3}(3^2)(i^3) + \binom{5}{4}(3)(i^4) + \binom{5}{5}(i^5)$$
$$= \frac{5!}{0!5!}(243) + \frac{5!}{1!4!}(81)(i) + \frac{5!}{2!3!}(27)(-1)+$$
$$\frac{5!}{3!2!}(9)(-i) + \frac{5!}{4!1!}(3)(1) + \frac{5!}{5!0!}(i)$$
$$= 243 + 405i - 270 - 90i + 15 + i$$
$$= -12 + 316i$$

39. We use factorial notation. Note that $a = \sqrt{2}$, $b = -i$, and $n = 4$.

$$(\sqrt{2}-i)^4 = \binom{4}{0}(\sqrt{2})^4 + \binom{4}{1}(\sqrt{2})^3(-i)+$$
$$\binom{4}{2}(\sqrt{2})^2(-i)^2 + \binom{4}{3}(\sqrt{2})(-i)^3+$$
$$\binom{4}{4}(-i)^4$$
$$= \frac{4!}{0!4!}(4) + \frac{4!}{1!3!}(2\sqrt{2})(-i)+$$
$$\frac{4!}{2!2!}(2)(-1) + \frac{4!}{3!1!}(\sqrt{2})(i)+$$
$$\frac{4!}{4!0!}(1)$$
$$= 4 - 8\sqrt{2}i - 12 + 4\sqrt{2}i + 1$$
$$= -7 - 4\sqrt{2}i$$

41.
$$(a-b)^n = \binom{n}{0}a^n(-b)^0 + \binom{n}{1}a^{n-1}(-b)^1+$$
$$\binom{n}{2}a^{n-2}(-b)^2 + \cdots +$$
$$\binom{n}{n-1}a^1(-b)^{n-1} + \binom{n}{n}a^0(-b)^n$$
$$= \binom{n}{0}(-1)^0 a^n b^0 + \binom{n}{1}(-1)^1 a^{n-1}b^1+$$
$$\binom{n}{2}(-1)^2 a^{n-2}b^2 + \cdots +$$
$$\binom{n}{n-1}(-1)^{n-1}a^1 b^{n-1}+$$
$$\binom{n}{n}(-1)^n a^0 b^n$$
$$= \sum_{k=0}^{n} \binom{n}{k}(-1)^k a^{n-k}b^k$$

43.
$$\frac{(x+h)^n - x^n}{h}$$
$$= \frac{\binom{n}{0}x^n + \binom{n}{1}x^{n-1}h + \cdots + \binom{n}{n}h^n - x^n}{h}$$
$$= \binom{n}{1}x^{n-1} + \binom{n}{2}x^{n-2}h + \cdots + \binom{n}{n}h^{n-1}$$
$$= \sum_{k=1}^{n} \binom{n}{k}x^{n-k}h^{k-1}$$

45. $(fg)(x) = f(x)g(x) = (x^2+1)(2x-3) = 2x^3 - 3x^2 + 2x - 3$

47. $(g \circ f)(x) = g(f(x)) = g(x^2+1) = 2(x^2+1) - 3 = 2x^2 + 2 - 3 = 2x^2 - 1$

49. $\displaystyle\sum_{k=0}^{4} \binom{4}{k}(-1)^k x^{4-k}6^k = \sum_{k=0}^{4} \binom{4}{k}x^{4-k}(-6)^k$, so

the left side of the equation is sigma notation for $(x-6)^4$. We have:

$$(x-6)^4 = 81$$

$x - 6 = \pm 3$ Taking the 4th root on both sides

$x - 6 = 3$ or $x - 6 = -3$

$x = 9$ or $x = 3$

The solutions are 9 and 3.

If we also observe that $(3i)^4 = 81$, we also find the imaginary solutions $6 \pm 3i$.

51. The $(k+1)$st term of $\left(\sqrt[3]{x} - \dfrac{1}{\sqrt{x}} \right)^7$ is

$\binom{7}{k} (\sqrt[3]{x})^{7-k} \left(-\dfrac{1}{\sqrt{x}} \right)^k$. The term containing $\dfrac{1}{x^{1/6}}$ is the term in which the sum of the exponents is $-1/6$. That is,

$$\left(\dfrac{1}{3} \right)(7-k) + \left(-\dfrac{1}{2} \right)(k) = -\dfrac{1}{6}$$

$$\dfrac{7}{3} - \dfrac{k}{3} - \dfrac{k}{2} = -\dfrac{1}{6}$$

$$-\dfrac{5k}{6} = -\dfrac{15}{6}$$

$$k = 3$$

Find the $(3+1)$st, or 4th term.

$$\binom{7}{3} (\sqrt[3]{x})^4 \left(-\dfrac{1}{\sqrt{x}} \right)^3 = \dfrac{7!}{4!3!} (x^{4/3})(-x^{-3/2}) =$$

$$-35x^{-1/6}, \ \text{ or } \ -\dfrac{35}{x^{1/6}}.$$

53. $_{100}C_0 + _{100}C_1 + \cdots + _{100}C_{100}$ is the total number of subsets of a set with 100 members, or 2^{100}.

55. $\displaystyle\sum_{k=0}^{23} \binom{23}{k} (\log_a x)^{23-k} (\log_a t)^k =$

$$(\log_a x + \log_a t)^{23} = [\log_a(xt)]^{23}$$

57. See the answer section in the text.

Exercise Set 8.8

1. a) We use Principle P.

For 1: $P = \dfrac{18}{100}$, or 0.18

For 2: $P = \dfrac{24}{100}$, or 0.24

For 3: $P = \dfrac{23}{100}$, or 0.23

For 4: $P = \dfrac{23}{100}$, or 0.23

For 5: $P = \dfrac{12}{100}$, or 0.12

b) Opinions may vary, but it seems that people tend not to select the first or last numbers.

3. $28.5\%(18,200) = 0.285(18,200) = 5187$ emails

5. a) Since there are 14 equally likely ways of selecting a marble from a bag containing 4 red marbles and 10 green marbles, we have, by Principle P,

$$P(\text{selecting a red marble}) = \dfrac{4}{14} = \dfrac{2}{7}.$$

b) Since there are 14 equally likely ways of selecting a marble from a bag containing 4 red marbles and 10 green marbles, we have, by Principle P,

$$P(\text{selecting a green marble}) = \dfrac{10}{14} = \dfrac{5}{7}.$$

c) Since there are 14 equally likely ways of selecting a marble from a bag containing 4 red marbles and 10 green marbles, we have, by Principle P,

$$P(\text{selecting a purple marble}) = \dfrac{0}{14} = 0.$$

d) Since there are 14 equally likely ways of selecting a marble from a bag containing 4 red marbles and 10 green marbles, we have, by Principle P,

$$P(\text{selecting a red or a green marble}) =$$

$$\dfrac{4+10}{14} = 1.$$

7. There are 6 possible outcomes. There are 3 numbers less than 4, so the probability is $\dfrac{3}{6}$, or $\dfrac{1}{2}$.

9. a) There are 4 queens, so the probability is $\dfrac{4}{52}$, or $\dfrac{1}{13}$.

b) There are 4 aces and 4 tens, so the probability is $\dfrac{4+4}{52}$, or $\dfrac{8}{52}$, or $\dfrac{2}{13}$.

c) There are 13 hearts, so the probability is $\dfrac{13}{52}$, or $\dfrac{1}{4}$.

d) There are two black 6's, so the probability is $\dfrac{2}{52}$, or $\dfrac{1}{26}$.

11. The number of ways of drawing 3 cards from a deck of 52 is $_{52}C_3$. The number of ways of drawing 3 aces is $_4C_3$. The probability is

$$\dfrac{_4C_3}{_{52}C_3} = \dfrac{4}{22,100} = \dfrac{1}{5525}.$$

13. The total number of people on the sales force is $10+10$, or 20. The number of ways to choose 4 people from a group of 20 is $_{20}C_4$. The number of ways of selecting 2 people from a group of 10 is $_{10}C_2$. This is done for both the men and the women.

$$P(\text{choosing 2 men and 2 women}) = \dfrac{_{10}C_2 \cdot _{10}C_2}{_{20}C_4} =$$

$$\dfrac{45 \cdot 45}{4845} = \dfrac{135}{323}$$

15. The number of ways of selecting 5 cards from a deck of 52 cards is $_{52}C_5$. Three sevens can be selected in $_4C_3$ ways and 2 kings in $_4C_2$ ways.

$$P(\text{drawing 3 sevens and 2 kings}) = \dfrac{_4C_3 \cdot _4C_2}{_{52}C_5}, \ \text{or}$$

$$\dfrac{1}{108,290}.$$

17. The number of ways of selecting 5 cards from a deck of 52 cards is $_{52}C_5$. Since 13 of the cards are spades, then 5 spades can be drawn in $_{13}C_5$ ways

$$P(\text{drawing 5 spades}) = \dfrac{_{13}C_5}{_{52}C_5} = \dfrac{1287}{2,598,960} =$$

$$\dfrac{33}{66,640}$$

19. a) HHH, HHT, HTH, HTT, THH, THT, TTH, TTT

b) Three of the 8 outcomes have exactly one head. Thus, $P(\text{exactly one head}) = \dfrac{3}{8}$.

c) Seven of the 8 outcomes have exactly 0, 1, or 2 heads. Thus, $P(\text{at most two heads}) = \dfrac{7}{8}$.

d) Seven of the 8 outcomes have 1, 2, or 3 heads. Thus, $P(\text{at least one head}) = \dfrac{7}{8}$.

e) Three of the 8 outcomes have exactly two tails. Thus, $P(\text{exactly two tails}) = \dfrac{3}{8}$.

21. The roulette wheel contains 38 equally likely slots. Eighteen of the 38 slots are colored black. Thus, by Principle P,

$P(\text{the ball falls in a black slot}) = \dfrac{18}{38} = \dfrac{9}{19}$.

23. The roulette wheel contains 38 equally likely slots. Only 1 slot is numbered 0. Then, by Principle P,

$P(\text{the ball falls in the 0 slot}) = \dfrac{1}{38}$.

25. The roulette wheel contains 38 equally likely slots. Thirty-six of the slots are colored red or black. Then, by Principle P,

$P(\text{the ball falls in a red or a black slot}) = \dfrac{36}{38} = \dfrac{18}{19}$.

27. The roulette wheel contains 38 equally likely slots. Eighteen of the slots are odd-numbered. Then, by Principle P,

$P(\text{the ball falls in a an odd-numbered slot}) =$
$\dfrac{18}{38} = \dfrac{9}{19}$.

29. zero

31. function; domain; range; domain; range

33. combination

35. factor

37. a) There are $\dbinom{13}{2}$ ways to select 2 denominations from the 13 denominations. Then in each denomination there are $\dbinom{4}{2}$ ways to choose 2 of the 4 cards. Finally there are $\dbinom{44}{1}$ ways to choose the fifth card from the 11 remaining denominations ($4 \cdot 11$, or 44 cards). Thus the number of two pairs hands is

$\dbinom{13}{2} \cdot \dbinom{4}{2} \cdot \dbinom{4}{2} \cdot \dbinom{44}{1}$, or 123,552.

b) $\dfrac{123,552}{_{52}C_5} = \dfrac{123,552}{2,598,960} \approx 0.0475$

39. a) There are 13 ways to select a denomination and then $\dbinom{4}{3}$ ways to choose 3 of the 4 cards in that denomination. Now there are $\dbinom{48}{2}$ ways to choose 2 cards from the 12 remaining denominations ($4 \cdot 12$, or 48 cards). But these combinations include the 3744 hands in a full house like Q-Q-Q-4-4 (Exercise 38), so these must be subtracted. Thus the number of three of a kind hands is $13 \cdot \dbinom{4}{3} \cdot \dbinom{48}{2} - 3744$, or 54,912.

b) $\dfrac{54,912}{_{52}C_5} = \dfrac{54,912}{2,598,960} \approx 0.0211$

Chapter 8 Review Exercises

1. The statement is true. See page 526 in the text.

3. The statement is true. See page 556 in the text.

5. $a_n = (-1)^n \left(\dfrac{n^2}{n^4 + 1} \right)$

$a_1 = (-1)^1 \left(\dfrac{1^2}{1^4 + 1} \right) = -\dfrac{1}{2}$

$a_2 = (-1)^2 \left(\dfrac{2^2}{2^4 + 1} \right) = \dfrac{4}{17}$

$a_3 = (-1)^3 \left(\dfrac{3^2}{3^4 + 1} \right) = -\dfrac{9}{82}$

$a_4 = (-1)^4 \left(\dfrac{4^2}{4^4 + 1} \right) = \dfrac{16}{257}$

$a_{11} = (-1)^{11} \left(\dfrac{11^2}{11^4 + 1} \right) = -\dfrac{121}{14,642}$

$a_{23} = (-1)^{23} \left(\dfrac{23^2}{23^4 + 1} \right) = -\dfrac{529}{279,842}$

7. $\displaystyle\sum_{k=1}^{4} \dfrac{(-1)^{k+1}3^k}{3^k - 1}$

$= \dfrac{(-1)^{1+1}3^1}{3^1 - 1} + \dfrac{(-1)^{2+1}3^2}{3^2 - 1} + \dfrac{(-1)^{3+1}3^3}{3^3 - 1} + \dfrac{(-1)^{4+1}3^4}{3^4 - 1}$

$= \dfrac{3}{2} - \dfrac{9}{8} + \dfrac{27}{26} - \dfrac{81}{80}$

$= \dfrac{417}{1040}$

9. $a_n = a_1 + (n-1)d$

$a_1 = \dfrac{3}{4}, d = \dfrac{13}{12} - \dfrac{3}{4} = \dfrac{1}{3}$, and $n = 10$

$a_{10} = \dfrac{3}{4} + (10-1)\dfrac{1}{3} = \dfrac{3}{4} + 3 = \dfrac{15}{4}$

11. $a_n = a_1 + (n-1)d$

$a_{18} = 4 + (18-1)3 = 4 + 51 = 55$

$S_n = \dfrac{n}{2}(a_1 + a_n)$

$S_{18} = \dfrac{18}{2}(4 + 55) = 531$

13. $a_1 = 5$, $a_{17} = 53$

$53 = 5 + (17 - 1)d$; $d = 3$

$a_3 = 5 + (3 - 1)3 = 11$

15. $a_1 = -2$, $r = 2$, $a_n = -64$

$a_n = a_1 r^{n-1}$

$-64 = -2 \cdot 2^{n-1}$

$-64 = -2^n$; $n = 6$

$S_n = \dfrac{a_1(1 - r^n)}{1 - r}$

$S_n = \dfrac{-2(1 - 2^6)}{1 - 2} = 2(1 - 64) = -126$

17. Since $|r| = \left| \dfrac{27.5}{2.5} \right| = |1.1| = 1.1 > 1$, the sum does not exist.

19. Since $|r| = \left| \dfrac{-\frac{1}{6}}{\frac{1}{2}} \right| = \left| -\dfrac{1}{3} \right| = \dfrac{1}{3} < 1$, the series has a sum.

$S_\infty = \dfrac{\frac{1}{2}}{1 - \left(-\frac{1}{3} \right)} = \dfrac{\frac{1}{2}}{\frac{4}{3}} = \dfrac{1}{2} \cdot \dfrac{3}{4} = \dfrac{3}{8}$

21. $5, m_1, m_2, m_3, m_4, 9$

We look for m_1, m_2, m_3, and m_4, such that $5, m_1, m_2, m_3, m_4, 9$ is an arithmetic sequence. In this case, $a_1 = 5$, $n = 6$, and $a_6 = 9$. First we find d:

$a_n = a_1 + (n - 1)d$

$9 = 5 + (6 - 1)d$

$4 = 5d$

$\dfrac{4}{5} = d$

Then we have:

$m_1 = a_1 + d = 5 + \dfrac{4}{5} = 5\dfrac{4}{5}$

$m_2 = m_1 + d = 5\dfrac{4}{5} + \dfrac{4}{5} = 6\dfrac{3}{5}$

$m_3 = m_2 + d = 6\dfrac{3}{5} + \dfrac{4}{5} = 7\dfrac{2}{5}$

$m_4 = m_3 + d = 7\dfrac{2}{5} + \dfrac{4}{5} = 8\dfrac{1}{5}$

23. $S_n = \dfrac{a_1(1 - r^n)}{1 - r}$

$S_{18} = \dfrac{2000\left[1 - (1.028)^{18}\right]}{1 - 1.028} \approx \$45,993.04$

25. $a_1 = 24,000,000,000$; $r = 0.73$

$S_\infty = \dfrac{24,000,000,000}{1 - 0.73} \approx \$88,888,888,889$

27. $S_n : 1 + 3 + 3^2 + \ldots + 3^{n-1} = \dfrac{3^n - 1}{2}$

$S_1 : 1 = \dfrac{3^1 - 1}{2}$

$S_k : 1 + 3 + 3^2 + \ldots + 3^{k-1} = \dfrac{3^k - 1}{2}$

$S_{k+1} : 1 + 3 + 3^2 + \ldots + 3^{(k+1)-1} = \dfrac{3^{k+1} - 1}{2}$

1. *Basis step*: $\dfrac{3^1 - 1}{2} = \dfrac{2}{2} = 1$ is true.

2. *Induction step*: Assume S_k. Add 3^k to both sides.

$1 + 3 + \ldots + 3^{k-1} + 3^k$

$= \dfrac{3^k - 1}{2} + 3^k = \dfrac{3^k - 1}{2} + 3^k \cdot \dfrac{2}{2}$

$= \dfrac{3 \cdot 3^k - 1}{2} = \dfrac{3^{k+1} - 1}{2}$

29. $6! = 720$

31. $\begin{pmatrix} 15 \\ 8 \end{pmatrix} = \dfrac{15!}{8!(15 - 8)!} = 6435$

33. $\dfrac{9!}{1!4!2!2!} = 3780$

35. a) $_6P_5 = \dfrac{6!}{(6 - 5)!} = 720$

b) $6^5 = 7776$

c) $_5P_4 = \dfrac{5!}{(5 - 4)!} = 120$

d) $_3P_2 = \dfrac{3!}{(3 - 2)!} = 6$

37. $(m + n)^7$

Pascal's triangle method: Use the 8th row of Pascal's triangle.

1 7 21 35 35 21 7 1

$(m + n)^7 = m^7 + 7m^6n + 21m^5n^2 + 35m^4n^3 +$
$\qquad\qquad 35m^3n^4 + 21m^2n^5 + 7mn^6 + n^7$

Factorial notation method:

$(m+n)^7 = \begin{pmatrix} 7 \\ 0 \end{pmatrix} m^7 + \begin{pmatrix} 7 \\ 1 \end{pmatrix} m^6n + \begin{pmatrix} 7 \\ 2 \end{pmatrix} m^5n^2 + \begin{pmatrix} 7 \\ 3 \end{pmatrix} m^4n^3 +$

$\qquad \begin{pmatrix} 7 \\ 4 \end{pmatrix} m^3n^4 + \begin{pmatrix} 7 \\ 5 \end{pmatrix} m^2n^5 + \begin{pmatrix} 7 \\ 6 \end{pmatrix} mn^6 + \begin{pmatrix} 7 \\ 7 \end{pmatrix} n^7$

$= m^7 + 7m^6n + 21m^5n^2 + 35m^4n^3 +$
$\qquad\qquad 35m^3n^4 + 21m^2n^5 + 7mn^6 + n^7$

39. Expand: $(x^2 - 3y)^4$

Pascal's triangle method: Use the 5th row.

1 4 6 4 1

$(x^2 - 3y)^4 = (x^2)^4 + 4(x^2)^3(-3y) + 6(x^2)^2(-3y)^2 +$
$\qquad\qquad 4(x^2)(-3y)^3 + (-3y)^4$

$= x^8 - 12x^6y + 54x^4y^2 - 108x^2y^3 + 81y^4$

Factorial notation method:

$$(x^2 - 3y)^4 = \binom{4}{0}(x^2)^4 + \binom{4}{1}(x^2)^3(-3y) +$$

$$\binom{4}{2}(x^2)^2(-3y)^2 + \binom{4}{3}(x^2)(-3y)^3 +$$

$$\binom{4}{4}(-3y)^4$$

$$= x^8 - 12x^6y + 54x^4y^2 - 108x^2y^3 + 81y^4$$

41. Expand: $(1 + 5i)^6$

Pascal's triangle method: Use the 7th row.

1 6 15 20 15 6 1

$$(1 + 5i)^6 = 1^6 + 6(1)^5(5i) + 15(1)^4(5i)^2 + 20(1)^3(5i)^3 +$$

$$15(1)^2(5i)^4 + 6(1)(5i)^5 + (5i)^6$$

$$= 1 + 30i - 375 - 2500i + 9375 +$$

$$18,750i - 15,625$$

$$= -6624 + 16,280i$$

Factorial notation method:

$$(1 + 5i)^6 = \binom{6}{0}1^6 + \binom{6}{1}(1)^5(5i) + \binom{6}{2}(1)^4(5i)^2 +$$

$$\binom{6}{3}(1)^3(5i)^3 + \binom{6}{4}(1)^2(5i)^4 +$$

$$\binom{6}{5}(1)(5i)^5 + \binom{6}{6}(5i)^6$$

$$= 1 + 30i - 375 - 2500i + 9375 +$$

$$18,750i - 15,625$$

$$= -6624 + 16,280i$$

43. Find 12th term of $(2a - b)^{18}$.

$$\binom{18}{11}(2a)^7(-b)^{11} = -\binom{18}{11}128a^7b^{11}$$

45. Of 52 cards, 13 are clubs.

Probability $= \dfrac{13}{52} = \dfrac{1}{4}$.

47. A: $\dfrac{86}{86 + 97 + 23} = \dfrac{86}{206} \approx 0.42$

B: $\dfrac{97}{86 + 97 + 23} = \dfrac{97}{206} \approx 0.47$

C: $\dfrac{23}{86 + 97 + 23} = \dfrac{23}{206} \approx 0.11$

49. There are 3 pairs that total 4: 1 and 3, 2 and 2, 3 and 1. There are $6 \cdot 6$, or 36, possible outcomes. Thus, we have $\dfrac{3}{36}$, or $\dfrac{1}{12}$. Answer A is correct.

51. a) If all of the terms of $a_1, a_2, \ldots, a_n$ are positive or if they area all negative, then $b_1, b_2, \ldots, b_n$ is an arithmetic sequence whose common difference is $|d|$, where d is the common difference of $a_1, a_2, \ldots, a_n$.

b) Yes; if d is the common difference of $a_1, a_2, \ldots, a_n$, then it is also the common difference of $b_1, b_2, \ldots, b_n$ and consequently $b_1, b_2, \ldots, b_n$ is an arithmetic sequence.

c) Yes; if d is the common difference of $a_1, a_2, \ldots, a_n$, then each term of $b_1, b_2, \ldots, b_n$ is obtained by adding $7d$ to the previous term and $b_1, b_2, \ldots, b_n$ is an arithmetic sequence.

d) No (unless a_n is constant)

e) No (unless a_n is constant)

f) No (unless a_n is constant)

53. $r = -\dfrac{1}{3}$, $S_\infty = \dfrac{3}{8}$

$$S_\infty = \frac{a_1}{1 - r}$$

$$\frac{3}{8} = \frac{a_1}{1 - \left(-\dfrac{1}{3}\right)}$$

$$a_1 = \frac{1}{2}$$

$$a_2 = \frac{1}{2}\left(-\frac{1}{3}\right) = -\frac{1}{6}$$

$$a_3 = -\frac{1}{6}\left(-\frac{1}{3}\right) = \frac{1}{18}$$

55.

$$\sum_{k=0}^{10} (-1)^k \binom{10}{k}(\log x)^{10-k}(\log y)^k$$

$$= (\log x - \log y)^{10}$$

$$= \left(\log \frac{x}{y}\right)^{10}$$

57.

$$\binom{n}{n-1} = 36$$

$$\frac{n!}{(n-1)![n - (n-1)]!} = 36$$

$$\frac{n(n-1)!}{(n-1)!1!} = 36$$

$$n = 36$$

59. For each circular arrangement of the numbers on a clock face there are 12 distinguishable ordered arrangements on a line. The number of arrangements of 12 objects on a line is $_{12}P_{12}$, or 12!. Thus, the number of circular permutations is $\dfrac{_{12}P_{12}}{12} = \dfrac{12!}{12} = 11! = 39,916,800$.

In general, for each circular arrangement of n objects, there are n distinguishable ordered arrangements on a line. The total number of arrangements of n objects on a line is $_{n}P_{n}$, or $n!$. Thus, the number of circular permutations is $\dfrac{n!}{n} = \dfrac{n(n-1)!}{n} = (n-1)!$.

61. Order is considered in a combination lock.

Chapter 8 Test

1. $a_n = (-1)^n(2n + 1)$

$a_{21} = (-1)^{21}[2(21) + 1]$

$= -43$

2. $a_n = \dfrac{n+1}{n+2}$

$a_1 = \dfrac{1+1}{1+2} = \dfrac{2}{3}$

$a_2 = \dfrac{2+1}{2+2} = \dfrac{3}{4}$

$a_3 = \dfrac{3+1}{3+2} = \dfrac{4}{5}$

$a_4 = \dfrac{4+1}{4+2} = \dfrac{5}{6}$

$a_5 = \dfrac{5+1}{5+2} = \dfrac{6}{7}$

3. $\displaystyle\sum_{k=1}^{4} (k^2 + 1) = (1^2 + 1) + (2^2 + 1) + (3^2 + 1) + (4^2 + 1)$

$= 2 + 5 + 10 + 7$

$= 34$

4. $\displaystyle\sum_{k=1}^{6} 4k$

5. $\displaystyle\sum_{k=1}^{\infty} 2^k$

6. $a_{n+1} = 2 + \dfrac{1}{a_n}$

$a_1 = 2 + \dfrac{1}{1} = 2 + 1 = 3$

$a_2 = 2 + \dfrac{1}{3} = 2\dfrac{1}{3}$

$a_3 = 2 + \dfrac{1}{\frac{7}{3}} = 2 + \dfrac{3}{7} = 2\dfrac{3}{7}$

$a_4 = 2 + \dfrac{1}{\frac{17}{7}} = 2 + \dfrac{7}{17} = 2\dfrac{7}{17}$

7. $d = 5 - 2 = 3$

$a_n = a_1 + (n-1)d$

$a_{15} = 2 + (15-1)3 = 44$

8. $a_1 = 8,\ a_{21} = 108,\ n = 21$

$a_n = a_1 + (n-1)d$

$108 = 8 + (21-1)d$

$100 = 20d$

$5 = d$

Use $a_n = a_1 + (n-1)d$ again to find a_7.

$a_7 = 8 + (7-1)(5) = 8 + 30 = 38$

9. $a_1 = 17,\ d = 13 - 17 = -4,\ n = 20$

First find a_{20}:

$a_n = a_1 + (n-1)d$

$a_{20} = 17 + (20-1)(-4) = 17 - 76 = -59$

Now find S_{20}:

$S_n = \dfrac{n}{2}(a_1 + a_n)$

$S_{20} = \dfrac{20}{2}(17 - 59) = 10(-42) = -420$

10. $\displaystyle\sum_{k=1}^{25} (2k + 1)$

$a_1 = 2 \cdot 1 + 1 = 3$

$a_{25} = 2 \cdot 25 + 1 = 51$

$S_n = \dfrac{n}{2}(a_1 + a_n)$

$S_{25} = \dfrac{25}{2}(3 + 51) = \dfrac{25}{2} \cdot 54 = 675$

11. $a_1 = 10,\ r = \dfrac{-5}{10} = -\dfrac{1}{2}$

$a_n = a_1 r^{n-1}$

$a_{11} = 10\left(-\dfrac{1}{2}\right)^{11-1} = \dfrac{5}{512}$

12. $r = 0.2,\ S_4 = 1248$

$S_n = \dfrac{a_1(1 - r^n)}{1 - r}$

$1248 = \dfrac{a_1(1 - 0.2^4)}{1 - 0.2}$

$1248 = \dfrac{0.9984 a_1}{0.8}$

$a_1 = 1000$

13. $\displaystyle\sum_{k=1}^{8} 2^k$

$a_1 = 2^1 = 2,\ r = 2,\ n = 8$

$S_8 = \dfrac{2(1 - 2^8)}{1 - 2} = 510$

14. $a_1 = 18,\ r = \dfrac{6}{18} = \dfrac{1}{3}$

Since $|r| = \dfrac{1}{3} < 1$, the series has a sum.

$S_\infty = \dfrac{18}{1 - \dfrac{1}{3}} = \dfrac{18}{\dfrac{2}{3}} = 18 \cdot \dfrac{3}{2} = 27$

15. $0.\overline{56} = 0.56 + 0.0056 + 0.000056 + \ldots$

$|r| = \left|\dfrac{0.0056}{0.56}\right| = |0.01| = 0.01 < 1$, so the series has a sum.

$S_\infty = \dfrac{0.56}{1 - 0.01} = \dfrac{0.56}{0.99} = \dfrac{56}{99}$

16. $a_1 = \$10,000$

$a_2 = \$10,000 \cdot 0.80 = \$8000$

$a_3 = \$8000 \cdot 0.80 = \$6400$

$a_4 = \$6400 \cdot 0.80 = \$5120$

$a_5 = \$5120 \cdot 0.80 = \$4096$

$a_6 = \$4096 \cdot 0.80 = \$3276.80$

17. We have an arithmetic sequence \$12.25, \$12.55, \$12.85, \$13.15, and so on with $d = \$0.30$. Each year there are 12/3, or 4 raises, so after 4 years the sequence will have the original hourly wage plus the $4 \cdot 4$, or 16, raises for a total of 17 terms. We use the formula $a_n = a_1 + (n-1)d$ with $a_1 = \$12.25$, $d = \$0.30$, and $n = 17$.

$a_{17} = \$12.25 + (17 - 1)(\$0.30) = \$12.25 + 16(\$0.30) = \$12.25 + \$4.80 = \$17.05$

At the end of 4 years William's hourly wage will be \$17.05.

18. We use the formula $S_n = \dfrac{a_1(1-r^n)}{1-r}$ with $a_1 = \$2500$, $r = 1.056$, and $n = 18$.

$$S_{18} = \frac{2500[1-(1.056)^{18}]}{1-1.056} = \$74,399.77$$

19.

$$S_n: \quad 2+5+8+\ldots+(3n-1) = \frac{n(3n+1)}{2}$$

$$S_1: \quad 2 = \frac{1(3 \cdot 1 + 1)}{2}$$

$$S_k: \quad 2+5+8+\ldots+(3k-1) = \frac{k(3k+1)}{2}$$

$$S_{k+1}: \quad 2+5+8+\ldots+(3k-1)+[3(k+1)-1] = \frac{(k+1)[3(k+1)+1]}{2}$$

1) *Basis step:* $\dfrac{1(3 \cdot 1 + 1)}{2} = \dfrac{1 \cdot 4}{2} = 2$, so S_1 is true.

2) *Induction step:*

$$\underbrace{2+5+8+\ldots+(3k-1)}+[3(k+1)-1]$$

$$= \quad \underbrace{\frac{k(3k+1)}{2}} \quad + [3k+3-1] \text{ By } S_k$$

$$= \frac{3k^2}{2} + \frac{k}{2} + 3k + 2$$

$$= \frac{3k^2}{2} + \frac{7k}{2} + 2$$

$$= \frac{3k^2 + 7k + 4}{2}$$

$$= \frac{(k+1)(3k+4)}{2}$$

$$= \frac{(k+1)[3(k+1)+1]}{2}$$

20. $_{15}P_6 = \dfrac{15!}{(15-6)!} = 3,603,600$

21. $_{21}C_{10} = \dfrac{21!}{10!(21-10)!} = 352,716$

22. $\begin{pmatrix} n \\ 4 \end{pmatrix} = \dfrac{n!}{4!(n-4)!}$

$$= \frac{n(n-1)(n-2)(n-3)(n-4)!}{4!(n-4)!}$$

$$= \frac{n(n-1)(n-2)(n-3)}{24}$$

23. $_6P_4 = \dfrac{6!}{(6-4)!} = 360$

24. a) $6^4 = 1296$

    b) $_5P_3 = \dfrac{5!}{(5-3)!} = 60$

25. $_{28}C_4 = \dfrac{28!}{4!(28-4)!} = 20,475$

26. $_{12}C_8 \cdot _8C_4 = \dfrac{12!}{8!(12-8)!} \cdot \dfrac{8!}{4!(8-4)!} = 34,650$

27. Expand: $(x+1)^5$.

Pascal's triangle method: Use the 6th row.

1   5   10   10   5   1

$(x+1)^5 = x^5 + 5x^4 \cdot 1 + 10x^3 \cdot 1^2 + 10x^2 \cdot 1^3 + 5x \cdot 1^4 + 1^5$

$$= x^5 + 5x^4 + 10x^3 + 10x^2 + 5x + 1$$

Factorial notation method:

$$(x+1)^5 = \begin{pmatrix} 5 \\ 0 \end{pmatrix} x^5 + \begin{pmatrix} 5 \\ 1 \end{pmatrix} x^4 \cdot 1 + \begin{pmatrix} 5 \\ 2 \end{pmatrix} x^3 \cdot 1^2 +$$

$$\begin{pmatrix} 5 \\ 3 \end{pmatrix} x^2 \cdot 1^3 + \begin{pmatrix} 5 \\ 4 \end{pmatrix} x \cdot 1^4 + \begin{pmatrix} 5 \\ 5 \end{pmatrix} 1^5$$

$$= x^5 + 5x^4 + 10x^3 + 10x^2 + 5x + 1$$

28. Find 5th term of $(x-y)^7$.

$$\begin{pmatrix} 7 \\ 4 \end{pmatrix} x^3(-y)^4 = 35x^3y^4$$

29. $2^9 = 512$

30. $\dfrac{8}{6+8} = \dfrac{8}{14} = \dfrac{4}{7}$

31. $\dfrac{_6C_1 \cdot _5C_2 \cdot _4C_5}{_{15}C_6} = \dfrac{6 \cdot 10 \cdot 4}{5005} = \dfrac{48}{1001}$

32. $a_n = 2_n - 2$

Only integers $n \geq 1$ are inputs.

$a_1 = 2 \cdot 1 - 2 = 0$, $a_2 = 2 \cdot 2 - 2 = 2$, $a_3 = 2 \cdot 3 - 2 = 4$, $a_4 = 2 \cdot 4 - 2 = 6$

Some points on the graph are $(1,0)$, $(2,2)$, $(3,4)$, and $(4,6)$. Thus the correct answer is B.

33.

$$_nP_7 = 9 \cdot _nP_6$$

$$\frac{n!}{(n-7)!} = 9 \cdot \frac{n!}{(n-6)!}$$

$$\frac{n!}{(n-7)!} \cdot \frac{(n-6)!}{n!} = 9 \cdot \frac{n!}{(n-6)!} \cdot \frac{(n-6)!}{n!}$$

$$\frac{(n-6)(n-7)!}{(n-7)!} = 9$$

$$n-6 = 9$$

$$n = 15$$

Just-in-Time Review

1. Real Numbers

1. Rational numbers: $\frac{2}{3}$, 6, -2.45, $18.\overline{4}$, -11, $\sqrt[3]{27}$, $5\frac{1}{6}$, $-\frac{8}{7}$, 0, $\sqrt{16}$

2. Rational numbers but not integers: $\frac{2}{3}$, -2.45, $18.\overline{4}$, $5\frac{1}{6}$, $-\frac{8}{7}$

3. Irrational numbers: $\sqrt{3}$, $\sqrt[6]{26}$, $7.151551555\ldots$, $-\sqrt{35}$, $\sqrt[5]{3}$
   (Although there is a pattern in $7.151551555\ldots$, there is no repeating block of digits.)

4. Integers: 6, -11, $\sqrt[3]{27}$, 0, $\sqrt{16}$

5. Whole numbers: 6, $\sqrt[3]{27}$, 0, $\sqrt{16}$

6. Real numbers: All of them

2. Properties of Real Numbers

1. $-24 + 24 = 0$ illustrates the additive inverse property.

2. $7(xy) = (7x)y$ illustrates the associative property of multiplication.

3. $9(r - s) = 9r - 9s$ illustrates a distributive property.

4. $11 + z = z + 11$ illustrates the commutative property of addition.

5. $-20 \cdot 1 = -20$ illustrates the multiplicative identity property.

6. $5(x + y) = (x + y)5$ illustrates the commutative property of multiplication.

7. $q + 0 = q$ illustrates the additive identity property.

8. $75 \cdot \frac{1}{75} = 1$ illustrates the multiplicative inverse property.

9. $(x + y) + w = x + (y + w)$ illustrates the associative property of addition.

10. $8(a + b) = 8a + 8b$ illustrates a distributive property.

3. Absolute Value

1. $|-98| = 98$ $(|a| = -a, \text{ if } a < 0.)$

2. $|0| = 0$ $(|a| = a, \text{ if } a \geq 0.)$

3. $|4.7| = 4.7$ $(|a| = a, \text{ if } a \geq 0.)$

4. $\left|-\frac{2}{3}\right| = \frac{2}{3}$ $(|a| = -a, \text{ if } a < 0.)$

5. $|-7 - 13| = |-20| = 20$, or
   $|13 - (-7)| = |13 + 7| = |20| = 20$

6. $|2 - 14.6| = |-12.6| = 12.6$, or
   $|14.6 - 2| = |12.6| = 12.6$

7. $|-39 - (-28)| = |-39 + 28| = |-11| = 11$, or
   $|-28 - (-39)| = |-28 + 39| = |11| = 11$

8. $\left|-\frac{3}{4} - \frac{15}{8}\right| = \left|-\frac{6}{8} - \frac{15}{8}\right| = \left|-\frac{21}{8}\right| = \frac{21}{8}$, or
   $\left|\frac{15}{8} - \left(-\frac{3}{4}\right)\right| = \left|\frac{15}{8} + \frac{6}{8}\right| = \left|\frac{21}{8}\right| = \frac{21}{8}$

4. Operations with Real Numbers

1. $8 - (-11) = 8 + 11 = 19$

2. $-\frac{3}{10} \cdot \left(-\frac{1}{3}\right) = \frac{3 \cdot 1}{10 \cdot 3} = \frac{3}{3} \cdot \frac{1}{10} = 1 \cdot \frac{1}{10} = \frac{1}{10}$

3. $15 \div (-3) = -5$

4. $-4 - (-1) = -4 + 1 = -3$

5. $7 \cdot (-50) = -350$

6. $-0.5 - 5 = -0.5 + (-5) = -5.5$

7. $-3 + 27 = 24$

8. $-400 \div -40 = 10$

9. $4.2 \cdot (-3) = -12.6$

10. $-13 - (-33) = -13 + 33 = 20$

11. $-60 + 45 = -15$

12. $\frac{1}{2} - \frac{2}{3} = \frac{1}{2} + \left(-\frac{2}{3}\right) = \frac{3}{6} + \left(-\frac{4}{6}\right) = -\frac{1}{6}$

13. $-24 \div 3 = -8$

14. $-6 + (-16) = -22$

15. $-\frac{1}{2} \div \left(-\frac{5}{8}\right) = -\frac{1}{2} \cdot \left(-\frac{8}{5}\right) = \frac{1 \cdot 8}{2 \cdot 5} = \frac{1 \cdot \cancel{2} \cdot 4}{\cancel{2} \cdot 5} = \frac{4}{5}$

5. Order on the Number Line

1. 9 is to the right of -9 on the number line, so it is false that $9 < -9$.

2. -10 is to the left of -1 on the number line, so it is true that $-10 \leq -1$.

3. $-5 = -\sqrt{25}$, and $-\sqrt{26}$ is to the left of $-\sqrt{25}$, or -5, on the number line. Thus it is true that $-\sqrt{26} < -5$.

4. $\sqrt{6} = \sqrt{6}$, so it is true that $\sqrt{6} \le \sqrt{6}$.

5. -30 is to the left of -25 on the number line, so it is false that $-30 > -25$.

6. $-\dfrac{4}{5} = -\dfrac{16}{20}$ and $-\dfrac{5}{4} = -\dfrac{25}{20}$; $-\dfrac{16}{20}$ is to the right of $-\dfrac{25}{20}$, so it is true that $-\dfrac{4}{5} > -\dfrac{5}{4}$.

6. Interval Notation

1. This is a closed interval, so we use brackets. Interval notation is $[-5, 5]$.

2. This is a half-open interval. We use a parenthesis on the left and a bracket on the right. Interval notation is $(-3, -1]$.

3. This interval is of unlimited extent in the negative direction, and the endpoint -2 is included. Interval notation is $(-\infty, -2]$.

4. This interval is of unlimited extent in the positive direction, and the endpoint 3.8 is not included. Interval notation is $(3.8, \infty)$.

5. $\{x | 7 < x\}$, or $\{x | x > 7\}$.

   This interval is of unlimited extent in the positive direction and the endpoint 7 is not included. Interval notation is $(7, \infty)$.

6. The endpoints -2 and 2 are not included in the interval, so we use parentheses. Interval notation is $(-2, 2)$.

7. The endpoints -4 and 5 are not included in the interval, so we use parentheses. Interval notation is $(-4, 5)$.

8. The interval is of unlimited extent in the positive direction, and the endpoint 1.7 is included. Internal notation is $[1.7, \infty)$.

9. The endpoint -5 is not included in the interval, so we use a parenthesis before -5. The endpoint -2 is included in the interval, so we use a bracket after -2. Interval notation is $(-5, -2]$.

10. This interval is of unlimited extent in the negative direction, and the endpoint $\sqrt{5}$ is not included. Interval notation is $(-\infty, \sqrt{5})$.

7. Integers as Exponents

1. $3^{-6} = \dfrac{1}{3^6}$ Using $a^{-m} = \dfrac{1}{a^m}$

2. $\dfrac{1}{(0.2)^{-5}} = (0.2)^5$ Using $a^{-m} = \dfrac{1}{a^m}$

3. $\dfrac{w^{-4}}{z^{-9}} = \dfrac{z^9}{w^4}$ Using $\dfrac{a^{-m}}{b^{-n}} = \dfrac{b^n}{a^m}$

4. $\left(\dfrac{z}{y}\right)^2 = \dfrac{z^2}{y^2}$ Raising a quotient to a power

5. $100^0 = 1$ Using $a^0 = 1$, $a \ne 0$

6. $\dfrac{a^5}{a^{-3}} = a^{5-(-3)} = a^{5+3} = a^8$ Using the quotient rule

7. $(2xy^3)(-3x^{-5}y) = 2(-3)x \cdot x^{-5} \cdot y^3 \cdot y$
   $$= -6x^{1+(-5)}y^{3+1}$$
   $$= -6x^{-4}y^4, \text{ or } -\dfrac{6y^4}{x^4}$$

8. $x^{-4} \cdot x^{-7} = x^{-4+(-7)} = x^{-11}$, or $\dfrac{1}{x^{11}}$

9. $(mn)^{-6} = m^{-6}n^{-6}$, or $\dfrac{1}{m^6n^6}$

10. $(t^{-5})^4 = t^{-5 \cdot 4} = t^{-20}$, or $\dfrac{1}{t^{20}}$

8. Scientific Notation

1. Convert $18{,}500{,}000$ to scientific notation.

   We want the decimal point to be positioned between the 1 and the 8, so we move it 7 places to the left. Since $18{,}500{,}000$ is greater than 10, the exponent must be positive.
   $$18{,}500{,}000 = 1.85 \times 10^7$$

2. Convert 0.000786 to scientific notation.

   We want the decimal point to be positioned between the 7 and the 8, so we move it 4 places to the right. Since 0.000786 is between 0 and 1, the exponent must be negative.
   $$0.000786 = 7.86 \times 10^{-4}$$

3. Convert 0.0000000023 to scientific notation.

   We want the decimal point to be positioned between the 2 and the 3, so we move it 9 places to the right. Since 0.0000000023 is between 0 and 1, the exponent must be negative.
   $$0.0000000023 = 2.3 \times 10^{-9}$$

4. Convert $8{,}927{,}000{,}000$ to scientific notation.

   We want the decimal point to be positioned between the 8 and the 9, so we move it 9 places to the left. Since $8{,}927{,}000{,}000$ is greater than 10, the exponent must be positive.
   $$8{,}927{,}000{,}000 = 8.927 \times 10^9$$

5. Convert 4.3×10^{-8} to decimal notation.

   The exponent is negative, so the number is between 0 and 1. We move the decimal point 8 places to the left.
   $$4.3 \times 10^{-8} = 0.000000043$$

6. Convert 5.17×10^6 to decimal notation.

   The exponent is positive, so the number is greater than 10. We move the decimal point 6 places to the right.
   $$5.17 \times 10^6 = 5{,}170{,}000$$

7. Convert 6.203×10^{11} to decimal notation.

The exponent is positive, so the number is greater than 10. We move the decimal point 11 places to the right.

$$6.203 \times 10^{11} = 620,300,000,000$$

8. Convert 2.94×10^{-5} to scientific notation.

The exponent is negative, so the number is between 0 and 1. We move the decimal point 5 places to the left.

$$2.94 \times 10^{-5} = 0.0000294$$

9. Order of Operations

1. $3 + 18 \div 6 - 3 = 3 + 3 - 3$ Dividing

$\qquad\qquad\qquad\quad = 6 - 3 = 3$ Adding and subtracting

2. $\quad = 5 \cdot 3 + 8 \cdot 3^2 + 4(6 - 2)$

$\quad = 5 \cdot 3 + 8 \cdot 3^2 + 4 \cdot 4$ Working inside parentheses

$\quad = 5 \cdot 3 + 8 \cdot 9 + 4 \cdot 4$ Evaluating 3^2

$\quad = 15 + 72 + 16$ Multiplying

$\quad = 87 + 16$ Adding in order

$\quad = 103$ from left to right

3. $\quad 5[3 - 8 \cdot 3^2 + 4 \cdot 6 - 2]$

$\quad = 5[3 - 8 \cdot 9 + 4 \cdot 6 - 2]$

$\quad = 5[3 - 72 + 24 - 2]$

$\quad = 5[-69 + 24 - 2]$

$\quad = 5[-45 - 2]$

$\quad = 5[-47]$

$\quad = -235$

4. $\quad 16 \div 4 \cdot 4 \div 2 \cdot 256$

$\quad = 4 \cdot 4 \div 2 \cdot 256$ Multiplying and dividing

$\qquad\qquad\qquad\qquad$ in order from left to right

$\quad = 16 \div 2 \cdot 256$

$\quad = 8 \cdot 256$

$\quad = 2048$

5. $\quad 2^6 \cdot 2^{-3} \div 2^{10} \div 2^{-8}$

$\quad = 2^3 \div 2^{10} \div 2^{-8}$

$\quad = 2^{-7} \div 2^{-8}$

$\quad = 2$

6. $\quad \dfrac{4(8-6)^2 - 4 \cdot 3 + 2 \cdot 8}{3^1 + 19^0}$

$\quad = \dfrac{4 \cdot 2^2 - 4 \cdot 3 + 2 \cdot 8}{3 + 1}$ Calculating in the numerator and in the denominator

$\quad = \dfrac{4 \cdot 4 - 4 \cdot 3 + 2 \cdot 8}{4}$

$\quad = \dfrac{16 - 12 + 16}{4}$

$\quad = \dfrac{4 + 16}{4}$

$\quad = \dfrac{20}{4}$

$\quad = 5$

10. Introduction to Polynomials

1. $5 - x^6$

The term of highest degree is $-x^6$, so the degree of the polynomial is 6.

2. $x^2 y^5 - x^7 y + 4$

The degree of $x^2 y^5$ is $2 + 5$, or 7; the degree of $-x^7 y$ is $7 + 1$, or 8; the degree of 4 is 0 ($4 = 4x^0$). Thus the degree of the polynomial is 8.

3. $2a^4 - 3 + a^2$

The term of highest degree is $2a^4$, so the degree of the polynomial is 4.

4. $-41 = -41x^0$, so the degree of the polynomial is 0.

5. $4x - x^3 + 0.1x^8 - 2x^5$

The term of highest degree is $0.1x^8$, so the degree of the polynomial is 8.

6. $x - 3$ has two terms. It is a binomial.

7. $14y^5$ has one term. It is a monomial.

8. $2y - \dfrac{1}{4}y^2 + 8$ has three terms. It is a trinomial.

11. Add and Subtract Polynomials

1. $\quad (8y - 1) - (3 - y)$

$\quad = (8y - 1) + (-3 + y)$

$\quad = (8 + 1)y + (-1 - 3)$

$\quad = 9y - 4$

2. $\quad (3x^2 - 2x - x^3 + 2) - (5x^2 - 8x - x^3 + 4)$

$\quad = (3x^2 - 2x - x^3 + 2) + (-5x^2 + 8x + x^3 - 4)$

$\quad = (3 - 5)x^2 + (-2 + 8)x + (-1 + 1)x^3 + (2 - 4)$

$\quad = -2x^2 + 6x - 2$

3. $\quad (2x + 3y + z - 7) + (4x - 2y - z + 8) +$

$\qquad\qquad\qquad (-3x + y - 2z - 4)$

$\quad = (2 + 4 - 3)x + (3 - 2 + 1)y + (1 - 1 - 2)z +$

$\qquad\qquad\qquad (-7 + 8 - 4)$

$\quad = 3x + 2y - 2z - 3$

4. $\quad (3ab^2 - 4a^2b - 2ab + 6) +$

$\qquad\qquad\qquad (-ab^2 - 5a^2b + 8ab + 4)$

$\quad = (3 - 1)ab^2 + (-4 - 5)a^2b + (-2 + 8)ab + (6 + 4)$

$\quad = 2ab^2 - 9a^2b + 6ab + 10$

5. $\quad (5x^2 + 4xy - 3y^2 + 2) - (9x^2 - 4xy + 2y^2 - 1)$

$\quad = (5x^2 + 4xy - 3y^2 + 2) + (-9x^2 + 4xy - 2y^2 + 1)$

$\quad = (5 - 9)x^2 + (4 + 4)xy + (-3 - 2)y^2 + (2 + 1)$

$\quad = -4x^2 + 8xy - 5y^2 + 3$

12. Multiply Polynomials

1. $(3a^2)(-7a^4) = [3(-7)](a^2 \cdot a^4)$
$$= -21a^6$$

2. $(y-3)(y+5)$
$= y^2 + 5y - 3y - 15$ Using FOIL
$= y^2 + 2y - 15$ Collecting like terms

3. $(x+6)(x+3)$
$= x^2 + 3x + 6x + 18$ Using FOIL
$= x^2 + 9x + 18$ Collecting like terms

4. $(2a+3)(a+5)$
$= 2a^2 + 10a + 3a + 15$ Using FOIL
$= 2a^2 + 13a + 15$ Collecting like terms

5. $(2x+3y)(2x+y)$
$= 4x^2 + 2xy + 6xy + 3y^2$ Using FOIL
$= 4x^2 + 8xy + 3y^2$

6. $(x+3)^2$
$= x^2 + 2 \cdot x \cdot 3 + 3^2$
$$[(A+B)^2 = A^2 + 2AB + B^2]$$
$= x^2 + 6x + 9$

7. $(5x-3)^2$
$= (5x)^2 - 2 \cdot 5x \cdot 3 + 3^2$
$$[(A-B)^2 = A^2 - 2AB + B^2]$$
$= 25x^2 - 30x + 9$

8. $(2x+3y)^2$
$= (2x)^2 + 2(2x)(3y) + (3y)^2$
$$[(A+B)^2 = A^2 + 2AB + B^2]$$
$= 4x^2 + 12xy + 9y^2$

9. $(n+6)(n-6)$
$= n^2 - 6^2$ $[(A+B)(A-B) = A^2 - B^2]$
$= n^2 - 36$

10. $(3y+4)(3y-4)$
$= (3y)^2 - 4^2$ $[(A+B)(A-B) = A^2 - B^2]$
$= 9y^2 - 16$

13. Factor Polynomials

1. $3x + 18 = 3 \cdot x + 3 \cdot 6 = 3(x+6)$

2. $2z^3 - 8z^2 = 2z^2 \cdot z - 2z^2 \cdot 4 = 2z^2(z-4)$

3. $3x^3 - x^2 + 18x - 6$
$= x^2(3x-1) + 6(3x-1)$
$= (3x-1)(x^2+6)$

4. $t^3 + 6t^2 - 2t - 12$
$= t^2(t+6) - 2(t+6)$
$= (t+6)(t^2-2)$

5. $w^2 - 7w + 10$

We look for two numbers with a product of 10 and a sum of -7. By trial, we determine that they are -5 and -2.
$$w^2 - 7w + 10 = (w-5)(w-2)$$

6. $t^2 + 8t + 15$

We look for two numbers with a product of 15 and a sum of 8. By trial, we determine that they are 3 and 5.
$$t^2 + 8t + 15 = (t+3)(t+5)$$

7. $2n^2 - 20n - 48 = 2(n^2 - 10n - 24)$

Now factor $n^2 - 10n - 24$. We look for two numbers with a product of -24 and a sum of -10. By trial, we determine that they are 2 and -12. Then $n^2 - 10n - 24 = (n+2)(n-12)$. We must include the common factor, 2, to have a factorization of the original trinomial.
$$2n^2 - 20n - 48 = 2(n+2)(n-12)$$

8. $y^4 - 9y^3 + 14y^2 = y^2(y^2 - 9y + 14)$

Now factor $y^2 - 9y + 14$. Look for two numbers with a product of 14 and a sum of -9. The numbers are -2 and -7. Then $y^2 - 9y + 14 = (y-2)(y-7)$. We must include the common factor, y^2, in order to have a factorization of the original trinomial.
$$y^4 - 9y^3 + 14y^2 = y^2(y-2)(y-7)$$

9. $2n^2 + 9n - 56$

We use the FOIL method.

1. There is no common factor other than 1 or -1.
2. The factorization must be of the form $(2n+ \quad)(n+ \quad)$.
3. Factor the constant term, -56. The possibilities are $-1 \cdot 56$, $1(-56)$, $-2 \cdot 28$, $2(-28)$, $-4 \cdot 16$, $4(-16)$, $-7 \cdot 8$, and $7(-8)$. The factors can be written in the opposite order as well: $56(-1)$, $-56 \cdot 1$, $28(-2)$, $-28 \cdot 2$, $16(-4)$, $-16 \cdot 4$, $8(-7)$, and $-8 \cdot 7$.
4. Find a pair of factors for which the sum of the outside and the inside products is the middle term, $9n$. By trial, we determine that the factorization is $(2n-7)(n+8)$.

10. $2y^2 + y - 6$

We use the grouping method.

1. There is no common factor other than 1 or -1.
2. Multiply the leading coefficient and the constant: $2(-6) = -12$.
3. Try to factor -12 so that the sum of the factors is the coefficient of the middle term, 1. The factors we want are 4 and -3.
4. Split the middle term using the numbers found in step (3):
$$y = 4y - 3y$$

5. Factor by grouping.

$$2y^2 + y - 6 = 2y^2 + 4y - 3y - 6$$
$$= 2y(y + 2) - 3(y + 2)$$
$$= (y + 2)(2y - 3)$$

11. $z^2 - 81 = z^2 - 9^2 = (z + 9)(z - 9)$

12. $16x^2 - 9 = (4x)^2 - 3^2 = (4x + 3)(4x - 3)$

13. $7pq^4 - 7py^4 = 7p(q^4 - y^4)$
$$= 7p[(q^2)^2 - (y^2)^2]$$
$$= 7p(q^2 + y^2)(q^2 - y^2)$$
$$= 7p(q^2 + y^2)(q + y)(q - y)$$

14. $x^2 + 12x + 36 = x^2 + 2 \cdot x \cdot 6 + 6^2$
$$= (x + 6)^2$$

15. $9z^2 - 12z + 4 = (3z)^2 - 2 \cdot 3z \cdot 2 + 2^2 = (3z - 2)^2$

16. $a^3 + 24a^2 + 144a$
$$= a(a^2 + 24a + 144)$$
$$= a(a^2 + 2 \cdot a \cdot 12 + 12^2)$$
$$= a(a + 12)^2$$

17. $x^3 + 64 = x^3 + 4^3$
$$= (x + 4)(x^2 - 4x + 16)$$

18. $m^3 - 216 = m^3 - 6^3$
$$= (m - 6)(m^2 + 6m + 36)$$

19. $3a^5 - 24a^2 = 3a^2(a^3 - 8)$
$$= 3a^2(a^3 - 2^3)$$
$$= 3a^2(a - 2)(a^2 + 2a + 4)$$

20. $t^6 + 1 = (t^2)^3 + 1^3$
$$= (t^2 + 1)(t^4 - t^2 + 1)$$

14. Equation-Solving Principles

1. $7t = 70$

$t = 10$ Dividing by 7

The solution is 10.

2. $x - 5 = 7$

$x = 12$ Adding 5

The solution is 12.

3. $3x + 4 = -8$

$3x = -12$ Subtracting 4

$x = -4$ Dividing by 3

The solution is −4.

4. $6x - 15 = 45$

$6x = 60$ Adding 15

$x = 10$ Dividing by 6

The solution is 10.

5. $7y - 1 = 23 - 5y$

$12y - 1 = 23$ Adding $5y$

$12y = 24$ Adding 1

$y = 2$ Dividing by 12

The solution is 2.

6. $3m - 7 = -13 + m$

$2m - 7 = -13$ Subtracting m

$2m = -6$ Adding 7

$m = -3$ Dividing by 2

The solution is −3.

7. $2(x + 7) = 5x + 14$

$2x + 14 = 5x + 14$

$-3x + 14 = 14$ Subtracting $5x$

$-3x = 0$ Subtracting 14

$x = 0$

The solution is 0.

8. $5y - (2y - 10) = 25$

$5y - 2y + 10 = 25$

$3y + 10 = 25$ Collecting like terms

$3y = 15$ Subtracting 10

$y = 5$ Dividing by 3

The solution is 5.

15. Inequality-Solving Principles

1. $p + 25 \geq -100$

$p \geq -125$ Subtracting 25

2. $-\dfrac{2}{3}x > 6$

$x < -\dfrac{3}{2} \cdot 6$ Multiplying by $-\dfrac{3}{2}$ and reversing the inequality symbol

$x < -9$

3. $9x - 1 < 17$

$9x < 18$ Adding 1

$x < 2$ Dividing by 9

4. $-x - 16 \geq 40$

$-x \geq 56$ Adding 6

$x \leq -56$ Multiplying by -1 and reversing the inequality symbol

5. $\dfrac{1}{3}y - 6 < 3$

$\dfrac{1}{3}y < 9$ Adding 6

$y < 27$ Multiplying by 3

6. $8 - 2w \le -14$

$\qquad -2w \le -22$ Subtracting 8

$\qquad\quad w \ge 11$ Dividing by -2 and
reversing the inequality symbol

16. Principle of Zero Products

1. $(a + 7)(a - 1) = 0$

$\quad a + 7 = 0 \quad or \quad a - 1 = 0$

$\qquad a = -7 \quad or \qquad a = 1$

The solutions are -7 and 1.

2. $(5y + 3)(y - 4) = 0$

$\quad 5y + 3 = 0 \quad or \quad y - 4 = 0$

$\qquad 5y = -3 \quad or \qquad y = 4$

$\qquad y = -\dfrac{3}{5} \quad or \qquad y = 4$

The solutions are $-\dfrac{3}{5}$ and 4.

3. $6x^2 + 7x - 5 = 0$

$\quad (3x + 5)(2x - 1) = 0$

$\quad 3x + 5 = 0 \quad or \quad 2x - 1 = 0$

$\qquad 3x = -5 \quad or \qquad 2x = 1$

$\qquad x = -\dfrac{5}{3} \quad or \qquad x = \dfrac{1}{2}$

The solutions are $-\dfrac{5}{3}$ and $\dfrac{1}{2}$.

4. $t(t - 8) = 0$

$\quad t = 0 \quad or \quad t - 8 = 0$

$\quad t = 0 \quad or \qquad t = 8$

The solutions are 0 and 8.

5. $x^2 - 8x - 33 = 0$

$\quad (x + 3)(x - 11) = 0$

$\quad x + 3 = 0 \quad or \quad x - 11 = 0$

$\qquad x = -3 \quad or \qquad x = 11$

The solutions are -3 and 11.

6. $x^2 + 13x = 30$

$\quad x^2 + 13x - 30 = 0$

$\quad (x + 15)(x - 2) = 0$

$\quad x + 15 = 0 \quad or \quad x - 2 = 0$

$\qquad x = -15 \quad or \qquad x = 2$

The solutions are -15 and 2.

17. Principle of Square Roots

1. $x^2 - 36 = 0$

$\qquad x^2 = 36$

$\quad x = \sqrt{36} \quad or \quad x = -\sqrt{36}$

$\quad x = 6 \qquad or \quad x = -6$

The solutions are 6 and -6, or ± 6.

2. $2y^2 - 20 = 0$

$\qquad 2y^2 = 20$

$\qquad y^2 = 10$

$\quad y = \sqrt{10} \quad or \quad y = -\sqrt{10}$

The solutions are $\sqrt{10}$ and $-\sqrt{10}$, or $\pm\sqrt{10}$.

3. $6z^2 = 18$

$\qquad z^2 = 3$

$\quad z = \sqrt{3} \quad or \quad z = -\sqrt{3}$

The solutions are $\sqrt{3}$ and $-\sqrt{3}$, or $\pm\sqrt{3}$.

4. $3t^2 - 15 = 0$

$\qquad 3t^2 = 15$

$\qquad t^2 = 5$

$\quad t = \sqrt{5} \quad or \quad t = -\sqrt{5}$

The solutions are $\sqrt{5}$ and $-\sqrt{5}$, or $\pm\sqrt{5}$.

5. $z^2 - 1 = 24$

$\qquad z^2 = 25$

$\quad z = \sqrt{25} \quad or \quad z = -\sqrt{25}$

The solutions are 5 and -5, or ± 5.

6. $5x^2 - 75 = 0$

$\qquad 5x^2 = 75$

$\qquad x^2 = 15$

$\quad x = \sqrt{15} \quad or \quad x = -\sqrt{15}$

The solutions are $\sqrt{15}$ and $-\sqrt{15}$, or $\pm\sqrt{15}$.

18. Simplify Rational Expressions

1. $\dfrac{3x - 3}{x(x - 1)}$

The denominator is 0 when the factor $x = 0$ and also when $x - 1 = 0$, or $x = 1$. The domain is the set of all real numbers except 0 and 1.

2. $\dfrac{y + 6}{y^2 + 4y - 21} = \dfrac{y + 6}{(y + 7)(y - 3)}$

The denominator is 0 when $y = -7$ or $y = 3$. The domain is the set of all real numbers except -7 and 3.

3. $\dfrac{x^2 - 4}{x^2 - 4x + 4} = \dfrac{(x + 2)(x - 2)}{(x - 2)(x - 2)} = \dfrac{x + 2}{x - 2}$

4. $\dfrac{x^2 + 2x - 3}{x^2 - 9} = \dfrac{(x-1)(\cancel{x+3})}{(\cancel{x+3})(x-3)} = \dfrac{x-1}{x-3}$

5. $\dfrac{x^3 - 6x^2 + 9x}{x^3 - 3x^2} = \dfrac{x(x^2 - 6x + 9)}{x^2(x-3)}$

$= \dfrac{\cancel{x}(\cancel{x-3})(x-3)}{\cancel{x} \cdot x(\cancel{x-3})}$

$= \dfrac{x-3}{x}$

6. $\dfrac{6y^2 + 12y - 48}{3y^2 - 9y + 6} = \dfrac{6(y^2 + 2y - 8)}{3(y^2 - 3y + 2)}$

$= \dfrac{2 \cdot \cancel{3} \cdot (y+4)(\cancel{y-2})}{\cancel{3}(y-1)(\cancel{y-2})}$

$= \dfrac{2(y+4)}{y-1}$

19. Multiply and Divide Rational Expressions

1. $\dfrac{r-s}{r+s} \cdot \dfrac{r^2 - s^2}{(r-s)^2} = \dfrac{(r-s)(r^2 - s^2)}{(r+s)(r-s)^2}$

$= \dfrac{(\cancel{r-s})(\cancel{r-s})(\cancel{r+s}) \cdot 1}{(\cancel{r+s})(\cancel{r-s})(\cancel{r-s})}$

$= 1$

2. $\dfrac{m^2 - n^2}{r+s} \div \dfrac{m-n}{r+s}$

$= \dfrac{m^2 - n^2}{r+s} \cdot \dfrac{r+s}{m-n}$

$= \dfrac{(m+n)(\cancel{m-n})(\cancel{r+s})}{(\cancel{r+s})(\cancel{m-n})}$

$= m+n$

3. $\dfrac{4x^2 + 9x + 2}{x^2 + x - 2} \cdot \dfrac{x^2 - 1}{3x^2 + x - 2}$

$= \dfrac{(4x+1)(\cancel{x+2})(\cancel{x+1})(\cancel{x-1})}{(\cancel{x+2})(\cancel{x-1})(3x-2)(\cancel{x+1})}$

$= \dfrac{4x+1}{3x-2}$

4. $\dfrac{3x+12}{2x-8} \div \dfrac{(x+4)^2}{(x-4)^2}$

$= \dfrac{3x+12}{2x-8} \cdot \dfrac{(x-4)^2}{(x+4)^2}$

$= \dfrac{3(\cancel{x+4})(\cancel{x-4})(x-4)}{2(\cancel{x-4})(\cancel{x+4})(x+4)}$

$= \dfrac{3(x-4)}{2(x+4)}$

5. $\dfrac{a^2 - a - 2}{a^2 - a - 6} \div \dfrac{a^2 - 2a}{2a + a^2}$

$= \dfrac{a^2 - a - 2}{a^2 - a - 6} \cdot \dfrac{2a + a^2}{a^2 - 2a}$

$= \dfrac{(\cancel{a-2})(a+1)(\cancel{a})(2+a)}{(a-3)(\cancel{a+2})(\cancel{a})(\cancel{a-2})}$

$= \dfrac{a+1}{a-3}$

6. $\dfrac{x^2 - y^2}{x^3 - y^3} \cdot \dfrac{x^2 + xy + y^2}{x^2 + 2xy + y^2}$

$= \dfrac{(x+y)(x-y)(x^2 + xy + y^2)}{(x-y)(x^2 + xy + y^2)(x+y)(x+y)}$

$= \dfrac{1}{x+y} \cdot \dfrac{(x+y)(x-y)(x^2 + xy + y^2)}{(x+y)(x-y)(x^2 + xy + y^2)}$

$= \dfrac{1}{x+y} \cdot 1 \qquad \text{Removing a factor of 1}$

$= \dfrac{1}{x+y}$

20. Add and Subtract Rational Expressions

1. $\dfrac{a-3b}{a+b} + \dfrac{a+5b}{a+b} = \dfrac{2a + 2b}{a+b}$

$= \dfrac{2(\cancel{a+b})}{1 \cdot (\cancel{a+b})}$

$= 2$

2. $\dfrac{x^2 - 5}{3x^2 - 5x - 2} + \dfrac{x+1}{3x - 6}$

$= \dfrac{x^2 - 5}{(3x+1)(x-2)} + \dfrac{x+1}{3(x-2)}$

$= \dfrac{x^2 - 5}{(3x+1)(x-2)} \cdot \dfrac{3}{3} + \dfrac{x+1}{3(x-2)} \cdot \dfrac{3x+1}{3x+1}$

$= \dfrac{3(x^2 - 5) + (x+1)(3x+1)}{3(3x+1)(x-2)}$

$= \dfrac{3x^2 - 15 + 3x^2 + 4x + 1}{3(3x+1)(x-2)}$

$= \dfrac{6x^2 + 4x - 14}{3(3x+1)(x-2)}$

3. $\dfrac{a^2 + 1}{a^2 - 1} - \dfrac{a-1}{a+1}$

$= \dfrac{a^2 + 1}{(a+1)(a-1)} - \dfrac{a-1}{a+1}, \text{ LCD is } (a+1)(a-1)$

$= \dfrac{a^2 + 1 - (a-1)(a-1)}{(a+1)(a-1)}$

$= \dfrac{a^2 + 1 - a^2 + 2a - 1}{(a+1)(a-1)}$

$= \dfrac{2a}{(a+1)(a-1)}$

4.
$$\frac{9x+2}{3x^2-2x-8}+\frac{7}{3x^2+x-4}$$

$$=\frac{9x+2}{(3x+4)(x-2)}+\frac{7}{(3x+4)(x-1)},$$
$$\text{LCD is }(3x+4)(x-2)(x-1)$$

$$=\frac{9x+2}{(3x+4)(x-2)}\cdot\frac{x-1}{x-1}+\frac{7}{(3x+4)(x-1)}\cdot\frac{x-2}{x-2}$$

$$=\frac{9x^2-7x-2}{(3x+4)(x-2)(x-1)}+\frac{7x-14}{(3x+4)(x-1)(x-2)}$$

$$=\frac{9x^2-16}{(3x+4)(x-2)(x-1)}$$

$$=\frac{\cancel{(3x+4)}(3x-4)}{\cancel{(3x+4)}(x-2)(x-1)}$$

$$=\frac{3x-4}{(x-2)(x-1)}$$

5.
$$\frac{y}{y^2-y-20}-\frac{2}{y+4}$$

$$=\frac{y}{(y+4)(y-5)}-\frac{2}{y+4},\ \text{LCD is }(y+4)(y-5)$$

$$=\frac{y}{(y+4)(y-5)}-\frac{2}{y+4}\cdot\frac{y-5}{y-5}$$

$$=\frac{y}{(y+4)(y-5)}-\frac{2y-10}{(y+4)(y-5)}$$

$$=\frac{y-(2y-10)}{(y+4)(y-5)}$$

$$=\frac{y-2y+10}{(y+4)(y-5)}$$

$$=\frac{-y+10}{(y+4)(y-5)}$$

6.
$$\frac{3y}{y^2-7y+10}-\frac{2y}{y^2-8y+15}$$

$$=\frac{3y}{(y-2)(y-5)}-\frac{2y}{(y-5)(y-3)},$$
$$\text{LCD is }(y-2)(y-5)(y-3)$$

$$=\frac{3y(y-3)-2y(y-2)}{(y-2)(y-5)(y-3)}$$

$$=\frac{3y^2-9y-2y^2+4y}{(y-2)(y-5)(y-3)}$$

$$=\frac{y^2-5y}{(y-2)(y-5)(y-3)}$$

$$=\frac{y\cancel{(y-5)}}{(y-2)\cancel{(y-5)}(y-3)}$$

$$=\frac{y}{(y-2)(y-3)}$$

---

21. Simplify Complex Rational Expressions

1.
$$\frac{\dfrac{x}{y}-\dfrac{y}{x}}{\dfrac{1}{y}+\dfrac{1}{x}}=\frac{\dfrac{x}{y}-\dfrac{y}{x}}{\dfrac{1}{y}+\dfrac{1}{x}}\cdot\frac{xy}{xy},\ \text{LCM is }xy$$

$$=\frac{\left(\dfrac{x}{y}-\dfrac{y}{x}\right)(xy)}{\left(\dfrac{1}{y}+\dfrac{1}{x}\right)(xy)}$$

$$=\frac{x^2-y^2}{x+y}$$

$$=\frac{\cancel{(x+y)}(x-y)}{\cancel{(x+y)}\cdot1}$$

$$=x-y$$

2.
$$\frac{\dfrac{a-b}{b}}{\dfrac{a^2-b^2}{ab}}=\frac{a-b}{b}\cdot\frac{ab}{a^2-b^2}$$

$$=\frac{a-b}{b}\cdot\frac{ab}{(a+b)(a-b)}$$

$$=\frac{a\cancel{b}\cancel{(a-b)}}{\cancel{b}(a+b)\cancel{(a-b)}}$$

$$=\frac{a}{a+b}$$

3.
$$\frac{w+\dfrac{8}{w^2}}{1+\dfrac{2}{w}}=\frac{w\cdot\dfrac{w^2}{w^2}+\dfrac{8}{w^2}}{1\cdot\dfrac{w}{w}+\dfrac{2}{w}}$$

$$=\frac{\dfrac{w^3+8}{w^2}}{\dfrac{w+2}{w}}$$

$$=\frac{w^3+8}{w^2}\cdot\frac{w}{w+2}$$

$$=\frac{\cancel{(w+2)}(w^2-2w+4)\cancel{w}}{\cancel{w}\cdot w\cancel{(w+2)}}$$

$$=\frac{w^2-2w+4}{w}$$

4.
$$\frac{\dfrac{x^2-y^2}{xy}}{\dfrac{x-y}{y}}=\frac{x^2-y^2}{xy}\cdot\frac{y}{x-y}$$

$$=\frac{(x+y)\cancel{(x-y)}\cancel{y}}{x\cancel{y}\cancel{(x-y)}}$$

$$=\frac{x+y}{x}$$

5. $\dfrac{\dfrac{a}{b} - \dfrac{b}{a}}{\dfrac{1}{a} - \dfrac{1}{b}} = \dfrac{a^2 - b^2}{b - a}$ Multiplying by $\dfrac{ab}{ab}$

$\qquad = \dfrac{(a+b)(a-b)}{b-a}$

$\qquad = \dfrac{(a+b)(a - b)}{-1 \cdot (a - b)}$

$\qquad = -a - b$

22. Simplify Radical Expressions

1. $\sqrt{(-21)^2} = |-21| = 21$

2. $\sqrt{9y^2} = \sqrt{(3y)^2} = |3y| = 3y$

3. $\sqrt{(a-2)^2} = a - 2$

4. $\sqrt[3]{-27x^3} = \sqrt[3]{(-3x)^3} = -3x$

5. $\sqrt[4]{81x^8} = \sqrt[4]{(3x^2)^4} = 3x^2$

6. $\sqrt[5]{32} = \sqrt[5]{2^5} = 2$

7. $\sqrt[4]{48x^6y^4} = \sqrt[4]{16x^4y^4 \cdot 3x^2} = 2xy\sqrt[4]{3x^2} =$
$2xy\sqrt[4]{3x^2}$

8. $\sqrt{15}\sqrt{35} = \sqrt{15 \cdot 35} = \sqrt{3 \cdot 5 \cdot 5 \cdot 7} = \sqrt{5^2 \cdot 3 \cdot 7} =$
$\sqrt{5^2} \cdot \sqrt{3 \cdot 7} = 5\sqrt{21}$

9. $\dfrac{\sqrt{40xy}}{\sqrt{8x}} = \sqrt{\dfrac{40xy}{8x}} = \sqrt{5y}$

10. $\dfrac{\sqrt[3]{3x^2}}{\sqrt[3]{24x^5}} = \sqrt[3]{\dfrac{3x^2}{24x^5}} = \sqrt[3]{\dfrac{1}{8x^3}} = \dfrac{1}{2x}$

11. $\sqrt{x^2 - 4x + 4} = \sqrt{(x-2)^2} = x - 2$

12. $\sqrt{2x^3y}\sqrt{12xy} = \sqrt{24x^4y^2} = \sqrt{4x^4y^2 \cdot 6} = 2x^2y\sqrt{6}$

13. $\sqrt[3]{3x^2y}\sqrt[3]{36x} = \sqrt[3]{108x^3y} = \sqrt[3]{27x^3 \cdot 4y} = 3x\sqrt[3]{4y}$

14. $5\sqrt{2} + 3\sqrt{32} = 5\sqrt{2} + 3\sqrt{16 \cdot 2}$
$\qquad\qquad\qquad = 5\sqrt{2} + 3 \cdot 4\sqrt{2}$
$\qquad\qquad\qquad = 5\sqrt{2} + 12\sqrt{2}$
$\qquad\qquad\qquad = (5 + 12)\sqrt{2}$
$\qquad\qquad\qquad = 17\sqrt{2}$

15. $7\sqrt{12} - 2\sqrt{3} = 7 \cdot 2\sqrt{3} - 2\sqrt{3} = 14\sqrt{3} - 2\sqrt{3} = 12\sqrt{3}$

16. $2\sqrt{32} + 3\sqrt{8} - 4\sqrt{18} = 2 \cdot 4\sqrt{2} + 3 \cdot 2\sqrt{2} - 4 \cdot 3\sqrt{2} =$
$8\sqrt{2} + 6\sqrt{2} - 12\sqrt{2} = 2\sqrt{2}$

17. $6\sqrt{20} - 4\sqrt{45} + \sqrt{80} = 6\sqrt{4 \cdot 5} - 4\sqrt{9 \cdot 5} + \sqrt{16 \cdot 5}$
$\qquad\qquad\qquad\qquad = 6 \cdot 2\sqrt{5} - 4 \cdot 3\sqrt{5} + 4\sqrt{5}$
$\qquad\qquad\qquad\qquad = 12\sqrt{5} - 12\sqrt{5} + 4\sqrt{5}$
$\qquad\qquad\qquad\qquad = (12 - 12 + 4)\sqrt{5}$
$\qquad\qquad\qquad\qquad = 4\sqrt{5}$

18. $\quad (2 + \sqrt{3})(5 + 2\sqrt{3})$
$\qquad = 2 \cdot 5 + 2 \cdot 2\sqrt{3} + \sqrt{3} \cdot 5 + \sqrt{3} \cdot 2\sqrt{3}$
$\qquad = 10 + 4\sqrt{3} + 5\sqrt{3} + 3 \cdot 2$
$\qquad = 10 + 9\sqrt{3} + 6$
$\qquad = 16 + 9\sqrt{3}$

19. $\quad \left(\sqrt{8} + 2\sqrt{5}\right)\left(\sqrt{8} - 2\sqrt{5}\right)$
$\qquad = \left(\sqrt{8}\right)^2 - \left(2\sqrt{5}\right)^2$
$\qquad = 8 - 4 \cdot 5$
$\qquad = 8 - 20$
$\qquad = -12$

20. $(1 + \sqrt{3})^2 = 1^2 + 2 \cdot 1 \cdot \sqrt{3} + (\sqrt{3})^2$
$\qquad\qquad\quad = 1 + 2\sqrt{3} + 3$
$\qquad\qquad\quad = 4 + 2\sqrt{3}$

23. Rationalizing Denominators

1. $\dfrac{4}{\sqrt{11}} = \dfrac{4}{\sqrt{11}} \cdot \dfrac{\sqrt{11}}{\sqrt{11}} = \dfrac{4\sqrt{11}}{11}$

2. $\sqrt{\dfrac{3}{7}} = \sqrt{\dfrac{3}{7} \cdot \dfrac{7}{7}} = \sqrt{\dfrac{21}{49}} = \dfrac{\sqrt{21}}{\sqrt{49}} = \dfrac{\sqrt{21}}{7}$

3. $\dfrac{\sqrt[3]{7}}{\sqrt[3]{2}} = \dfrac{\sqrt[3]{7}}{\sqrt[3]{2}} \cdot \dfrac{\sqrt[3]{4}}{\sqrt[3]{4}} = \dfrac{\sqrt[3]{28}}{\sqrt[3]{8}} = \dfrac{\sqrt[3]{28}}{2}$

4. $\sqrt[3]{\dfrac{16}{9}} = \sqrt[3]{\dfrac{16}{9} \cdot \dfrac{3}{3}} = \sqrt[3]{\dfrac{48}{27}} = \dfrac{\sqrt[3]{48}}{\sqrt[3]{27}} =$
$\dfrac{\sqrt[3]{8 \cdot 6}}{3} = \dfrac{2\sqrt[3]{6}}{3}$

5. $\dfrac{3}{\sqrt{30} - 4} = \dfrac{3}{\sqrt{30} - 4} \cdot \dfrac{\sqrt{30} + 4}{\sqrt{30} + 4}$
$\qquad = \dfrac{3\sqrt{30} + 12}{30 - 16}$
$\qquad = \dfrac{3\sqrt{30} + 12}{14}$

6. $\dfrac{1 - \sqrt{2}}{\sqrt{3} - \sqrt{6}} = \dfrac{1 - \sqrt{2}}{\sqrt{3} - \sqrt{6}} \cdot \dfrac{\sqrt{3} + \sqrt{6}}{\sqrt{3} + \sqrt{6}}$
$\qquad = \dfrac{\sqrt{3} + \sqrt{6} - \sqrt{6} - \sqrt{12}}{3 - 6}$
$\qquad = \dfrac{\sqrt{3} + \sqrt{6} - \sqrt{6} - 2\sqrt{3}}{3 - 6}$
$\qquad = \dfrac{-\sqrt{3}}{-3} = \dfrac{\sqrt{3}}{3}$

7. $\dfrac{6}{\sqrt{m} - \sqrt{n}} = \dfrac{6}{\sqrt{m} - \sqrt{n}} \cdot \dfrac{\sqrt{m} + \sqrt{n}}{\sqrt{m} + \sqrt{n}}$
$\qquad = \dfrac{6(\sqrt{m} + \sqrt{n})}{(\sqrt{m})^2 - (\sqrt{n})^2}$
$\qquad = \dfrac{6\sqrt{m} + 6\sqrt{n}}{m - n}$

24. Rational Exponents

1. $y^{5/6} = \sqrt[6]{y^5}$

2. $x^{2/3} = \sqrt[3]{x^2}$

3. $16^{3/4} = (16^{1/4})^3 = \left(\sqrt[4]{16}\right)^3 = 2^3 = 8$

4. $4^{7/2} = (\sqrt{4})^7 = 2^7 = 128$

5. $125^{-1/3} = \dfrac{1}{125^{1/3}} = \dfrac{1}{\sqrt[3]{125}} = \dfrac{1}{5}$

6. $32^{-4/5} = \left(\sqrt[5]{32}\right)^{-4} = 2^{-4} = \dfrac{1}{16}$

7. $\sqrt[12]{y^4} = y^{4/12} = y^{1/3}$

8. $\sqrt{x^5} = x^{5/2}$

9. $x^{1/2} \cdot x^{2/3} = x^{1/2+2/3} = x^{3/6+4/6} = x^{7/6} = \sqrt[6]{x^7} = x\sqrt[6]{x}$

10. $(a-2)^{9/4}(a-2)^{-1/4} = (a-2)^{9/4+(-1/4)} =$
    $(a-2)^{8/4} = (a-2)^2$

11. $(m^{1/2}n^{5/2})^{2/3} = m^{\frac{1}{2}\cdot\frac{2}{3}}n^{\frac{5}{2}\cdot\frac{2}{3}} = m^{1/3}n^{5/3} =$
    $\sqrt[3]{m}\,\sqrt[3]{n^5} = \sqrt[3]{mn^5} = n\sqrt[3]{mn^2}$

25. Pythagorean Theorem

1. $\begin{aligned} a^2 + b^2 &= c^2 \\ 8^2 + 15^2 &= c^2 \\ 64 + 225 &= c^2 \\ 289 &= c^2 \\ 17 &= c \end{aligned}$

2. $\begin{aligned} a^2 + b^2 &= c^2 \\ 4^2 + 4^2 &= c^2 \\ 16 + 16 &= c^2 \\ 32 &= c^2 \\ \sqrt{32} &= c \\ 5.657 &\approx c \end{aligned}$

3. $\begin{aligned} a^2 + b^2 &= c^2 \\ 5^2 + b^2 &= 13^2 \\ 25 + b^2 &= 169 \\ b^2 &= 144 \\ b &= 12 \end{aligned}$

4. $\begin{aligned} a^2 + b^2 &= c^2 \\ a^2 + 12^2 &= 13^2 \\ a^2 + 144 &= 169 \\ a^2 &= 25 \\ a &= 5 \end{aligned}$

5. $\begin{aligned} a^2 + b^2 &= c^2 \\ (\sqrt{5})^2 + b^2 &= 6^2 \\ 5 + b^2 &= 36 \\ b^2 &= 31 \\ b &= \sqrt{31} \approx 5.568 \end{aligned}$